The Vent and Seep Biota

Aims and Scope
Topics in Geobiology Book Series

Topics in Geobiology series treats geobiology – the broad discipline that covers the history of life on Earth. The series aims for high quality, scholarly volumes of original research as well as broad reviews. Recent volumes have showcased a variety of organisms including cephalopods, corals, and rodents. They discuss the biology of these organisms-their ecology, phylogeny, and mode of life – and in addition, their fossil record – their distribution in time and space.
Other volumes are more theme based such as predator-prey relationships, skeletal mineralization, paleobiogeography, and approaches to high resolution stratigraphy, that cover a broad range of organisms. One theme that is at the heart of the series is the interplay between the history of life and the changing environment. This is treated in skeletal mineralization and how such skeletons record environmental signals and animal-sediment relationships in the marine environment.
The series editors also welcome any comments or suggestions for future volumes.

Series Editors
Neil H. Landman, landman@amnh.org
Peter Harries, harries@shell.cas.usf.edu

For other titles published in this series, go to
http://www.springer.com/series/6623

The Vent and Seep Biota

Aspects from Microbes to Ecosystems

Steffen Kiel
Editor

Springer

Editor
Steffen Kiel
Institut für Geowissenschaften
Christian-Albrechts-Universität
Ludewig-Meyn-Str. 10
24118 Kiel
Germany

Geowissenschaftliches Zentrum
Georg-August-Universitat Göttingen
Abteilung Geobiologie
Goldschmidtstr. 3
37077 Göttingen
steffen.kiel@gmx.de

ISBN 978-94-007-3339-8 978-90-481-9572-5 (eBook)
DOI 10.1007/978-90-481-9572-5
Springer Dordrecht Heidelberg London New York

Softcover reprint of the hardcover 1st edition 2010

Cover illustrations:
Main image: An association of the mussel *Bathymodiolus platifrons*, the galatheid crab *Shinkaia crosnieri*, as well as shrimps and gastropod limpets at a vent site in the Okinawa Trough. Photo taken by Mutsunori Tokeshi (Kyushu University), © by JAMSTEC.

Small images: False color SEM images of symbiotic bacteria in the gills of a solemyid (fig. 1), a thyasirid (fig. 2), and a lucinid (fig. 3). Images courtesy of JD Taylor and EA Glover (London).

Printed on acid-free paper

Springer is part of Springer Science+Business Media (www.springer.com)

Foreword

The discovery of hydrothermal vent communities at the Galapagos Rift in 1977 turned several paradigms on deep-sea biology upside down. Deep-sea animals were considered as small and slow growing, but the tube worm *Riftia pachyptila* reaches nearly 2 m in length and turned out to be one of the fastest growing invertebrates on Earth. Likewise, diversity in the deep was considered high but low in abundance, yet most of the biomass at hydrothermal vents consists of a few highly specialized and superabundant species. But the perhaps biggest surprise was that these entire ecosystems rely on nutrients provided by chemoautotrophic bacteria, rather than on photosynthesis.

How and when have animals adapted to life in such an extreme environment, and how do they survive there? To answer these questions, deep-sea investigation following the initial discovery at the Galapagos Rift found similar communities on mid-oceanic ridges in all other ocean basins, at hydrocarbon seeps on continental slopes, and on whale carcasses as well as an assortment of other sites of organic enrichment in the deep sea. The ecology of these ecosystems and their inhabitants is now increasingly well understood, biogeographic patterns and adaptational pathways are emerging, and paleontology provides insights into the origin and evolutionary history of these fascinating communities.

This book attempts to provide summaries of key aspects of the vent and seep biota across disciplines that emerged during the past 10 years. The ecology and evolution of vent and seep communities is characterized by complex interactions between and among fluids, microbes, and macrofauna. Consequently, the chapters of this book cover a range of topics including evolutionary and ecologic interactions between invertebrate hosts and their bacterial symbionts (Chapters 2, 3, and 6), the physiology, life strategies and geologic history of major vent and seep invertebrates (Chapters 4, 7–9), how life at vents and seeps is shaped by the animals that inhabit them (Chapter 10), the peculiarities of shallow-water vent and seep communities (Chapter 11), as well as insights on regions with exceptional vent and seep faunas (Chapters 10, 12, and 14).

It is hoped that this book raises and maintains interest in these fantastic faunas and that it inspires research for the coming years.

Steffen Kiel

Acknowledgments

This book would not have been possible with the help and support of many people. I would like to thank, in no particular order, Peter Harris for inviting me to edit this volume, all the authors for their fine chapters and the reviewers for their efforts to improve them, Jim Goedert (Wauna) for opening the door into seep paleontology, Sven N. Nielsen (Kiel) for providing access to his ftp server, the ChEss office for sponsoring the color appendix, Tamara Welschot and Judith Terpos at Springer for guiding me through the editorial process, and last but not least my wife Silke for her patience.

Contents

Contributors

Martin Blumenberg
Department of Geobiology, Geoscience Centre (GZG), Georg-August-University of Göttingen, Goldschmidtstraße 3, 37077 Göttingen, Germany
martin.blumenberg@geo.uni-goettingen.de

Erik E. Cordes
Biology Department, Temple University, 1900 N 12th St, Philadelphia, PA 19122, USA
ecordes@temple.edu

Paul R. Dando
Marine Biological Association of the United Kingdom, Citadel Hill, Plymouth PL1 2PB, UK
pdando@mba.ac.uk

Sébastien Duperron
UMR 7138 (UPMC CNRS IRD MNHN), Systématique Adaptation Evolution – Adaptation aux Milieux Extrêmes, Université Pierre et Marie Curie, Bât. A, 7 quai St Bernard, 75005 Paris, France
sebastien.duperron@snv.jussieu.fr

Katsunori Fujikura
Japan Agency for Marine-Earth Science and Technology, 2-15 Natsushima, Yokosuka 237-0061, Japan
fujikura@jamstec.go.jp

Yoshihiro Fujiwara
Japan Agency for Marine-Earth Science and Technology, 2-15 Natsushima-cho, Yokosuka 237-0061, Japan
fujiwara@jamstec.go.jp

Françoise Gaill
CNRS, Ecology and Environment Institute, 3 rue Michel Ange, 75016, Paris, France
francoise.gaill@snv.jussieu.fr

Emily A. Glover
Department of Zoology, The Natural History Museum, London SW7 5BD, United Kingdom
emily.glover@dial.pipex.com

Breea Govenar
Biology Department, Woods Hole Oceanographic Institution, Woods Hole, MA 02543, USA
bgovenar@whoi.edu

Stèphane Hourdez
Station Biologique de Roscoff, CNRS-UPMC, Roscoff, France
hourdez@sb-roscoff.fr

Yasunori Kano
Atmosphere and Ocean Research Institute, The University of Tokyo, 5-1-5 Kashiwanoha, Kashiwa, Chiba 277-8564, Japan
kano@ori.u-tokyo.ac.jp

Steffen Kiel
Institut für Geowissenschaften, Christian-Albrechts-Universität, Ludewig-Meyn-Str. 10, 24118 Kiel, Germany
steffen.kiel@gmx.de

Shigeaki Kojima
Atmosphere and Ocean Research Institute, University of Tokyo, 5-1-5 Kashiwanoha, Kashiwa, Chiba 277-8564, Japan
kojima@ori.u-tokyo.ac.jp

Nadine Le Bris
Département Etude des Ecosystèmes Profonds, Ifremer, BP70, 29280, Plouzané, France
and
CNRS-UPMC FRE 3350, Laboratoire d'Ecogéochimie des Environnements Benthiques, 66650 Banyuls-sur-Mer, France
lebris@obs-banyuls.fr

Jun-Ichi Miyazaki
Yamanashi University, 4-4-37 Takeda, Kofu, Yamanashi 400-8510, Japan
miyazaki@yamanashi.ac.jp

Takashi Okutani
Japan Agency for Marine-Earth Science and Technology, 2-15 Natsushima, Yokosuka 237-0061, Japan
okutani@jamstec.go.jp

Harry H. Roberts
Coastal Studies Institute, Louisiana State University, Baton Rouge, LA, USA
hrober3@lsu.edu

Michael R. Sandy
Department of Geology, University of Dayton, 300 College Park, Dayton, Ohio 45469-2364, USA
michael.sandy@notes.udayton.edu

Takenori Sasaki
The University Museum, The University of Tokyo,
7-3-1 Hongo, Bunkyo-ku, Tokyo, 113-0033, Japan
sasaki@um.u-tokyo.ac.jp

John D. Taylor
Department of Zoology, The Natural History Museum,
London SW7 5BD, Uinted Kingdom
j.taylor@nhm.ac.uk

Paul A. Tyler
National Oceanography Centre, Southampton, University of Southampton Waterfront Campus, European Way, Southampton SO14 3ZH, UK
pat8@noc.soton.ac.uk

Robert C. Vrijenhoek
Monterey Bay Aquarium Research Institute, 7700 Sandholdt Road,
Moss Landing, CA 95039, USA
vrijen@mbari.org

Anders Warén
Swedish Museum of Natural History, Box 50007,
SE-10405 Stockholm, Sweden
anders.waren@nrm.se

Hiromi Watanabe
Japan Agency for Marine-Earth Science and Technology,
2-15 Natsushima-cho, Yokosuka 237-0061, Japan
hwatanabe@jamstec.go.jp

Chapter 1
Chemosynthetically-Driven Ecosystems in the Deep Sea

Steffen Kiel and Paul A. Tyler

Contents

1.1 Introduction

The process of chemosynthesis, by which reduced inorganic chemicals are used as the primary energy source rather than sunlight, has been known for many years (Winogradski 1887), the most obvious expression in the marine environment being the 'black layer' found in marine sediments. However, this was considered a very minor ecosystem that contributed very little to the carbon budget of the marine environment. All this was to change in 1977.

Elder (1965) predicted the existence of thermal springs along mid-ocean ridges. Data on the anomalous heat flow measured along selected spreading centres (Talwani et al. 1971), and changes in basalt chemistry and metalliferous sediments (Bostrom and Peterson 1966; Deffreyes 1970) supported this suggestion. A mechanism of active convective circulation through newly-formed oceanic crust was proposed by Sclater and Klitgord (1973) and supported by heat flow measurements (Williams et al. 1974). To test the hypothesis that there was convective heat flow at mid-ocean ridges, Corliss et al. (1979) used the submersible *Alvin* to explore the Galapagos Rift at ~2,500 m depth in 1977. As predicted they found thermal springs, now called hydrothermal vents, but the outstanding serendipitous discovery were

S. Kiel (✉)
Institut für Geowissenschaften, Christian-Albrechts-Universität, Ludewig-Meyn-Str. 10, 24118 Kiel, Germany
e-mail: steffen.kiel@gmx.de

P.A. Tyler
National Oceanography Centre, Southampton, University of Southampton Waterfront Campus, European Way, Southampton SO14 3ZH, UK

S. Kiel (ed.), *The Vent and Seep Biota*, Topics in Geobiology 33,
DOI 10.1007/978-90-481-9572-5_1,

the substantial fauna associated with these vents and the subsequent elucidation that the dominants in these ecosystems harboured chemosynthetic bacteria capable of using the hydrogen sulphide emitted from the hydrothermal vent. Thus was the discovery of a major marine ecosystem driven by chemosynthesis rather than by plant material. This unexpected discovery changed one of the major paradigms in deep-sea biology and has been a constant source of exploration, analysis and not a little wonder ever since.

These initial discoveries were rapidly followed by discoveries of vents along the East Pacific Rise (Grassle 1985) and Guymas Basin, the NE Pacific (Juan de Fuca, Gorda and Explorer Ridges) (Tunnicliffe 1988; Rona et al. 1990) (Fig. 1.1, Plate 1). Vents were predicted along the Trans-Atlantic Geotraverse (TAG) region of the Mid-Atlantic Ridge and in 1984 hydrothermal sediments and shrimp were collected by dredge at what became known as the TAG site (Rona et al. 1986). This led to the first concept of hydrothermal biogeography in that Pacific vents were dominated by tubeworms and alvinellid polychaetes whilst Atlantic vents were dominated by the caridean shrimp *Rimicaris exoculata*. Common to both biogeographic regions were vent mussels of the genus *Bathymodiolus*. A common physical parameter to all vents discovered by the mid-1980s was elevated temperature.

1.2 Diversity of Chemosynthetic Ecosystems

More recently, the hottest vents yet discovered, at 407°C have been discovered on the MAR at 5°S (C. German 2006, personal communication, doi:10.1038/news060522-15). An anomaly in the Atlantic was the discovery of the 15 km-off-axis Lost City hydrothermal vent (Kelley et al. 2001). This site was believed to be driven by heat derived from the exothermic serpentinization reactions between mantle rocks and seawater, but more recently Allen and Seyfried (2004) have proposed that the vent uses magmatic heat sources from the nearby mid-ocean ridge. The chimneys are steep-side, formed of carbonate and vent fluids have a maximum temperature of 75°C and an alkaline pH of 9–9.8 (Kelley et al. 2001).

Contemporaneously with the discovery of vents in the Atlantic was a chemosynthetic ecosystem driven by hydrogen sulphide but not at elevated temperatures. These systems became known at 'cold seeps' because there was no elevation in the ambient temperature. The first was found at the base of the Florida Escarpment in the Gulf of Mexico where sulphide-rich water emanated from the limestone platform and supported a chemosynthetically driven community (Paull et al. 1984) (Fig. 1.2, Plate 2). The discovery of a sulphide-based ecosystem was rapidly supplemented by the discovery of cold seeps along the northern side of the Gulf of Mexico where communities were supported by both methane and hydrogen sulphide. The surprising feature about these cold seeps is that they had tubeworms closely related to those found along vents in the Pacific.

The late 1980s saw an increased interest in back-arc basins where short stretches of mid-ocean ridge are isolated from other ridge segments (Fig. 1.1, Plate 1).

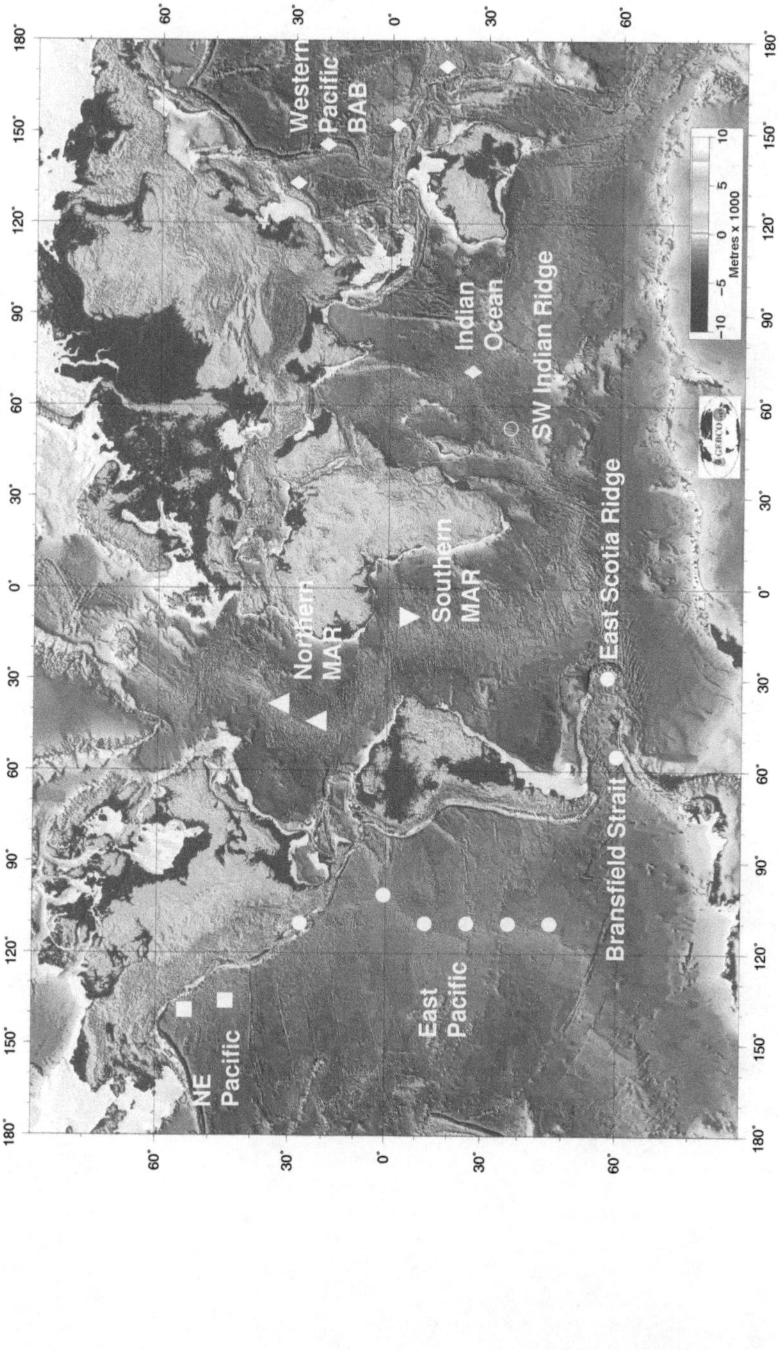

Fig. 1.1 Putative biogeographic provinces of hydrothermal vents in the global ocean. Individual vents are not marked. Open circles represent sites of known hydrothermal activity but vent ecosystems are yet to be found (as of November 2008). Bransfield Strait (BS) is yet to be assigned to a biogeographic province. A color plate of this figure can be found in Appendix (Plate 1)

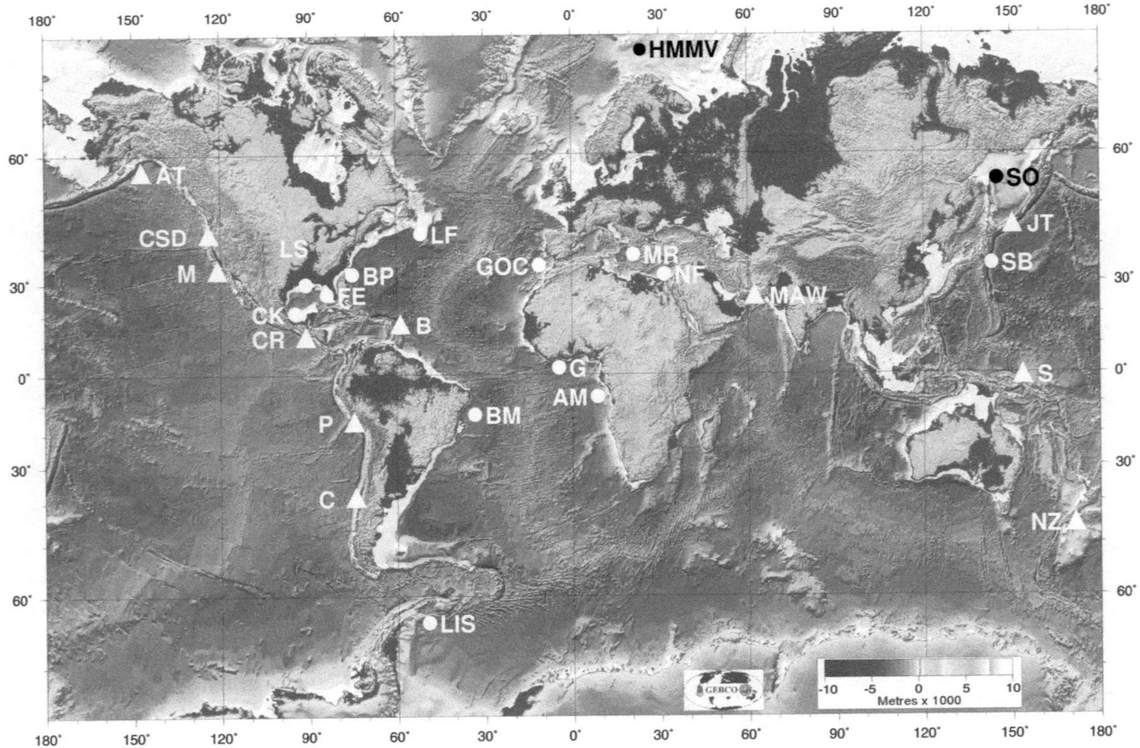

Fig. 1.2 Known cold seeps in the global ocean. Seeps on active margins (triangles) include: Aleutian Trench (AT), Barbados Accretionary Prism (B), Chile (C), Cascadia Margin/Oregon (CSD), Costa Rica (CR), Japan Trench (JT), Monterey Canyon (strictly a strike slip margin)(M), Makran Accretionary Wedge (MAW), New Zealand (NZ), Peru (P), Sissan (S). Seeps on passive margins (circles): Angola Margin (AM), Brazil Margin (BM), Blake Plateau (BP), Haakon Mosby Mud Volcano (HMMV), Campeche Knolls (CK), Florida Escarpment (FE), Guinea (G), Gulf of Cadiz (GOC), Laurentian Fan (LF), Larsen Ice Shelf (LIS), Louisiana Slope (LS), Mediterranean Ridge (MR), Nile Fan (NF), Sagami Bay (SB), Sea of Okhotsk (SO). A color plate of this figure can be found in Appendix (Plate 2)

The first analysis of these back arc basins was the Mariana Trough (Hessler and Lonsdale 1991). Subsequent exploration has revealed faunas in the North Fiji, Manus and Lau basins (Hashimoto and Jollivet 1989; Tufar 1989; Desbruyères et al. 1994), often dominated by snails rather than the more readily recognised tubeworms or shrimp. To the south of Japan vent systems were recognised in the Okinawa Trough (Kimura et al. 1988; Tsuchida et al. 2000). These vents differed from other back arc basins in that the vent chimneys were carbonate rather than acidic (Izawa et al. 1991).

By the late 1990s, the only regions where vents had not been discovered were in polar latitudes and the Indian Ocean. In the Indian Ocean, vents sites were found close to the Rodriguez Triple Junction and preliminary analysis of their fauna suggested they were dominated by the shrimp and anemone typical of the Mid-Atlantic Ridge. More detailed analysis of their less obvious fauna suggested a link to the back-arc basins of the western Pacific (Hashimoto et al. 2001; Van Dover et al. 2001). Hydrothermal plumes have been recognised along the SW Indian Ridge (German et al. 1998) but at the time of writing no vents had been directly observed along this section of mid-ocean ridge (Fig. 1.1, Plate 1). Further south there was evidence from vent plume studies that hydrothermal vents would be found along the East Scotia Ridge (German et al. 2000), possibly the most isolated back-arc basin in the global ocean. Similar evidence of hydrothermal plumes has been found along the Gakkel Ridge in the Arctic (Edmonds et al. 2003), and the East Scotia Ridge (German et al. 1998). Initial observations in 2009 (http://www.noc.soton.ac.uk/chess/) report a vent fauna along the East Scotia Ridge, but a vent fauna has yet to be observed along the Gakkel Ridge. In the Bransfield Strait there is strong evidence of hydrothermal venting (Klinkhammer et al. 2001) but the tube worms present appear to avoid the warm waters (Sahling et al. 2005).

The discoveries of cold seeps were no less active than the discovery of vents during the late 1980s and 1990s (Fig. 1.2, Plate 2). Whereas vents were confined to spreading centres, cold seeps have been found along many passive and active margins, and generally over a much wider depth range than vents. Along active margins in the E Pacific, cold seeps have been found down to ~5,000 m depth associated with scarps on the landward side of the subduction zone off Peru (Olu et al. 1996a). Similar cold seeps sites have been found along this active margin off Costa Rica (Kahn and Orange 1994) and north to Monterey Bay (technically a strike/slip margin) (Embley et al. 1990; Barry et al. 1996) and the Oregon Margin (Suess et al. 1985; Sahling et al. 2002). There is strong evidence from trawlings of chemosynthetically-driven bivalve faunas off Chile that cold seeps occur in this region also (Stuardo and Valdovinos 1988; Sellanes et al. 2004). On the other side of the Pacific off Japan, cold seeps have been discovered in Sagami Bay and down to more than 7,000 m on the subduction zone off Japan (Laubier et al. 1986; Sibuet et al. 1988; Fujikura et al. 1999). In the SW Pacific, the Papua-New Guinea earthquake of 1998 resulted in landslides that gave rise to small isolated chemosynthetically-driven ecosystems dominated by *Bathymodiolus* (Tappin et al. 2001). This site is close to the vent systems of the Bismarck Basin (Tufar 1989). Active margin sites in the Atlantic are much less rare than in the

Pacific but cold seeps have been found on the Barbados Accretionary Prism in the western Atlantic where mud volcanoes are the main expression of cold seep activity (Olu et al. 1996b).

In the last 10 years numerous cold seeps have been found on passive margins, particularly in the Atlantic (Fig. 1.2, Plate 2). Sites now known include the Blake Plateau off the eastern seaboard of N. America (Van Dover et al. 2003), the Brazil Margin, the Angolan and Congo River margin (Ondréas et al. 2005; Gay et al. 2006; Olu-Le Roy et al. 2007) and extensive mud volcanoes in the Gulf of Cadiz (Génio et al. 2008; P. Tyler 2007, personal observation). Determining the age of cold seeps is often a problem but one cold seep, the Laurentian Fan, is readily dateable as it was formed by a land slide associated with the 1929 Grand Banks earthquake (Mayer et al. 1988). The well-established vents along the northern border of the Gulf of Mexico were complemented by cold seeps in an asphalt community in the Campeche Knolls in the SW of the Gulf (MacDonald et al. 2004). In this community the asphalt had the appearance of lava flows and the tubeworm community developed where the asphalt cracked.

In the southern hemisphere cold seep exploration is developing rapidly. Cold seeps were recognised off New Zealand in the late 1990s (Lewis and Marshall 1996) and have recently been the subject of intense study (Baco et al. 2010). Cold seeps are known from the Arctic. The discovery of the Haakon Mosby mud volcano (Hjelsteun et al. 1999) presaged one of the most intense studies of a mud volcano that continues to this day. The HMMV is also one of the larger mud volcanoes and is showing temporal as well as spatial variation. As the other end of the Atlantic in the Weddell Sea there is evidence of cold seeps close to the Larsen Ice Shelf (Domack et al. 2005). Cold seeps bivalves were observed by cameras lowered from a surface vessel but subsequent ROV dives have failed to locate an active cold seep.

A region of current active interest is the Mediterranean. Cold seep faunas have been found along the Mediterranean Ridge (Corselli and Brasso 1996; Salas and Woodside 2002; Olu et al. 2004) and there is active research under the auspices of the EU-HERMES programme (www.eu-hermes.org) examining the microbial populations of reducing environments along the Nile delta and in the Black Sea.

Not all cold seeps are natural. The *Francois Vieljeux* was a coastal vessel carrying copper ingots, coffee and tins of pineapples when she sank in 1,200 m of water off the coast of Spain. In the salvage operation to recover the copper, the grab brought up siboglinid tubeworms typical of cold seeps. It would appear that the hold containing the coffee and pineapples had rotted and formed a reducing environment producing hydrogen sulphide that supported a dense siboglinid population (Dando et al. 1992).

Possibly the most unexpected cold seeps are those associated with whale falls. In the late 1980s Craig Smith discovered a gray whale carcass off the coast of California. The whale was skeletonised but the whale bones were covered in a dense 'fur' of filamentous bacteria and supported a population of small mussels of the genus *Idas* on the bone and one or more species of vesicomyid in the surrounding reduced sediment (Smith et al. 1989; Bennett et al. 1994; Smith and Baco 2003). Since this discovery of a natural whale fall there have been implanted whale 'falls' and the discovery of natural ones especially at bathyal depths off California and Smith and Baco (2003) suggest a distinct sequence of events leading to a 'sulphophilic' stage based on sulphate reduction in the vertebrae of the whales.

At the time of writing (late 2008) research into hydrothermal vents and cold seeps ecosystem communities continues apace. Exploration is less of a focus now and process analysis tends to dominate. However, scientists have become more skilled in recognising vents and seeps and there is a battery of sophisticated technologies to aid us in discovery. AUVs and ROVs are at the forefront of these technologies, especially in combination with dynamic positioning. At vents there is the miniaturisation of sensors that can detect plumes in situ in real time. Bottom seismic reflectors are used to locate areas of gas hydrates and thus cold seeps. Areas of the most intense activity are the polar sea, both north and south, as well as areas where there is active oil exploration such as off West Africa.

1.3 Fossil History

Shortly after the discoveries of vent, seep, and whale fall communities in the modern oceans, geologists and paleontologists found fossil analogs of these ecosystems (Fig. 1.3). Five years after Corliss et al. (1979) reported dense tube worm clusters from vents on the Galapagos Ridge, tube-like structures were discovered in volcanogenic massive sulphide (VMS) deposits of the Cretaceous Samail ophiolite in

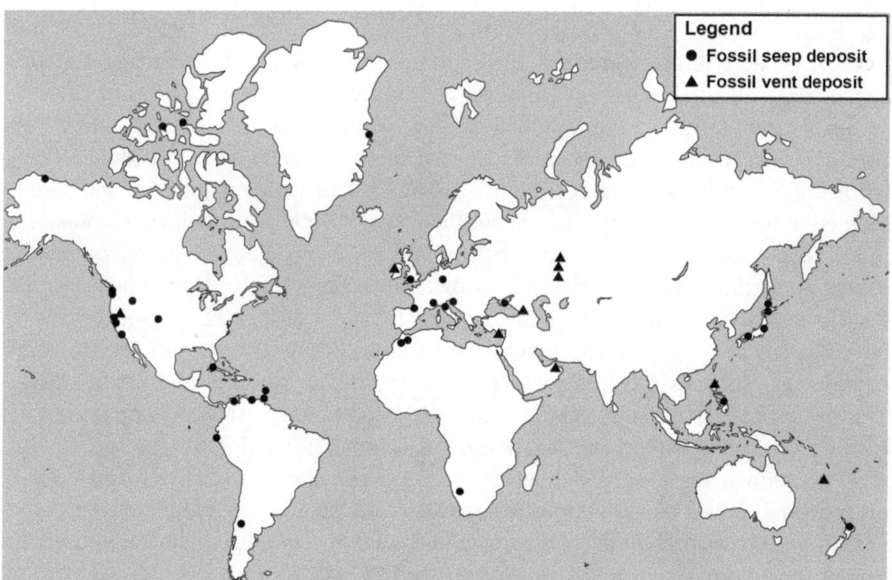

Fig. 1.3 Occurrences of fossil vent and seep deposits

Oman (Haymon et al. 1984). Only one year after the discovery of seep communities on the Florida escarpment (Paull et al. 1984) fossil seep communities were recognized by Gaillard and Rolin (1986) in the Late Jurassic of southern France and the Late Cretaceous of the United States. And it took two years from Craig Smith's finding of chemosymbiotic animals on whale remains in 1989 until fossil examples were reported from Oligocene deep-water sediments exposed on the Olympic peninsula in western Washington (Squires et al. 1991). Recently – after many years of speculation whether large marine reptile carcasses in the Mesozoic were also inhabited by such communities – Kaim et al. (2008b) discovered two Late Cretaceous plesiosaur skeletons associated with provannid and skeneimorph gastropods in deep-water sediments on Hokkaido.

Further vent fossils were discovered during the 1980s, expanding the geologic range of these ecosystems from the Carboniferous to the Eocene (Banks 1985; Oudin et al. 1985; Boirat and Fouquet 1986). Because all these fossils were tubes, it was suggested that worms similar or perhaps ancestral to the modern vestimentiferan vent worms also inhabited this type of ecosystem in the past. Better preserved and more diverse fossil vent communities were discovered during the 1990s. These included Paleozoic communities from the Ural mountains with large mollusks, brachiopods, and worm tubes (Kuznetsov et al. 1993; Little et al. 1997), an Early Jurassic vent community from California hosting gastropods, brachiopods and worm tubes (Little et al. 1999b), and Late Cretaceous examples from the Troodos ophiolite on Cyprus that contain a diverse gastropod fauna along with the ubiquitous worm tubes (Little et al. 1999a). In addition to these Phanerozoic macrofaunal communities, a deep-time perspective of life at ancient vents was provided by Rasmussen (2000) who reported filamentous microfossils in a 3.235-million-year-old VMS deposit in Australia, that were interpreted as sulphur-oxidizing bacteria. Research on fossil vents has slowed down during the past decade and revealed only a single new site with tubular fossil (Little et al. 2007).

Seep communities have a much higher potential of preservation than vent communities, because increased pH values at seeps enhances both carbonate precipitation and the preservation of carbonaceous shells. Consequently, the fossil record of seep communities is significantly richer than that of vent communities. Hundreds of fossil seep deposits are now known from 29 countries situated on all continents except Australia. The great majority of them are subduction-related deposits on the Pacific coasts of Japan and the United States (Majima et al. 2005; Campbell 2006), but there are also various occurrences on passive continental margins and epicontinental seas, such as the famous Tepee Buttes in the Western Interior Seaway, a shallow seaway that connected the Gulf of Mexico and the Arctic Ocean during the Late Cretaceous (Gaillard and Rolin 1986; Kauffman et al. 1996).

Most known fossil seep faunas are of Cenozoic age. Unrivalled world-wide in diversity and preservation are the seep faunas of the deep-water sediments in western Washington, which have been used to address taxonomic, geochemical and evolutionary questions since their discovery (Goedert and Squires 1990; Goedert et al. 2003; Kiel 2006; Kiel and Goedert 2006; Schweitzer and Feldmann 2008). The Miocene seep deposits of Italy – known as '*Calcari a Lucina*' – continue to be in the focus of researchers. Recent work unraveled the internal structure of mud volcanoes, showed the growth stages encompassing the entire geologic life-time of individual seep deposits, and outlined the evolution of seeps in relation to the orogeny of the Apennine Mountains (Clari et al. 2004; Conti and Fontana 2005; Conti et al. 2007, 2008). Clusters of vesicomyid clams have long been recognized in the Cenozoic sediments of Japan (Oinomikado and Kanehara 1938) and Niitsuma et al. (1989) recognized such a cluster of *Calyptogena* clams in Pliocene sediments in Japan as fossil analogs of cold seep faunas in the subduction zone off Japan. In addition to subduction-related seeps found in the Cenozoic sediments along the Pacific coast of Japan, fossil seep faunas are also known from the Japan Sea borderland (Amano et al. 2001); see also the review by Majima et al. (2005).

Geologically older seep-bearing strata of current active interest are sediments of the Yezo Group that outcrop in a North–south trending belt on Hokkaido, the northernmost island of Japan. These Cretaceous deep-water sediments yield a wealth of well-preserved seep faunas (Hikida et al. 2003; Jenkins et al. 2007a, b; Kaim et al. 2008a) along with similar faunas found on plesiosaurs as mentioned earlier, and on sunken wood (Kiel et al. 2009). The rich mollusc fauna of Late Jurassic to Early Cretaceous seep deposits of the Great Valley Group in California has recently received attention (Kiel and Campbell 2005; Campbell et al. 2008; Kiel et al. 2008). Many of these taxa are part of a distinctive seep fauna that apparently had a world-wide distribution during this age (Kiel and Peckmann 2008). Even older seep deposits have been located in eastern Oregon, where the sediments of numerous terranes and subduction zones were accreted to the western margin of the North American continent during the Triassic and Jurassic (Peckmann, Kiel, Taylor and Sandy 2008, personal communication).

Perhaps the most interesting discovery in the fossil record of seep communities is that many Mesozoic and Paleozoic sites are dominated by brachiopods – animals that are virtually absent from modern vents and seeps. Even more surprising is that these brachiopods tend to be much larger than their shallow-water relatives and that they appear to be restricted to seep deposits. Mass occurrences of the Early Cretaceous rhynchonellid brachiopod *Peregrinella* in hemi-pelagic sediments in southern France were the first to be discussed in this context (Lemoine et al. 1982). The idea was later supported by Campbell and Bottjer (1993, 1995) who found the carbonates enclosing *Peregrinella* to show features characteristic for seep-carbonates. Sandy (1995) identified a number of Paleozoic and Mesozoic brachiopod genera that could potentially be seep-related, based on their isolated and highly disjunct mass occurrences. One of these brachiopods was the late Devonian rhynchonellid *Dzieduszyckia*; recent work on such an occurrence in Morocco showed that it indeed lived at an ancient cold seep (Peckmann et al. 2007).

A fascinating line of research into the nature of ancient seeps that developed during the past 10 years focuses on geochemical markers and molecular fossils to identify the intensity and composition of fluids at ancient seeps. Fossilized oil has been found at a Carboniferous seep in the Harz Mountains in Germany, providing evidence for oil seepage (Peckmann et al. 2001). The discovery of biomarkers (molecular fossils) characteristic for anaerobic methane-oxidizing Archaea was a major step forward to provide evidence for methane oxidation – and therewith the base of the food-chain – at ancient seeps (Peckmann et al. 1999; Thiel et al. 1999). Such molecular fossil can now be traced back into the Carboniferous (Birgel et al. 2008).

The rising interest in taxonomic identifications of vent and seep fossils will help paleontologists to use the direct evidence provided by the fossil record to test hypotheses on the evolutionary history of vent and seep faunas put forward by their colleagues from the biology departments. Likewise, ongoing work on biogeochemical proxies will help to understand how and under which circumstances animals lived at chemosynthetic ecosystems in the geologic past.

Acknowledgments We would like to thank the ChEss (http://www.noc.soton.ac.uk/chess/) and Census of Marine Life programs (http://www.coml.org) that kindly sponsored the color appendix of this book.

References

Allen DE, Seyfried WE (2004) Serpentinization and heat generation: constraints from Lost City and Rainbow hydrothermal systems. Geochim Cosmochim Acta 68:1347–1354

Amano K, Hamuro T, Hamuro M, Fujii S (2001) The oldest vesicomyid bivalves from the Japan Sea borderland. Venus 60:189–198

Baco AR, Rowden AA, Levin LA, Smith CR et al (2010) Initial characterization of cold seep faunal communities on the New Zealand Hikurangi margin. Mar Geol 272:251–259

Banks DA (1985) A fossil hydrothermal worm assemblage from the Tynagh lead-zinc deposit in Ireland. Nature 313:128–131

Barry JP, Greene G, Orange DL, Baxter CH et al (1996) Biologic and geologic characteristics of cold seeps in Monterey Bay, California. Deep Sea Res I 43:1739–1762

Bennett BA, Smith CR, Glaser B, Maybaum HL (1994) Faunal community structure of a chemotrophic assemblage on whale bones in the deep northeast Pacific Ocean. Mar Ecol Prog Ser 108:205–223

Birgel D, Himmler T, Freiwald A, Peckmann J (2008) A new constraint on the antiquity of anaerobic oxidation of methane: late Pennsylvanian seep limestones from southern Namibia. Geology 36:543–546

Boirat J-M, Fouquet Y (1986) Découverte de tubes de vers hydrothermaux fossiles dans un amas sulfuré de l'Éocène supérieur (Barlo, ophiolite de Zambalès, Philippines). CR Acad Sci Paris, Sci Terre 302:941–946

Bostrom K, Peterson MN (1966) Precipitaes from hydrothermal exhalations on the East Pacific Rise. Econ Geol 61:1258–1265

Campbell KA (2006) Hydrocarbon seep and hydrothermal vent paleoenvironments and paleontology: past developments and future research directions. Palaeogeogr Palaeocl 232:362–407

Campbell KA, Bottjer DJ (1993) Fossil cold seeps. Natl Geogr Res Explor 9:326–343

Campbell KA, Bottjer DJ (1995) *Peregrinella*: an Early Cretaceous cold-seep-restricted brachiopod. Paleobiology 24:461–478

Campbell KA, Peterson D, Alfaro AC (2008) Two new species of *Retiskenea*? (Gastropoda: Neomphalidae) from Lower Cretaceous hydrocarbon seep-carbonates of northern California. J Paleontol 82:140–153

Clari P, Cavagna S, Martire L, Hunziker J (2004) A Miocene mud volcano and its plumbing system: a chaotic complex revisited (Monferrato, NW Italy). J Sediment Res 74:662–676

Conti S, Fontana D (2005) Anatomy of seep-carbonates: ancient examples from the Miocene of the northern Apennines (Italy). Palaeogeogr Palaeocl 227:156–175

Conti C, Artoni A, Piola G (2007) Seep-carbonates in a thrust-related anticline at the leading edge of an orogenic wedge: the case of the middle–late Miocene Salsomaggiore Ridge (Northern Apennines, Italy). Sediment Geol 199:233–251

Conti S, Fontana D, Lucente CC (2008) Authigenic seep-carbonates cementing coarse-grained deposits in a fan-delta depositional system (middle Miocene, Marnoso-arenacea Formation, central Italy). Sedimentology 55:471–486

Corliss JB, Dymond J, Gordon LI, Edmond JM et al (1979) Submarine thermal springs on the Galápagos Rift. Science 203:1073–1083

Corselli C, Brasso D (1996) First evidence of benthic communities based on chemosynthesis on the Mediterranean Ridge (Eastern Mediterranean). Mar Geol 132:227–239

Dando PR, Southward AJ, Southward EC, Dixon DR et al (1992) Shipwrecked tube worms. Nature 356:667

Deffreyes KL (1970) The axial valley: a steady state feature of the terraine. In: Johnson H, Smith BL (eds) Megatectonics of continents and oceans. Rutgers University Press, New Brunswick, NJ

Desbruyères D, Alayse-Danet A-M, Ohta S, Scientific Parties of BIOLAU and STARMER Cruises (1994) Deep-sea hydrothermal communities in southwestern Pacific back-arc basins (the North Fiji and Lau Basins): composition, microdistribution and food web. Mar Geol 116:227–242

Domack E, Ishman S, Leventer A, Sylva S et al (2005) A chemotrophic ecosystem found beneath Antarctic ice shelf. Eos 86(269):271–272

Edmonds HN, Michael PJ, Baker ET, Connelly DP et al (2003) Discovery of abundant hydrothermal venting on the ultraslow-spreading Gakkel Ridge in the Arctic Ocean. Nature 421:252–256

Elder JW (1965) Physical processes in geothermal areas. AGU Monogr 8:211–239

Embley RW, Eittreim SL, McHugh CH, Normark WR et al (1990) Geological setting of chemosynthetic communities in the Monterey Fan Valley system. Deep Sea Res 37:1651–1667

Fujikura K, Kojima S, Tamaki K, Maki Y et al (1999) The deepest chemosynthesis-based community yet discovered from the hadal zone, 7326 m deep, in the Japan Trench. Mar Ecol Prog Ser 190:17–26

Gaillard C, Rolin Y (1986) Paléobiocœnoses susceptibles d'être liées à des sources sous-marines en milieu sédimentaire. L'exemple des pseudobiohermes des Terres Noires (S. E. France) et des tepee buttes de la Pierre Shale Formation (Colorado, U.S.A.). CR Acad Sci Paris 303: 1503–1508

Gay A, Lopez M, Ondréas H, Charlou JL et al (2006) Seafloor facies related to upward methane flux within a giant pockmark of the lower Congo Basin. Mar Geol 226:81–95

Génio L, Johnson SB, Vrijenhoek RC, Cunha MR et al (2008) New record of *Bathymodiolus mauritanicus* Cosel from the Gulf of Cadiz (NE Atlantic) mud volcanoes. J Shell Res 27:53–61

German CR, Baker ET, team tFS (1998) Hydrothermal activity along the southwest Indian Ridge. Nature 395:490–493

German CR, Livermore RA, Baker ET, Bruguier NI et al (2000) Hydrothermal plumes above the East Scotia Ridge: an isolated high-latitude back-arc spreading centre. Earth Planet Sci Lett 184:241–250

Goedert JL, Squires RL (1990) Eocene deep-sea communities in localized limestones formed by subduction-related methane seeps, southwestern Washington. Geology 18:1182–1185

Goedert JL, Thiel V, Schmale O, Rau WW et al (2003) The late Eocene 'Whiskey Creek' methane-seep deposit (western Washington State) Part I: geology, palaeontology, and molecular geobiology. Facies 48:223–240

Grassle JF (1985) Hydrothermal vent animals: distribution and biology. Science 229:713–725

Hashimoto J, Jollivet D (1989) The hydrothermal vent communities in the North Fiji Basin: results of Japan-France cooperative research on board Kaiyo 88. La mer Tokyo 27:62–71

Hashimoto J, Ohta S, Gamo T, Chiba H et al (2001) First hydrothermal vent communities from the Indian Ocean discovered. Zool Sci 18:717–721

Haymon RM, Koski R, Sinclair C (1984) Fossils of hydrothermal vent worms from Cretaceous sulfide ores of the Samail ophiolite, Oman. Science 223:1407–1409
Hessler RR, Lonsdale P (1991) Biogeography of Mariana Trough hydrothermal vent communities. Deep Sea Res 38:185–199
Hikida Y, Suzuki S, Togo Y, Ijiri A (2003) An exceptionally well-preserved seep community from the Cretaceous Yezo forearc basin in Hokkaido, northern Japan. Paleontol Res 7:329–342
Hjelsteun BO, Eldholm O, Faleide JI, Vogt PR (1999) Regional settings of Hakon Mosby Mud Volcano, SE Barents Sea margin. Geo Mar Lett 19:22–28
Izawa E, Motomura Y, Tanaka T, Kimura M (1991) Hydrothermal carbonate chimneys in the Iheya Ridge of the Okinawa Trough. JAMSTEC J Deep Sea Res 7:185–192
Jenkins RG, Kaim A, Hikida Y (2007a) Antiquity of the substrate choice among acmaeid limpets from the Late Cretaceous chemosynthesis-based communities. Acta Palaeontol Pol 52:369–373
Jenkins RG, Kaim A, Hikida Y, Tanabe K (2007b) Methane-flux-dependent lateral faunal changes in a Late Cretaceous chemosymbiotic assemblage from the Nakagawa area of Hokkaido, Japan. Geobiol 5:127–139
Kahn LM, Orange D (1994) Evidence of fluid venting from Costa Rica accretionary prism. Eos, 1994 Spring Meeting Supplement, p 324
Kaim A, Jenkins RG, Warén A (2008a) Provannid and provannid-like gastropods from Late Cretaceous cold seeps of Hokkaido (Japan) and the fossil record of the Provannidae (Gastropoda: Abyssochrysoidea). Zool J Linn Soc 154:421–436
Kaim A, Kobayashi Y, Echizenya H, Jenkins RG et al (2008b) Chemosynthesis-based associations on Cretaceous plesiosaurid carcasses. Acta Palaeontol Pol 53:97–104
Kauffman EG, Arthur MA, Howe B, Scholle PA (1996) Widespread venting of methane-rich fluids in Late Cretaceous (Campanian) submarine springs (Tepee Buttes), Western Interior seaway, U.S.A. Geology 24:799–802
Kelley DS, Karson JA, Blackmann DK, Früh-Green GL et al (2001) An off-axis hydrothermal vent field near the Mid-Atlantic Ridge at 30 degree N. Nature 412:145–149
Kiel S (2006) New records and species of mollusks from Tertiary cold-seep carbonates in Washington State, USA. J Paleontol 80:121–137
Kiel S, Campbell KA (2005) *Lithomphalus enderlini* gen. et sp. nov. from cold-seep carbonates in California - a Cretaceous neomphalid gastropod? Palaeogeogr Palaeoclim Palaeoeco 227:232–241
Kiel S, Goedert JL (2006) Deep-sea food bonanzas: early Cenozoic whale-fall communities resemble wood-fall rather than seep communities. Proc R Soc B 273:2625–2631
Kiel S, Peckmann J (2008) Paleoecology and evolutionary significance of an Early Cretaceous *Peregrinella*-dominated hydrocarbon-seep deposit on the Crimean Peninsula. Palaios 23:751–759
Kiel S, Campbell KA, Elder WP, Little CTS (2008) Jurassic and Cretaceous gastropods from hydrocarbon-seeps in forearc basin and accretionary prism settings, California. Acta Palaeontol Pol 53:679–703
Kiel S, Amano K, Hikida Y, Jenkins RG (2009) Wood-fall associations from Late Cretaceous deep-water sediments of Hokkaido, Japan. Lethaia 42:74–82
Kimura M, Uyeda S, Kato Y, Tanaka T et al (1988) Active hydrothermal mounds in the Okinawa Trough backarc basin, Japan. Tectonophysics 145:319–324
Klinkhammer GP, Chin CS, Dählmann A, Sahling H et al (2001) Discovery of new hydrothermal vent sites in Bransfield Strait, Antarctica. Earth Planet Sci Lett 193:395–407
Kuznetsov AP, Maslennikov VV, Zaikov VV (1993) The near-hydrothermal fauna of the Silurian paleoocean in the south Ural. Izv Akad Nauk Ser Biol 4:525–534 (in Russian)
Laubier L, Ohta S, Sibuet M (1986) Decouverte de communautes animales profondes durant la campagne franco-japonaise KAIKO de plongees dans les fosses de subduction autour du Japon. CR Acad Sci Paris Ser III 303:25–29
Lemoine M, Arnaud-Vanneau A, Arnaud H, Létolle R et al (1982) Indices possibles de paléo-hydrothermalisme marin dans le Jurassique et le Crétacé des Alpes occidentales

(océan téthysien et sa marge continetale européenne): essai d'inventaire. Bull Soc Géol Fr 24:641–647

Lewis KB, Marshall BA (1996) Seep faunas and other indicators of methane-rich dewatering on New Zealand convergent margins. NZ J Geol Geophys 39:181–200

Little CTS, Herrington RJ, Maslennikov VV, Morris NJ et al (1997) Silurian hydrothermal vent community from the southern Urals, Russia. Nature 385:146–148

Little CTS, Cann JR, Herrington RJ, Morisseau M (1999a) Late Cretaceous hydrothermal vent communities from the Troodos ophiolite, Cyprus. Geology 27:1027–1030

Little CTS, Herrington RJ, Haymon RM, Danelian T (1999b) Early Jurassic hydrothermal vent community from the Franciscan Complex, San Rafael Mountains, California. Geology 27:167–170

Little CTS, Magalashvili AG, Banks DA (2007) Neotethyan Late Cretaceous volcanic arc hydrothermal vent fauna. Geology 35:835–838

MacDonald IR, Bohrmann G, Escobar E, Abegg F et al (2004) Asphalt volcanism and chemosynthetic life in the Campeche Knolls, Gulf of Mexico. Science 304:999–1002

Majima R, Nobuhara T, Kitazaki T (2005) Review of fossil chemosynthetic assemblages in Japan. Palaeogeogr Palaeocl 227:86–123

Mayer LA, Shor AN, Huges Clarke J, Piper DJW (1988) Dense biological communities at 3850 m on the Laurentian Fan and their relationship to the deposits of the 1929 Grand Banks earthquake. Deep Sea Res 35:1235–1246

Niitsuma N, Matsushima Y, Hirata D (1989) Abyssal molluscan colony of *Calyptogena* in the Pliocene strata of the Miura peninsula, Central Japan. Palaeogeogr Palaeocl 71:193–203

Oinomikado T, Kanehara K (1938) A new species of *Calyptogena* from the Higasiyama Oil Field, Niigata-ken, Japan. J Geol Soc Jpn 45(93–95):677–679

Olu K, Duperret A, Sibuet M, Foucher J-P et al (1996a) Structure and distribution of cold seep communities along the Peruvian active margin: relationship to geological and fluid patterns. Mar Ecol Prog Ser 132:109–125

Olu K, Sibuet M, Harmegnies F, Foucher J-P et al (1996b) Spatial distribution of diverse cold seep communities living on various diapiric structures of the southern Barbados prism. Prog Oceanogr 38:347–376

Olu K, Sibuet M, Fiala-Médoni A, Gofas S et al (2004) Cold seep communities in the deep eastern Mediterranean Sea: composition, symbiosis and spatial distribution on mud volcanoes. Deep Sea Res I 51:1915–1936

Olu-Le Roy K, Caprais J-C, Fifis A, Fabri M-C et al (2007) Cold-seep assemblages on a giant pockmark off West Africa: spatial patterns and environmental control. Mar Ecol 28:115–130

Ondréas H, Olu K, Fouquet Y, Charlou JL et al (2005) ROV study of a giant pockmark on the Gabon continental margin. Geo Mar Lett 25:281–292

Oudin E, Bouladon J, Paris J-P (1985) Vers hydrothermaux fossils dans une minéralisation sulfurée des ophiolites de Nouvelle-Calédonie. CR Acad Sci Paris 301:157–162

Paull CK, Hecker B, Commeau R, Freeman-Lynde RP et al (1984) Biological communities at the Florida Escarpment resemble hydrothermal vent taxa. Science 226:965–967

Peckmann J, Thiel V, Michaelis W, Clari P et al (1999) Cold seep deposits of Beauvoisin (Oxfordian; southeastern France) and Marmorito (Miocene; northern Italy): microbially induced authigenic carbonates. Int J Earth Sci 88:60–75

Peckmann J, Gischler E, Oschmann W, Reitner J (2001) An Early Carboniferous seep community and hydrocarbon-derived carbonates from the Harz Mountains, Germany. Geology 29:271–274

Peckmann J, Campbell KA, Walliser OH, Reitner J (2007) A Late Devonian hydrocarbon-seep deposit dominated by dimerelloid brachiopods, Morocco. Palaios 22:114–122

Rasmussen B (2000) Filamentous microfossils in a 3.235-million-year-old volcanogenic massive sulphide deposit. Nature 405:676–679

Rona PA, Klinkhammer GP, Nelsen TA, Trefry JH et al (1986) Black smokers, massive sulphides and vent biota at the Mid-Atlantic Ridge. Nature 321:33–37

Rona PA, Denlinger RP, Fisk MR, Howard KJ et al (1990) Major off-axis hydrothermal activity on the northern Gorda Ridge. Geology 18:493–496
Sahling H, Rickert D, Lee RW, Linke P et al (2002) Macrofaunal community structure and sulfide flux at gas hydrate deposits from the Cascadia convergent margin, NE Pacific. Mar Ecol Prog Ser 231:121–138
Sahling H, Wallmann K, Dählmann A, Schmaljohann R et al (2005) The physiochemical habitat of *Sclerolinum* sp at Hook Ridge hydrothermal vent, Bransfield Strait, Antarctica. Limnol Oceanogr 50(2):598–606
Salas C, Woodside J (2002) *Lucinoma kazani* n. sp (Mollusca: Bivalvia): evidence of a living benthic community associated with a cold seep in the Eastern Mediterranean Sea. Deep Sea Res I 49:991–1005
Sandy MR (1995) A review of some Palaeozoic and Mesozoic brachiopods as members of cold seep chemosynthetic communities: "unusual" palaeoecology and anomalous palaeobiogeographic pattern explained. Foldtani Kozlony 125:241–258
Schweitzer CE, Feldmann RM (2008) New Eocene hydrocarbon seep decapod crustacean (Anomura: Galatheidae: Shinkaiinae) and its paleobiology. J Paleontol 82:1021–1029
Sclater JG, Klitgord KD (1973) A detailed heat flow, topographic and magnetic survey across the Galapagos Spreading Center at 86°W. J Geophys Res 78:6951–6975
Sellanes J, Quiroga E, Gallardo VA (2004) First direct evidence of methane seepage and associated chemosynthetic communities in the bathyal zone off Chile. J Mar Biol Assoc UK 84:1065–1066
Sibuet M, Juniper SK, Pautot G (1988) Cold-seep benthic communities in the Japan subduction zones: geological control of community development. J Mar Res 46:333–348
Smith CR, Baco AR (2003) Ecology of whale falls at the deep-sea floor. Oceanogr Mar Biol 41:311–354
Smith CR, Kukert H, Wheatcroft RA, Jumars PA et al (1989) Vent fauna on whale remains. Nature 341:27–28
Squires RL, Goedert JL, Barnes LG (1991) Whale carcasses. Nature 349:574
Stuardo J, Valdovinos C (1988) A new bathyal *Calyptogena* from off the coast of Central Chile (Bivalvia: Vesicomyidae). Venus 47:241–250
Suess E, Carson B, Ritger SD, Moore JC et al (1985) Biological communities at vent sites along the subduction zone off Oregon. Bull Biol Soc Wash 6:475–484
Talwani M, Windisch CC, Langseth MG (1971) Reykjanes Ridge Crest: a detailed geophysical study. J Geophys Res 76:473–517
Tappin DR, Watts P, McMurtry GM, Lafoy Y et al (2001) The Sissano, Papua New Guinea tsunami of July 1998 - offshore evidence on the source mechanism. Mar Geol 175:1–23
Thiel V, Peckmann J, Seifert R, Wehrung P et al (1999) Highly isotopically depleted isoprenoids: molecular markers for ancient methane venting. Geochim Cosmochim Acta 63:3959–3966
Tsuchida S, Watanabe K, Ishibashi J, Miyake H et al (2000) Preliminary report of a biological, geological, and geochemical survey on hydrothermalism at the Hotoma and Minna Knolls. JAMSTEC J Deep Sea Res 17:35–42
Tufar W (1989) Modern hydrothermal activity, formation of complex massive sulfide deposits and associated vent communities in the Manus back-arc basin (Bismark Sea, Papua New Guinea). Mitt Österr Geol Ges 8:183–210
Tunnicliffe V (1988) Biogeography and evolution of hydrothermal-vent fauna in the eastern Pacific Ocean. Proc R Soc B B233:347–366
Van Dover CL, Humphris SE, Fornari D, Cavanaugh CM et al (2001) Biogeography and ecological setting of Indian Ocean hydrothermal vents. Science 294:818–823
Van Dover CL, Aharon P, Bernhard JM, Caylor E et al (2003) Blake Ridge methane seeps: characterization of a soft-sediment, chemosynthetically based ecosystem. Deep Sea Res I 50:281–300
Williams DL, Von Herzen RP, Sclater JG, Anderson RN (1974) The Galapagos spreading centre: lithospheric cooling and hydrothermal circulation. Geophys J Roy Astr S 38:587–608
Winogradski SN (1887) Über Schwefelbacterien. Bot Ztg 45:489–610

Chapter 2
Genetics and Evolution of Deep-Sea Chemosynthetic Bacteria and Their Invertebrate Hosts

Robert C. Vrijenhoek

Contents

R.C. Vrijenhoek (✉)
Monterey Bay Aquarium Research Institute, 7700 Sandholdt Road,
Moss Landing, CA 95039, USA
e-mail: vrijen@mbari.org

S. Kiel (ed.), *The Vent and Seep Biota*, Topics in Geobiology 33,
DOI 10.1007/978-90-481-9572-5_2, © Springer Science+Business Media B.V. 2010

2.1 Introduction

The earth scientists who discovered deep-sea hydrothermal vent communities along the Galápagos Rift in 1977 set the stage for revealing chemosynthetic symbioses in animals. They found high concentrations of hydrogen sulfide in the vent effluents and hypothesized that "a significant proportion of the organic carbon utilized within these hot spring regions could be produced by chemolithotrophic sulfur-oxidizing bacteria" (Corliss et al. 1979, p. 1079). Chemolithotrophic microbes are faced with the problem of extracting energy from narrow redox zones in marine environments. The free-living species typically occur in biofilms on sulfidic rocks or in filamentous mats like *Beggiatoa* spp. (Fig. 2.1a, Plate 3a), absorbing reduced gases from the substrate below and oxygen from the ambient water above. Symbiotic microbes, however, can span broader oxic-anoxic boundaries by exploiting the behavior, physiology and morphology of their animal hosts (Stewart et al. 2005). For example, the small mobile shrimp *Rimicaris exoculata* (Fig. 2.1b, Plate 3b), which lives at the interface between hot sulfidic vent water and ambient seawater, circulates vent fluids through a gill chamber that is packed with sulfur-oxidizing (thiotrophic) episymbionts. In contrast, the sedentary vesicomyid clam, *Calyptogena magnifica* (Fig. 2.1c, Plate 3c), spans a broad redox zone by growing up to 30 cm in length. The clam protrudes its highly vascularized and extensible foot deeply into small rocky fissures to absorb dissolved sulfides, and it uses its siphon to circulate ambient seawater to gills housing intracellular thiotrophic endosymbionts. The sessile vestimentiferan tubeworm, *Riftia pachyptila* (Fig. 2.1d, Plate 3d), grows up to 1.5 m in length. It absorbs dissolved sulfide and oxygen from the ambient bottom water with a feathery plume (the obturaculum) and delivers the gases through its circulatory system to the trophosome, a specialized organ housing thiotrophic endosymbionts. *Riftia* has a leathery tube that allows it to flex and relocate its plume among water masses that are variably sulfidic or oxygenated. Other species (Fig. 2.1e, Plate 3e) have rigid tubes that penetrate deeply into anoxic sediments, allowing absorption of sulfides through the worm's posterior end (Freytag et al. 2001). The vent mussel, *Bathymodiolus azoricus* (Fig. 2.1f, Plate 3f), is more versatile, because it absorbs sulfide and methane from venting waters and hosts thiotrophic and methanotrophic endosymbionts in its gills. Soon after the discovery of hydrothermal vents, related taxa were found at cold-water hydrocarbon seeps, in anoxic basins, on whale- and wood-falls, and in coastal reducing sediments (reviews in Sibuet and Olu 1998; Tunnicliffe et al. 1998; Smith and Baco 2003). I subsequently use the term "chemosynthetic" (as defined in Dubilier et al. 2008) to describe these ecosystems and the animals and microbes that are supported mainly by thiotrophic (i.e. chemolithoautotrophic) or methanotrophic (i.e. chemoorgano-autotrophic) primary production.

Marine chemosynthetic symbioses have been characterized now from seven invertebrate phyla (Dubilier et al. 2008). Though many invertebrate taxa have independently evolved chemosynthetic symbioses, six families dominate the biomass at deep-sea vents, seeps, wood- and whale-falls: polychaete annelids in the families Siboglinidae and Alvinellidae; bivalve molluscs in the families Vesicomyidae and Mytilidae; gastropod molluscs in the family Provannidae; and decapod crustaceans in

Fig. 2.1 Chemosynthetic taxa found at hydrothermal vents and seeps. (**a**) *Beggiatoa* mat growing on sulfides in the Gulf of California. (**b**) *Rimicaris* shrimp swarming at a hydrothermal vent on Central Indian Ridge. (**c**) *Calyptogena magnifica* clams clustering in basaltic cracks at 21°N latitude on the East Pacific Rise. (**d**) *Riftia pachyptila* cluster growing at the base of a hydrothermal vent chimney in the Gulf of California. (**e**) A tubeworm cluster composed of *Escarpia spicata* and *Lamellibrachia barhami* growing in soft sediments at cold seeps in the Gulf of California. (**f**) A cluster of *Bathymodiolus azoricus* mussels (courtesy of C. Van Dover) from the Snake Pit vent locality on the Mid-Atlantic Ridge. A color plate of this figure can be found in Appendix (Plate 3)

the shrimp family Alvinocarididae. A variety of ε-Proteobacteria dominate the episymbiont populations associated with decapod crustaceans, alvinellid worms and provannid snails. *Alvinella pompejana* worms host a complex array of ε-Proteobacteria on their bodies (Cary et al. 1997). These and other bacteria pack the gill chamber of *Rimicaris* shrimp (Zbinden et al. 2008). The "hairy" snail, *Alvinoconcha hessleri*, is also associated with external ε-Proteobacteria (Urakawa et al. 2005). Filamentous ε-Proteobacteria form dense tufts that attach to long hair-like structures on the legs of the Yeti crab, *Kiwa hirsuta* (Goffredi et al. 2008). The roles that these episymbionts play in providing nutritional support to their hosts remains unclear in most cases, and too little is known about their biogeographical diversity to consider these bacteria further in the present context. Current information about the phylogenetic diversity of bacterial epi- and endosymbionts associated with decapod crustaceans, the bivalve families Solemyidae, Lucinidae and Thyasiridae, the gastropods families Provannidae and Lepetodrilidae, and a variety of annelids, sponges and protists can be found elsewhere (Bright and Giere 2005; DeChaine and Cavanaugh 2005; Cavanaugh et al. 2006; Dubilier et al. 2008). Because more is known about their transmission modes, genetics and evolution, I focus instead on the chemosynthetic γ-Proteobacteria associated with three taxa, the vesicomyid clams, vestimentiferan tubeworms and bathymodiolin mussels. A comparison of genetic, demographic and evolutionary processes affecting these endosymbionts and their hosts should provide a foundation for assessing lesser-known symbioses in the other taxa.

The consequences of various symbiont transmission modes must be considered with respect to one of greatest challenges facing deep-sea chemosynthetic organisms – finding suitable island-like habitats in vast ocean basins. Hydrothermal vents are distributed intermittently along global mid-ocean ridge system, in back arc basins and on volcanically active seamounts (Van Dover et al. 2002). Distances between active vent fields are typically on the order of tens to hundreds of km, and isolation-by-distance affects some vent taxa with relatively limited dispersal distances (Vrijenhoek 1997). Topographical structures such as large transform faults displace adjacent spreading centers and limit opportunities for along-axis dispersal (Young et al. 2008). Hydrocarbon seeps are distributed as discrete localized patches, mostly along continental margins (Sibuet and Olu 1998). Whale carcasses occur at breeding and feeding grounds and along migratory pathways, and wood-falls that exist mostly along terrestrial margins (Smith and Baco 2003; Pailleret et al. 2007).

These chemosynthetic habitats are variably ephemeral. The vent fields found along rapidly spreading ridge axes persist for only a few decades before they are eliminated, but fields found at slow-spreading axes can persist for thousands of years (Lalou and Brichet 1982; Lalou et al. 1993). Seeps probably are not as ephemeral, but they tend to be patchy, even at local scales (Barry et al. 1996). Whale-falls last for varying periods depending apparently on depth, dissolved oxygen and the assemblage of decomposer organisms (Smith and Baco 2003; Braby et al. 2007). Metapopulation processes, governed by the frequency of local habitat extinctions, sources of colonizing species, and rates of dispersal will affect both the richness of species and the diversity of genotypes found in chemosynthetic communities (Vrijenhoek 1997). Perhaps many of the chemosynthetic organisms that live at ephemeral vents and whale-fall habitats should be considered "weedy species" (sensu Baker 1965) that must grow fast, reproduce early and disperse effectively to colonize new habitats before their local habitat is extinguished. On the other hand, some seep habitats appear to support chemosynthetic animals that grow very slowly and live long, raising the possibility that these animals and their associated microbes are subject to very different metapopulation processes (Berquist et al. 2000).

2.2 Some General Consequences of Symbiont Transmission Modes

The modes by which symbionts are transmitted to offspring will profoundly affect the demography and evolution of the microbes and their hosts. Vertical transmission from parent to offspring (Fig. 2.2a) can occur via large gametes such as eggs (transovarial), by subsequent inoculation of retained zygotes or larvae in brooding organisms, or by feeding as in some social insects (Douglas 1989). Horizontal transmission (Fig. 2.2b) can occur directly via host-to-host transfers or indirectly via infections by a free-living stage of the microbe. Among the chemosynthetic invertebrates that dominate vent, seep, and whale-fall communities, transovarial transmission has been convincingly documented only for vesicomyid clams (Endow and Ohta 1990; Cary and Giovannoni 1993), though it is suspected in solemyid bivalves (Krueger et al. 1996). Horizontal transmission via an apparently free-living infectious phase (also known as environmental acquisition) has been convincingly documented in vesti-

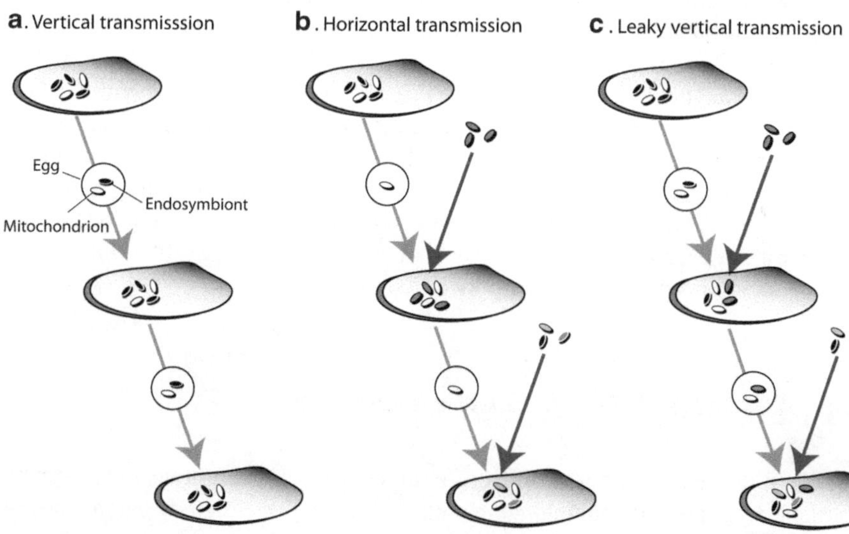

Fig. 2.2 Transmission modes that have been documented or inferred in various chemosynthetic organisms, illustrated for a hypothetical bivalve mollusc. The white ovals represent mitochondria. The black and shaded ovals represent endosymbiotic bacteria. (**a**) Under obligately vertical transmission, the symbiont is inherited maternally along with the mitochondria, leading to genetic coupling of mitochondrial and endosymbiont genotypes. (**b**) Under horizontal transmission, the bacteria are acquired anew in each generation, decoupling the cytoplasmic organelles from bacterial genotypes. Furthermore, infections involving environmental bacteria will likely lead to mixed symbiont genotypes (strains) within a host. (**c**) Leaky vertical transmission is predominantly vertical with occasional environmental acquisition, or vertical with massive environmental swamping. In either case, the vertical transmission component will create a small lag-time in the decoupling of host mitochondrial and symbiont genotypes. The horizontal component, depending how prevalent, can create mixed-strain infections. A color plate of this figure can be found in Appendix (Plate 4)

mentiferan tubeworms. Mixed transmission modes (Fig. 2.2c) also exist, in which vertical transmission is supplemented by host-to-host transfers or de novo environmental infections with free-living bacteria. Before critically evaluating what we know about symbiont transmission in deep-sea chemosynthetic organisms, it will be helpful to consider some general consequences of various transmission modes.

2.2.1 Symbiont Assurance Versus Locally Optimal Symbionts for the Host

Vertical transmission provides the host with symbiont assurance. Dispersing propagules arriving at a nascent vent, seep or whale-fall habitat already carry the right kind of chemosynthetic symbiont needed to grow and prosper. Nonetheless, symbiont assurance carries a hidden cost because dispersing propagules might not carry the optimal bacterial strain for the particular habitat in which they settle. In contrast, host species that acquire symbionts after settlement have the potential to adopt bacterial strains that are optimally adapted to the local environment (Won et al. 2003b). Yet, environmental acquisition poses a number of risks. Perhaps most significantly, dispersing host propagules might fail to be infected altogether. Environmental acquisition can also lead to mixed symbiont genotypes within a host individual creating opportunities for the evolution of selfish “cheaters” that exploit the host by providing less nutrition (Frank 1996). Finally, environmental acquisition creates potential pathways for infection that can be exploited by pathogens. Host species employing leaky vertical systems of transmission still face the risks of infection by cheaters and pathogens, but they also enjoy the benefits of symbiont assurance and the capacity to acquire locally optimal strains.

2.2.2 A Dispersal Benefit for the Symbionts

If microbial dispersal is limited in deep-ocean basins, vertically transmitted symbionts might benefit by hitchhiking along with host propagules, thereby gaining access to new chemosynthetic habitats. A persistent dictum in microbiology is that “everything is everywhere, and nature selects” (Beijerinck 1913), but molecular studies have made it apparent that microbial populations are often spatially subdivided (Papke et al. 2003; Whitaker et al. 2003). Local-scale differentiation can result from limited dispersal abilities of free-living microbe (Harvey and Garabedian 1991; Schloter et al. 2000). For example, isolation-by-distance occurs on very small scales for *Pseudomonas* (Cho and Tiedje 2000), but local differences might not be adaptively neutral. The fitness of *Bacillus* strains decays exponentially when strains are grown in soil extracts taken only meters away from their home site (Belotte et al. 2003). Living in many places at the same time decreases the risk of extinction due to local catastrophes, so the dispersal benefit provides an advantage to facultatively vertical symbionts that recycle to the ambient environment. The evolutionary fate of obligately vertical symbionts, however, is linked inexorably to the fate of their host.

2.2.3 Genetic Consequences of Transmission Modes

Obligately vertical transmission (Fig. 2.2a) is expected to result in genetic homogeneity of the symbiont lineages found within host individuals. The theoretical framework developed to explain the clonality of transovarial symbionts was developed for the maternally inherited cytoplasmic genomes of the chloroplasts and mitochondria in eukaryotes (Birky et al. 1983, 1989). Individual animals typically possess only one haploid mitochondrial genotype (haplotype) that was inherited from the mother, a state referred to as homoplasmic. Countless numbers of mitochondria exist among the cells of an individual, but only the germ-line mitochondria are transmitted to eggs. A series of intra-individual bottlenecks occur as germ-line cells divide and sort a subset of the mitochondria into daughter cells. For example, after about 25 cell divisions in the germ-line of a *Drosophila* fly, genetic drift among its mitochondrial genomes will result in eggs that are homoplasmic. The rate at which homoplasmy is reached in a species will be a function of the mean number of mitochondria transmitted during each cell division and the number of germ-line cell divisions. Transovarial endosymbionts will similarly achieve clonal homogeneity within a host lineage at a rate determined primarily by its transmission numbers through each egg generation of the host (Rispe and Moran 2000).

Horizontal transmission, on the other hand, can retain symbiont heterogeneity (mixed infections), depending on several factors. The diversity of mixed strains within host individuals will be a function of the diversity of infectious strains in the ambient environment or in the co-occurring pool of hosts if transmission is contagious. Intra-host symbiont diversity should also be affected by length of time that a window for infection remains open during host development. In the squid-*Vibrio* symbiosis this window for infection is open for a very brief "permissive" phase during early embryogenesis, which limits the number of bacteria establishing a symbiotic association (reviewed by Nyholm and McFall-Ngai 2004). A brief permissive phase also occurs in the larvae of vestimentiferan tubeworms, but the entire lifespan of bathymodiolin mussels might be permissive to infections (discussed below).

Vertical and horizontal symbionts are expected to differ greatly in their capacity to acquire foreign genetic material through lateral gene transfer or homologous recombination. The clonality of obligately vertical symbionts greatly limits these opportunities unless occasional leakage of symbionts occurs between hosts (Stewart et al. 2008). Nonetheless, natural selection can only discriminate among the recombinational variants that are permissible within the obligate host's environment. This limitation on the scope for selection also exists for horizontally infectious symbionts that lack a free-living phase altogether. In contrast, facultatively vertical and horizontal symbionts that retain a free-living phase will be exposed to mixed infections and thereby have greater opportunities for recombination, increasing the scope for selection (Jiggins et al. 2001).

2.2.4 Demographic Consequences of Transmission Modes for the Symbionts

Transovarial transmission dramatically reduces the genetically effective size (N_e) of the symbiont population. The bottlenecking process reduces N_e to N_{fem}, the effective number of breeding females in the host population (Rispe and Moran 2000). This reduction has profound evolutionary consequences. According to the "Nearly Neutral" theory of molecular evolution, genetic drift in small populations leads to the fixation of slightly deleterious mutations and a concomitant acceleration of nucleotide substitution rates (Ohta 1987). Bacterial genomes are predominantly clonal, so they also are subject to Muller's ratchet – i.e. mutational decay due to an accumulation of slightly deleterious mutations in small populations (Muller 1964). Indeed, nucleotide substitution rates are faster in the transovarial *Buchnera* endosymbionts of aphids than in their free-living counterparts (Moran 1996). The observed mutations are considered slightly deleterious because they are expected to destabilize the secondary structure of rRNA molecules (Lambert and Moran 1998) and increase amino acid substitutions in protein-coding loci (Wernegreen and Moran 1999).

Conversely, horizontal symbionts are expected to exhibit lower substitution rates than vertical symbionts because they have potentially much larger population sizes. If the bacteria recycle between host-associated and free-living phases, then N_e will be equal the sum of the effective number of bacteria transmitted to the next generation through each phase ($N_{sym} + N_{free}$). However, if the symbionts are entirely devoured by the host and fail to recycle, then $N_e = N_{free}$. Alternatively, if the bacteria do not reproduce in a free-living phase, but only occur as spores or dormant infectious stages, then $N_e = N_{sym}$. These demographic considerations warrant further investigation because the fitness contributions of these bacterial life phases will likely determine possible paths of coevolution (see for example Genkai-Kato and Yamamura 1999).

2.2.5 Cytoplasmic Cotransmission and Genetic Hitchhiking

Transovarial symbionts of animals are inherited maternally in tandem with other cytoplasmic factors such as mitochondria (Funk et al. 2000). Cytoplasmic cotransmission effectively couples the host and symbiont genomes (Fig. 2.2a), which is manifested as covariance (gametic phase disequilibrium) in the frequencies of host mitochondrial and symbiont genetic polymorphisms (Hurtado et al. 2003). In contrast, meiosis randomly shuffles the nuclear genome in sexually reproducing populations; consequently, nuclear polymorphisms should not exhibit associations with cytoplasmic genes (cytonuclear disequilibrium). Natural selection can tighten associations between cytoplasmically cotransmitted factors. A beneficial new mutation arising in a symbiont genome might lead to a selective sweep that carries along a cotransmitted mitochondrial haplotype (Hurst and Jiggins 2005), a process known as genetic hitchhiking (Maynard Smith and Haigh 1974). Conversely, a selective sweep involving a beneficial mitochondrial variant will cotransmit a coupled symbiont genome. In either case, one might expect lower genetic diversity among the cytoplasmic factors of species with vertically transmitted obligate symbionts (Hurst and Jiggins 2005).

2.2.6 Host-Symbiont Cospeciation

As a host species spreads geographically and splits during cladogenic events, the obligately vertical (heritable) symbiont lineages will split, as well. Cladogenic events in the host and symbiont lineages will be congruent and their phylogenetic trees should exhibit parallel topologies and congruent evolutionary ages of the internal nodes (Huelsenbeck et al. 1997). With time and increasing divergence, hybridization becomes improbable, and cospeciation is inevitable, unless occasional leakage of symbionts occur via host transfers. Once reproductive isolation completely blocks opportunities for hybridization and recombination among the nuclear genomes of the host species, these genes too will begin to covary with changes in the genes of heritable symbionts.

Though it is more likely, cospeciation does not require vertical transmission. Obligate horizontal symbionts can also speciate in synchrony if splitting events involve geographic isolation. For example, chewing lice are obligate parasites of pocket gophers (Hafner and Nadler 1988). The lice spread by host-to-host transfers, so allopatric speciation events in the host will also lead to corresponding isolation of the parasites. Tight adaptive constraints forced on obligate horizontal symbionts by their specific hosts also contributes to cospeciation (Clayton et al. 2003). Caution is warranted, however, in considering reported examples of cospeciation in horizontal symbionts. For example, Atlantic and Pacific species of sepiolid squid host corresponding Atlantic and Pacific strains of luminescent *Vibrio* endosymbionts that are acquired infectiously (Nishiguchi et al. 1998). Though extraordinary evidence for coevolution has been documented in the squid-*Vibrio* association, parallel phylogenies might result from underlying geographic covariance, which should always be factored-out of studies on host–symbiont cospeciation (Nieberding et al. 2008; Won et al. 2008).

2.2.7 Symbiont Genome Reduction

Genome reduction is expected when endosymbionts become vertical and lose their free-living phase entirely (Andersson 2006; Pál et al. 2006). Vertical symbionts are enslaved in the host environment and face no selective pressures to maintain unnecessary or redundant functions. Mitochondria, for example, are cytoplasmic organelles descended from proteobacterial endosymbionts (Gray et al. 1999). Animal mitochondria retain a small number of functional genes that encode a minimal set of transfer RNAs, large and small subunit ribosomal RNAs and some essential respiratory chain proteins. The genes required for replication of mitochondrial DNA and metabolic functions exist in the cell's nucleus. When free-living bacteria make the transition to obligate intracellular symbionts, genes that were needed to survive in the external environment may no longer be useful in the intracellular environment and can therefore degenerate or be deleted. Adaptively neutral gene losses will accumulate because of genetic drift, and slightly deleterious deletions can accrue due to the actions of Muller's ratchet in small populations, but deletion mutations might benefit the symbionts if they lead to more efficient replication.

Significant reductions in genome size are well documented in a number of obligate, intracellular pathogens and mutualists (reviewed in Moran 2002). Obligate nutritional endosymbionts of insects provide extreme examples among the γ-Proteobacteria. *Buchnera aphidicola* strain BCc, which infects aphids, has only ~416 Kb of DNA encoding ~362 proteins (Perez-Brocal et al. 2006), and the *Carsonella rudii* strain that infects psyllids is even smaller, having only ~160 Kb of DNA encoding 182 proteins (Nakabachi et al. 2006). For comparison, free-living *Escherichia coli* contain an order of magnitude more DNA, ~4,600 Kb and ~4,300 protein-coding genes (Blattner et al. 1997). The highly reduced insect endosymbionts obtain most of the compounds needed for intermediate metabolism from their host cells, and many of the genes responsible for DNA replication and cell membrane components are missing. Shifts in nucleotide composition to higher AT content and a loss of genes involved in DNA repair are hypothesized to accelerate genomic erosion in these obligate symbionts (Moran et al. 2009). Nonetheless, *Buchnera* and *Carsonella* retain subsets of genes responsible for the biosynthesis of critical amino acids that their hosts are unable to obtain from strict diets of plant saps.

In contrast, horizontally transmitted and facultative endosymbionts that retain a metabolically active free-living stage are expected to retain the genes that control motility, cell division, and cellular metabolism. Indeed, some horizontal symbionts have evolved enlarged genomes to deal with the diverse contingencies of living in the host-associated and ambient environments. For example, a horizontally transmitted nitrogen-fixing rhizobium associated with legumes contains an enlarged chromosome and plasmids that carry accessory genes involved in nitrogen fixation, nodule formation and proteins that facilitate the infection of root tissues (reviewed in Downie and Young 2001).

2.3 Vesicoymid Clams and Vertical Transmission

The vesicomyid clam, *Calyptogena magnifica* Boss and Turner (1980), has a reduced digestive system and obtains its nutrition from the thiotrophic γ-proteobacterial endosymbiont, Candidatus *Ruthia magnifica* Newton et al. (2007), which lives in the clam's gills. Endow and Ohta (1990) were first to suggest that vesicomyid symbionts are vertically transmitted. They identified microscopic, rod-shaped, bacterial inclusions in the primary oocytes and follicle cells of *C. soyoae*. Earlier oogonial stages did not contain the inclusions; so, the authors hypothesized that a transmission stage of the gill symbiont is delivered to follicle cells and eventually to eggs during a narrow developmental window. Soon after, Cary and Giovannoni (1993) examined bacterial 16S rRNA sequences from vesicomyid symbionts and verified the presence of these sequences in ovarian tissues of *C. magnifica* and several other vesicomyids. Their symbiont-specific 16S probes hybridized with the bacterial inclusions in the follicle cells, but unlike Endow and Ohta, they saw no inclusions in the oocytes. Cary and Giovannoni (ibid. pp. 5698–5699) suggested, "nutritive [follicle] cells would provide the ideal mechanism for inoculation of the developing eggs." To date, the mechanisms that these clams use to inoculate their eggs or adhering cells remain a mystery, as do pathways they might use to translocate a putative transmission stage of the symbiont from bacteriocytes in the gill to follicle cells in the ovary.

2.3.1 Symbiont Homogeneity and Genetic Drift

The number of symbionts transmitted through a clam's eggs also is unknown, but the transmitted bacteria are undoubtedly a miniscule subset of the number that lived in the clam's gills. Bottlenecking and genetic drift during translocation to the ova are inevitable; so symbiont clonality is expected. Conversely, an observation of mixed symbiont genotypes within a host individual suggests horizontal transmission or occasional symbiont leakage. As predicted, the vertically transmitted symbionts of vesicomyid clams appear to be clonal within individuals of nearly all the vesicomyid species examined to date. Hurtado et al. (2003) examined *R. magnifica* symbionts from five hydrothermal vent areas along the Galápagos Rift (GAR) and East Pacific Rise (EPR) between 21°N and 17°S latitude. All the clams examined from this 4,800 km range hosted a single 16S rRNA phylotype, but the ITS region of the ribosomal operon exhibited 11 subtypes that varied within and among the sampled geographical localities. Nonetheless, each individual clam hosted only a single subtype. Five ITS subtypes were found in different clams from the 17°S EPR locality, so mixed genotypic infections should have been detectable had they occurred there. Goffredi et al. (2003) similarly examined the symbionts hosted by three morphologically indistinguishable species of the *Vesicomya* (=*Calyptogena*) *pacifica* cryptic-species complex. Though these clams occupy a variety of vent and cold seep habitats along the western American margin, the species segregate mostly according to depth. Each of the clams examined in this study was associated with only one host-specific ITS genotype. A caveat exists, however, regarding the detection of ITS heterogeneity via the polymerase chain reaction (PCR), a run-away process that depends on primer specificity and the starting numbers of various target sequences in a potential mixture of sequences. If a minority genotype occurs in a mixture, it might not amplify in sufficient quantity to be evident in DNA sequence traces or in studies that involve cloning of PCR products (e.g., Stewart et al. 2008).

As previously argued, rates of Nearly Neutral nucleotide substitution should increase in small populations. Consequently, vertical symbionts that have very low N_e are expected to exhibit accelerated evolutionary rates compared to free-living bacteria. Peek et al. (1998a) found that vesicomyid endosymbionts have significantly accelerated substitution rates for 16S rRNA sequences. Nearly Neutral theory also predicts that mutations should be more pronounced in regions of molecules that are subject to lower selective constraints. As expected, substitution rates were consistently greater in the unconstrained loop regions of the rRNA molecule versus the pair-bonded stem regions. Various domains of the rRNA molecule also experience different selective constraints. Domain II for example is highly constrained and no significant difference in substitution rates existed between vertical symbionts and free-living bacteria. Domains I and III, however, are less constrained and there the majority of substitutions fell into the loop regions.

2.3.2 Cytoplasmic Cotransmission and Rare Leakage

Obligately vertical symbionts are inherited maternally along with other cytoplasmic factors such as mitochondria, which should lead to gametic phase disequilibrium between these factors. Nuclear polymorphisms, on the other hand, should not exhibit cytonuclear disequilibrium in a randomly mating population. Hurtado et al. (2003) examined these predictions in a *Calyptogena magnifica* population from 17°S along the EPR (Table 2.1). As predicted, the symbiont ITS polymorphisms were significantly associated with mitochondrial COI haplotypes, whereas the nuclear alleles varied independently of both cytoplasmic factors.

Cytoplasmically cotransmitted genomes can be decoupled, however, if vertical transmission is leaky (Fig. 2.2c). Stewart et al. (2008) reported that an unnamed member of the *V. pacifica* cryptic species complex (sp. mt-II) was infected with a foreign vesicomyid symbiont. They examined 118 clams from the Endeavor segment (47°58′N) of the Juan de Fuca Ridge and all but two of the clams hosted symbiont type A (*symA*), which occurs elsewhere in *V.* sp. mt-II. The two unusual clams hosted a highly divergent strain (*symB*). Recent evidence based on strain-specific

Table 2.1 Likelihood ratio contingency tests of gametic phase disequilibrium between host and symbiont genetic markers. Three tests were conducted: (i) symbiont versus host mitochondria; (ii) host nuclear versus host mitochondria; and (iii) host nuclear versus symbiont. Expected numbers appear in parentheses. Each gametic combination is labeled with *S* representing 17°S-endemic variants and *N* representing alternative variants occurring in northern populations (From Hurtado et al. 2003)

		Frequency of gametic combinations					
Test	Contrast	*N/N*	*N/S*	*S/N*	*S/S*	$G_{(1\ df)}^2$	*P*
(i)	*sym*ITS/*mt*COI	8 (6.1)	0 (1.9)	21 (22.9)	9 (7.1)	4.951	0.026*
(ii)	*Cmg24/mt*COI	5 (5.9)	3 (2.1)	47 (46.1)	15 (16.0)	0.612	0.434
(iii)	*Cmg24/sym*ITS	2 (1.6)	6 (6.4)	12 (12.4)	50 (49.6)	0.134	0.714

PCR primers and pyrosequencing methods indicates that mixed (*symA/symB*) infections are more frequent than reported in the earlier study (Stewart and Cavanaugh 2009). The source of *symB* infections and the mechanism of host transfer remain unknown, but the evidence of symbiont leakage in vesicomyids is intriguing. Rare horizontal transfers create possibilities for genetic recombination among endosymbiont strains. Recombination could obscure patterns of cospeciation and generate variability that might act to retard genome reduction in these symbionts (Stewart et al. 2009). Nonetheless, recent genomic studies reveal significant reductions genome size and strongly support the hypothesis that vertical transmission is the predominant mode of symbiont acquisition in vesicomyids (summarized below).

2.3.3 *Cospeciation*

Despite the potential for symbiont leakage in some vesicomyids a general pattern of host-symbiont cospeciation is observed. Peek et al. (1998b) found that the 16S rRNA gene tree derived from the bacterial endosymbionts was not significantly different in topology from the mitochondrial COI gene tree derived from nine vesicomyid species (Fig. 2.3). Some minor discrepancies exist

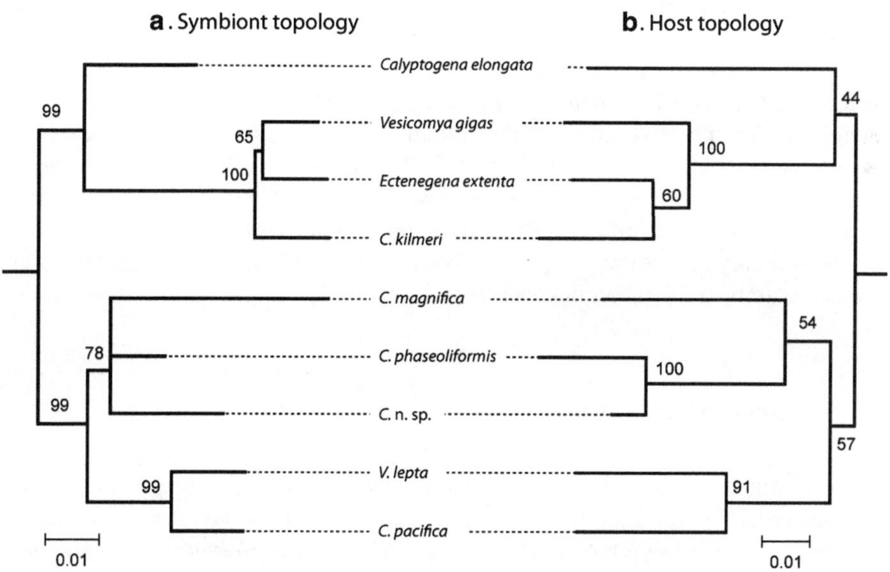

Fig. 2.3 Cospeciation between endosymbionts and nine vesicomyid clams species (Redrawn from Peek et al. 1998b). (**a**) The maximum likelihood (ML) phylogram for the bacteria was based on 16S rRNA sequences. (**b**) The ML phylogram for the vesicomyids was based on combined data from portions of the mitochondrial COI gene and the large subunit 16S rRNA gene. The numbers at nodes represent bootstrap support values. See original publication for statistical tests of cospeciation

between the two trees, but the differences can be more parsimoniously explained by statistical uncertainties in the inferred trees rather than symbiont leakage. Additional gene sequences involving more clam species are currently being examined to produce a more robust host tree (A. Audzijonyte, personal communication, 2010). All vesicomyid symbionts that have been examined to date comprise a discrete monophyletic clade that excludes related γ-Proteobacteria found in other bivalves or local marine environments (Dubilier et al. 2008). So, if occasional leakage occurs it must result from host-to-host transfers within the Vesicomyidae and not from environmental infections by free-living strains or host transfers from other molluscan families.

This vesicomyid-endosymbiont association appears to be young compared to the 200 million year old association between aphids and *Buchnera* endosymbionts (Moran et al. 1993). The oldest confirmed vesicomyid shells (late middle Eocene, 45 million years ago) are attributable to the genus *Archivesica* (summarized in Amano and Kiel 2007). Two other genera also appeared during the late Eocene, so the last common ancestor of all vesicomyids must have lived at least 45 million years ago. These fossil data are consistent with molecular estimates for a relatively young common ancestor for vesicomyids, 22–44 million years ago (Peek et al. 1997; Little and Vrijenhoek 2003). Previously, these estimates of evolutionary age appeared to be at odds with reports of "vesicomyid" shells from Cretaceous cold seep deposits in Japan (Kanie et al. 1993; Kanie and Nishida 2000), but this fossil evidence for ~100 million year old vesicomyids is now disputed (Amano et al. 2008). New fossil specimens from Hokkaido, Japan, revealed that a putative Cretaceous "*Vesicomya*" lacks the characteristic hinge morphology distinguishing vesicomyids from other heterodonts and is instead a member of the family Lucinidae (Amano et al. 2008). Other putative Cretaceous vesicomyids appear to be misidentified solemyid shells (Kiel et al. 2008). For now, the fossil and molecular evidence together suggest that a conservative estimate for the origin of vesicomyid clams is about 45 million years ago. A molecular clock method for estimating the evolutionary age of the vesicomyid symbionts is difficult to develop, but a current estimate broadly encompasses this young age (Peek et al. 1998b).

2.3.4 Genome Reduction

An evolutionary history of obligately vertical transmission is expected to result in genomic reduction for the symbionts. Two complete genomes have been sequenced, to date – the endosymbionts *Ruthia magnifica* from *Calyptogena magnifica* (Newton et al. 2007) and *Vesicomyiosocious okutanii* from *Calyptogena okutanii* (Kuwahara et al. 2007). When compared to the free-living thiotroph, *Thiomicrospira crunogena*, the two symbiont genomes are roughly half the size, have less than half of the protein-coding genes, and have reduced GC contents (Newton et al. 2008). Genes for externally expressed structures such as flagella and pili are missing in the symbionts. The absence of *ftsZ* and related genes in *V. okutanii* (Kuwahara et al.

2007) suggests that the bacterium has ceded control of cell division to the host. Nonetheless, the symbionts retain most of the metabolic functions of *T. crunogena* and have added genes that encode various sulfur oxidation pathways. Both the number of protein-coding genes and order of these genes in the genome appear to be highly conserved between *R. magnifica* and *V. okutanii*. Differences exist, however, for genes that encode cell envelope components and proteins involved in nitrate reduction. Comparative genomics involving additional vesicomyid endosymbionts will help to resolve questions regarding the phylogenetic pathways of genome reduction and gene retention in these bacteria. Whole genome studies might also help to resolve whether the apparent "leakage" reported for *Vesicomya* cf. *pacifica* sp II has contributed to recombination among the symbiont lineages in different hosts, thereby providing some resistance to the genomic degradation of these symbionts (Stewart et al. 2009).

2.4 *Riftia* and Environmental Infections

Adults of the vestimentiferan tubeworm *Riftia pachyptila* Jones (1981) lack a digestive system entirely and have instead a specialized organ, the trophosome, that houses the sulfur-oxidizing, intracellular, γ-Proteobacteria endosymbiont, Candidatus *Endoriftia persephone* Robidart et al. (2008). *Riftia* acquires its *E. persephone* symbionts horizontally from the local environment in which the worm larvae settle. The rod-shaped symbionts are not observed in *Riftia*'s eggs, sperm or freshly settled larvae (Cavanaugh et al. 1981; Cary et al. 1989, 1993; Nussbaumer et al. 2006). Molecular probing suggests that bacteria bearing very similar, if not identical, 16S rRNA sequences occur in sea-water and biofilms from the hydrothermal vent environment (Harmer et al. 2008), but direct linkages to a competent infectious phase have not been made. In a landmark field experiment, Nussbaumer and coworkers (2006) found that freshly settled vestimentiferan larvae are free of endosymbionts (aposymbiotic) and have a rudimentary gut. Unlike frenulate siboglinids that appear to acquire endosymbionts via the digestive tract (Southward 1988), *Riftia* larvae are infected by rod-shaped bacteria that penetrate the epidermis. Then the bacteria migrate to the dorsal mesentery where they are enclosed in vacuoles within mesodermal cells, a process that initiates development of the trophosome and metamorphosis to a juvenile stage, when subsequent infections are halted by massive apoptosis of the skin.

2.4.1 Absence of Cospeciation

If the environmental infection model of Nussbaumer et al. (2006) can be generalized to all vestimentiferans, symbiont phylotypes should be associated with local environments where the worm larvae settle rather than a particular host species.

Indeed, no evidence is found for cospeciation between vestimentiferan hosts and their symbionts (Feldman et al. 1997; Nelson and Fisher 2000; McMullin et al. 2003). Host and symbiont phylogenies are broadly incongruent (Fig. 2.4). Instead, the symbionts are associated with types of habitats (basaltic vents versus sedimented seeps) and broad biogeographical regions. The bacterial 16S sequences amplified from the trophosomes of all the vestimentiferan species examined to date reveal two primary phylotypes of the thiotrophic symbiont (Feldman et al. 1997; Di Meo et al. 2000; Nelson and Fisher 2000; McMullin et al. 2003; Vrijenhoek et al. 2007). Sequence divergence between phylotypes I and II is 4.3% on average, so they might be considered legitimate bacterial species according to the criteria of many microbial systematists (e.g., Stackebrandt and Goebel 1994). This degree of sequence divergence suggests that phylotypes I and II diverged more than 200 million years ago (Feldman et al. 1997), long before the radiation of their vestimentiferan hosts, most probably about 60 million years ago (Black et al. 1997; Chevaldonné et al. 2002; Little and Vrijenhoek 2003). Previous evidence for "vestimentiferan" tubes in Jurassic hydrothermal vent deposits (Little et al. 2004) is now disputed (Kiel and Dando 2009), so it seems likely that these worms have radiated during the Cenozoic Era.

Phylotype I is globally widespread, occurring in five vestimentiferan genera (*Lamellibrachia*, *Escarpia*, *Seepiophila*, *Arcovestia* and *Alaysia*) that occupy cold seep environments in the Atlantic, Gulf of Mexico and Pacific, and occupy vents only in the western Pacific (S. Johnson, MBARI, 2010). Minor 16S subtypes of phylotype I (Fig. 2.4a) appear to segregate according to depth of the seep habitats in the Gulf of Mexico (McMullin et al. 2003). Phylotype II, or *E. persephone* Robidart et al. (2008), is less variable and more narrowly distributed, occurring in four vestimentiferan genera: *Riftia*, *Tevnia*, *Oasisia* and *Ridgeia*, found at eastern Pacific hydrothermal vents. Phylotypes I and II occur in close proximity at vents and cold seeps in the Gulf of California, but were never found together in the same habitat (Vrijenhoek et al. 2007). On the other hand, some vestimentiferan species will switch symbiont phylotypes based on the habitats in which their larvae settle. The eastern Pacific seep worm *Escarpia spicata* ordinarily hosts phylotype I, but specimens found adjacent to a hydrothermal vent in the Gulf of California hosted phylotype II, just like its *Riftia* neighbors (Di Meo et al. 2000).

2.4.2 Shared Hosts and Mixed Symbionts

According to the environmental infection model, different vestimentiferan species that settle together in the same habitat should be infected by the same local strains of the symbiont. Furthermore, mixed-strain infections are expected within individual hosts if the window for infection is open for an appreciable period during larval development. Indeed, *Riftia*, *Tevnia*, and *Oasisia* sampled from the same eastern Pacific vent fields appear to share the same symbiont phylotype (Feldman et al.

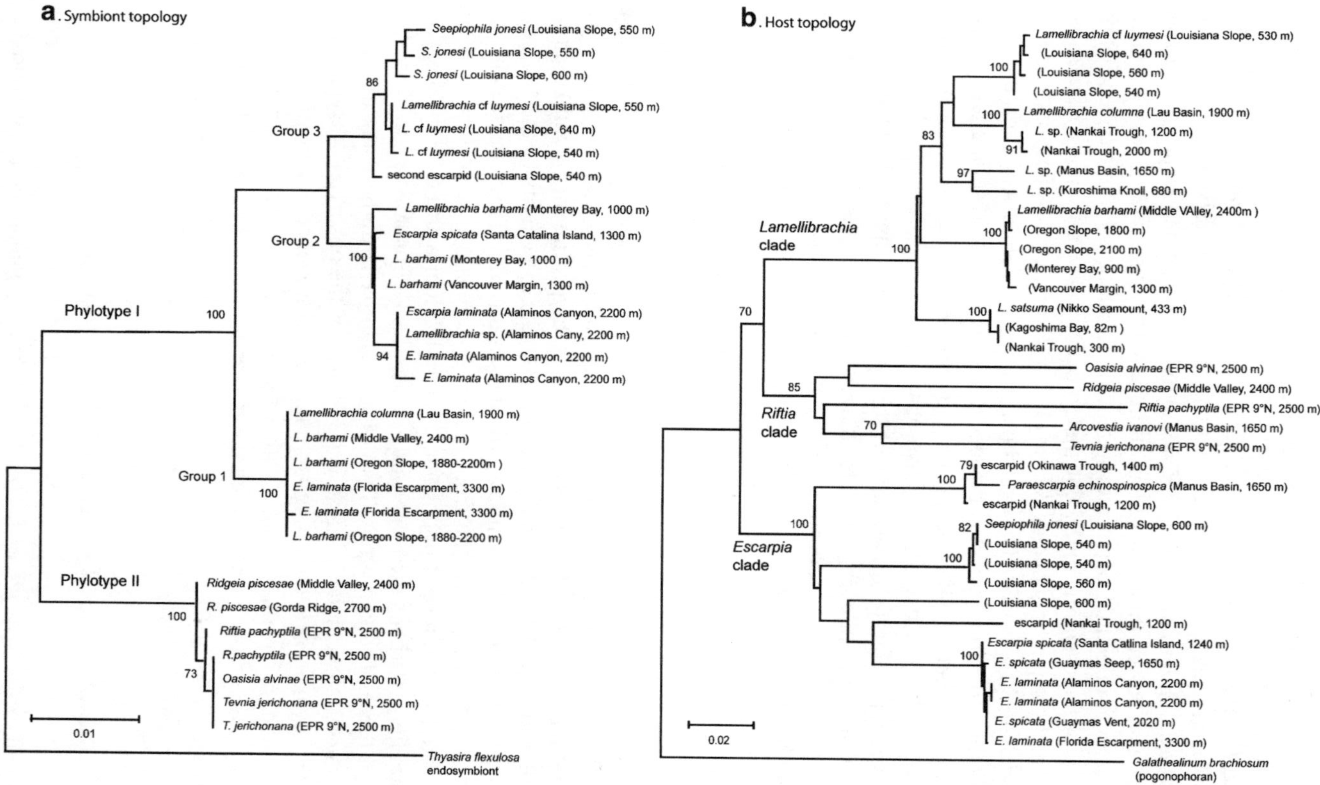

Fig. 2.4 Absence of cospeciation between endosymbionts and vestimentiferan tubeworm species (Redrawn from McMullin et al. 2003). (**a**) Neighbor-joining (NJ) phylogram for the bacteria was based on 16S rRNA sequences. (**b**) NJ phylogram for the worms was based on mitochondrial COI sequences. The numbers at nodes represent bootstrap support values from original publication. Note that the taxa do not correspond between the two trees so no attempt was made to connect hosts and symbionts. Eastern Pacific members of the *Riftia* clade only host symbiont phylotype II. Phylotype I symbionts can be further subdivided into three subgroups

1997; McMullin et al. 2003), but the 16S sequences examined in these studies are too conservative to reflect differences that might exist among co-infecting symbiont strains (Di Meo et al. 2000). Nucleotide sequences from a portion of the RuBisCO gene (*Rbc*) distinguish subtype strains of phylotype I that occur in *Lamellibrachia barhami* and *Escarpia spicata* from the Gulf of California (Vrijenhoek et al. 2007). These two species are commonly found living together in mixed aggregations. Nearly half of the *L. barhami* and *E. spicata* individuals sampled from one of these aggregations also hosted mixed symbiont strains. When aggregations were sampled from discrete environmental patches only tens of meters apart, the frequencies of *Rbc* subtypes differed significantly, but the co-occurring worms hosted the same local strains in the same frequencies. Worms living 7 km apart hosted a distinct *Rbc* subtype, which again was shared by both host species. *Lamellibrachia barhami* and *E. spicata* belong to two very distinct clades that were separated soon after the origin of vestimentiferan siboglinids, yet no evidence was found for discrimination among these subtypes by these highly divergent worm genera.

Though vestimentiferans do not appear to differentiate among local strains of the symbiont, it would be wrong to infer that they are not selective regarding the kinds of bacteria that they incorporate. Only phylotype I bacteria were detected in the *L. barhami* and *E. spicata* specimens examined in our study, so the worm larvae must be highly effective at sequestering the right mutualistic symbiont and rejecting potential pathogens. It seems improbable that the *Rbc* alleles assessed in our study might affect external phenotypes that could be detected by the host during bacterial infections. Nevertheless, the nucleotide substitutions distinguishing these RuBisCO subtypes differ by as many as four amino acids and might not be adaptively neutral if they affect enzyme activities and bacterial maintenance during the symbiotic phase. Additionally, the *Rbc* subtypes might hitchhike along with variation in linked genes that confer local adaptive benefits to the bacteria. Various geochemical factors might favor distinct strains in different environmental patches or at different times. The frequencies of various symbiont strains might also vary spatially within the trophosome of an individual worm as it spans a geochemical gradient. Perhaps the footprints of natural selection still reside in the *Rbc* sequences or those of linked genes. Tests of this adaptive hypothesis are underway in my laboratory. Nonetheless, it also is possible that the observed subtype variation is adaptively neutral and nothing more than a consequence of genetic drift and small-scale subdivision of microbes with very limited dispersal capacity. These horizontal symbionts would not obtain the dispersal benefits afforded by vertical transmission.

2.4.3 Population Size of Horizontal Symbionts

The genetically effective size of horizontal symbiont populations is composed of two components, $N_e = N_{sym} + N_{free}$. A free-living form of *E. persephone* (phylotype II) settles in biofilms on basaltic surfaces and has been filtered from seawater sampled up to 100 m from an eastern Pacific vent, which "suggests a potentially large environmental pool of symbionts" (Harmer et al. 2008, p. 3897). Genomic evidence (Robidart et al. 2008) also suggests that *E. persephone* can persist as a heterotroph while living in the ambient environment, but the contribution of heterotrophic nutrition to N_{free} is unknown. In addition, no mechanism is known for vestimentiferan endosymbionts to escape the host-associated environment and invade the ambient environment. If the endosymbionts are completely exploited by the host (incorporated, cultivated, and entirely digested), N_{sym} is zero and the host-associated phase contributes nothing to N_e and bacterial fitness. Bacterial co-evolution should not occur under a scenario of complete exploitation, but mathematical models involving very low fitness payoffs ($N_{sym} > 0$) suggest that coevolution is possible for rapidly evolving facultative symbionts (Frean and Abraham 2004). To avoid complete exploitation, some of the host-associated bacteria must escape and re-inoculate the ambient environment or infect other hosts (see Section 2.4.4, below). Perhaps molecular footprints for the evolution of cooperation versus coercion can be found in the genomic and proteomic sequences of *E. persephone* (Robidart et al. 2008). Excellent models for the evolution of cooperation with very low payoffs exist in the plant-rhizobium and aphid-*Buchnera* systems (reviewed by Simms and Taylor 2002; Moya et al. 2008).

Though we do not know how N_{sym} or N_{free} contribute to the total population size of *E. persephone*, a tangential approach offers some insight. Very slow rates of nucleotide substitution are observed in the horizontal endosymbionts hosted mostly by vestimentiferans (Peek et al. 1998a). This deceleration relative to vertically transmitted symbionts could be interpreted as evidence for a greatly enhanced N_e, according to the Nearly Neutral model of molecular evolution. The distribution of mutations observed in various domains of the 16S rRNA molecule is consistent with expectations of the model. An alternative model, however, suggests that infectious horizontal symbionts should decelerate evolutionary rates because they are subject to severe constraints that limit any form of change. According to the Red King model, the slowest runner wins a coevolutionary race involving mutualists (Bergstrom and Lachmann 2003). Changes in a mutualist are perceived as potential threats (an invasive pathogen for example) by the host, so they will be purged by purifying selection. The Red King model is the converse of the Red Queen model (Van Valen 1973) for antagonistic relationships, wherein evolutionary rates are accelerated due to an arms race between the antagonists. Perhaps a good example of a Red Queen process is hypervariability in the externally expressed, central portion of the flagellin gene (*fliC*) of pathogenic *Escherichia coli* (Reid et al. 1999). The pathogen must change its external appearance frequently to stay ahead of the host's capacity to recognize and develop mechanisms to exclude the pathogen. Mutualists, on the other hand, must avoid any changes that might be recognized as threatening; so decelerated nucleotide substitution rates are expected. Though the 16S sequences examined by Peek et al. (1998a) evolved more slowly in the horizontal endosymbionts, these data alone are not sufficient to discriminate between the Nearly Neutral and Red King models. Future studies should examine the evolutionary rates of genes that are externally expressed and potentially involved in host–symbiont signaling and recognition. Now that the genomes from several vertical and horizontal symbionts have been sequenced, it may be possible to target a number of genes that might shed light on this matter.

2.4.4 Escape Strategies and the Absence of Genome Reduction

As evidenced by the enlarged genomes of nitrogen fixing rhizobium (Downie and Young 2001), horizontal transmission should not result in genome reduction if the symbionts face conflicting selective pressures imposed by host-associated and free-living environments. Present genomic evidence suggests that both the free-living and host-associated phases contribute to the overall fitness of *E. persephone*. Early research revealed that genome size (~3.3 Mb) and GC (~58%) content of the *Riftia* endosymbiont are about what is expected for free-living bacteria (Nelson et al. 1984). A recent metagenomic analysis reveals that *E. persephone* has retained the functional versatility needed to survive and reproduce in the ambient environment and also adapt to the symbiont lifestyle (Robidart et al. 2008). *E. persephone* has retained a full suite of genes needed for heterotrophic metabolism, including all the genes needed for glycolysis, fructose degradation and the Kreb's cycle. The genome also contains the requisite components for autotrophic carbon fixation via partial Calvin-Benson and reverse tri-carboxylic acid (rTCA) cycles. However, several key enzymes of the Kreb's cycle were not detected in a proteomic analysis of host-associated *E. persephone*, suggesting that the bacterium can adaptively regulate heterotrophic metabolism when it occurs in the symbiont phase (Markert et al. 2007). *E. persephone* retains suites of genes involved in cellular motility, signal transduction and specific genes that control cell division, a critical fitness-related function. In most respects, the symbiont's genome resembles that of the free-living thiotroph *Thiomicrospira crunogena* in the functional categories of genes it contains; however, it appears to have elevated contents of genes involved in combating host defenses against pathogens and in energy production, critical functions needed to invade and prosper in the host environment. These apparently coevolved changes suggest that the host-associated phase contributes significantly to overall fitness in *E. Persephone*, because the evolution of such differences seems improbable if the host-associated phase is just a demographic dead-end.

A search for exit strategies from vestimentiferan hosts is warranted. Perhaps a transmissive phase of *E. persephone* invades the host's circulatory system to exit via respiratory structures, through excretory pores or along with gametes through gonopores. We do not know what happens to *E. persephone* when *Riftia* dies. I have observed that the trophosome, and presumably its content of symbionts, shrinks notably in size when the worms are starved of sulfides at waning vents (unpublished data). Damage to the host due to predation by bythograeid crabs (Micheli et al. 2002) and polynoid annelids (personal observations) might also create opportunities for the bacteria to re-inoculate the environment.

2.5 Bathymodiolin Mussels and Dual Symbiosis

Mussels in the genus *Bathymodiolus* Kenk and Wilson (1985) and related genera of the subfamily Bathymodiolinae (Bivalvia: Mytilidae) are also among the dominant constituents of chemosynthetic environments (Jones et al. 2006). Bathymodiolins appear to be more versatile than vesicomyids and vestimentiferans, because the mussels are mixotrophic, retaining a functional digestive tract while hosting nutritional endosymbionts (Page et al. 1990). The majority of bathymodiolin species host only intracellular thiotrophic γ-Proteobacteria in their gill tissues, but several species are known to host only intracellular methanotrophic γ-Proteobacteria, and other species can host both types simultaneously, a phenomenon known as dual symbiosis (Fisher et al. 1993). Some members of the genus *Idas* host a high diversity of extracellular sulfur-oxidizing symbionts on their gill filaments (Duperron et al. 2008a). The distribution of symbiont types among various mussel hosts has been summarized elsewhere (DeChaine and Cavanaugh 2005; Won et al. 2008). Recently *B. heckerae* was reported to host four bacterial phylotypes (Duperron et al. 2007) and a newly discovered *Idas* species from the Mediterranean was reported to host six (Duperron et al. 2008b). Little is known about the genetics and evolution of these bacteria, and no genome sequences are available from any of the mussel endosymbionts.

The preponderance of evidence strongly suggests that bathymodiolins acquire their thiotrophic endosymbionts locally from the environment in which they live; nonetheless, evidence for vertical transmission also exists. Cytological investigations of *Bathymodiolus thermophilus* sperm and eggs revealed no evidence for bacteria (Herry and Le Pennec 1986); consequently Le Pennec et al. (1988) were first to suggest that *B. thermophilus* acquires its endosymbionts horizontally by endocytosis through gill epithelium. Subsequent weak inferences, however, generated momentum for an alternative hypothesis that the symbionts were vertically transmitted. Cary and Giovannoni (1993, p. 5699) concluded their seminal publication on vesicomyid symbiont transmission with the statement, "Recent preliminary studies have revealed a similar transovarial transmission mechanism in the symbiont of the mytilid bivalve *Bathymodiolus thermophilus*." Their molecular probes hybridized with bacteria in the gonad-bearing mantle tissue of *B. thermophilus*, but the background signal was too noisy to localize the bacteria in eggs or ovarian nurse cells (S.C. Cary, personal communication, 2003). Unlike vesicomyids, the bathymodiolin symbionts are not restricted to gills (Salerno et al. 2005). Cary and Giovannoni's concluding statement and a subsequent report of apparent congruence between host and symbiont phylogenies among several bivalve taxa including vesicomyids, lucinids, thyasirids, solemyids and *B. thermophilus* (Distel et al. 1994, p. 540), led other researchers to infer vertical transmission in mussels (Nelson and Fisher 1995; Peek et al. 1998a; Trask and Van Dover 1999). The methanotrophic endosymbionts associated with bathymodiolins have not been studied in similar depth, so even less is known about their potential modes of transmission.

2.5.1 Mixed Transmission and Cytonuclear Decoupling in a Hybrid Zone

A fortuitous discovery allowed us to test some predictions of vertical and horizontal transmission hypotheses for bathymodiolins. Closely related northern (*N*) and southern (*S*) species (*B. azoricus* and *B. puteoserpentis*, respectively) that occupy the Mid-Atlantic Ridge host dual symbionts (Fisher et al. 1993). These mussels hybridize along an intermediate portion of the ridge axis (O'Mullan et al. 2001; Won et al. 2003a). Host mitochondrial and symbiont genetic polymorphisms would be decoupled in the hybrid zone if the symbionts were acquired horizontally from the local environment. Alternatively, cytoplasmic co-transmission would maintain symbiont-mitochondrial coupling in hybrids if symbiont transmission were strictly vertical. *Bathymodiolus azoricus* females would co-transmit *mtN* mitochondria and *symN* bacteria to their offspring and *B. puteoserpentis* females would cotransmit *mtS/symS* to their offspring. Cytoplasmic co-transmission would also occur in the hybrid females as long as it is strictly maternal, so the *mtN/symN* and *mtS/symS* cytotypes should remain coupled in F_1 hybrids and subsequent hybrid generations. Won et al. (2003b) were able to test these predictions, because the thiotrophs segregated into distinct *N* and *S* subtypes based on ITS sequences. A sample of mussels from the hybrid zone included 24 individuals with the southern cytotype *mtS/symS*, 18 with a recombinant cytotype *mtN/symS*, and five mixed infections with the northern mitotype *mtN*/(*symN* + *symS*). The 18 recombinant cytotypes and five mixed infections were not consistent with predictions of strictly vertical transmission. Environmental acquisition is the simplest explanation for these data, but anomalies existed. No individuals had the *mtN/symN* coupling type or the *mtS/symN* recombinant type. If the *symS* and *symN* bacterial strains both occurred locally as free-living infectious agents, a few *mtN/symN* and *mtS/symN* cytotypes would be expected, but their absence might be a consequence of sample size ($n = 47$ mussels from the hybrid zone). We tested for PCR bias against the *symN* type as a likely explanation for the absence of *mtN/symN* and *mtS/symN* by probing the mussels with sensitive molecular hybridization methods, but without success. An alternative hypothesis emerged, however, upon examination of nuclear-encoded allozyme polymorphisms: the latter could recombine with the cytotypes of subsequent hybrids and gradually eliminate cytonuclear coupling in an old demographically stable hybrid population. Apparent F_1 hybrids and beyond were identified in the sample, but the five *mtN*/(*symN* + *symS*) mussels had nuclear genotypes that would be expected for first generation immigrants from the north. Consequently, Won et al. (2003b) hypothesized that these putative immigrants might have transported the *symN* thiotrophs from their natal sites, before emigrating to the south, where they were subsequently infected by local *symS* thiotrophs. Electron microscopy and molecular evidence are consistent with this hypothesis, as thiotrophic bacteria were found in pediveliger larvae and the earliest juveniles of *B. azoricus* settling at Mid-Atlantic vents (Salerno et al. 2005).

Several lines of evidence support the hypothesis that mussel symbionts can be replaced throughout the host's lifespan. Cytological evidence exists for the apparent endocytosis of free-living bacteria through the gill membranes of adult mussels (Le Pennec et al. 1988; Won et al. 2003b; Salerno et al. 2005) and other bivalves (Gros et al. 1996, 1998). A bacterial exit strategy for re-inoculation of the local environment may also exist. Salerno et al. (2005) suggested that the pit-like structures in gill epithelium might involve exocytosis of the bacteria. Exo- and endocytosis are suggested in a recent experiment reported by Kádár et al. (2005). They "cleared" *B. azoricus* mussels of thiotrophic endosymbionts by exposing them to sulfide-free seawater for a period of 30 days in the laboratory. A combination of microscopy and the application of symbiont-specific PCR primers indicated that the mussels appeared to be aposymbiotic, though the authors could not be entirely sure of this matter. Then they exposed the "cleared" mussels to sulfide-enriched seawater and newly acquired symbiont-bearing mussels. Thiotrophic bacteria soon grew again in gills of the "cleared" mussels. Kádár et al. (2005) hypothesized that the bacteria exited the newly acquired mussels and entered the gills of "cleared" mussels through pit-like structures in the epithelium. They interpreted the new infections as evidence that the thiotrophs have the capacity to exit the host environment and revert to an infectious stage. Though research that is more definitive is needed to resolve this matter, present evidence suggests that bathymodiolin symbionts might obtain a significant dispersal benefit if their hosts engage in some degree of vertical transmission. Dispersing along with the host larvae would allow the symbionts to "seed" bacterial populations at nascent chemosynthetic sites and possibly leave a historical imprint that covaries with the history of host dispersal.

2.5.2 Absence of Cospeciation in Thiotrophs

Environmental acquisition alone, however, appears to provide the simplest explanation for the phylogeographic distribution of the thiotrophic endosymbionts (Won et al. 2008). The gene tree constructed for the thiotrophs associated with 15 host species was not congruent with the phylogeny of corresponding host species (Fig. 2.5). Pairwise genetic distances among the 15 thiotrophs were not correlated with genetic distances among host species, but they were positively correlated with geographical distances among host localities. This evidence for isolation-by-distance among the thiotrophs was not observed among the mussels – genetic distances did not correspond at all with geographical distances. In some cases closely related hosts like *B. azoricus* and *B. puteoserpentis* live near one another, yet other related species like "*Bathymodiolus*" *tangaroa* and "*B.*" *mauritanicus* occur in different ocean basins on the opposite sides of the globe.[1] Conversely,

[1] The quotes denote the dubious assignment of this genus name to these species (Jones and Vrijenhoek 2006).

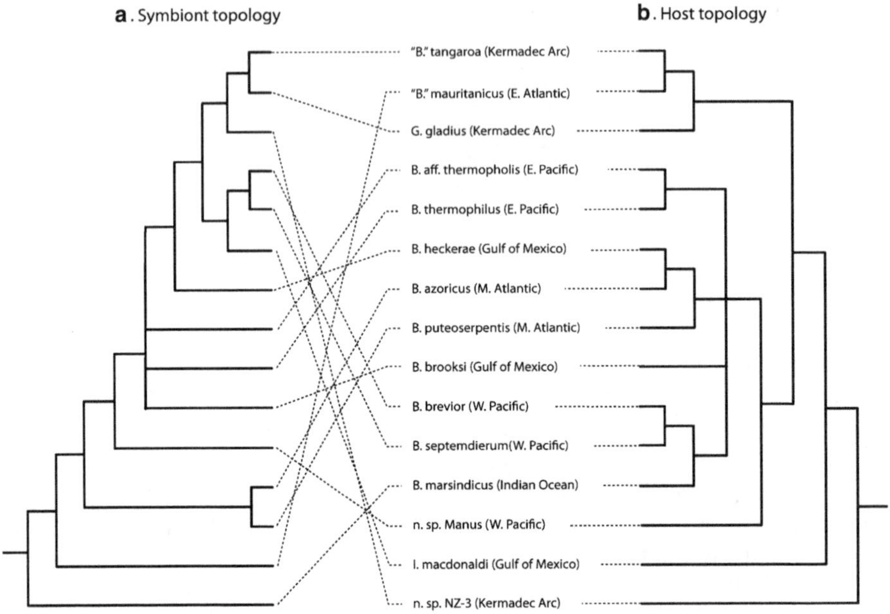

Fig. 2.5 Absence of cospeciation between endosymbionts and 15 species of bathymodiolin mussels (Redrawn from Won et al. 2008). (**a**) Bayesian cladogram of the bacteria was based on 16S rRNA. (**b**) Bayesian cladogram of the host species was based on combined data from three genes: mitochondrial COI, ND4 and nuclear encoded 18S rRNA. See original publication for the bootstrap support values and statistical tests of cospeciation

very distantly related hosts *Idas macdonaldi*, *B. brooksi* and *B. heckerae* all occur in the Gulf of Mexico. Nominal species like *B. brevior*, *B. septemdierum* and *B. marisindicus* appear to be synonymous (Jones et al. 2006). *Bathymodiolus brevior* and *B. septemdierum* occur in the western Pacific and share closely related symbionts, but *B. marisindicus* occurs in the Indian Ocean and hosts the most divergent symbiont seen in this study.

Won et al. (2008) hypothesized that the mussel thiotrophs began to diverge and spread around the globe approximately 112–160 million years ago, long before the continents reached their present locations about 50 million years ago. They would have passed through more continuous ocean basins than exist today. The host phylogeny, on the other hand, suggests that bathymodiolins probably began to radiate about 60 million years ago, so their paths to spread globally were probably very different from the ancient paths used by the symbionts. Regardless, it may be safe to conclude that vertical transmission, if it occurs at all in these mussels, has left no historical imprint on the phylogeographic distribution of the thiotrophic symbionts. Similar phylogeographic studies involving 16S rRNA and other genetic markers remain to be conducted with the methanotrophs, but these studies will be more difficult because the methanotrophs exhibit a greater diversity of genotypes than found in the mixed thiotroph hosted by individual mussels.

2.6 Conclusions

A large number of observed and expected differences exist between the obligately vertical endosymbionts hosted by vesicomyid clams and the horizontal endosymbionts hosted by vestimentiferans tubeworms (summarized in Table 2.2). The situation in bathymodiolin mussels is more complex, however. The mussels acquire thiotrophic symbionts from the local environment, but a component of vertical transmission might also exist. Transmission modes of methanotrophs and other bacterial endosymbionts remain unknown (Duperron et al. 2008b).

Symbiont infections involving multiple strains as seen in vestimentiferans and bathymodiolins appear to be a good indicator of horizontal transmission. Environmental infections are hypothesized to create opportunities for exploiting locally optimal symbiont strains (Won et al. 2003b), despite the associated risks. Pathogenic microorganisms might evolve to exploit the host acquisition pathways, requiring investments by the host in surveillance and active defense mechanisms that do not discriminate against mutualists. *Riftia* opens a time-limited window for environmental infections and then closes it by destroying subsequently infected epidermal tissues (Nussbaumer et al. 2006). *Bathymodiolus*, on the other hand, might be susceptible to environmental infections throughout its adult life (Kádár et al. 2005). To assess whether horizontal transmission carries a significant *pathogen penalty*, it might be valuable to compare the diversities and densities of non-mutualist microbes associated with horizontal and vertical hosts. Evidence exists for a variety of other microbes associated with the tissues of chemosynthetic organisms (e.g., Naganuma et al. 1997; Elsaied et al. 2002; Goffredi et al. 2004), but their roles are unknown. Pathogenic fungi are known to infect *Bathymodiolus* mussels (Van Dover et al. 2007). Horizontal symbionts should bear a significant cost associated with enhanced host surveillance of potential pathogens. The *Riftia* endosymbiont, *E. persephone*, has an elevated content of genes that appear to be involved in combating host defenses, as in many pathogenic bacteria (Robidart et al. 2008). Similar elevations are not evident in the vertical symbionts *Ruthia magnifica* and *Vesicomyiosocious okutanii*. Studies of the trade-offs between acquisition of locally optimal symbionts and the risk of pathogens deserve more attention.

Mixed environmental infections also engender risks of within-host symbiont competition. To provide the host with nutrition, chemosynthetic symbionts must be devoured via intracellular autophagy or they must "leak" nutrients to the host, or both (reviewed in Cavanaugh et al. 2006). Competition among multiple symbiont genotypes within a host should favor cheaters that contribute less to the host while gaining access to a broad redox zone. How can honest symbionts evolve in such mixtures unless their contribution to the host also increases their own fitness? Trying to envision the evolution of cooperative mutualists in mixed genotypic infections is difficult, but a number of theoreticians have attempted to address this problem (e.g., Frank 1996; Doebeli and Knowlton 1998; Frean and Abraham 2004). Cheaters who exploit the host too intensively will decrease their own fitness if the host dies and the bacteria fail to escape to the ambient environment

Table 2.2 Three model systems involving deep-sea chemosynthetic endosymbionts and a summary of the genetic, demographic and evolutionary consequences of their transmission modes. For lines 16–22, designations of "lost", "retained", "reduced" or "enhanced" pertain to comparisons with the free-living thiotroph *Thiomicrospira crunogena*

Host	*Calyptogena magnifica*[a]	*Riftia pachyptila*	*Bathymodiolus azoricus*
1. Endosymbiont	*Ruthia magnifica*	*Endoriftia persephone*	Undescribed
2. Chemosynthetic type	Thiotroph	Thiotroph	Dual symbiosis[b]
3. Symbiont location	Gill bacteriocytes	Trophosome bacteriocytes	Gill bacteriocytes[c]
4. Transmission mode	Transovarial	Environmental	Environmental[d]
5. Symbiont assurance for the host	Yes	No	Potentially[e]
6. Dispersal benefit for the symbiont	Yes	No	Potentially[e]
7. Locally optimal symbionts	No	Yes	Yes
8. Mixed symbionts	No	Yes	Yes
9. Symbiont recombination[f]	Unlikely[g]	Possible	Possible
10. Maternal inheritance	Yes	No	Unclear
11. Expectation of symbiont N_e	Very small	Potentially very large	Large
12. Expectation of mutational decay	Yes	Unlikely	Unlikely
13. Observed nucleotide substitution rates	Very fast	Very slow	Unclear
14. Evidence for cytoplasmic coupling	Yes	No	Unclear
15. Host-symbiont cospeciation in family	Yes	No	No
16. Symbiont genome sequenced	Yes	Yes, but not closed[h]	No
17. Symbiont genome reduction	By about one-half	No	?
18. Symbiont motility genes	Lost	Retained	?
19. Symbiont central metabolism genes	Retained	Retained	?
20. Symbiont energy production genes	Enhanced	Enhanced	?

21. Symbiont DNA replication genes	Reduced	Retained	?
22. Symbiont GC content	Reduced	Enhanced	?
23. Symbiont competition	No	Yes	?
24. Risk of cheaters	Limited	Yes	Yes

[a] *Calyptogena okutanii* and Candidatus *Vesicomyiosocious okutanii* provide an equally valid model.

[b] γ-proteobacterial thiotrophs and methanotrophs in gill.

[c] Bacteriocytes also occur in other tissues including mantle and gonad.

[d] A vertical component of transmission might also exist.

[e] Potentially if a vertical component of transmission does exists.

[f] Recombination might occur through transformation, transduction and conjugation in mixed symbiont infections. It might also occur in vertical symbionts if they occasionally are leaky.

[g] Exceptions appear to exist (Stewart et al. 2009).

[h] A consequence of mixed infections involving multiple symbiont genotypes.

(Genkai-Kato and Yamamura 1999). On the other hand, if reproduction in the host and re-inoculation of the ambient environment enhances bacterial fitness even slightly, cooperation with the host might be expected. Experimental studies with legume-rhizobium mutualists reveal that the plant host exerts sanctions against bacterial cheaters by controlling the distribution of critical resources to less productive modular tissues (root nodules) (Kiers et al. 2003). Similar experiments are impossible with most deep-sea vent and seep mutualists, but they are warranted for chemosynthetic mutualists that could be cultured from shallower environments.

Enslavement of chemosynthetic symbionts through obligately vertical transmission avoids of the problems of mixed symbiont genotypes and intra-host symbiont competition. The relatively recent (~45 million year old) enslavement of intracellular γ-Proteobacteria by vesicomyid clams might be considered analogous to the symbiogenic origin of organelles in eukaryotes (Vetter 1991). The origins of mitochondria and chloroplasts by symbiogenesis are now well established though the order of events leading to eukaryotes is still debated (Cavalier-Smith and Lee 1985; Embley and Martin 2006). The extreme genome reduction reported for the insect endosymbiont *Carsonella rudii* suggest that this bacterium exists in a state "between living cell and organelles" (Tamames et al. 2007). Nonetheless, "full-fledged" organelles like mitochondria and chloroplasts have evolved sophisticated mechanisms to import proteins encoded by the nuclear genome and synthesized in the cellular cytoplasm (Cavalier-Smith and Lee 1985; Theissen and Martin 2006). It will be interesting to see if the vertically transmitted vesicomyid endosymbionts have evolved even rudimentary mechanisms for protein transport. Clearly there is much to be learned about evolution from the genome sequencing and proteomic analyses of additional chemosynthetic endosymbionts.

Acknowledgements I wish to thank the Frank Stewart, Shana Goffredi, Monika Bright, Julie Robidart, Steffen Kiel, Julio Harvey, Shannon Johnson and an anonymous reviewer for providing information and criticisms that improved the scope of this manuscript. Funding was provided by grants from the David and Lucile Packard Foundation to the Monterey Bay Research Institute and the National Science Foundation (OCE 0241613).

References

Amano K, Kiel S (2007) Fossil vesicomyid bivalves from the North Pacific region. Veliger 49:270–293

Amano K, Jenkins RG, Kurihara Y, Kiel S (2008) A new genus for *Vesicomya inflata* Kanie & Nishida, a lucinid shell convergent with that of vesicomyids, from Cretaceous strata of Hokkaido, Japan. Veliger 50:255–262

Andersson SGE (2006) The bacterial world gets smaller. Science 314:259–260

Baker HG (1965) Characteristics and modes of origin of weeds. In: Baker HG, Stebbins GL (eds) Genetics of colonizing species. Academic, New York, pp 147–172

Barry JP, Greene HG, Orange DL, Baxter CH, Robison BH, Kochevar RE, Nybakken JW, Reed DL, McHugh CM (1996) Biologic and geologic characteristics of cold seeps in Monterey Bay, California. Deep Sea Res I 43:1739–1762

Beijerinck MW (1913) Jaarboek van de Koninklijke Akademie v. Wetenschoppen. Muller, Amsterdam, The Netherlands

Belotte D, Curien J-B, Maclean RC, Bell G (2003) An experimental test of local adaptation in soil bacteria. Evolution 57:27–36

Bergstrom CT, Lachmann M (2003) The Red King effect: when the slowest runner wins the coevolutionary race. Proc Natl Acad Sci USA 100:593–598

Berquist DC, Williams FM, Fisher CR (2000) Longevity record for deep-sea invertebrate. Nature 403:499–500

Birky CWJ, Maruyama T, Fuerst P (1983) An approach to population and evolutionary genetic theory for genes in mitochondria and chloroplasts, and some results. Genetics 103: 513–527

Birky CW Jr, Fuerst P, Maruyama T (1989) Organelle diversity under migration, mutation, and drift: equilibrium expectations, approach to equilibrium, effects of heteroplasmic cells, and comparison to nuclear genes. Genetics 121:613–627

Black MB, Halanych KM, Maas PAY, Hoeh WR, Hashimoto J, Desbruyères D, Lutz RA, Vrijenhoek RC (1997) Molecular systematics of vestimentiferan tube worms from hydrothermal vents and cold-water seeps. Mar Biol 130:141–149

Blattner FR, Plunkett G III, Bloch CA, Perna NT, Burland V, Riley M, Collado-Vides J, Glasner JD, Rode CK, Mayhew GF, Gregor J, Davis NW, Kirkpatrick HA, Goeden MA, Rose DJ, Mau B, Shao Y (1997) The complete genome sequence of *Escherichia coli* K-12. Science 277: 1453–1462

Boss KJ, Turner RD (1980) The giant white clam from the Galápagos rift, *Calyptogena magnifica* species novum. Malacologia 20:161–194

Braby CE, Rouse GW, Johnson SB, Jones WJ, Vrijenhoek RC (2007) Bathymetric and temporal variation among *Osedax* boneworms and associated megafauna on whale-falls in Monterey Bay, California. Deep Sea Res I 54:1773–1791

Bright M, Giere O (2005) Microbial symbiosis in Annelida. Symbiosis 38:1–45

Cary SC, Giovannoni SJ (1993) Transovarial inheritance of endosymbiotic bacteria in clams inhabiting deep-sea hydrothermal vents and cold seeps. Proc Natl Acad Sci USA 90:5695–5699

Cary SC, Felbeck H, Holland ND (1989) Observations on the reproductive biology of the hydrothermal vent tube worm *Riftia pachyptila*. Mar Ecol Prog Ser 52:89–94

Cary SC, Warren W, Anderson E, Giovannoni SJ (1993) Identification and localization of bacterial endosymbionts in hydrothermal vent taxa with symbiont-specific polymerase chain reaction amplification and *in situ* hybridization techniques. Mol Mar Biol Biotechnol 2:51–62

Cary SC, Cottrell MT, Stein JT, Camacho F, Desbruyères D (1997) Molecular identification and localization of filamentous symbiotic bacteria associated with the hydrothermal vent annelid *Alvinella pompejana*. Appl Environ Microbiol 63:1124–1130

Cavalier-Smith T, Lee JJ (1985) Protozoa as hosts for endosymbioses and the conversion of symbionts into organelles. J Eukaryot Microbiol 32:376–379

Cavanaugh CM, Gardiner SL, Jones ML, Jannasch HW, Waterbury JB (1981) Prokaryotic cells in the hydrothermal vent tube worm *Riftia pachyptila* Jones: possible chemoautotrophic symbionts. Science 213:340–342

Cavanaugh CM, McKinness AP, Newton ILG, Stewart FJ (2006) Marine chemosynthetic symbiosis. In: Dworkin M, Falkow S, Rosenberg E, Schleifer K-H, Stackebrandt E (eds) The prokaryotes. Springer, New York, pp 475–507

Chevaldonné P, Jollivet D, Desbruyères D, Lutz RA, Vrijenhoek RC (2002) Sister-species of eastern Pacific hydrothermal-vent worms (Ampharetidae, Alvinelidae, Vestimentifera) provide new mitochondrial clock calibration. Cah Biol Mar 43:367–370

Cho J-C, Tiedje JM (2000) Biogeography and degree of endemicity of fluorescent *Pseudomonas* strains in soil. Appl Environ Microbiol 66:5448–5456

Clayton DH, Bush SE, Goates BM, Johnson KP (2003) Host defense reinforces host-parasite cospeciation. Proc Natl Acad Sci USA 100:15694–15699

Corliss JB, Dymond J, Gordon LI, Edmond JM, Von Herzen RP, Ballard RD, Green K, Williams D, Bainbridge A, Crane K, Van Andel TH (1979) Submarine thermal springs on the Galápagos Rift. Science 203:1073–1083

DeChaine EG, Cavanaugh CM (2005) Symbioses of methanotrophs and deep-sea mussels (Mytilidae: Bathymodiolinae). In: Overmann J (ed) Progress in molecular and subcellular biology: molecular basis of symbiosis. Springer, Berlin, Heidelberg, pp 227–249
Di Meo CA, Wilbur AE, Holben WE, Feldman RA, Vrijenhoek RC, Cary SC (2000) Genetic variation among endosymbionts of widely distributed vestimentiferan tubeworms. Appl Environ Microbiol 66:651–658
Distel DL, Felbeck H, Cavanaugh CM (1994) Evidence for phylogenetic congruence among sulfur-oxidizing chemoautotrophic bacterial endosymbionts and their bivalve hosts. J Mol Evol 38:533–542
Doebeli M, Knowlton N (1998) The evolution of interspecific mutualisms. Proc Natl Acad Sci USA 95:8676–8680
Douglas AE (1989) Mycetocyte symbiosis in insects. Biol Rev 64:409–434
Downie JA, Young JPW (2001) Genome sequencing: the ABC of symbiosis. Nature 412: 597–598
Dubilier N, Bergin C, Lott C (2008) Symbiotic diversity in marine animals: the art of harnessing chemosynthesis. Nat Rev Microbiol 6:725–740
Duperron S, Sibuet M, MacGregor BJ, Kuypers MMM, Fisher CR, Dubilier N (2007) Diversity, relative abundance and metabolic potential of bacterial endosymbionts in three *Bathymodiolus* mussel species from cold seeps in the Gulf of Mexico. Environ Microbiol 9:1423–1438
Duperron S, Laurent MCZ, Gaill F, Gros O (2008a) Sulphur-oxidizing extracellular bacteria in the gills of Mytilidae associated with wood falls. FEMS Microbiol Ecol 63:338–349
Duperron S, Halary S, Lorion J, Sibuet M, Gaill F (2008b) Unexpected co-occurrence of six bacterial symbionts in the gills of the cold seep mussel *Idas* sp. (Bivalvia: Mytilidae). Environ Microbiol 10:433–445
Elsaied H, Kimura H, Naganuma T (2002) Molecular characterization and endosymbiotic localization of the gene encoding D-ribulose 1, 5-bisphosphate carboxylase–oxygenase (RuBisCO) form II in the deep-sea vestimentiferan trophosome. Microbiology 148:1947–1957
Embley TM, Martin W (2006) Eukaryotic evolution, changes and challenges. Nature 440:623–630
Endow K, Ohta S (1990) Occurrence of bacteria in the primary oocytes of vesicomyid clam *Calyptogena soyoae*. Mar Ecol Prog Ser 64:309–311
Feldman RA, Black MB, Cary CS, Lutz RA, Vrijenhoek RC (1997) Molecular phylogenetics of bacterial endosymbionts and their vestimentiferan hosts. Mar Mol Biol Biotech 6: 268–277
Fisher CR, Brooks JM, Vodenichar JS, Zande JM, Childress JJ, Burke RA Jr (1993) The co-occurance of methanotrophic and chemoautotrophic sulfur-oxidizing bacterial symbionts in a deep-sea mussel. Mar Ecol 14:277–289
Frank SA (1996) Host-symbiont conflict over mixing of symbiotic lineages. Proc R Soc Lond B 263:33–344
Frean MR, Abraham ER (2004) Adaptation and enslavement in endosymbiont-host associations. Phys Rev E 69:051913
Freytag JK, Girguis PR, Bergquist DC, Andras JP, Childress JJ, Fisher CR (2001) A paradox resolved: sulfide acquisition by roots of seep tubeworms sustains net chemoautotrophy. Proc Natl Acad Sci USA 98:13408–13413
Funk DJ, Helbling L, Wernegreen JJ, Moran NA (2000) Intraspecific phylogenetic congruence among multiple symbiont genomes. Proc R Soc Lond B 267:2517–2521
Genkai-Kato M, Yamamura N (1999) Evolution of mutualistic symbiosis without vertical transmission. Theor Popul Biol 55:309–323
Goffredi SK, Hurtado LA, Hallam S, Vrijenhoek RC (2003) Evolutionary relationships of deep-sea vent and seep clams (Mollusca: Vesicomyidae) of the '*pacifica/lepta*' species complex. Mar Biol 142:311–320
Goffredi SK, Warén A, Orphan VJ, Van Dover CL, Vrijenhoek RC (2004) Novel forms of structural integration between microbes and a vent gastropod from the Indian Ocean. Appl Environ Microbiol 70:3082–3090

Goffredi SK, Jones WJ, Erhlich H, Springer A, Vrijenhoek RC (2008) Epibiotic bacteria associated with the recently discovered Yeti crab, *Kiwa hirsuta*. Environ Microbiol 10: 2623–2634

Gray MW, Burger G, Lang BF (1999) Mitochondrial evolution. Science 283:1476–1481

Gros O, Darrasse A, Durand P, Frenkiel L, Moueza M (1996) Environmental transmission of sulfur-oxidizing bacterial gill endosymbiont in the tropical lucinid bivalve *Codakia orbicularis*. Appl Environ Microbiol 62:2324–2330

Gros O, Frenkiel L, Moueza M (1998) Gill filament differentiation and experimental colonization by symbiotic bacteria in aposymbiotic juveniles of *Codakia orbicularis* (Bivalvia: Lucinidae). Invertebr Reprod Dev 34:219–231

Hafner MS, Nadler SA (1988) Phylogenetic trees support the coevolution of parasites and their hosts. Nature 332:258–259

Harmer TL, Rotjan RD, Nussbaumer AD, Bright M, Ng AW, DeChaine EG, Cavanaugh CM (2008) Free-living tube worm endosymbionts found at deep-sea vents. Appl Environ Microbiol 74:3895–3898

Harvey RW, Garabedian SP (1991) Use of colloid filtration theory in modeling movement of bacteria through a contaminated sandy aquifer. Environ Sci Technol 25:178–185

Herry A, Le Pennec M (1986) Ultrastructure de la gonade d'un Mytilidae hydrothermal profond de la ride du Pacifique oriental. Haliotis 16:295–307

Huelsenbeck JP, Rannala B, Yang Z (1997) Statistical tests of host-parasite cospeciation. Evolution 51:410–419

Hurst GDD, Jiggins FM (2005) Problems with mitochondrial DNA as a marker in population, phylogeographic and phylogenetic studies: the effects of inherited symbionts. Proc R Soc Lond B 272:1525–1534

Hurtado LA, Mateos M, Lutz RA, Vrijenhoek RC (2003) Coupling of bacterial endosymbiont and host mitochondrial genomes in the hydrothermal vent clam *Calyptogena magnifica*. Appl Environ Microbiol 69:2058–2064

Jiggins FM, JHGvd S, Hurst GDD, Majerus MEN (2001) Recombination confounds interpretations of *Wolbachia* evolution. Proc R Soc Lond B 268:1423–1427

Jones ML (1981) *Riftia pachyptila*, new genus, new species, the vestimentiferan worm from the Galápagos Rift geothermal vents (Pogonophora). Proc Biol Soc Wash 93:1295–1313

Jones WJ, Vrijenhoek RC (2006) Evolutionary relationships within the "*Bathymodiolus*" *childressi* group. Cah Biol Mar 47:403–407

Jones WJ, Won YJ, Maas PAY, Smith PJ, Lutz RA, Vrijenhoek RC (2006) Evolution of habitat use by deep-sea mussels. Mar Biol 148:841–851

Kádár E, Bettencourt R, Costa V, Serrão Santos R, Lobo-da-Cunha A, Dando P (2005) Experimentally induced endosymbiont loss and re-acquirement in the hydrothermal vent bivalve *Bathymodiolus azoricus*. J Exp Mar Biol Ecol 318:99–110

Kanie Y, Nishida T (2000) New species of chemosynthetic bivalves, *Vesicomya* and *Acharax*, from the Cretaceous deposits of northwestern Hokkaido. Sci Rep Yokosuka City Mus 47:79–84

Kanie Y, Yoshikawa Y, Sakai T, Takahash T (1993) The Cretaceous chemosynthetic cold water-dependent molluscan community discovered from Mikasa City, Central Hokkaido. Sci Rep Yokosuka City Mus 41:31–132

Kenk VC, Wilson BR (1985) A new mussel (Bivalvia, Mytilidae) from hydrothermal vents in the Galápagos Rift zone. Malacologia 26:253–271

Kiel S, Dando PR (2009) Chaetopterid tubes from vent and seep sites: implications for fossil record and evolutionary history. Acta Palaeontol Pol 54(3):443–448

Kiel S, Amano K, Jenkins RG (2008) Bivalves from Cretaceous cold-seep deposits on Hokkaido, Japan. Acta Palaeontol Pol 53:525–537

Kiers ET, Rousseau RA, West SA, Denison RF (2003) Host sanctions and the legume-rhizobium mutualism. Nature 425:78–81

Krueger DM, Gustafson RG, Cavanaugh CM (1996) Vertical transmission of chemoautotrophic symbionts in the bivalve *Solemya velum* (Bivalvia: Protobranchia). Biol Bull 190:195–202

Kuwahara H, Yoshida T, Takaki Y, Shimamura S, Nishi S, Harada M, Matsuyama K, Takishita K, Kawato M, Uematsu K, Fujiwara Y, Sato T, Kato C, Kitagawa M, Kato I, Maruyama T (2007) Reduced genome of the thioautotrophic intracellular symbiont in a deep-sea clam, *Calyptogena okutanii*. Curr Biol 17:881–886

Lalou C, Brichet E (1982) Ages and implications of East Pacific Rise sulfide deposits at 21°N. Nature 300:169–171

Lalou C, Reyss J-L, Brichet E, Arnold M, Thompson G, Fouquet Y, Rona P (1993) New age data for Mid-Atlantic Ridge hydrothermal sites: TAG and Snakepit chronology revisited. J Geophys Res 98:9705–9713

Lambert JD, Moran NA (1998) Deleterious mutations destabilize ribosomal RNA in endosymbiotic bacteria. Proc Natl Acad Sci USA 95:4458–4462

Le Pennec M, Diouris M, Herry A (1988) Endocytosis and lysis of bacteria in gill epithelium of *Bathymodiolus thermophilus*, *Thyasira flexuosa* and *Lucinella divaricata* (Bivalve, Molluscs). J Shell Res 7:483–489

Little CTS, Vrijenhoek RC (2003) Are hydrothermal vent animals living fossils? Trends Ecol Evol 18:582–588

Little CTS, Danelian T, Herrington RJ, Haymon R (2004) Early Jurassic hydrothermal vent community from the Franciscan complex, California. J Paleontol 78:542–559

Markert S, Arndt C, Felbeck H, Becher D, Sievert SM, Hugler M, Albrecht D, Robidart J, Bench S, Feldman RA, Hecker M, Schweder T (2007) Physiological proteomics of the uncultured endosymbiont of *Riftia pachyptila*. Science 315:247–250

Maynard Smith J, Haigh J (1974) The hitch-hiking effect of a favorable gene. Gene Res (Camb) 23:23–35

McMullin E, Hourdez S, Schaeffer SW, Fisher CR (2003) Phylogeny and biogeography of deep sea vestimentiferans and their bacterial symbionts. Symbiosis 34:1–41

Micheli F, Peterson CH, Mullineaux LS, Fisher CR, Mills SW, Sancho G, Johnson GA, Lenihan HS (2002) Predation structures communities at deep-sea hydrothermal vents. Ecol Monogr 72:365–382

Moran NA (1996) Accelerated evolution and Muller's ratchet in endosymbiotic bacteria. Proc Natl Acad Sci USA 93:2873–2878

Moran NA (2002) Microbial minimalism: genome reduction in bacterial pathogens. Cell 108: 583–586

Moran NA, Munson MA, Baumann P, Ishikawa H (1993) A molecular clock in endosymbiotic bacteria is calibrated using the insect hosts. Proc R Soc Lond B 253:167–171

Moran NA, McLaughlin HJ, Sorek R (2009) The dynamics and time scale of ongoing genomic erosion in symbiotic bacteria. Science 323:379–382

Moya A, Pereto J, Gil R, Latorre A (2008) Learning how to live together: genomic insights into prokaryote-animal symbioses. Nat Rev Genet 9:218–229

Muller HJ (1964) The relation of mutation to mutational advance. Mutat Res 1:2–9

Naganuma T, Naka J, Okayama Y, Minami A, Horikoshi K (1997) Morphological diversity of the microbial population in a vestimentiferan tubeworm. J Mar Biotech 53:193–197

Nakabachi A, Yamashita A, Toh H, Ishikawa H, Dunbar HE, Moran NA, Hattori M (2006) The 160-kilobase genome of the bacterial endosymbiont *Carsonella*. Science 314:267

Nelson DC, Fisher CR (1995) Chemoautotrophic and methanotrophic endosymbiotic bacteria at deep-sea vents and seeps. In: Karl DM (ed) Microbiology of deep sea hydrothermal vent habitats. CRC Press, Boca Raton, FL, pp 125–167

Nelson K, Fisher CR (2000) Absence of cospeciation in deep-sea vestimentiferan tube worms and their bacterial endosymbionts. Symbiosis 28:1–15

Nelson DC, Waterbury JB, Jannasch HW (1984) DNA base composition and genome size of the prokaryotic symbiont in *Riftia pachyptila* (Pogonophora). FEMS Microbiol Lett 24: 267–271

Newton ILG, Woyke T, Auchtung TA, Dilly GF, Dutton RJ, Fisher MC, Fontanez KM, Lau E, Stewart FJ, Richardson PM, Barry KW, Saunders E, Detter JC, Wu D, Eisen JA, Cavanaugh CM

(2007) The *Calyptogena magnifica* chemoautotrophic symbiont genome. Science 315:998–1000
Newton ILG, Girguis PR, Cavanaugh CM (2008) Comparative genomics of vesicomyid clam (Bivalvia: Mollusca) chemosynthetic symbionts. BMC Genomics 9:585
Nieberding CM, Durette-Desset M-C, Vanderpoorten A, Casanova JC, Ribas A, Deffontaine V, Feliu C, Morand S, Libois R, Michaux JR (2008) Geography and host biogeography matter for understanding the phylogeography of a parasite. Mol Phylogenet Evol 47:538–554
Nishiguchi MK, Ruby EG, McFall-Ngai MJ (1998) Competitive dominance among strains of luminous bacteria provides an unusual evidence for parallel evolution in sepiolid squid-vibrio symbioses. Appl Environ Microbiol 64:3209–3213
Nussbaumer AD, Fisher CR, Bright M (2006) Horizontal endosymbiont transmission in hydrothermal vent tubeworms. Nature 441:345–348
Nyholm SV, McFall-Ngai MJ (2004) The winnowing: establishing the squid–*Vibrio* symbiosis. Nat Rev Microbiol 2:632–642
O'Mullan GD, Maas PAY, Lutz RA, Vrijenhoek RC (2001) A hybrid zone between hydrothermal vent mussels (Bivalvia: Mytilidae) from the Mid-Atlantic Ridge. Mol Ecol 10:2819–2831
Ohta T (1987) Very slightly deleterious mutations and the molecular clock. J Mol Evol 26:1–6
Page HM, Fisher CR, Childress JJ (1990) Role of filter-feeding in the nutritional biology of a deep-sea mussel with methanotrophic symbionts. Mar Biol 104:251–257
Pailleret M, Haga T, Petit P, Prive-Gill C, Saedlou N, Gaill F, Zbinden M (2007) Sunken wood from the Vanuatu Islands: identification of wood substrates and preliminary description of associated fauna. Mar Ecol 28:233–241
Pál C, Papp B, Lercher MJ, Csermely P, Oliver SG, Hurst LD (2006) Chance and necessity in the evolution of minimal metabolic networks. Nature 440:667–670
Papke RT, Ramsing NB, Bateson MM, Ward DM (2003) Geographical isolation in hot spring cyanobacteria. Environ Microbiol 5:650–659
Peek A, Gustafson R, Lutz R, Vrijenhoek R (1997) Evolutionary relationships of deep-sea hydrothermal vent and cold-water seep clams (Bivalvia: Vesicomyidae): results from the mitochondrial cytochrome oxidase subunit I. Mar Biol 130:151–161
Peek AS, Vrijenhoek RC, Gaut BS (1998a) Accelerated evolutionary rate in sulfur-oxidizing endosymbiotic bacteria associated with the mode of symbiont transmission. Mol Biol Evol 15:1514–1523
Peek AS, Feldman RA, Lutz RA, Vrijenhoek RC (1998b) Cospeciation of chemoautotrophic bacteria and deep-sea clams. Proc Natl Acad Sci USA 95:9962–9966
Perez-Brocal V, Gil R, Ramos S, Lamelas A, Postigo M, Michelena JM, Silva FJ, Moya A, Latorre A (2006) A small microbial genome: the end of a long symbiotic relationship? Science 314:312–313
Reid SD, Selander RK, Whittam TS (1999) Sequence diversity of flagellin (fliC) alleles in pathogenic *Escherichia coli*. J Bacteriol 181:153–160
Rispe C, Moran NA (2000) Accumulation of deleterious mutations in endosymbionts: Muller's ratchet with two levels of selection. Am Nat 156:425–441
Robidart JC, Benc SR, Feldman RA, Novoradovsky A, Podell SB, Gaasterl T, Allen EE, Felbeck H (2008) Metabolic versatility of the *Riftia pachyptila* endosymbiont revealed through metagenomics. Environ Microbiol 10:727–737
Salerno JL, Macko SA, Hallam SJ, Bright M, Won Y-J, McKiness Z, Van Dover CL (2005) Characterization of symbiont populations in life-history stages of mussels from chemosynthetic environments. Biol Bull 208:145–155
Schloter M, Lebuhn M, Heulin T, Hartmann A (2000) Ecology and evolution of bacterial microdiversity. FEMS Microbiol Rev 24:647–660
Sibuet M, Olu K (1998) Biogeography, biodiversity and fluid dependence of deep-sea cold-seep communities at active and passive margins. Deep Sea Res II 45:517–567
Simms EL, Taylor DL (2002) Partner choice in nitrogen-fixation mutualisms of legumes and rhizobia. Integr Comp Biol 42:369–380

Smith CR, Baco AR (2003) Ecology of whale falls at the deep-sea floor. Oceanogr Mar Biol Annu Rev 41:311–354

Southward EC (1988) Development of the gut and segmentation of newly settled stages of *Ridgeia* (Vestimentifera): implications for relationship between Vestimentifera and Pogonophora. J Mar Biol Assoc UK 68:465–487

Stackebrandt E, Goebel BM (1994) Taxonomic note: a place for DNA: DNA reassociation and 16S rRNA sequence analysis in the present species definition in bacteriology. Int J Syst Bacteriol 44:846–849

Stewart FJ, Cavanaugh CM (2009) Pyrosequencing analysis of endosymbiont population structure: co-occurrence of divergent symbiont lineages in a single vesicomyid host clam. Environ Microbiol 11(8):2136–2147

Stewart FJ, Newton ILG, Cavanaugh CM (2005) Chemosynthetic endosymbioses: adaptations to oxic-anoxic interfaces. Trends Microbiol 13:439–448

Stewart FJ, Young CR, Cavanaugh CM (2008) Lateral symbiont acquisition in a maternally transmitted chemosynthetic clam endosymbiosis. Mol Biol Evol 25:673–687

Stewart FJ, Young CR, Cavanaugh CM (2009) Evidence for homologous recombination in intracellular chemosynthetic clam symbionts. Mol Biol Evol 26:1391–1404

Tamames J, Gil R, Latorre A, Pereto J, Silva F, Moya A (2007) The frontier between cell and organelle: genome analysis of Candidatus *Carsonella ruddii*. BMC Evol Biol 7:181

Theissen U, Martin W (2006) The difference between organelles and endosymbionts. Curr Biol 16:R1016–R1017

Trask JL, Van Dover CL (1999) Site-specific and ontogenetic variations in nutrition of mussels (*Bathymodiolus sp.*) from the Lucky Strike hydrothermal vent field, Mid-Atlantic Ridge. Limnol Oceanogr 44:334–343

Tunnicliffe V, McArthur AG, Mchugh D (1998) A biogeographical perspective of the deep-sea hydrothermal vent fauna. Adv Mar Biol 34:353–442

Urakawa H, Dubilier N, Fujiwara Y, Cunningham DE, Kojima S, Stahl DA (2005) Hydrothermal vent gastropods from the same family (Provannidae) harbour epsilon- and gamma-proteobacterial endosymbionts. Environ Microbiol 7:750–754

Van Dover CL, German CR, Speer KG, Parson LM, Vrijenhoek RC (2002) Evolution and biogeography of deep-sea vent and seep invertebrates. Science 295:1253–1257

Van Dover CL, Ward ME, Scott JL, Underdown J, Anderson B, Gustafson C, Whalen M, Carnegie RB (2007) A fungal epizootic in mussels at a deep-sea hydrothermal vent. Mar Ecol 28:54–62

Van Valen L (1973) A new evolutionary law. Evol Theor 1:1–30

Vetter RD (1991) Symbiosis and the evolution of novel trophic strategies: thiotrophic organisms at hydrothermal vents. In: Margulis L, Fester R (eds) Symbiosis as a source of evolutionary innovation. MIT Press, Cambridge, MA, pp 219–245

Vrijenhoek RC (1997) Gene flow and genetic diversity in naturally fragmented metapopulations of deep-sea hydrothermal vent animals. J Hered 88:285–293

Vrijenhoek RC, Duhaime M, Jones WJ (2007) Subtype variation among bacterial endosymbionts of tubeworms (Annellida: Siboglinidae) from the Gulf of California. Biol Bull 212: 180–184

Wernegreen JJ, Moran NA (1999) Evidence for genetic drift in endosymbionts (*Buchnera*): analyses of protein coding genes. Mol Biol Evol 16:83–97

Whitaker RJ, Grogan DW, Taylor JW (2003) Geographic barriers isolate endemic populations of hyperthermophilic Archaea. Science 301:976–978

Won Y-J, Hallam SJ, O'Mullan GD, Vrijenhoek RC (2003a) Cytonuclear disequilibrium in a hybrid zone involving deep-sea hydrothermal vent mussels of the genus *Bathymodiolus*. Mol Ecol 12:3185–3190

Won Y-J, Hallam SJ, O'Mullan GD, Pan IL, Buck KR, Vrijenhoek RC (2003b) Environmental acquisition of thiotrophic endosymbionts by deep-sea mussels of the genus *Bathymodiolus*. Appl Environ Microbiol 69:6785–6792

Won Y-J, Jones WJ, Vrijenhoek RC (2008) Absence of co-speciation between deep-sea mytilids and their thiotrophic endosymbionts. J Shell Res 27:129–138

Young CR, Fujio S, Vrijenhoek RC (2008) Directional dispersal between mid-ocean ridges: deep-ocean circulation and gene flow in *Ridgeia piscesae*. Mol Ecol 17:1718–1731

Zbinden M, Shillito B, Le Bris N, de Villardi de Montlaur C, Roussel E, Guyot F, Gaill F, Cambon-Bonavita M-A (2008) New insights on the metabolic diversity among the epibiotic microbial community of the hydrothermal shrimp *Rimicaris exoculata*. J Exp Mar Biol Ecol 359:131–140

Chapter 3
Microbial Habitats Associated with Deep-Sea Hydrothermal Vent Invertebrates: Insights from Microanalysis and Geochemical Modeling

Nadine Le Bris and Françoise Gaill

Contents

3.1 Introduction

Symbioses between hydrothermal vent invertebrates and chemosynthetic microbes have been recognized to form some of the most productive marine communities (Lutz et al. 1994; Halbach et al. 2003). Endosymbiotic organisms, like the giant

N. Le Bris (✉)
Département Etude des Ecosystèmes Profonds, Ifremer, BP70, 29280, Plouzané, France
and
CNRS-UPMC FRE 3350, Laboratoire d'Ecogéochimie des Environnements Benthiques, 66650 Banyuls-sur-Mer, France
e-mail: lebris@obs-banyuls.fr

F. Gaill
CNRS, Ecology and Environment Institute, 3 rue Michel Ange, 75016, Paris, France

S. Kiel (ed.), *The Vent and Seep Biota*, Topics in Geobiology 33,
DOI 10.1007/978-90-481-9572-5_3, © Springer Science+Business Media B.V. 2010

tubeworm *Riftia pachyptila* or the *Bathymodiolus* spp. mussels, flourish in diffuse flow areas of the seafloor. A key to their success is the ability to supply the autotrophic bacteria hosted in their tissues with electron donors and acceptors from the surrounding fluids (see Dubilier et al. 2008 for review). Communities colonizing the walls of hydrothermal smokers (i.e. focused vent habitats) are mostly devoid of these endosymbiotic organisms but also yield remarkably high biomasses. Only a few specialized metazoans are adapted to the sharp physico-chemical gradients and fluctuating conditions encountered in these habitats (Van Dover 2000). These species associate with abundant free-living microbes, sometimes attached to their body surface or colonizing extracellular matrix like tubes (Gaill and Hunt 1991; Gebruk et al. 1993). Morphological adaptations supporting the growth of dense filamentous bacterial epibioses have been described, suggesting mutualistic relationships between free-living microbes and their hosts (Gaill et al. 1988; Van Dover et al. 1988; Goffredi et al. 2004, 2008). Although the nutritional benefit for the host is still not demonstrated, the biomass of these invertebrate assemblages suggests that particularly suited physico-chemical conditions for chemoautotrophic primary producers may be found through these associations.

Although endosymbionts have not yet been isolated into pure culture, metabolic requirements and carbon fixation rates of endosymbiotic vent invertebrates have been qualitatively and quantitatively constrained on live organisms in pressurized aquaria (e.g., Childress and Fisher 1992; Girguis and Childress 2006; Riou et al. 2008). In comparison, the mechanisms underlying inorganic carbon fixation by microbes colonizing the cuticle or tube of invertebrate species are almost unknown, even though this question has held the attention of vent biologists for long (see reviews in Desbruyères et al. 1998; Le Bris and Gaill 2007; Schmidt et al. 2008a). One of the main difficulties is that very few representatives of these epibionts were successfully cultured and none of them belongs to the dominant phylotypes (Campbell and Cary 2001). Furthermore, in vivo experimentations on vent invertebrates carrying epibionts are still scarce (Ravaux et al. 2003; Shillito et al. 2004), and have been only recently dedicated to the symbiosis itself (Zbinden et al. 2008). The lack of accurate information concerning the physico-chemical conditions experienced by epibiotic microbes have long prevented researchers to address their metabolic activities and rates in relevant culture or in vivo conditions. First studies of Desbruyères et al. (1998, 2000) provided a general picture of the environment of invertebrate aggregations on chimney walls, but were unable to address epibiont environmental conditions. The volume of conventional water sampling devices (e.g. 200–500 mL titanium syringes e.g., Sarradin et al. 1998) exceeds by at least two order of magnitude the volume occupied by an individual and is even inadequate to resolve the steep gradients at the colony scale. Furthermore, large-volume samples not only mix a wide range of fractions with different chemical characteristics, but also favor the reaction of electron donors and acceptors during recovery, with a high risk of underestimating their in situ content (Le Bris et al. 2006).

In situ measurement techniques and micro-volume sampling allowed researchers to characterize more accurately the gradients experienced at the individual organism scale within animal aggregations (Di Meo-Savoie et al. 1999, 2004; Le Bris et al. 2001, 2005). Since the precision of submersible manipulations is not better than a few millimeters, a direct characterization of environment gradients at the scale of epibiont mats is still out of reach, but these data can be used to validate geochemical models. Indirect approaches based on microanalysis of mineralogical proxies combined with downscaling of geochemical models provide an interesting

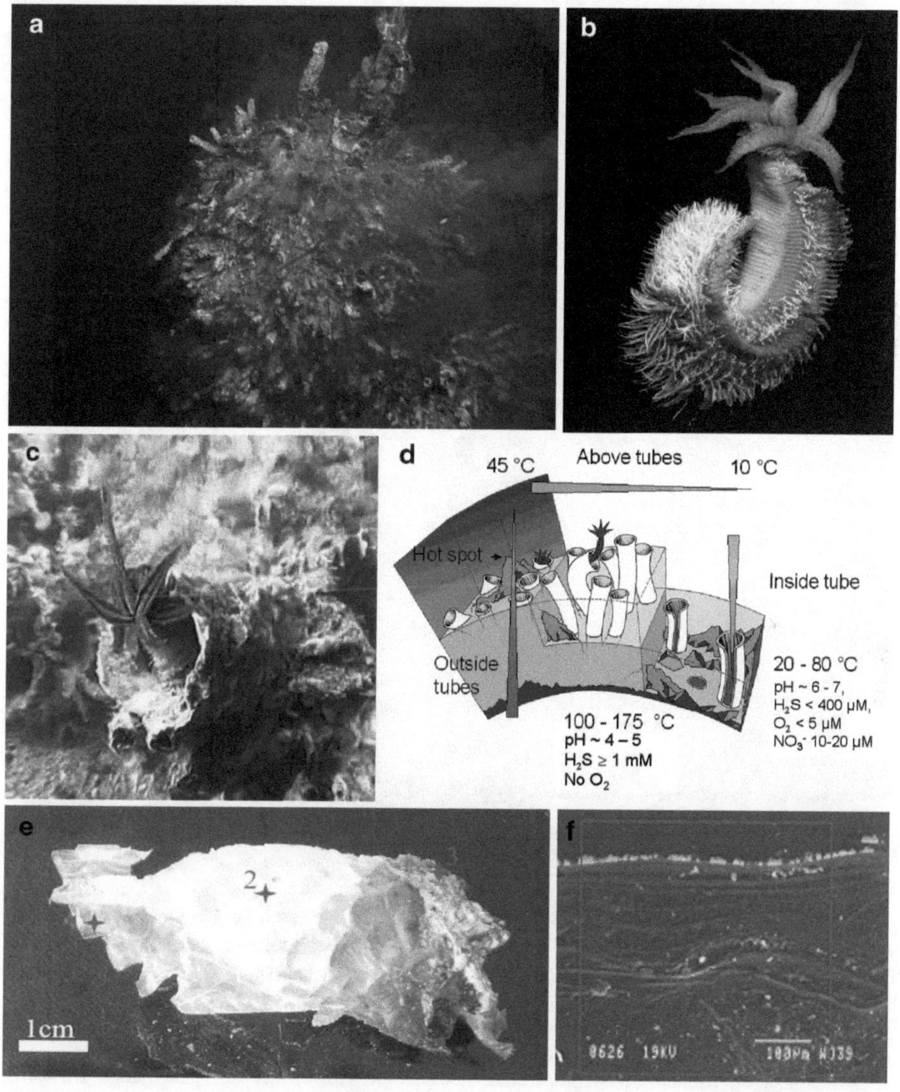

Fig. 3.1 *Alvinella pompejana*. (**a**) Pompeii worm colony on the wall of a hydrothermal smoker (EPR 13°N, Ifremer, PHARE cruise). (**b**) View of an individual and its dorsal filamentous epibiosis. (**c**) Brief appearance of a worm at tube opening. (**d**) Summary of temperature ranges measured at colony and animal scales (Adapted from Le Bris and Gaill 2007) (**e**) Tube: 1. Newly secreted portion at tube opening, 2. White central portion, 3. Mineralized aged portion (initially black, then rusty as the minerals oxidize in contact with air). (**f**) Thin sections of the tube wall showing successive organic layers and mineral deposits on the outside of the tube. A color plate of this figure can be found in Appendix (Plate 5)

Fig. 3.2 *Rimicaris exoculata*. (**a**) Swarm on smoker walls (Rainbow, Ifremer, ATOS cruise). (**b**) Positioning of shrimps in the thermal gradient, (**c**) A *Rimicaris exoculata* exuviate surrounded by live individuals from another species of shrimp feeding on it. Accumulated iron oxides in the branchial chamber are visible through the transparent carapace, (**d**) Semi-thin section of the cuticle covered by the epibiotic bacterial mat and iron oxide deposits. (**e**) TEM image showing a fraction of a bacterial filament (1) and a rod-shaped bacterium coated with iron oxides (2). (**f**) Scheme of the *Rimicaris exoculata* branchial chamber: lower pre-branchial compartment (1), true gill compartment in which respiration takes place (2), upper post-branchial compartment where iron oxides mostly accumulate on Rainbow individuals (3). The exopodite (Ex) delineate the lower and upper compartments. Scaphognatite (Sc) beats drive the water circulation along the three compartments (arrows). A color plate of this figure can be found in Appendix (Plate 6)

alternative to investigate the processes structuring physico-chemical gradients in these environments (Zbinden et al. 2003, 2004; Schmidt et al. 2008b). The following sections synthesize the main outcomes of these studies for the first two models of epibiosis discovered at hydrothermal vents: *Alvinella pompejana* and *Rimicaris exoculata*. Both species colonize the wall of active hydrothermal smokers with high population densities, but they largely differ in their geographic distribution and behavior. While the annelid polychaete, *Alvinella pompejana*, is endemic of the East Pacific Rise from 21°S to 13°N (Desbruyères et al. 1998), the alvinocarid shrimp, *Rimicaris exoculata*, has been described on five of the known mid-Atlantic ridge vent fields (Desbruyères et al. 2000). The first species builds mineral and organic structures through which hot fluids diffuse from the wall of hydrothermal smokers (Gaill and Hunt 1991) (Fig. 3.1a, Plate 5). The second is a mobile shrimp forming dense swarms on the wall of hydrothermal chimneys (Van Dover et al. 1988) (Fig. 3.2a, Plate 6). These two model organisms have been the subject of integrated multidisciplinary approaches in recent years, allowing the physico-chemical characteristics of epibiont micro-environments to be better constrained (Di-Meo et al. 2004; Le Bris et al. 2005; Schmidt et al. 2008b). In addition, the description of associated microbial communities and mineral deposits over time have allowed the temporal dynamics of these microhabitats to be accounted for (Zbinden et al. 2003, 2004, 2008; Corbari et al. 2008a, b). Conceptual and numerical models have been proposed to describe the mechanisms governing available energy resources for associated microbes, and serve as a basis for future experimental approaches and investigation of epibiont metabolic activities (Le Bris et al. 2008; Schmidt et al. 2009).

3.2 *Alvinella pompejana*

3.2.1 The Pompeï Worm Environment

Alvinella pompejana forms thick aggregations of proteinaceous tubes on the flanks of active hydrothermal smokers of the East Pacific Rise (Fig. 3.1a, Plate 5). This animal is characterized by, both, a dense dorsal filamentous epibiosis and abundant microbial communities colonizing the inner face and intra-layer space of their tubes (Gaill and Hunt 1991) (Fig. 3.1b, c, Plate 5). While hydrothermal end-member fluids exiting the main fluid conduits of EPR black smokers reach 330–405°C in temperature, subsurface mixing with cold seawater generates white or transparent fluids with a wide thermal range (Tivey 2007). *Alvinella pompejana* colonies associate with fluid emissions lying in the upper part of this range, as reflected by temperature maxima of 105–250°C reported in their immediate surroundings (see Le Bris and Gaill 2007 for review) (Fig. 3.1d, Plate 5). Temperature extremes as high as 175°C have even been measured by inserting a probe through the thickness of the tube layer (Di Meo et al. 2004). These measurements raised the assumption that this species might be one of the most thermotolerant metazoans on Earth (Chevaldonné et al. 1992). To the exception of the related species *Paralvinella sulfincola* from the Juan de Fuca Ridge, such a preference for high temperature habitats is unprecedented among vent invertebrates (Girguis and Lee 2006).

The most important characteristic of this habitat however relies in the steepness of temperature change (see Le Bris and Gaill 2007 for review). Temperature decreases to <10–45°C at tube openings, at the surface of the decimeter thick layer formed by tubes above the mineral wall (Le Bris et al. 2005) (Fig. 3.1d, Plate 5). As illustrated in Le Bris and Gaill (2007), the readings of two attached probes with a distance of less than 2–3 cm may differ by up to 20°C in this environment. It is therefore particularly challenging to precisely define temperature in this habitat from direct measurements. Chevaldonné et al. (1991) provided another illustration of this by reporting 105°C on a bare mineral wall, while a live Pompeï worm from the adjacent colony coiled around the temperature sensor obviously indicating that a much lower temperature was experienced only a few centimeter from the substratum.

Chemical features also distinguish these habitats from diffuse flow areas on the basaltic seafloor where most hydrothermal vent invertebrates proliferate. Almost millimolar concentrations of sulfide have been reported above *Alvinella pompejana* colonies from a number of vent sites (Luther et al. 2001; Le Bris et al. 2003; Le Bris and Gaill 2007), while sulfide levels only exceptionally exceed 100 μM in diffuse vent habitats (Johnson et al. 1988, 1994; Le Bris et al. 2003, 2006). In addition to numerous other potential stresses, including exposure to toxic heavy metals, low pH and high pCO_2 (see review Le Bris and Gaill 2007), Luther et al. (2001) and Di Meo et al. (2004) have shown that oxygen is largely depleted in this environment.

Not only are the temperature and chemical gradients extreme in amplitude, but they are also subject to significant temporal variability. Substantial temperature fluctuations have been reported at second to hour scales (Desbruyères et al. 1998; Di Meo et al. 2004; Le Bris et al. 2005). pH, which is a good proxy for the hydrothermal contribution, varies accordingly, indicating substantial chemical variability over time in this environment (Le Bris et al. 2005). The amplitude of these fluctuations might be particularly large in the turbulent plumes of local sources (Le Bris et al. 2005). On daily basis, Chevaldonné et al. (1991) additionally demonstrated that the tidal cycle may significantly influence this variability.

3.2.2 *Chemolithoautotrophic Microbes and Biomass Production in* Alvinella pompejana *Colonies*

Despite these a priori hostile conditions, in situ experiments using dedicated colonization devices (TRAC for Titanium Rings for *Alvinella* Colonization) have revealed that *Alvinella pompejana* colonies grow remarkably fast. The first worms settle within only a few days on such newly open habitats (Taylor et al. 1999; Pradillon et al. 2005, 2010). Within less than 3 months, biomineral assemblages are formed in which tubes and sulfide concretions are intimately entangled

(Zbinden et al. 2003). Microbes are pioneer colonizers in this environment, forming visible mats in only a few days on artificial substrates (Taylor et al. 1999; Alain et al. 2004). Epsilon (ε)–protobacteria dominate pioneer communities, consistent with the idea that this versatile microbial sub-group is particularly adapted to sulfur rich redox-clines (Nakagawa et al. 2005; Campbell et al. 2006). They were also among the most dominant phylotypes forming the dense *Alvinella pompejana* epibiosis (Grzymskia et al. 2008) (Fig. 3.1b, Plate 5). A number of strains have been isolated from the associated microflora (Campbell and Cary 2001; Campbell et al. 2003; Alain et al. 2004), but the dominant filamentous morphotypes forming the epibiosis eluded cultivation. Although no strict chemolithoautotrophs have been cultivated to date from the epibiotic community, Campbell and Cary (2004) revealed that the rTCA cycle is expressed and suggested this metabolic pathway could play a more important role in chemoautotrophic carbon fixation in this environment than the Calvin cycle. As discussed in Le Bris and Gaill (2007), primary producers using this metabolic pathway may have found appropriate conditions to sustain the exceptional rate of organic matter production, as quantitatively documented in a recent study (Pradillon et al. 2010). This carbon fixation pathway might indeed be particularly efficient in the high sulfide and low oxygen environment of *Avinella pompejana*, as shown for the free living shallow water *Arcobacter* strain (Wirsen et al. 2002). To what extent this productivity relies on the abundant epibiotic and free-living microbes colonizing the interior of tubes, in comparison to those colonizing the outer surface of tubes and the mineral substrate, is however unknown.

3.2.3 *Microhabitat Partitioning by* Alvinella pompejana *Tubes*

In an attempt to define growth conditions for the microbial communities living in association with *Alvinella pompejana*, dedicated investigations have focused on the physico-chemical characteristics experienced inside the tube. It was first suggested that the tube channels the hot fluid diffusing from the mineral substrate (Cary et al. 1998). The worms would then be exposed to similar conditions in this micro-environment rather than in the turbulent fluid-seawater mixing gradient that is established at the surface of chimney walls. A sharp contrast in mineral deposits from the exterior face of the tube to the interior face was however demonstrated from TEM and element microanalysis on thin sections of the tube wall (Zbinden et al. 2001, 2003) (Fig. 3.1f, Plate 5). Mineral particles associated with tube portions of 2.5 month-age, as indicated by their basal position in a colonization device, revealed a marked transition between an FeS_2-dominated mineral assemblage outside the tube to a ZnS-dominated assemblage inside the tube (Zbinden et al. 2003). These data provided first unambiguous evidence of distinct microhabitat conditions inside and outside the tube, rejecting the hypothesis of hot fluid flowing through it from the chimney wall. An increased dilution of the hydrothermal source fluid inside tubes, compared to the outside conditions, was furthermore inferred from these data. The observed correlation of ZnS and FeS_2–rich mineral zonation at the colony scale with the temperature gradient indeed suggested that their relative abundance could serve as a proxy for the fluid dilution rate (Zbinden et al. 2003).

Confirming the hypothesis of mineralogical studies, direct approaches using in situ measurements and micro-volume sampling have demonstrated that the tube is filled with a mixture dominated by seawater while the fluid circulating outside the tube is mostly of hydrothermal origin (Di Meo-Savoie et al. 2004; Le Bris et al. 2005). The abundance of magnesium, a tracer of seawater absent from hydrothermal end-members, and of electron donors like nitrate or sulfate (Di Meo et al. 2004) is consistent with a high seawater content inside tubes. The near neutral to slightly acidic pH measured inside tubes (Le Bris et al. 2005) furthermore contrasted with the more strongly acidic values measured among tubes (pH 4–5). Rather than being passively exposed to environmental gradients, *Alvinella pompejana* was therefore suspected by these authors to control its micro-environment, providing important insights to the understanding of *Alvinella pompejana*'s ability to farm its associated microflora and to colonize extremely hot substrates.

3.2.4 Thermal Buffering Within and Surrounding **Alvinella** *Tubes*

Despite the controversial interpretation of these data regarding the biology of *Alvinella pompejana* (Chevaldonné et al. 2000), Cary et al. (1998) and Di Meo et al. (2004) have repeatedly recorded mean temperatures above 50°C inside tubes hosting live *Alvinella pompejana*. These measurements were in apparent contradiction with a predominant contribution of cold ambient seawater in the tube. Yet, while temperature ranges overlap, distinct pH ranges were recorded inside and outside tubes (Le Bris et al. 2005). Geochemical modeling helped to clarify this apparent discrepancy, by suggesting that the seawater filling the tube is heated by conduction through the tube wall (Le Bris et al. 2005). Renewal of the fluids inside the tube resulting from active ventilation by the animal or passive convection would thus allow *Alvinella pompejana* to avoid excessive heating of its microenvironment and to stand at the limits authorized for its biological machinery (Le Bris and Gaill 2007).

The Pompeï worm colony itself appears to act as a thermal exchanger, gradually dissipating the heat of the fluid before it flows out of the tube assemblage (Le Bris et al. 2005). While temperatures well exceeding 100°C are found a few centimeters beneath the tube, the fluids that mix with seawater above colonies are indeed much lower in temperature (Le Bris and Gaill 2007 for review), preserving these animals and their associated microflora from deleterious temperature spikes. The organo-mineral assemblage of tubes formed by *Alvinella* colonies thus provides a mosaic of buffered micro-habitats which are expected to be much more suitable for microbial life than those associated with the turbulent mixing of high temperature fluids venting from bare chimney walls (Le Bris and Gaill 2007).

3.2.5 Temporal Dynamics of Electron Donor and Acceptor Supply to the Epibiotic Microflora

Despite the important seawater dilution of the fluid inside tubes, sulfide concentrations of several hundreds of micromolar have been measured inside tubes (Di Meo-Savoie et al. 2004). These values are consistent with a substantial dilution of source fluids escaping chimneys walls with almost millimolar sulfide contents (Luther et al. 2001; Le Bris et al. 2003; Le Bris and Gaill 2007). In addition to the advection of sulfidic waters from the tube opening, the diffusion of H_2S through the tube wall is expected to contribute significantly to the flux of sulfide to this microhabitat. The transparent extracellular material newly secreted by *Alvinella pompejana* to form its tube was shown to be highly permeable to hydrogen sulfide (Le Bris et al. 2008). From the first Fick's Law, these authors quantified the permeability of different portions of the tube along its length. The data revealed that the tube permeability to hydrogen sulfide is modulated by its progressive mineralization (Fig. 3.1d, e, Plate 5). The white tube portion in the upper layer of the colony is highly permeable to sulfide (mean permeability $14 \pm 2\ 10^{-5}$ cm s^{-1}, for an average thickness of 199 ± 20 μm, $n = 8$), while the diffusion of sulfide is prevented in the older mineralized part of the tube. Distinct microhabitat conditions should therefore be experienced within the same tube portion over time, as it ages. At the scale of a tube this results in the succession of different microhabitats, from the youngest part of the tube at the opening to the oldest mineralized part closer to the mineral wall (Fig. 3.1d, Plate 5).

The abundance of elemental sulfur in the tube matrix (12–25% according to Gaill and Hunt 1986) indicates a high rate of sulfide oxidation in the tube. Consistent with the depletion of oxygen in the tube documented by in situ voltammetry (Di Meo-Savoie et al. 2004), the main limitation to sulfide-oxidizing autotrophic carbon fixation in these conditions should be the availability of an electron acceptor. The renewal rate of the fluid inside the tube thus appears as a key factor in the electron acceptor supply (Le Bris et al. 2005, 2008). The brief appearances of Pompeii worms at tube openings (Fig. 3.1c, Plate 5), with intervals of 10 min or more according to statistical analysis of videos (Chevaldonné and Jollivet 1993), allowed researchers to consider that fluids in the anterior part of the tube is intermittently renewed through these fast in and out movements (Desbruyères et al. 1998; Le Bris et al. 2005). Le Bris et al. (2008) simulated the variation of sulfide in the tube over time from both H_2S diffusion through the tube wall and its abiotic reaction with oxygen as function of the ventilation rate. This simplified model, which did not account for biological consumption suggested that the energy available per volume unit for autotrophic microbial primary producers using oxygen and sulfide might increase over time, if not significantly balanced by sulfide-oxidizing microbe activity. These calculations also suggest that, within a 10 min-period, abiotic sulfide oxidation would significantly compete with microbial processes for oxygen (Le Bris et al. 2008). Both oxygen depletion and the abundance of nitrate in the *Alvinella pompejana* tubes reported by Di Meo-Savoie et al. (2004) indicated that this anion may be used as a secondary electron acceptor by the epibionts and associated free-living microbes. As simulated in Le Bris et al. (2008), the availability of this electron acceptor combined with sulfide diffusion through the tube may even constitute a substantial energy source as time increases following the intermittent renewal of the fluid.

The reduced permeability of the mineralized tube wall limits the diffusive flux of sulfide to the anterior part of the tube (Le Bris et al. 2008). As well, the availability of oxygen and nitrate is likely to be limited to the anterior portion of the tube where the fluid is periodically renewed. Microbial growth in the posterior part of the tube should therefore be less dependent on the outer environment and more directly reliant on the tube itself, with a lower contribution to autotrophic primary production. While elemental sulfur still constitutes a large stock of electron acceptor for microbes throughout the tube (Gaill and Hunt 1986), sulfur-reducing mixotrophs as reported in Campbell et al. (2006) for the *Alvinella pompejana* tube may therefore be favored by the abundance of organic electron donors resulting from the degradation of the tube material.

Although these distinct micro-niches appear much less chaotic in their chemical and thermal variations than the free-mixing zone on bare substrate, they are expected to undergo significant variation over timescales of weeks, in relation to the successive stages of tube secretion, mineralization and, ultimately, degradation and fossilization (Pradillon et al. 2010). Le Bris et al. (2008) suggested that optimal steady-state conditions for sulfide-oxidizing autotrophs are maintained in the anterior part of the tube in the dynamic environment associated with the rapidly growing chimney. This further supports the idea that *Alvinella pompejana* is not only a heat-tolerant pioneer colonizer but also acts as a new type of ecosystem bioengineer allowing the subsequent colonization of these extreme environments by various primary producers and consumers (Le Bris et al. 2005, 2008).

3.3 *Rimicaris exoculata*

3.3.1 The Environment of Rimicaris exoculata *Swarms*

Rimicaris exoculata shrimp gather in dense aggregations above and beside hot fluid issues on the walls of active chimney complexes (Van Dover et al. 1988) (Fig. 3.2a, Plate 6). *Rimicaris exoculata* is highly mobile, exceeding the motility of other shrimp species, *Mirocaris fortuna* and *Chorocaris chacei*, which have been described in association to *Rimicaris* swarms (Gebruk et al. 2000). Emphasizing previous observations, Copley et al. (1997) furthermore indicate that, when displaced from the substratum, shrimp immediately seek to re-establish themselves near the source of hydrothermal fluid. Similarly to *A. pompejana*, the availability of a substratum exposed to the flow of hydrothermal fluids appears to be a prerequisite for the development of these dense aggregations.

Rimicaris exoculata thrives within the turbulent mixing interface of hot fluids and seawater along chimney walls (Schmidt et al. 2008a). The temperature of the hot fluid source supplying the swarms is unknown, but values as high as 40°C have been reported in their close surroundings (see review in Schmidt et al. 2008a). Gebruk et al. (1993) reported that these high values refer to shimmering water surrounding the swarm and not directly to the swarm itself. The dedicated survey based on close-up video control, with a deported camera mounted on the temperature probe, allowed Schmidt et al. (2008b) to precisely define the *Rimicaris* thermal range. Over the seven swarms surveyed by these authors in 2001, 2005 and 2007 on the Rainbow and TAG vent fields, none exceeded 18°C in temperature. Temperature in the swarm ranged from this maximum down to the background seawater value (3.65 at Rainbow, 2.8°C at TAG). From this study, the habitat of *R. exoculata* thus appears quite homogeneous and reproducible in temperature, even though large temperature gradients and strongly fluctuating mixing conditions characterize their surroundings (Fig. 3.2b, Plate 6). This result is consistent with the fact that a thermal stress response is induced below 25°C (Ravaux et al. 2003), suggesting that these swarms developed a particularly reactive escape behavior. Although the possibility of higher temperatures deeper within swarms cannot be ruled out, it is likely that the temperature maxima around 40°C reported in the first studies of Van Dover et al. (1988) and Segonzac et al. (1993) have overestimated the habitat range. These measurements may have suffered from poor spatial accuracy, as it is particularly challenging to accurately define the position of a sensor tip with respect to steep swarm boundaries without a deported camera.

Although they are both physically associated to the walls of active smokers, *Alvinella pompejana* and *Rimicaris exoculata* then markedly differ in their positioning within the thermal gradient. While Pompeï worms experience a wide range of temperature throughout their colonies, *Rimicaris exoculata* shrimp occupy a much narrower thermal range. Contrasting with the sessile behavior of *Alvinella pompejana*, the strategy of the shrimp to select an optimal thermal range, avoiding exposure to deleterious temperatures likely relies on its ability to rapidly change its position within the hot fluid-seawater mixing gradient along the chimney wall (Copley et al. 1997).

3.3.2 *Available Energy Sources for Chemolithoautotrophs in the* Rimicaris exoculata *Environment*

One of the most interesting features of the *Rimicaris exoculata* habitat derives from the geochemical diversity of the mid-Atlantic Ridge vent fields. The composition of high-temperature fluids fueling hydrothermal smokers varies widely over the five vent fields where *Rimicaris* swarms have been described (see table 3 in Schmidt et al. 2008a). As reviewed by these authors, not only the diversity of dominant electron donors available to chemoautotrophic microbes is large but their relative abundance remarkably changes from one vent field to the other. In an attempt to evaluate the dominant energy sources for primary production within *Rimicaris* swarms, Schmidt et al. (2008b) compared TAG and Rainbow in terms of energy budgets available in the oxic part of the mixing gradient. Substantial enrichments in methane and ferrous iron have been reported in *Rimicaris* habitats at Rainbow, respectively from discrete sampling and in situ measurements (Desbruyères et al. 2000; Zbinden et al. 2004; Schmidt et al. 2008b). While sulfide appears much more depleted in the shrimp environment at Rainbow, it was suggested to dominate the electron donor pool at TAG (Schmidt et al. 2008b). These differences reflect different geological settings hosting the two vent fields. The interaction of peridotite rocks with water (called serpentinization) results in a large hydrogen enrichment in the Rainbow fluids, which subsequently reacts with CO_2 to form methane (Charlou et al. 2002). Hydrogen has not been precisely quantified in the environment of *Rimicaris*, but its abundance in end-member fluids (Charlou et al. 2002) suggests a significant role of this electron donor at Rainbow.

Despite the enrichment of the environment in electron donors rapidly reacting with oxygen, such as hydrogen or sulfide, oxygen concentration appears to be quite high in the surrounding of *Rimicaris exoculata* (Zbinden et al. 2004; Schmidt et al. 2008b). This is consistent with, both, the low temperature of this environment indicating a large seawater contribution and fast turbulent mixing conditions which prevent oxygen to be removed by abiotic processes. In contrast to the *Alvinella pompejana* microenvironment which was described as hypoxic to fully anoxic, oxygen is permanently renewed and maintained to a substantial level within swarms. Quantitative estimates of the chemical energy budgets available from different electron donors in the oxic part of the vent fluid-seawater mixing gradient at TAG and Rainbow have been proposed by Schmidt et al. (2008b). Their modeling approach was based on mixing simulation and Gibbs free energy calculations. It confirmed that ferrous iron is the most abundant energy source available for aerobic chemolithotrophs in the fluid-seawater mixing gradient at Rainbow below 30°C, exceeding methane and sulfide which are the most common electron donors fueling chemosynthetic communities. According to these authors, hydrogen may constitute another primary energy source at this vent field, but this hypothesis still needs to be quantitatively assessed. Consistent with previous assumptions (Wirsen et al. 1993), sulfide was confirmed as the most abundant energy source for TAG, while iron may constitute a secondary source. These models also revealed that the maximum energy available per kilogram of mixed fluid is limited by the availability of electron donors in the environment of *Rimicaris exoculata*, contrasting with the *Alvinella pompejana* model. Oxygen becomes limited only above the maximum temperature measured in shrimp swarms (Schmidt et al. 2008b). Beyond this temperature threshold, the energy available using this electron acceptor progressively decreases as temperature is increasing, supporting the assumption that *Rimicaris* shrimps may farm their epibionts by positioning themselves in optimal conditions for chemolithotrophic growth within the sharp mixing gradient while avoiding deleterious temperatures.

3.3.3 *The Diversity of* Rimicaris exoculata *Epibiosis*

One of the most characteristic specializations of these shrimp relies on the abundant bacterial communities colonizing in the inner surface of the gill chamber formed by its enlarged carapace (Van Dover et al. 1988; Segonzac et al. 1993). These authors described dense mats of filamentous bacteria covering the inner face of the carapace as well as the exopodite and scaphognatite appendages. Anderson et al. (2008) identified three bacterial morphotypes: thin and thick filaments and rod shape cells (Fig. 3.2e, Plate 6). In a more recent study, Zbinden et al. (2008) even reported three types of filaments (two thin types and one large) and two types of rods. Using TEM (Transmission Electron Microscopy), Anderson et al. (2008) pointed out that thin filaments are found closest to the cuticle in great numbers, whereas the larger, multi-chambered filamentous bacteria are fewer in number, with the rod-shape bacteria filling the spaces around them. The filamentous morphotypes were suggested to reflect at least two different ε proteobacteria phylotypes (Dubilier et al. 2008), contrasting with first studies, which reported a single phylotype among *Rimicaris exoculata* epibionts (Poltz and Cavanaugh 1995). Similar to the *Alvinella pompejana* epibiosis, the *Rimicaris* epibiosis may be in fact much more diverse than previously considered (Zbinden et al. 2008), with at least a third dominant epibiont belonging to γ proteobacteria (Dubilier et al. 2008).

This diverse epibiosis was assumed to be the main energy source for the dense shrimp swarms (Gebruk et al. 2000). Direct evidence of this symbiotic nutritional relationship has not been provided to date, but these authors invoke a series of clues to support this assumption. First, no visible free-living bacterial crops were observed on the mineral substratum surrounding swarms. Second the isotopic carbon signature of the shrimp tissues is inconsistent with the uptake of organic carbon from the mineral walls despite the fact that over 60% of the rRNA of the microbial community on sulfide surfaces seemed identical to that of the epibionts (Poltz and Cavanaugh 1995; Rieley et al. 1999). An alternative hypothesis involving a gut bacterial symbiosis using iron sulfide for growth has also been proposed (Zbinden and Cambon 2003). Whatever the mode of nutrition for the shrimp, its epibionts are likely to contribute to the chemoautotrophic primary production in these environments. Using ^{13}C incubation, Wirsen et al. (1993) have supported this hypothesis and confirmed Rubisco activity in samples from the shrimp epibiosis. ^{13}C isotopic ratios reported in Gebruk et al. (2000) are however significantly higher than those of *Bathymodiolus azoricus* mussels relying on sulfide oxidizing symbionts using Rubisco type I (Le Bris and Duperron in press). As already done for the *Alvinella pompejana* epibiosis, further analysis of alternative metabolic pathways, including those relying on the rTCA cycle, remains necessary to define the main carbon fixation pathways sustained by *Rimicaris* swarms.

The chemical energy source used for autototrophic carbon fixation remains puzzling. A predominant role of sulfide oxidizers has been primarily inferred for *Rimicaris* (Wirsen et al. 1993; Gebruk et al. 1993). The presence of intracellular granules, presumed to be elemental sulfur, supported this hypothesis

(Gebruk et al. 1993; Zbinden et al. 2008). Uncertainty in the composition of these granules inherent to TEM sample preparation however makes this observation conspicuous in demonstrating the use of this metabolic pathway. According to Poltz et al. (1998), no statistically significant stimulation of carbon fixation by the epibionts was observed in the presence of reduced sulfur compounds in Snake Pit samples. In a recent molecular survey, Zbinden et al. (2008) revealed that APS (sulfide oxidation) encoding genes were successfully amplified from DNA extracted from the scaphognatite of Rainbow specimens but their approach did not provide any clues on the effective expression of these genes. Furthermore, these authors did not observe any positive influence of sulfide exposure on the epibiotic bacteria of live *Rimicaris* from Rainbow repressurized in aquaria. Conversely, they identified bacteria with stacks of intracytoplasmic membranes typical of methanotrophs and genes encoding for pmoA (methane oxidation) were successfully amplified, indicating that methane can be used as an electron donor. These cells are however not among the predominant morphotypes of epibiotic bacteria at Rainbow. The lack of a gene sequence encoding for iron-oxidizer enzymatic pathways prevented these authors to confirm their occurrence among the epibionts. Furthermore, iron oxidizers broadly distribute among the Proteobacteria and therefore cannot be efficiently tracked from their phylogenetic affiliation in one or the other group of proteobacteria (Edwards et al. 2003). To date, there is no more evidence that H_2 might be used by epibionts, even though this is a common pathway in ε proteobacteria and this electron acceptor has been reported as a potentially dominant energy source at Rainbow. Although they have provided useful indications on the diversity of the epibiosis, molecular approaches are therefore insufficient to discriminate among the diverse chemical energy sources that can be exploited by *Rimicaris* on different vent fields.

3.3.4 Partitioning of the Shrimp Branchial Chamber into Distinct Micro-habitats

Most relevant clues suggesting the presence of active iron oxidizers in the branchial chamber of *Rimicaris exoculata* came from the investigation of relationships between microhabitat conditions within the gill cavity and the dynamics of the shrimp molting cycle. The hypertrophied branchial chamber of *Rimicaris exoculata* has been described in detail in Segonzac et al. (1993). Beating of the scaphognatites at rates of 50–100 beats min^{-1} drives water circulation within this cavity (Gebruk et al. 1993). Normal ventilatory currents in free swimming shrimps allows the surrounding water to enter the post-ventral border of the carapace, to pass over the gill in the central part of the cavity before emerging below the antenna bases (Gebruk et al. 2000) (Fig. 3.2f, Plate 6). The large mineral particles characteristic of smoker plume particles that were evidenced by Zbinden et al. (2004) in the lower compartment of the branchial chamber confirmed a similar direction of the flow for *Rimicaris exoculata*. Further information on the distribution of minerals and bacteria, obtained by electron microscopy and EDX microanalysis, has demonstrated distinct microenvironments in the *Rimicaris exoculata* branchial cavity (Zbinden et al. 2004): (1) the lower pre-branchial chamber, which houses bacteria but few minerals, (2) the true branchial chamber, which contains the gills but is free of bacteria and minerals, and (3) the upper post-branchial chamber, housing most of the bacteria and associated minerals (Fig. 3.2f, Plate 6). Corbari et al. (2008a) confirmed this distribution on individuals from the Rainbow vent field, but showed distinct mineral deposition features on individuals of the TAG vent field. While rusty colored iron(III) oxides dominate the mineral deposit at Rainbow, both rusty, grey and black deposits were observed on TAG shrimps, indicating that iron sulfides were relatively more abundant in shrimp from this vent field. Contrasting with iron oxides in Rainbow shrimps, iron sulfides appear much more uniform in the branchial chamber of TAG shrimp, covering the inner face of the carapace in both pre- and post-branchial compartments (Corbari et al. 2008a).

The reduced occurrence of iron oxide in the pre-branchial compartment of Rainbow shrimps lead Zbinden et al. (2004) to reject the hypothesis of an environmental contamination and lead to the conclusion that its deposition is favored in the post-branchial compartment (Zbinden et al. 2008; Corbari et al. 2008a). The distribution and composition of minerals in the branchial chamber of TAG shrimps suggest a reverse situation (Corbari et al. 2008a). Contamination from the surrounding environment is a likely origin for iron sulfide minerals consistent with the black smokers surrounding of *Rimicaris* swarms at TAG (Schmidt et al. 2008b). The occurrence of iron oxide in this case, also reported by Wirsen et al. (1993), may then result from the oxidation of FeS, abiotically or through bacterial processes. The non-uniform distribution of oxides in Rainbow shrimps led Zbinden et al. (2004) to hypothesize different micro-habitat conditions, potentially hosting different microbial communities. The tight association of these minerals with bacterial cells further supported the idea of a microbial influence on the deposition of these oxides. Gloter et al. (2004) mentioned that a significant part of the minerals are found attached to the bacterial cell walls. Anderson et al. (2008) further documented their three-dimensional (3D) distribution around cells in the search for further proof of microbially mediated mineral deposition. Minerals appeared preferentially associated with rod-shaped bacteria. These authors described different types of association between iron oxide and the surface of these cells, but the more frequently observed was a close association between the bacterial membrane and mineral with no visible space between them. Yet, another type of association showing indented 'teeth-like' mineral growths around a cell draws the attention on the potential link between exopolymeric secretion on discrete sites of the cell membrane and mineral deposition. The biogenic origin of the oxide was additionally supported by the mineralogy of these deposits. Gloter et al. (2004) identified a nearly single-phase mineralogy for these oxides composed of two-line ferrihydrite (Gloter et al. 2004; Corbari et al. 2008a). This mineralogical composition is characteristic of biogenic iron oxides formed in the presence of microbial cells, although it cannot be considered as a definitive proof of their biogenic origin (Fortin and Langley 2005).

3.3.5 *Dynamics of Iron Oxide Deposition in the Branchial Cavity*

Several mechanisms have been proposed to explain the surface precipitation of poorly ordered iron oxide on the cell-wall surface (see Fortin and Langley 2005 for review). Active surface sites may decrease the activation energy barrier by supplying a template for ferrihydrite nucleation in super-saturated conditions. Such a passive mechanism however fails to explain the accumulation of iron in the sole post-branchial chamber, unless surface properties of cell membranes are considered to differ significantly between the microbial communities colonizing the two microhabitats (Zbinden et al. 2004). Furthermore, Schmidt et al. (2008b) reported that dissolved ferrous iron accounts for most of the iron present in dissolved or colloid forms in the mixed fluids surrounding swarms. The accumulation of iron oxide in the post-branchial chamber thus suggests that this Fe(II) is transported and oxidized locally into Fe(III)(Zbinden et al. 2004). These authors pointed out that the kinetics of abiotic iron oxidation at neutral pH and moderate temperature (~10°C) encountered in the environment of *Rimicaris* should be much slower than usually considered for seawater at pH 8 and ambient temperature around 20°C. This may allow iron oxidizers to compete to a certain extent with the abiotic reaction pathway as described in Druschel et al. (2008). Iron oxidizers have developed specific strategies to facilitate electron exchange from oxygen and ferrous iron ions and subsequently benefit from the energy gained from redox disequilibria relaxation (Chan et al. 2004). They are yet competing with their own by-products (i.e. iron oxides) which catalyze iron oxidation (Rentz et al. 2007). Despite this competition, iron oxidizing bacteria are able to produce thick mats combining cells and iron oxides in association with iron-rich fluid emissions, such as on the flanks of the Loihi Hawaii submarine volcano (Emerson and Moyer 2002).

The development of this ectosymbiosis in relation with the progressive accumulation of iron oxides was investigated on Rainbow shrimps, using microscopy techniques, over the duration of the moulting cycle (Corbari et al. 2008b). By comparison with moulting stages in littoral shrimp species, these authors estimated the duration of this cycle to be about 10 days. They also revealed that bacterial re-colonization occurs within 2 days after each moult. Iron oxyhydroxide precipitation starts shortly after this first stage and thickening of the mineral layer progresses rapidly in time, until its elimination with the exuviate (Corbari et al. 2008b) (Fig. 3.2c, Plate 6). The high moulting frequency hence allows the shrimp to get rid of the dense mineral crust forming after a week or so within its branchial chamber. An equivalent of 7 mg of iron oxyhydroxide per day per shrimp may be removed from the fluid and deposited locally at the base of chimney (Schmidt et al. 2009). The catalysis of ferrous iron oxidation on iron oxyhydroxide is however known as particularly a fast process (Park and Dempsey 2005) and is likely to compete significantly with microbial use. Autocatalytic oxidation may explain the formation of the dense and thick crust in the later stages of the moulting cycle. The initiation and enhancement of the iron oxide deposition process in the post-branchial chamber, compared to the pre-branchial chamber which is more exposed to environmental contamination, support the assumption that iron-oxidizing bacteria may be active in this micro-environment. The growth of these microbes would be favored during initial moulting stages when the competition with their by-products is limited. If this iron oxidation was effectively used to fix carbon, this process could contribute at least partly to the energetic cost of the moult. Here again, the biology of the host regarding secretion of extracellular material appears as a key element in governing the growth of epibionts.

3.4 Conclusion

In a widely variable environment, invertebrate surfaces offer much more reproducible conditions for episymbiotic microbes than the mineral surface exposed to the turbulent mixing of vent fluid and seawater. The two examples considered in this chapter illustrate the diverse strategies implemented by host invertebrates to buffer environmental variability. Much remains to be known on the functioning of these associations with respect to energy acquisition and metabolic pathways, but combining mineral proxies and geochemical modeling helped to constrain the physico-chemical properties of some of these micro-niches. Metabolic requirements inferred from future molecular approaches may then be confronted with these characteristics to appreciate the potential importance of these metabolisms in local primary production. A variety of new types of epibioses have been described in recent years, among those the outstanding scaly foot gastropod (Goffredi et al. 2004) and Yeti crab (Goffredi et al. 2008). Each of these epibiotic associations presumably relies on specific microhabitat conditions governed by the host behavior. Indirect approaches such as those presented here constitute a relevant strategy to address the mutual relationships between the microflora and their host.

Acknowledgements The authors would like to acknowledge the CNRS, Ifremer and the MoMARNet EC contract MCRTN-CT-2004–552703.

References

Alain K, Zbinden M, Le Bris N, Lesongeur F, Querellou J, Gaill F, Cambon-Bonavita MA (2004) Early steps in microbial colonisation processes at deep-sea hydrothermal vents. Environ Microbiol 6(3):227–241

Anderson LM, Halary S, Lechaire JP, Boudier T, Frebourg G, Marco S, Zbinden M, Gaill F (2008) Tomography of bacteria-mineral associations within the deep sea hydrothermal vent shrimp *Rimicaris exoculata*. CR Acad Sci Chim 11(3):268–280

Campbell BJ, Cary SC (2001) Characterization of a novel spirochete associated with the hydrothermal vent polychaete annelid, *Alvinella pompejana*. Appl Environ Microbiol 67:110–117

Campbell BJ, Cary SC (2004) Abundance of reverse tricarboxylic acid cycle genes in free-living microorganisms at deep-sea hydrothermal vents. Appl Environ Microbiol 70:6282–6289

Campbell BJ, Stein JL, Cary SC (2003) Evidence of chemolithoautotrophy in the bacterial community associated with *Alvinella pompejana*, a hydrothermal vent polychaete. Appl Environ Microbiol 67:4566–4572

Campbell BJ, Summers EA, Porter ML, Takai K (2006) The versatile ε-proteobacteria: key players in sulphidic habitats. Environ Microbiol 69:5070–5078

Cary SC, Shank T, Stein J (1998) Worms bask in extreme temperatures. Nature 391:545–546

Chan CS, De Stasio G, Welch SA, Girasole M, Frazer BH, Nesterova MV, Fakra S, Banfield JF (2004) Microbial polysaccharides template assembly of nanocrystal fibers. Science 303:1656–1658

Charlou JL, Donval JP, Fouquet Y, Jean-Baptiste P, Holm N (2002) Geochemistry of high H_2 and CH_4 vent fluids issuing from ultramafic rocks at the Rainbow hydrothermal field (36°14′N, MAR). Chem Geol 191:345–359

Chevaldonné P, Desbruyères D, Le Haître M (1991) Time-series of temperature from three deep-sea hydrothermal vent sites. Deep-Sea Res 38:1417–1430

Chevaldonné P, Desbruyères D, Childress JJ (1992) And some even hotter. Nature 359:593

Chevaldonné P, Jollivet D (1993) Videoscopic study of deep-sea hydrothermal vent alvinellid polychaete populations: biomass estimation and behaviour. Mar Ecol Prog Ser 95:251–262

Chevaldonné P, Childress JJ, Fisher CR, Desbruyères D, Jollivet D, Zal F, Toulmond A (2000) Thermotolerance and the "Pompeii worms". Mar Ecol Prog Ser 208:293–295

Childress JJ, Fisher CR (1992) The biology of hydrothermal vent animals: physiology, biochemistry, and autotrophic symbioses. Oceanogr Mar Biol Annu Rev 30:337–441

Copley JTP, Tyler PA, Van Dover CL, Murton BJ (1997) Spatial and interannual variation in the faunal distribution at Broken Spur vent field (29°N, Mid-Atlantic Ridge). Mar Biol 129: 723–733

Corbari L, Cambon-Bonavita MA, Long GJ, Grandjean F, Zbinden M, Gaill F, Compère P (2008a) Iron oxide deposits associated with the ectosymbiotic bacteria in the hydrothermal vent shrimp *Rimicaris exoculata*. Biogeoscience 5:1825–1865

Corbari L, Zbinden M, Gaill F, Compère P (2008b) Moult-related settlement of bacterial symbionts and mineral deposits in the branchial chamber of the MAR hydrothermal vent shrimp *Rimicaris exoculata*. Aquat Biol 1:225–238

Desbruyères D, Chevaldonne P, Alayse-Danet AM, Caprais JC, Cosson R, Gaill F, Guezennec J, Hourdez S, Jollivet D, Jouin-Toulmond C, Lallier FH, Laubier L, Riso R, Sarradin PM, Toulmond A, Zal F (1998) Burning subjects: biology and ecology of the "Pompei worm" (*Alvinella pompejana* Desbruyères et Laubier), a normal dweller of an extreme deep-sea environment. Deep Sea Res II 45:383–422

Desbruyères D, Almeida A, Biscoito M, Comtet T, Khripounoff A, Le Bris N, Sarradin PM, Segonzac M (2000) A review of the distribution of hydrothermal vent communities along the Northern Mid Atlantic Ridge: dispersal vs environmental controls. Hydrobiol 440: 201–216

Di Meo-Savoie CA, Wakefield JR, Cary SC (1999) A new device for sampling small volumes of water from marine microenvironments. Deep-Sea Res I 46:1279–1287

Di Meo-Savoie CA, Luther GW III, Cary SC (2004) Physico-chemical characterization of the microhabitat of the epibionts associated with *Alvinella pompejana*, a hydrothermal vent annelid. Geochim Cosmochim Acta 68:2055–2066

Druschel GK, Emerson D, Sutka R, Suchecki P, Luther GW III (2008) Low oxygen and chemical kinetic constraints on the geochemical niche of neutrophilic iron(II) oxidizing microorganisms. Geochim Cosmochim Acta 72(14):3358–3370

Dubilier N, Bergin C, Lott C (2008) Symbiotic diversity in marine animals: the art of harnessing chemosynthesis. Nat Rev Microbiol 6:725–740

Edwards KJ, Rogers DR, Wirsen CO, McCollom TM (2003) Isolation and characterization of novel psychrophilic, neutrophilic, Fe-oxidizing, chemolithoautotrophic ε– and γ-Proteobacteria from the deep sea. Appl Environ Microbiol 69(5):2906–2913

Emerson D, Moyer CL (2002) Neutrophilic Fe-oxidizing bacteria are abundant at the Loihi seamount hydrothermal vents and play a major role in Fe oxide deposition. Appl Environ Microbiol 68:3085–3093

Fortin D, Langley S (2005) Formation and occurrence of biogenic iron-rich minerals. Earth Sci Rev 72:1–19

Gaill F, Hunt S (1986) Tubes of deep-sea hydrothermal vent worms *Riftia pachyptila* (Vestimentifera) and *Alvinella pompejana* (Annelida). Mar Ecol Prog Ser 34:267–274

Gaill F, Hunt S (1991) The biology of annelid worms from high temperature hydrothermal vent regions. Rev Aquat Sci 4:107–137

Gaill F, Desbruyeres D, Laubier L (1988) Relationships between the "Pompeii worms" and their epibiotic bacteria. Oceanol Acta 8:147–155

Gebruk AV, Pimenov NV, Savvichev AS (1993) Feeding specialization of bresiliid shrimps in the TAG site hydrothermal community. Mar Ecol Prog Ser 98:247–253

Gebruk AV, Southward EG, Kennedy H, Southward AJ (2000) Food sources, behaviour, and distribution of hydrothermal vent shrimps at the Mid-Atlantic Ridge. J Mar Biol Assoc UK 80: 485–499

Girguis PR, Childress JJ (2006) Metabolite uptake, stoichiometry and chemoautotrophic function of the hydrothermal vent tubeworm *Riftia pachyptila*: responses to environmental variations in substrate concentrations and temperature. J Exp Biol 209:3516–3528

Girguis PR, Lee RW (2006) Thermal preference and tolerance of alvinellids. Science 312:231

Gloter A, Zbinden M, Guyot F, Gaill F, Colliex C (2004) TEM-EELS study of natural ferrihydrite from geological-biological interactions in hydrothermal systems. Earth Planet Sci Lett 222:947–957

Goffredi SK, Warén A, Orphan VJ, Van Dover CL, Vrijenhoek RC (2004) Novel forms of structural integration between microbes and a hydrothermal vent gastropod from the Indian Ocean. Appl Environ Microbiol 70:3082–3090

Goffredi SK, Jones WJ, Erhlich H, Springer A, Vrijenhoek RC (2008) Epibiotic bacteria associated with the recently discovered Yeti crab, *Kiwa hirsuta*. Environ Microbiol 10(10): 2623–2634

Grzymskia JJ, Murraya AE, Campbell BJ, Kaplarevic M, Gaoc GR, Lee G, Daniele R, Ghadirif A, Feldman RA, Cary SC (2008) Metagenome analysis of an extreme microbial symbiosis reveals eurythermal adaptation and metabolic flexibility. Proc Natl Sci Acad USA 105(45): 17516–17521

Halbach P, Tunnicliffe V, Hein J (eds) (2003) Mass and energy flux through hydrothermal systems. Dahlem Press, Berlin

Johnson KS, Childress JJ, Hessler RR, Sakamoto-Arnold CM, Beelher CL (1988) Chemical and biological interactions in the Rose Garden hydrothermal vent field, Galapagos spreading center. Deep Sea Res I 35:1723–1744

Johnson KS, Childress JJ, Beelher CL, Sakamoto CM (1994) Biogeochemistry of hydrothermal vent mussel communities: the deep sea analogue to the intertidal zone. Deep Sea Res I 41: 993–1011

Le Bris N, Sarradin PM, Pennec S (2001) A new deep-sea probe for *in situ* pH measurement in the environment of hydrothermal vent biological communities. Deep Sea Res I 48:1941–1951

Le Bris N, Sarradin PM, Caprais JC (2003) Contrasted sulphide chemistries in the environment of 13°N EPR vent fauna. Deep Sea Res I 50:737–747

Le Bris N, Zbinden M, Gaill F (2005) Processes controlling the physico-chemical micro-environments associated with Pompeii worms. Deep Sea Res I 52(6):1071–1083

Le Bris N, Rodier P, Sarradin PM, Le Gall C (2006) Is temperature a good proxy for sulfide in hydrothermal vent habitats? Cah Biol Mar 47:465–470

Le Bris N, Gaill F (2007) How does the annelid *Alvinella pompejana* deal with an extreme hydrothermal environment? Rev Environ Sci Biotech 6:197–221

Le Bris N, Anderson L, Chever F, Gaill F (2008) Sulfide diffusion and chemoautotrophy requirements in an extremophilic worm tube. In: Mertens LP (ed) Biological oceanography research trends. Nova Science, New York, pp 157–175

Le Bris N, Duperron D (2010) Chemosynthetic communities and biogeochemical energy pathways along the MAR: the case of *Bathymodiolus azoricus*. In: Rona P, Devey C, Dyment J, Murton B (eds) Diversity of hydrothermal systems on slow-spreading ridges. AGU Geophysical Monograph 188:409–429

Luther GW, Rozan TF, Taillefer M, Nuzzio DB, Meo CD, Shank TM, Lutz RA, Cary SC (2001) Chemical speciation drives hydrothermal vent ecology. Nature 410:813–816

Lutz RA, Shank TM, Fornari DJ, Haymon RM, Lilley MD, Von Damm KL, Desbruyères D (1994) Rapid growth at deep-sea vents. Nature 371:663–664

Nakagawa S, Takai K, Inagaki F, Hirayama H, Nunoura T, Horikoshi K, Sako Y (2005) Distribution, phylogenetic diversity and physiological characteristics of ε-proteobacteria in a deep-sea hydrothermal field. Environ Microbiol 7:1619–1632

Park B, Dempsey BA (2005) Heterogeneous oxidation of Fe(II) on ferric oxide at neutral pH and a low partial pressure of O2. Environ Sci Technol 39:6494–6500

Poltz MF, Cavanaugh CM (1995) Dominance of one bacterial phylotype at a Mid-Atlantic Ridge hydrothermal vent site. Proc Natl Acad Sci USA 92:7232–7236

Poltz MF, Robinson JJ, Cavanaugh CM, Dover CL (1998) Trophic ecology of massive shrimp aggregations at a Mid-Atlantic Ridge hydrothermal vent site. Limnol Oceanogr 43:1631–1638

Pradillon F, Zbinden M, Mullineaux LS, Gaill F (2005) Colonisation of newly-opened habitat by a pioneer species, *Alvinella pompejana* (Polychaeta:Alvinellidae), at East Pacific Rise vent sites. Mar Ecol Prog Ser 302:147–157

Pradillon F, Zbinden M, Le Bris N, Hourdez S, Barnay AS, Gaill F (2010) Morphogenesis of high temperature Alvinellid-dominated biogeoassemblages at East Pacific Rise hydrothermal vents. Deep-Sea Res II 56:1622–1631

Ravaux J, Gaill F, Le Bris N, Sarradin PM, Jollivet D, Shillito B (2003) Heat-shock response and temperature resistance in the deep-sea vent shrimp *Rimicaris exoculata*. J Exp Biol 206: 2345–2354

Rentz JA, Kraiya C, Luther GW III, Emerson D (2007) Control of ferrous iron oxidation within circumneutral microbial iron mats by cellular activity and autocatalysis. Environ Sci Technol 41:6084–6089

Rieley G, Van Dover CL, Hedrick DB, Eglinton G (1999) Trophic ecology of *Rimicaris exoculata*: a combined lipid abundance/stable isotope approach. Mar Biol 133:495–499

Riou V, Halary S, Duperron S, Bouillon S, Elskens M, Bettencourt R, Santos RS, Dehairs F, Colaco A (2008) Influence of CH_4 and H_2S availability on symbiont distribution, carbon assimilation and transfer in the dual symbiotic vent mussel *Bathymodiolus azoricus*. Biogeosciences 5:1681–1691

Sarradin PM, Caprais JC, Briand P, Gaill F, Shillito B, Desbruyères D (1998) Chemical and thermal description of the Genesis hydrothermal vent community environment (13°N, EPR). Cah Biol Mar 39:159–167

Schmidt C, Le Bris N, Gaill F (2008a) Interactions of deep sea vent invertebrates with their environment: the case of *Rimicaris exoculata*. J Shell Res 27(1):1–12

Schmidt C, Vuillemin R, Le Gall C, Gaill F, Le Bris N (2008b) Geochemical energy sources for microbial primary production in the environment of hydrothermal vent shrimps. Mar Chem 108:18–31

Schmidt C, Corbari L, Gaill F, Le Bris N (2009) Biotic and abiotic controls on iron oxyhydroxide formation in the gill chamber of the hydrothermal vent shrimp *Rimicaris exoculata*. Geobiology 7:1–11

Segonzac M, de Saint LM, Casanova B (1993) L'énigme du comportement trophique des crevettes Alvinocarididae des sites hydrothermaux de la dorsale medio-atlantique. Cah Biol Mar 34:535–571

Shillito B, Le Bris N, Gaill F, Rees JF, Zal F (2004) First access to live Alvinellas. High Press Res 24(1):169–172

Taylor CD, Wirsen CO, Gaill F (1999) Rapid microbial production at filamentous sulfur mats at hydrothermal vents. Appl Environ Microbiol 35:2253–2255

Tivey MK (2004) Environmental conditions within active seafloor vent structures: sensitivity to vent fluid composition and fluid flow. In: Wilcock W, Cary C, DeLong E, Kelley D, Baross J (eds.) Subseafloor Biosphere at Mid-Ocean Ridges, AGU, Washington, DC, pp 137–152

Tivey MK (2007) Generation of seafloor hydrothermal vent fluids and associated mineral deposits. Oceanography 20:50–65

Van Dover CL (2000) The ecology of deep-sea hydrothermal vents. Princeton University Press, Princeton, NJ

Van Dover CL, Fry B, Grassle JF, Humphris S, Rona PA (1988) Feeding biology of the shrimp *Rimicaris exoculata* at hydrothermal vents on the Mid-Atlantic Ridge. Mar Biol 98:209–216

Wirsen CO, Jannasch HW, Molyneaux ST (1993) Chemosynthetic microbial activity at Mid-Atlantic Ridge hydrothermal vent sites. J Geophys Res 98:9693–9703

Wirsen CO, Sievert SM, Cavanaugh CM, Molyneaux SJ, Ahmad AR, Taylor LT, DeLong EF, Taylor CD (2002) Characterization of an autotrophic sulfide-oxidizing marine *Arcobacter* sp. that produces filamentous sulfur. Appl Environ Microbiol 68:316–325

Zbinden M, Martinez I, Guyot F, Cambon-Bonavita MA, Gaill F (2001) Zinc-iron sulphide mineralization in tubes of hydrothermal vent worms. Eur J Mineral 13:653–658

Zbinden M, Cambon-Bonavita MA (2003) Occurrence of Deferribacterales and Entomoplasmatales in the deep-sea shrimp *Rimicaris exoculata* gut. FEMS Microbiol Ecol 46:23–30

Zbinden M, Le Bris N, Compère P, Martinez I, Guyot F, Gaill F (2003) Mineralogical gradients associated with Alvinellids at deep-sea hydrothermal vents. Deep Sea Res I 50:269–280

Zbinden M, Le Bris N, Gaill F, Compère P (2004) Distribution of bacteria and associated minerals in the gill chamber of the vent shrimp *Rimicaris exoculata* and related biogeochemical processes. Mar Ecol Prog Ser 284:237–251

Zbinden M, Shillito B, Le Bris N, de Villardi de Montlaur C, Roussel E, Guyot F, Gaill F, Cambon-Bonavita MA (2008) New insights on the metabolic diversity among the epibiotic microbial community of the hydrothermal shrimp *Rimicaris exoculata*. J Exp Mar Biol Ecol 359:131–140

Chapter 4
Microbial Chemofossils in Specific Marine Hydrothermal and Methane Cold Seep Settings

Martin Blumenberg

Contents

4.1 Introduction

Marine hydrothermal vent and cold seep settings fuel phylogenetically and metabolically diverse microbial communities. In contrast to most organisms on earth, which depend directly (photosynthesis) or indirectly (heterotroph) on sunlight, many prokaryotes (archaea and bacteria; Fig. 4.1) at hydrothermal vents and cold seeps use energy from the reduced chemicals that are emitted at these sites (Fisher et al. 2007 and references cited therein). Consequently, these environments are considered as oases of life in oceanic deep sea deserts. Since the first discovery of deep sea hydrothermal vent systems (Corliss et al. 1979) such settings have become fascinating study sites for multidisciplinary research. Geobiologists in particular are attracted by these settings, because diffusive vents and smokers were considered to have offered the most favorable conditions for the development of early life on Earth (Nisbet and Sleep 2001). In hydrothermal settings, the

M. Blumenberg (✉)
Department of Geobiology, Geoscience Centre (GZG), Georg-August-University of Göttingen, Goldschmidtstraße 3, 37077, Göttingen, Germany
e-mail: martin.blumenberg@geo.uni-goettingen.de

S. Kiel (ed.), *The Vent and Seep Biota*, Topics in Geobiology 33,
DOI 10.1007/978-90-481-9572-5_4,

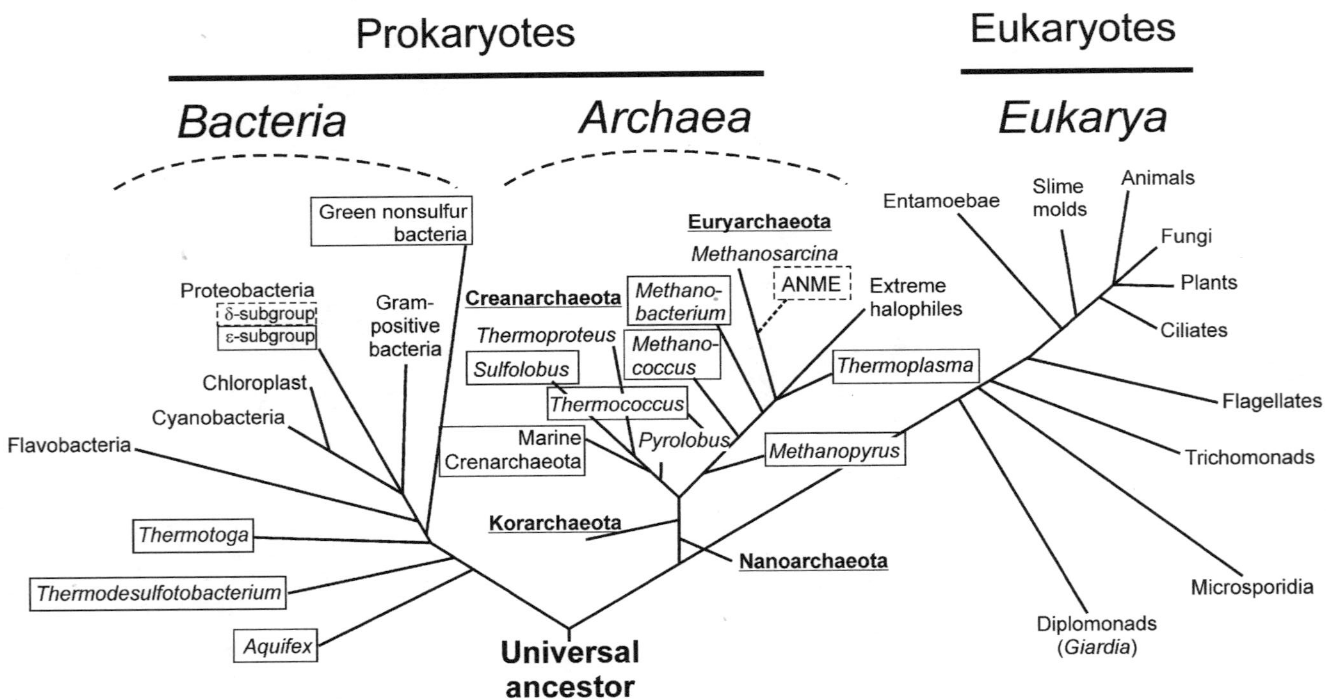

Fig. 4.1 Phylogenetic tree of life (Graphically modified after Woese et al. 1990; Konhauser 2006). ANME = anaerobic methanotrophic archaea, boxes = common prokaryotes at deep sea hydrothermal vents, dashed boxes = relevant prokaryotes at cold methane seeps

temperatures of the fluids may exceed 400°C (Haase et al. 2007), and mixing with cold sea water creates sharp physico-geochemical gradients with various microbiological niches. However, our knowledge of the functioning of microbial life in these environments is still very limited. Only recent technological developments and the routine operations of manned submersibles, autonomous underwater vehicles (AUVs), and remotely operated vehicles (ROVs) allow for detailed sampling, in situ experiments and visual inspections in the deep sea (e.g., Kelley et al. 2001, 2005; Teske et al. 2002; Nercessian et al. 2005; Inagaki et al. 2008; Haase et al. 2007; Perner et al. 2007; Nakagawa et al. 2008).

Unfortunately, no more than 0.1–1% of the archaeal and bacterial species in a microbial community can be cultivated (Madigan et al. 2002). Consequently, other molecular microbiological methods are being increasingly used to qualitatively and (semi)quantitatively describe microbial associations (e.g., staining of cells by fluorescence in situ hybridization (FISH) or extraction of DNA and subsequent quantitative polymerase chain reaction (qPCR)). Both methods have excellently broadened the existing knowledge of what living microbial communities in extreme environments are composed of, and in which biogeochemical niches they exist (see for reviews: Valentine 2002; Strous and Jetten 2004; Takai et al. 2006). DNA, in contrast to RNA, is relatively stable, so DNA extraction and subsequent identification also works for sediments that are several thousands of years old (Coolen and Overmann 2007). Another excellent tool to get information on recent as well as former microbial communities is the investigation of lipid biomarkers or chemofossils. Most biomarkers originate from molecules in cell membranes, have specific chemical structures so that their origins can be identified, and are resistant to diagenetic influences. Moreover, the stable isotope signatures of biomarkers contain information on substrates as well as the metabolisms of the source organisms.

This chapter describes, after briefly introducing hydrothermal vent and cold seep settings, the principles of this technique. Furthermore, the application of biomarkers for the understanding of microbial associations will be demonstrated for selected outstanding sites; hydrothermal settings on the Mid-Atlantic Ridge and methane cold seeps in the Black Sea.

4.1.1 Hydrothermal Vents

Hydrothermal settings in the deep sea are ubiquitous in oceanic regions characterized by sea floor spreading and support phylogenetically diverse macrofaunal communities (see Chapter 1; see also Little and Vrijenhoek 2003 for a review of fossil analogues). However, at the base of the food chain are very specific microorganisms (e.g. Fisher et al. 2007); bacterial and archaeal associations which are so far insufficiently understood.

One of the best studied sites with respect to microbial traces in oceanic hydrothermal rocks is the Lost City hydrothermal field (LCHF) on the Mid-Atlantic

Ridge (e.g. Kelley et al. 2005; Bradley et al. 2009). The ultramafic LCHF is unusual among hydrothermal vent settings because it is mostly driven by serpentinization reactions and cooling of mantle rocks, resulting in a reducing, high-pH (10), relatively low-temperature environment (~40–90°C), with fluids containing abundant hydrogen and methane. These conditions promote various methanogenic and sulfur-metabolizing microorganisms (Kelley et al. 2001, 2005; Brazelton et al. 2006). Hydrogen is produced directly during serpentinization and by coupling to the reduction of CO_2 and methane is mainly produced abiogenically (Schroeder et al. 2002). Due to the high alkalinity at the LCHF, carbonates precipitate, forming the building blocks of white chimneys.

In contrast to those at the LCHF, fluids at typical black smoker settings are characterized by higher temperatures and commonly higher acidity and are solute rich (metals, sulfide, etc.). Before the water is ejected from black smokers it had been drawn into the sea floor and heated by magma plumes in the shallow subsurface. During this heating, reduced compounds such as metal sulfides are dissolved and precipitate as "black smoke" when the hot fluid mixes up with the ambient, cold and oxygenated seawater (Fig. 4.2).

One recently discovered Black Smoker setting is the Turtle Pits Field at 5°S on the Mid-Atlantic Ridge (Haase et al. 2007; Fig. 4.3, Plate 7). The large sulfide chimneys at this site are populated by the vent mussel *Bathymodiolus puteoserpentis*,

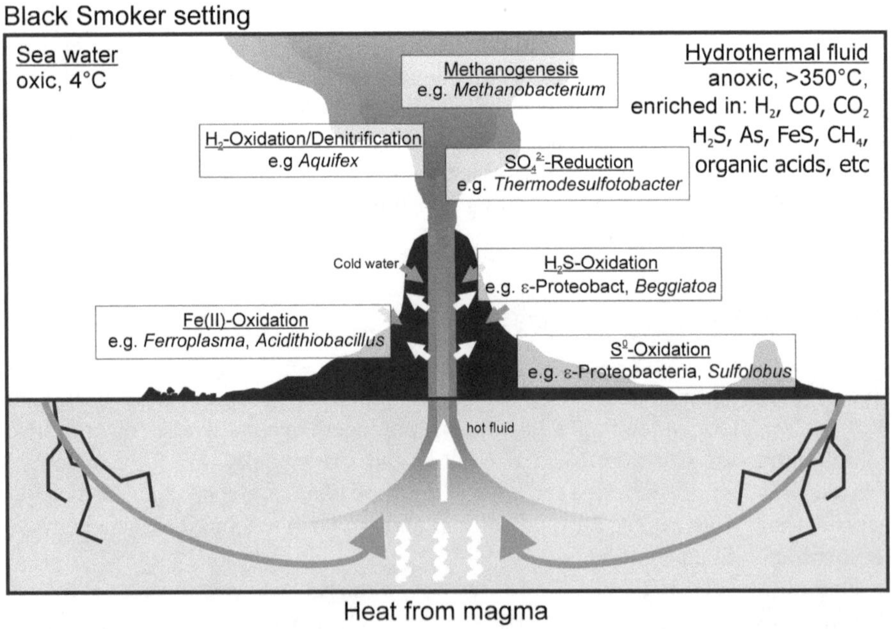

Fig. 4.2 Scheme of biogeochemical and geomicrobiological principles at a black smoker setting (simplified after Campbell et al. 2006)

Fig. 4.3 Vent and seep settings and associated bacteria. (**a**) Image of a black smoker in the Turtle Pits Field on the Mid-Atlantic Ridge (Photo courtesy of MARUM, University of Bremen and Richard Seifert, University of Hamburg); A color plate of this figure can be found in Appendix I (Plate 7); (**b**) Image of cold seeps in the Black Sea (microbial reef towers in the GHOSTDABS-field; 230 m water depth). The towers shown are about 1 m high (Photo courtesy of Walter Michaelis and Richard Seifert (University of Hamburg) and Karen Hissmann (JAGO team)); (**c**) Electron microscopy image of microorganisms (mainly ANME-1 archaea) from a microbial mat of the GHOSTDABS-field (Black Sea, 230 m water depth; Photo courtesy of Dietmar Keyser and Richard Seifert (University of Hamburg)

shrimps, and crabs. The pH of the fluid is at 3.1, and high concentrations of hydrogen and methane were observed (Haase et al. 2007). The highest fluid temperature ever recorded from any hydrothermal vent was found in one of the emanating fluids at the Turtle Pits Field (407°C; Haase et al. 2007). A weathered iron-sulfide from Turtle Pits was the object of a detailed lipid biomarker study focusing on the potential of microorganisms to be fossilized in the rock matrix (Blumenberg et al. 2007). The authors reported signals of primary, fluid associated microbes as well as secondary, leaching microbes living in cracks and fissure of the rock matrix (see Section 3.1.1).

4.1.2 Methane Cold Seeps

In contrast to hydrothermal settings, most cold seeps (sometimes also called cold vents) occur at continental margins and are characterized by relatively low temperatures, favoring mesophilic bacteria and archaea. At cold seeps and vents, mostly methane escapes into the water column, produced either by thermal cracking of organic matter in deeper sediments and/or by methanogenesis using organic matter as an electron donor (Schoell 1988). Consequently, most of these environments indirectly depend on sunlight, because the vast majority of organic matter buried in sediments was produced by photosynthesis. Methane from thermal cracking has a $\delta^{13}C$-value corresponding to the original organic matter (mostly between −20‰ and −45‰; Fig. 4.4). The second source of methane, methanogenesis, is more common in marine sediments. This methane is mostly generated by microbial CO_2-reduction with hydrogen or by acetate fermentation. Both processes are performed by strictly anaerobic archaea and lead to methane distinct from thermogenic methane, exhibiting a strong depletion in ^{13}C ($\delta^{13}C$ lower than −55‰; Whiticar 1999). Methane is present in almost all marine sediments. However, only if the methane concentration exceeds its solubility it migrates through the sedimentary column and forms cold seeps at the marine sea floor.

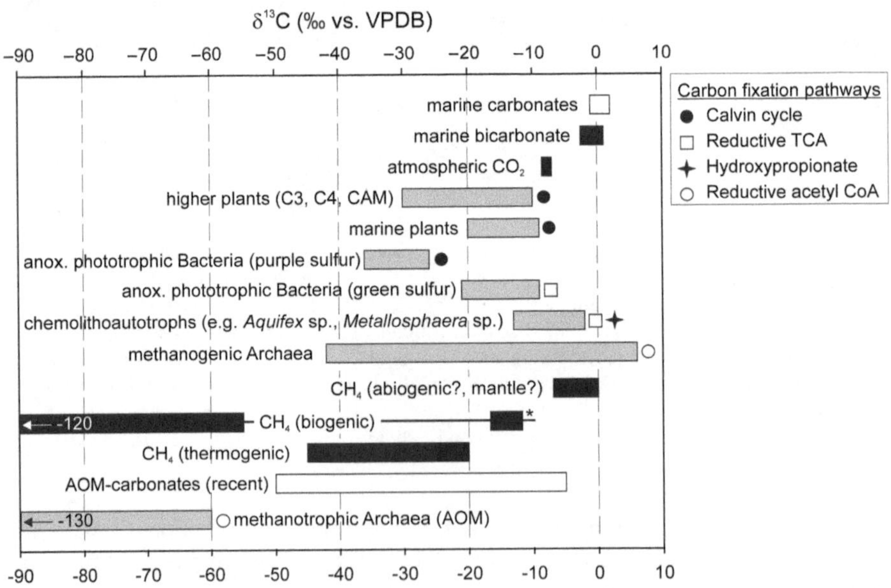

Fig. 4.4 Stable carbon isotopic composition of selected carbon pools and organisms (After Whiticar 1999; Hayes 2001; Schidlowski 2001; House et al. 2003; Hallam et al. 2004; Hoefs 2004; Killops and Killops 2005; Reitner et al. 2005; Niemann and Elvert 2008), *range of methane produced by a hyperthermophilic *Methanopyrus kandlerii* strain under high pressure conditions (40 MPa; Takai et al. 2008; assuming a (nearly common) $\delta^{13}C$ of the CO_2 used of −3‰). *Black boxes* indicate important carbon sources

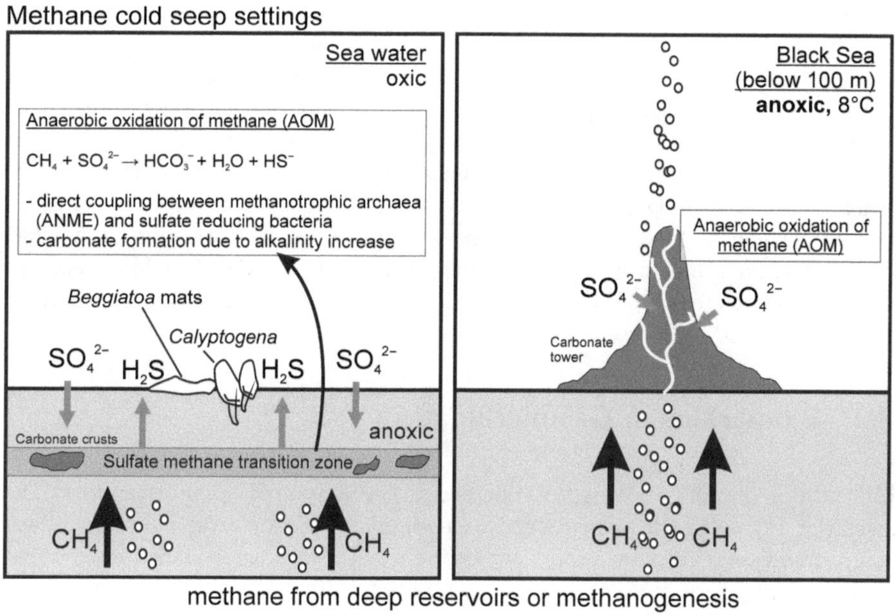

Fig. 4.5 Scheme of biogeochemical and geomicrobiological principles at AOM settings (*Left*: normal methane-rich marine environment and *right*: special situation present in the deep Black Sea). AOM = anaerobic oxidation of methane

Most of the methane in marine sediments (>90%) is converted to bicarbonate by the anaerobic oxidation of methane with sulfate (AOM; Barnes and Goldberg 1976; Reeburgh 1976, 2007; Fig. 4.5), thereby increasing the alkalinity and inducing carbonate precipitation (Ritger et al. 1987).

Although not understood in detail, AOM is suggested to be performed in a direct coupling between anaerobic methanotrophic archaea (ANME) and sulfate reducing bacteria (e.g. Boetius et al. 2000). It has been demonstrated that methanotrophic archaea are phylogenetically closely related to methanogenic archaea (e.g., Knittel et al. 2005). To date, three different phylogenetic groups of ANME archaea are known, with ANME-1 and ANME-2 being closely related to sulfate-reducing Bacteria (SRB) of the *Desulfosarcina*/*Desulfococcus*-cluster (Boetius et al. 2000; Orphan et al. 2001; Michaelis et al. 2002) and ANME-3 to *Desulfobulbus* spp. SRB (Niemann et al. 2006). However, aerobic microorganisms also indirectly gain energy from AOM. The methane-dependent sulfate reduction produces high amounts of hydrogen sulfide, which may then fuel free-living aerobic sulfide-oxidizing bacteria (e.g., *Beggiatoa* sp.) and those living in symbiosis with higher invertebrates (e.g. *Calyptogena* clam; Fig. 4.5 left; see Treude et al. 2003 for example). At many sites, carbonate rocks record AOM-activitiy in Earth history (Peckmann and Thiel 2004); to date, the oldest definite record of AOM dates back more than 300 million years (Birgel et al. 2008).

One unique AOM environment is located in the Black Sea, where permanently anoxic waters allow AOM communities to grow into the water column (Ivanov et al. 1991; Pimenov et al. 1997; Michaelis et al. 2002). In the Black Sea, below about 100 m water depth, a permanent anoxic water column was established in the aftermath of the last ice age (about 7,500 years before present; Jones and Gagnon 1994). In addition, in the Dnjepr paleodelta a very high methane flux exists (in 230 m water depth; GHOSTDABS-field; Michaelis et al. 2002). Most likely the combination of permanent anoxia, high methane escapement and the location on a ridge, well protected from mass-wastes, promoted the formation of 4 m high reef-like carbonate towers built by the very slow-growing AOM-microbes (Fig. 4.5 right).

4.2 Biomarkers in Geomicrobiology

Unicellular microorganisms without nuclei (prokaryotes; domains archaea and bacteria; Fig. 4.1) do not produce exo- or endoskeletons, but all macro- and microorganisms contain more or less specific carbonaceous biomolecules. Consequently, former microbial life can be reconstructed by findings of cellular remnants (e.g., Summons et al. 1988; Brocks et al. 1999; Brocks and Pearson 2005), in addition to the use of morphological indications, such as trace fossils in sedimentary structures (Cady et al. 2003; Furnes et al. 2004). Living organisms on Earth produce an enormous diversity of molecules from a few building blocks. This includes four bases in DNA, 20 amino acids of proteins and two kinds of lipid building blocks (Summons et al. 2008). However, since nucleic acids, proteins and carbohydrates are rapidly degraded after cell lysis and later diagenesis, the relatively stable lipids (cell membrane constituents) have the highest potential as biomarkers. The principal structure of most biological membranes is a phospholipid bilayer with hydrophobic hydrocarbon chains linked to a glycerol backbone (Madigan et al. 2002; Fig. 4.6). This chapter focuses on the use of biomarkers, because of their high quality to characterize former as well as recent microbial life. This organic-geochemical technique uses information encoded in these chemical structures. Ideally, stable isotopic compositions are also studied, since the combination of both contains information on organisms, substrates, carbon fixation pathways used, environmental conditions, and the trophic interrelationships in communities (e.g., Eglinton et al. 1964; Hayes 2001; Summons et al. 2008; see below). In the following, the terms biomarker, chemofossil and molecular biosignature are often used interchangeably, as is done in many publications. However, the term chemofossil is generally applied for the fossil remains of organisms and is used to describe structures that are diagenetically formed from lipid biomarkers (Fig. 4.7). Regardless of these definitions, several recent studies have demonstrated that the originally formed lipid biomarkers may also be preserved for millions of years without losing structural information.

The possibility of biomarkers being preserved over geological time scales and becoming a chemofossil depends on its structural characteristics and its resistance

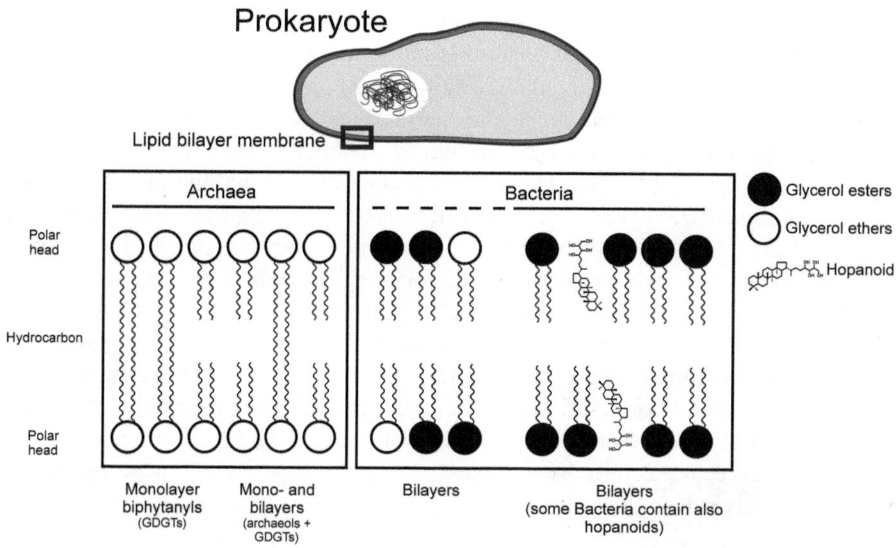

Fig. 4.6 Very simplified scheme of cell membranes of prokaryotes (archaea and bacteria) and the differences in lipid composition. Generally bacteria produce glycerol esters and many also contain hopanoids (not all). Glycerol ethers are rare in bacteria. Archaea contain only glycerol ethers, no hopanoids, and many produce monolayers consisting of biphytanyls (glycerol dialkyl glycerol tetraethers = GDGT)

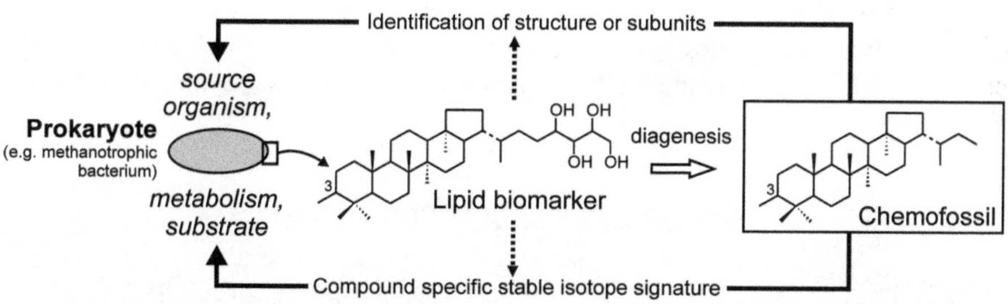

Fig. 4.7 Scheme of the biomarker (chemofossil) principle (hopanoid from a methanotrophic bacterium as example). Every organism contains more or less specific lipid membrane molecules (lipid biomarkers), from which after diagenesis structural subunits and stable isotope signatures are conserved (in particular a methyl group at position C-3). These chemofossils can be identified in rocks and sediments and include information on the respective source organism and metabolisms and substrates

to diagenetic stress. The conditions during deposition are also very important; i.e., anoxic settings favor the preservation of organic material. Many biomolecules decay within hours to days after burial, while other structures are very stable. The oldest chemofossils found so far date back to 2.77 billion years (Ga; Brocks et al. 1999), although the syngeneity of these structures has recently been questioned (Rasmussen et al. 2008).

More than only paleoenvironments can be reconstructed using biomarkers. In particular, the discovery of the anaerobic oxidation of methane as an important process is an excellent example to show how potent biomarkers are for recent geobiological purposes. The principles and thermodynamics of methane oxidation at the methane sulfate transition zone in anoxic sediments have long been known (Barnes and Goldberg 1976; Reeburgh 1976). However, only findings of ^{13}C-depleted microbial biomarkers (Elvert et al. 1999; Hinrichs et al. 1999; Thiel et al. 1999) and of those biomarkers in combination with molecular microbiological data (Elvert 1999; Hinrichs et al. 1999; Boetius et al. 2000; Orphan et al. 2001; Michaelis et al. 2002) directly proved the presence of AOM-performing consortia in marine sediments and microbial mats. In a similar manner, the combination of biomarkers and stable isotope signatures is also applied to gain information on the distribution of microbial groups, carbon fixation pathways and metabolisms at hydrothermal settings (Lein et al. 1998; Kelley et al. 2005; Pancost et al. 2006; Blumenberg et al. 2007; Bradley et al. 2009). Nevertheless, the understanding of these particular environments is still fragmentary.

4.2.1 *Structures of Biomarkers*

Archaeal and bacterial glycerolipids have two basic differences: First, most bacteria (and all eukaryotes) contain ester-bond fatty acids, whereas archaea have ether-bond alcohols (Fig. 4.6). Second, bacteria (and eukaryotes) have acetogenic and archaea isoprenoidal hydrocarbon chains (see examples in Fig. 4.8). One important exception from these rules is that several thermophilic, often phylogenetic deeply branching bacteria also contain ether-bonds, most likely to increase their membrane

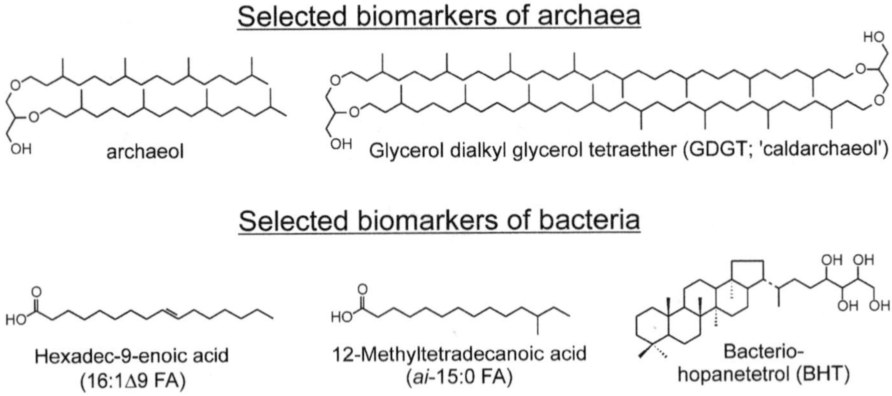

Fig. 4.8 Selected examples of common lipid constituents in cell membranes of archaea and bacteria. Archaea produce isoprenoid glycerol ethers (such as archaeol) and glycerol dialkyl glycerol tetraethers (GDGT) whereas bacteria contain mainly fatty acids which are linked to a glycerol backbone (not shown). Some bacteria produce also hopanoids such as bacteriohopanetetrol

rigidity (e.g., Langworthy et al. 1983; Huber et al. 1992). Among the bacterial and archaeal domains, and often dependent on environmental conditions, various structural modifications in the hydrocarbon chains are known (chain length difference, methyl branching, unsaturation, ring formation). For instance, bacteria produce high amounts of mono- to polyunsaturated fatty acids as a response to low temperatures, a mechanism to keep cell membranes more fluid (Russell 1990). However, despite the fact that such modifications are known to be triggered by varying temperatures, pH, etc., biomarkers also have chemotaxonomic relevance. In addition to glycerol-based lipids, many bacteria contain pentacyclic triterpenoids (hopanoids; Fig. 4.8), compounds that are not present in archaea.

4.2.2 Stable Isotope Signatures of Biomarkers

The distributions of naturally occurring isotopes in biomarkers provide information on the metabolic mode (e.g., autotrophy vs. heterotrophy), the substrate, and/or the carbon fixation pathway used by the source organisms. Carbon has two stable isotopes, with ^{12}C comprising on average 98.89% and the heavier ^{13}C 1.11%. However, there is a preference for ^{12}C during biological turnover, which leads to a ^{12}C enrichment in the products and a ^{13}C enrichment in the residue of the educts. This is due to the fact that molecules including ^{13}C isotopes exhibit higher bond energies and lower diffusion velocities than those with ^{12}C. Stable isotope fractionations are small and for better comparisons they are expressed in per mil as the difference from the $^{13}C/^{12}C$-ratio of a standard, which is defined as 0‰ (normally Vienna PeeDee Belemnite). For instance, an absolute ^{13}C depletion of 0.0011% represents a $\delta^{13}C$ of −1‰. Most biological carbon reservoirs are negative, spanning a range of slightly positive to less than −100‰ (Fig. 4.4). The extent of biological fractionation depends on the metabolic mode of the organisms and physical gradients and can be reconstructed by analyzing the $\delta^{13}C$ values of biomass and lipids (Hayes 2001; see Fig. 4.4).

The growth on inorganic carbon compounds is called autotrophy and is often accompanied by strong isotopic fractionations (for instance higher plants [C3, C4] using the Calvin cycle for carbon fixation fractionate between −10‰ and −22‰ with respect to the carbon source; land plants use atmospheric CO_2 with a $\delta^{13}C$ of −7‰ and consequently have a $\delta^{13}C$ between −17‰ and −29‰). Most autotrophic organisms on Earth use the Calvin-cycle for C-fixation, leading to uniform respective $\delta^{13}C$-values in recent and ancient bulk organic matter. This is because the consumption of large organic molecules by heterotrophic animals or microbes causes only a very minor isotopic fractionation. Therefore, the relationship between heterotrophic organisms and their food in terms of $\delta^{13}C$ values has been described as "you are what you eat (plus a few per mil)" (DeNiro and Epstein 1976).

For settings not based on photoautotrophy and where isotopically uncommon carbon sources are used, $\delta^{13}C$ values are powerful for geobiological characterizations. In particular, marine hydrothermal settings represent milieus where generally

^{13}C enriched carbon sources from magma/fluid reactions prevail. Moreover, in these particular settings uncommon carbon fixation pathways are used by the predominant bacteria and archaea (e.g., reductive TCA and the hydroxypropionate pathway; Nakagawa and Takai 2008; see Fig. 4.4).

Another process leading to characteristic stable carbon isotope signatures is the anaerobic oxidation of methane (AOM). Most methane in marine sediments is formed by biological methanogenesis in anoxic sediment layers and usually shows very negative $\delta^{13}C$ values (see biogenic methane in Fig. 4.4). Consequently, biomass and lipids of methane-consuming (synonym methanotrophic) prokaryotes are also depleted in ^{13}C. In anoxic environments these organisms are ANME-(anaerobic methanotrophic) archaea, as well as AOM-involved sulfate reducing bacteria. Both directly or indirectly take up the stable isotope signal of the methane carbon. In fact, the ^{13}C-depletion in archaeal lipids in these environments makes them distinguishable from the biosignatures of methanogenic archaea (due to their close phylogenetic relationship, both groups show a broad overlap in their lipid inventory).

The following section introduces specific biomarkers of prokaryotic organisms thriving at certain recent marine hydrothermal settings and cold seeps, and the information that can be deduced from their stable isotope signatures. Respective studies use biomarkers that are entrapped in the rock matrix and try to decipher which groups of bacteria and archaea. and which metabolisms are/were prevailing. As an example, the current knowledge of outstanding geobiological environments will be briefly presented, hydrothermal settings from the Mid-Atlantic ridge, with a special emphasis on black smoker settings and the unique cold seeps on the sea floor of the Black Sea. Notably, AOM-settings are much better studied for their lipid biomarker distributions and isotope signatures than hydrothermal smoker areas, so the picture of the latter is still fragmentary.

The scope of the following part is not to give a thorough overview of all work that has been published on these specific as well as other settings, but to present a brief overview and provide further readings. Both sections close with short subchapters that discuss what is known about fossilized cell remnants (chemofossils) of archaea and bacteria in old massive hydrothermal sulfides and cold seep methane-carbonates in the geological record.

4.3 Biomarkers of Microbes Thriving in Marine Hydrothermal Settings and Cold Seep Environments

4.3.1 Biomarkers in Hydrothermal Deep Sea Settings

Although bacteria and archaea are key-players at modern (e.g. Stetter 1996, 2006; Takai et al. 2006; Nakagawa and Takai 2008) and presumably also fossil hydrothermal settings, findings of biomarkers in hydrothermal rocks are scarce. Most respective studies have been performed at terrestrial hydrothermal settings such as hot springs (e.g., Jahnke et al. 2001; Pancost et al. 2006). Despite the limited knowledge on marine hydrothermal settings, biomarker distributions demonstrate that bacteria, archaea, and eukaryotes are not equally distributed in the walls of black smokers. For instance, by estimating the biomass using lipid concentrations, Schrenk et al. (2003) showed that microbial abundances are generally high in a sulfide chimney on the Juan de Fuca Ridge. The highest numbers, however, were found in the outer, cooler areas (>10^8 cells/g sulfide) and the lowest numbers in the interior zones (<10^5 cells/g sulfide). Higher amounts of bacterial biomarkers (fatty acids) in the exterior part were also found in a sulfide chimney from the Atlantic Rainbow field (Simoneit et al. 2004), suggesting that the outer, cooler parts of chimneys are the favored living zones for bacteria, whereas archaea dominate in the hotter parts of black smokers. It should be noted that (to date) the upper temperature limit of life is 121°C (shown for the archaeal species *Pyrodictium occultum*, Kashefi and Lovley 2003).

One major problem for biomarker studies at hydrothermal vents is the masking of biogenic signals by abiogenic aliphatics. The aliphatics are produced by Fischer-Tropsch type reactions (e.g. Proskurowski et al. 2008 and references cited therein). In addition, at hydrothermal settings carbon limitation may occur, which represses biological fractionation and thus may conflict with conventional interpretations of $\delta^{13}C$ values (Bradley et al. 2009). Despite these problems and although biomarker studies on the distribution of prokaryotes in active chimneys and altered hydrothermal deposits are rare, existing data indicate a high complexity of microbial associations.

In the following section, some of the biomarkers and stable isotope signatures found at the LCHF and Turtle Pits Field, so far two of the best studied settings, are presented and compared with other studies. However, it must be mentioned that these, as well as other relevent studies, present interesting but only very small pieces of the puzzle of microbe-mineral associations, since these are extraordinary heterogeneous with respect to age, alteration stage, locality, and space.

4.3.1.1 Biomarkers Preserved in Recent Hydrothermal Rocks from the Mid-Atlantic Ridge

Biomarkers of archaea. Biomarkers of methanogenic archaea often dominate the biosignatures present in hydrothermal rocks from the Mid-Atlantic ridge, underlining the importance of these archaea in these settings. For instance, findings of the nonspecific "phytanylglyceride" (commonly named archaeol, A II; Fig. 4.9 and Table 4.1) in sulfides from 14°45′N (Mid-Atlantic Ridge) have been suggested to be of methanogenic origin (Peresypkin et al. 1999). Of much higher specificity for methanogenic archaea are hydroxyarchaeols (A IIIa+IIIb). Hydroxyarchaeols were found in high concentrations in carbonate chimneys from the Lost City hydrothermal field (Kelley et al. 2005; Bradley et al. 2009). These studies also report 2,6,10,15,19-pentamethylicosane (PMI, A I) to be abundant in the carbonate, an isoprenoid hydrocarbon also indicative of methanogenic archaea (Tornabene et al. 1979). One very specific biomarker for hyperthermophilic methanogenic archaea is

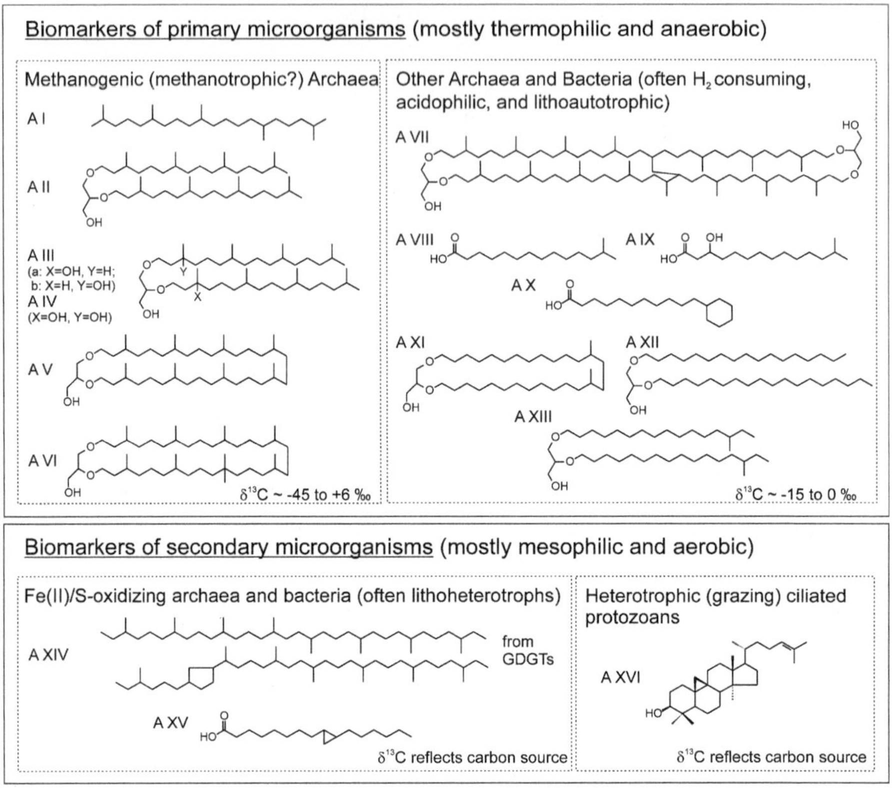

Fig. 4.9 Examples of lipids from microbes thriving at marine deep sea hydrothermal settings. Two communities are often present, primary, thermophilic archaea and bacteria (methanogens and other chemolithoautotrophs), which are directly linked to the fluids and show strong variations in $\delta^{13}C$-values and a secondary, mesophilic community, important in metal-sulfide oxidation. Names of the structures and further information are denoted in Table 4.1

a macrocyclic archaeol, which was found in massive sulfides from the Turtle Pits black smoker area (Blumenberg et al. 2007; AV). This compound has been found in cultured hyperthermophilic *Methanococcus jannaschii* (Comita and Gagosian 1983), underlining that temperatures of the fluids at the Turtle Pits area are much higher than at the LCHF, where this compound is lacking (Bradley et al. 2009). Moreover, at Turtle Pits, additional archaeal lipids with a most likely methanogenic origin were found, components that have not yet been described from cultured organisms (e.g., A VI). Interestingly, the $\delta^{13}C$ values of the lipids of putative methanogens differ strongly. The macrocyclic archaeol at the Turtle Pits field has ^{13}C-depleted $\delta^{13}C$ values around −50‰. In this high hydrogen environment, it

Table 4.1 Lipid biomarkers known from recent or sub-recent marine hydrothermal vents, potential source organisms, the quality as chemofossils and individual fossil records in ancient sulfides

Component	Structure	Source organisms at hydrothermal settings	Stability during diagenesis (chemofossil quality)	Record at fossil hydrothermal sites
PMI (2,6,10,15,19-pentamethylicosane)	A I (B II)	Archaea (methanogenic/ methanotrophic?)[1]	High	–
Archaeol	A II (B IV)	Archaea (various)[2, 3]	Medium	–
*sn*2/*sn*3-Hydroxyarchaeol	A IIIa+b	Archaea (methanogenic/ methanotrophic?)[1, 3]	Low	–
*sn*2-*sn*3-Dihydroxyarchaeol	A IV	Archaea (methanogenic/ methanotrophic?)[1, 4]	Low	–
Macrocyclic archaeol	A V	Thermophilic archaea (methanog./ methanotr.)[5, 6]	Medium	–
C_{41}-Macrocyclic archaeol	A VI	Thermophilic archaea (methanogenic)[3, 7]	Medium	–
'H-shaped' GDGT ('C_{80} tetraphytane')	A VII	Archaea (euryarchaeota, DHVE2-group)[8]	Medium	–
13-Methyltetradecanoic PLFA (*i*-15)[c]	A VIII	(chemolithoautotr.) Bacteria (e.g. *Myxobact*)[3, 9]	Medium	✓([a]indirect as hydrocarbon)
3-OH 13-Methyltetradec. PLFA (3OH-*i*-15)[c]	A IX	Various bacteria (*Thermus*?)[10]	Low	–
ω-Cyclohexylundecanoic PLFA	A X	Thermoacidophilic bacteria (Bacilli)[11]	Medium	✓([b]indirect as hydrocarbon)
C_{30} Macrocyclic non-isopr. DAGE	A XI	Thermoacidoph. bacteria (Acidob., ε-Proteob., Myxobat.?)[12, 3]	Medium	–
17/18-DAGE	A XII	Thermoacidoph. bacteria (*Aquifex, Ammonifex*)[13,14]	Medium	–

(continued)

Table 4.1 (continued)

Component	Structure	Source organisms at hydrothermal settings	Stability during diagenesis (chemofossil quality)	Record at fossil hydrothermal sites
*ai*17/*ai*17-DAGE	A XIII	Thermoacidoph. SRB[15]	Medium	–
Biphytanes (0–1 rings; from GDGTs)	A XIV	Archaea (*Ferroplasma*)[16,17,18]	High	–
ω7,8-Cyclopropylhexadecanoic PLFA	A XV	Fe/S-oxidizing bacteria (*Acidithiobacillus*?)[19]	Medium	–
Cycloartenol	A XVI	Ciliated protozoans[2]	High	–

For structures see Fig. 4.9. Please note that so far only two fossil sulfide deposit (see for review Reysenbach and Cady 2001) have been studied for chemofossils

[a] Yaman Kasy, Ural; Silurian; Lein et al. 1998

[b] Mt Isa. sulfide, Precambrian; Mycke et al. 1988

[c] If $\delta^{13}C \sim 0‰$. These papers report only hydrocarbons. GDGT = glycerol dialkyl glycerol tetraethers, PLFA = phospholipid fatty acid, DAGE = dialkyl glycerol ethers, SRB = sulfate reducing Bacteria

[1] Kelley et al. (2005); [2] Bradley et al. (2009); [3] Blumenberg et al. (2007); [4] Thiel et al. (2007); [5] Comita and Gagosian (1983); [6] Holzer et al. (1988); [7] Galliker (1990); [8] Schouten et al. (2008); [9] Eichorst et al. (2007); [10] Chung et al. (1997); [11] De Rosa et al. (1972); [12] Pancost et al. (2006); [13] Huber et al. (1992); [14] Huber et al. (1996); [15] Langworthy et al. (1983); [16] Simoneit et al. (2004); [17] Golyshina and Timmis (2005); [18] Glynn et al. (2006); [19] Kerger et al. (1986)

appears to be sourced by hydrogen-consuming, CO_2-reducing methanogens fixing carbon via the acetyl CoA pathway (Fig. 4.9, Blumenberg et al. 2007 and references therein). At the LCHF, the hydroxyarchaeols demonstrate very uncommon $\delta^{13}C$ values around 0‰, usually an indication of the use of abiotic carbon (Bradley et al. 2009). However, this might be related to carbon limitation, avoiding normal isotopic fractionations to be expressed by the methanogenic archaea at this site (Bradley et al. 2009).

The presence of methanotrophic archaea at the LCHF has been also suggested from findings of biomarkers. This was shown by the presence of archaeol with a strong depletion in ^{13}C ($\delta^{13}C$ about −77‰; Bradley et al. 2009), which is generally an indication of methane consumption. To summarize, biomarker findings at the LCHF suggest that archaea are clearly dominated by methanogens, while methanotrophs appear to also be present. With respect to metabolisms, the archaeal community in the studied Turtle Pits sulfide appears to be more heterogeneous, although biomarkers of methanogens are also present. In the massive, hydrothermal sulfide, a distribution of tetraether-derived biphytanes (A XIV) with 0–1 pentacylic rings was found, unlikely to be derived from methanogenic archaea (Blumenberg et al. 2007). High amounts of $C_{40:0}$ and $C_{40:1}$ biphytanes have also been described from sulfide samples of an inactive chimney at the Rainbow vent field (Simoneit et al. 2004) suggesting similar archaeal sources. These compounds have been described from cultured *Ferroplasma* relatives (Golyshina and Timmis 2005). *Ferroplasma* belong to the Thermoplasmata, are aerophilic, acidophilic, and are known as ferrous-iron oxidizing (Golyshina and Timmis 2005). Most likely, these archaea belong to the secondary community in the sulfides from black smokers, are not directly related to the fluids and gain their energy from the oxidation of the reduced metal sulfides (Blumenberg et al. 2007).

Many other archaeal groups (e.g. Thermococcales, Sulfolobaceae, Desulforococcales) are known from hydrothermal settings. However, these groups have not yet been unambiguously identified in the rock matrices using biomarkers, although some are known to contain very specific lipids. One good example is an H-shaped 'GDGT', which was recently discovered in obviously widely distributed thermoacidophilic Thermoplasmata of the DHVE2-group (Schouten et al. 2008; A VII). Further studies must show whether these or other characteristic lipids can also be identified in hydrothermal rocks from the deep sea.

Biomarkers of bacteria. Mesophilic and thermophilic bacteria are common at marine hydrothermal settings and may thrive very close to the hot fluids, using hydrogen or hydrogen sulfide as energy source. Furthermore, secondary bacteria gain energy from the oxidation of reduced compounds at rock surfaces (e.g., Konhauser 2006). As mentioned above, short chain fatty acids are the major building blocks of bacterial lipids (~12–20 carbon atoms). These fatty acids are often present in the rock matrix of hydrothermal sulfides and carbonates and are used as evidence for the presence of bacteria (Peresypkin et al. 1999; Lein et al. 2003). Unfortunately, most of these fatty acids are not very specific and can not be used to identify different groups of bacteria. From hydrothermal environments, less common structures and thus more sound biomarkers are also described. For instance, branched hydroxy (OH-*iso*) (A IX in Fig. 4.9 and Table 4.1; Peresypkin et al. 1999; Simoneit et al. 2004;

Blumenberg et al. 2007) and cyclohexyl fatty acids (A X) are often found at hydrothermal settings. Similar structures for the first have been found in thermophilic *Thermus* (Chung et al. 1997) and may be sourced by those or by phylogenetically related bacteria. The latter, cyclohexyl fatty acids, may originate from thermoacidophilic *Bacilli* (De Rosa et al. 1972). One other feature is the occurrence of thermally branched, *iso* and/or *anteiso*, fatty acids (*iso* = ω2-, *anteiso* = ω3-branched, where ω marks the position of the branching counted from the non-functionalized side of the structure). Findings of *anteiso* branched fatty acids, in particular pentadecanoic acid, are often used as an indication for sulfate-reducing bacteria (e.g., Kohring et al. 1994), but these structures appear to be not very abundant at hydrothermal settings studied so far. In contrast, *iso*-branched structures (fatty acids and ether lipids) are more abundant (Blumenberg et al. 2007). Unfortunately the source bacteria are still unknown. Similar structures have been described from Myxobacteria (e.g., *Myxococcus xanthus*; *Stigmatalla aurantiaca*; Fautz et al. 1979; Ring et al. 2006), which might be common at marine hydrothermal settings (Lopez-Garcia et al. 2003). Acidobacteria are also candidates as source organisms, since several members were recently found to contain high amounts of *iso*-15 fatty acids (Eichorst et al. 2007), and they are prominent in certain hydrothermal vent settings (Glamoclija et al. 2004). A set of very interesting structures recorded only from the Turtle Pits Field and one other site (Pancost et al. 2006), are macrocyclic non-isoprenoidal glycerol ethers (A XI). So far, these have not been described from any cultured bacteria, but the building blocks of these structures are also *iso*-branched pentadecanes. The probably diverse bacterial sources of the *iso*-branched ester and ether lipids at Turtle Pits, however, have yet to be identified. What can be concluded so far is that the source bacteria are thermoacidophilic and most likely chemolithoautotrophic, fixing the carbon via the reductive TCA or the hydroxypropionate cycle. This was concluded because the $\delta^{13}C$ values of the lipids are close to the marine bicarbonate presumably used as the carbon source (Blumenberg et al. 2007). Similar ^{13}C enriched carbon isotopic signatures (−8.6‰ to −14.2‰) were observed in lipids from other black smoker areas (Lein et al. 1997; Peresypkin et al. 1999; Vereshchaka et al. 2000), indicating that these carbon fixation pathways are common in chemoautotrophic bacteria living near black smokers on the Mid-Atlantic Ridge. *Iso*-branched lipids were not reported by these authors, although the distribution of "alkanes" in some of the sulfides shown by Peresypkin et al. (1999) appears to be similar to the long chain branched aliphatics that form the building blocks of the uncommon glycerol diethers at Turtle Pits (high concentrations of C_{30} to C_{35} carbon chains; Blumenberg et al. 2007; example A XI).

Another striking feature of hydrothermal settings is that bacterial dialkyl glycerol diether lipids (DAGE; A XII, A XIII) are much more common than in other environments (e.g., Jahnke et al. 2001; Pancost et al. 2006; Blumenberg et al. 2007; Bradley et al. 2009). (Hyper)thermophilic bacteria (e.g. *Aquifex*, *Ammonifex*, *Thermodesulfobacteria*; Langworthy et al. 1983; Huber et al. 1992; Huber et al. 1996) are potential sources, since glycerol ethers appear to be more rigid than glycerol esters and are thus advantageous in hot environments. Most of the above mentioned lipids are likely to originate from thermophilic, often anaerobic bacteria thriving in close vicinity to the hydrogen and/or sulfide rich fluids. But there are environments and energy sources for other clades of bacteria in these settings, too. A large amount of energy is stored in precipitated minerals, suitable for secondary, iron-sulfide oxidizing bacteria. Among these additional bacteria are *(Acidi)thiobacilli*, which may oxidize (aerobically) ferrous iron, elemental sulfur or reduced sulfur compounds. These bacteria grow on the cracks and fissures of metal sulfides and are known to produce a specific cyclopropylic fatty acid (A XV; Fig. 4.9 and Table 4.1; Kerger et al. 1986), a component that is abundant in the altered sulfide from the Turtle Pits Field (Blumenberg et al. 2007). The high concentration of this biomarker in the sulfide may be related to its highly weathered stage, suggesting that these bacteria are important in the leaching of iron-sulfides at Turtle Pits (Blumenberg et al. 2007). Nevertheless, the vast majority of bacteria from hydrothermal settings are still not well characterized, and many biomarkers are waiting to be assigned to their source bacteria.

Biomarkers of eukaryotes. Although the biosignatures of prokaryotes clearly prevail in hydrothermal settings, or are even the only signals, structures with an obvious eukaryotic origin have also been occasionally reported. For example, high concentrations of the steroid cycloartenol (A XVI; Fig. 4.9 and Table 4.1) were reported from the carbonate chimneys of the LCHF (Bradley et al. 2009) and indicate the presence of grazing, ciliated protozoa, as supported by molecular microbiological findings (Lopez-Garcia et al. 2007). It must be noted, however, that the majority of eukaryotic biomarkers in hydrothermal settings are allochthonous in origin (e.g., from algae or zooplankton). For instance, structural characteristics of most of the steranes found in sulfides from the Rainbow vent field suggest that they were produced by organisms living in the euphotic zone of the overlying water column (Simoneit et al. 2004).

4.3.1.2 Chemofossils in Ancient Hydrothermal Sulfides

Studies dealing with the biomarker content of fossil sulfides are scarce. Mycke et al. (1988) analyzed biomarkers in sedimentary sulfides of Precambrian age and found relatively high amounts of aliphatic hydrocarbons with cyclohexane rings in 1.5 Ga old sulfides from the Mt. Isa formation (Australia), as well as in sulfides from the Shungit formation (former USSR). They interpreted these as signals from thermoacidophilic bacteria due to the structural similarities to cyclohexylic fatty acids found in thermophilic *Bacilli* (De Rosa et al. 1972). The Yaman-Kasy and Taiskoe formation in the Ural are much younger sulfides formations (Silurian) and are suggested to represent fossil analogues of recent black smoker systems (Lein et al. 1998). Alla Yu Lein and colleagues, who put a lot of work and publications (many in Russian) on the (organic) geochemistry of recent and fossil hydrothermal systems, studied the biomarker contents of some of these ancient sulfides (e.g. Lein et al. 1998 and references cited therein). Lein et al. (1998) reported different suites of bacterial and archaeal biomarkers from these sulfides. However, more detailed investigations of fossil hydrothermal sulfides and enclosed chemofossils, including stable carbon isotope investigations, are still lacking and remain very interesting topics for future studies.

4.3.2 *Biomarkers at Methane Cold Seeps*

AOM-performing microbes produce specific lipid biomarkers that are clearly identifiable by their extremely negative carbon isotope signature, because this isotopic signal is transferred from the methane carbon to the microbial cells (the $\delta^{13}C$-values of the lipids are often lower than –70‰; Hinrichs et al. 1999; Thiel et al. 1999; Pancost et al. 2001b; Michaelis et al. 2002). The idea to get further insights into the AOM-process was spurred by findings of ^{13}C-depleted biomarkers of archaea and bacteria in marine sediments and microbial mats, suggesting close associations, and is thus an excellent example to demonstrate how powerful biomarker studies are in the field of geomicrobiology. The following section focuses on biomarker findings from microbial mats from the Black Sea. These structures were almost entirely formed by archaeal/bacterial associations performing the anaerobic oxidation of methane and serve as an excellent example to introduce the biomarker tool with respect to AOM-settings.

4.3.2.1 Biomarkers in Recent AOM-Mats and Carbonates from the Northwestern Black Sea

The presence of massive microbial mat-carbonate assemblages on the Black Sea seafloor was reported about 20 years ago (Ivanov et al. 1991) and gave clear evidence that these tower-like structures are formed by anaerobic methane oxidizing communities. Moreover, a biomarker study of such a carbonate clearly showed that biomarkers of AOM-performing microbes can be conserved in the carbonate matrix (Thiel et al. 2001a). Only in 2001, by using a manned submersible, were the extent, function, and shape of the carbonate towers understood in more detail (Michaelis et al. 2002; Fig. 4.3B, Plate 8). Michaelis et al. (2002) showed that in the so-called GHOSTDABS-field, almost the entire microbial community is fuelled by methane and that most of the microbes belong to a specific group of methanotrophic archaea, the ANME-1 cluster. These archaea have a cylindrical shape and a cell length of up to 10 μm (Fig. 4.3C). The ANME-1 archaea at this site are accompanied by sulfate-reducing bacteria of the *Desulfosarcina/Desulfococcus*-group (Michaelis et al. 2002). Subsequent studies demonstrated a higher complexity of microbes than previously suggested and a zonation of different AOM-communities in these microbial reefs, with additional niches for ANME-2 archaea which are closely associated to SRB of the *DSS*-group (Blumenberg et al. 2004). Interestingly, these bacteria include iron-sulfide minerals (greigite), and are obviously involved in the iron cycling in the mats (Reitner et al. 2005). In contrast to other AOM-sites, sulfide-oxidizing *Beggiatoa* bacteria (e.g., Treude et al. 2003) are not present at the GHOSTDABS-field due to the lack of oxygen in the waters overlying the sediment.

Because of the high microbial diversity, but particularly because the Black Sea reefs are unique in recent oceans, they offer the opportunity to study an analogue of a perhaps widely distributed biogeochemical scenario in the early oceans.

Consequently, many of the biomarkers present in Black Sea carbonates or their diagenetic products are also known from fossil carbonates and prove AOM as a key process during their precipitation (see for review Peckmann and Thiel 2004).

Biomarkers of archaea. As in hydrothermal settings, the presence of archaea at AOM-sites is indicated by isoprenoid hydrocarbons, glycerol diethers and/or glycerol dialkyl glycerol tetraethers (GDGT; Table 4.2 and Fig. 4.10; e.g. Schouten et al. 2003). At AOM sites like in the Black Sea, archaeal and bacterial biomarkers are strongly depleted in ^{13}C, clearly demonstrating methane carbon to be the major carbon source for the lipid biosynthesis (Figs. 4.4 and 4.10).

The isoprenoidal hydrocarbon (2,6,10,15,19-)PMI (Fig. 4.10; B II) has often been found at AOM sites and is also abundant in Black Sea mats and carbonates (Thiel et al. 2001a; Michaelis et al. 2002; Blumenberg et al. 2004). As mentioned above, PMI is also described from hydrothermal settings where it was assigned to methanogenic Archaea. In Black Sea mats, however, PMI most likely originates from methanotrophic archaea and serves as a hint supporting a close phylogenetic relationship between methanogenic and methanotrophic archaea. In fact, both archaeal groups produce several similar lipids, most of which can only be assigned to either methanotrophic or methanogenic archaea using their carbon isotopic signatures. However, the other very abundant hydrocarbon at AOM-sites, crocetane (B I), has so far not been found in methanogenic archaea. From detailed studies of Black Sea mats it was shown that crocetane is mostly present in zones with high numbers of ANME-2 archaea (Blumenberg et al. 2004) and may therefore be useful as an indication for this archaeal subgroup. Due to the highly structured distribution of archaeal/bacterial assemblages in the Black Sea mats, other ^{13}C-depleted archaeal biomarkers were also demonstrated to be applicable for a differentiation between ANME-1 and ANME-2 associations (e.g., *sn*2-hydroxyarchaeol (B V) indicative for ANME-2 and GDGT-derived biphytane hydrocarbons (B VII – IX) for ANME-1; Blumenberg et al. 2004). This study and follow-up investigations (see Niemann and Elvert (2008) and references cited therein) propose ratios and occurrences of distinct compounds to characterize archaeal communities at AOM-sites.

Although the biomarker distributions in the Black Sea mats are diverse, some compounds of AOM-performing microbes have only been described from other settings. For example, the recently discovered ANME-3 group appears to be characterized by a suite of unsaturated PMI derivatives (Niemann et al. 2006). Moreover, in several Mediterranean samples, a specific hydroxyarchaeol with one C_{20} and one C_{25} hydrocarbon side chain was very recently described (Stadnitskaia et al. 2008; B VI). It is known from halophilic Archaea as a non-hydroxylated analogue (Teixidor et al. 1993), and since *sn2*-hydroxyarchaeol is indicative for ANME-2, it is suggested to derive from yet unknown halophilic ANME-archaea, related to the ANME-2 cluster (Stadnitskaia et al. 2008).

It is still under discussion what controls the preferential growth of each community, but it is a common view that as yet unspecified physical and/or geochemical parameters are the controlling factors (e.g., gas partial pressure, presence of free gas, sulfide concentrations). Consequently, detailed characterizations of ANME-communities,

Table 4.2 Lipid biomarkers known from recent cold seep settings, potential source organisms, the quality as chemofossils, and individual fossil records in AOM carbonates

Component	Structure	Source organisms at AOM-sites[a]	Stability during diagenesis (chemofossil quality)	Record at fossil AOM-sites[b,c]
Crocetane	B I	Archaea (ANME-2?)[1–3]	High	✓
PMI (2,6,10,15,19-pentamethylicosane)	B II	Archaea (AOM-related)[1]	High	✓
PMI Δ4 and Δ5 (solely)		ANME-3[4]	Low	–
Squalene	B III	Archaea (AOM-related)[5]	Low	✓ (but mostly as Squalane)
Archaeol	B IV	Archaea (AOM-related)[3 and refs. therein]	Medium	✓
*sn*2-Hydroxyarchaeol	B V	ANME-2 or ANME-3 archaea [2, 4]	Low	–
C_{20},C_{25}-*sn*2-Hydroxyarchaeol	B VI	(halophilic?) AOM-related archaea[6]	Low	–
Biphytanes (0-2 rings) (from GDGTs)	B VII–IX	ANME-1 archaea[2, 7]	High	✓
12-Methyltetradecanoic PLFA (*ai*-15)	B X	SRB (*DSS*-group, ANME-1 related)[2, 8]	Medium	✓
Hexadec-11-enoic PLFA	B XI	SRB (*DSS*-group, ANME-2 related)[2, 9]	Low	–
ω-Cyclohexylundecanoic PLFA	B XII	Yet unknown AOM-related bacteria[10] (Bacilli?)	Medium	–
ω-5,6-cyclopropylhexadecanoic PLFA	B XIII	SRB (*DSS*-group, ANME-2 related)[9, 11]	Medium	–
Heptadec-11-enoic PLFA	B XIV	SRB (*Desulfobulbus* spp.)[6]	Low	–

Non-isopr. MAGE (e.g. 16-MAGE)	B XV	SRB? (ANME-2 related?)[11, 12]	High	✓
Non-isopr. DAGE (e.g. *ai*-15-DAGE)	B XVI	SRB (*DSS*-group, ANME-1 related)[2, 13]	High	✓
Tricos-10-ene	B XVII	Yet unknown (bacteria?)[14]	Low	(✓) (as ^{13}C-depl. tricosane)
Diploptene	B XVIII	Yet unknown bacteria (AOM-related, SRB?)[5, 15]	High	✓
Tetrahymanol	B XIX	Ciliated protozaons (anaerobic lifestyle)[16]	Low	(✓) (indirect as gammacerane)

The oldest chemofossil record of AOM goes back to Carboniferous times (Birgel et al. 2008). Structures are shown in Fig. 4.10

[a] If ^{13}C-depleted (<–50‰) GDGT = glycerol dialkyl glycerol tetraethers, PLFA = phospholipid fatty acid, MAGE = monoalkyl glycerol ethers, DAGE = dialkyl glycerol diethers, ANME = anaerobic methanotrophic archaea, SRB = sulfate reducing bacteria; DSS = *Desulfosarcina*/*Desulfococcus*

[b] = If ^{13}C-depleted

[c] Peckmann et al. 1999, 2004; Thiel et al. 1999; Burhan et al. 2002; Peckmann and Thiel 2004; Birgel et al. 2006a, b, 2008

[1] Elvert et al. (1999); [2] Blumenberg et al. (2004); [3] Niemann and Elvert (2008); [4] Niemann et al. (2006); [5] Elvert et al. (2000); [6] Stadnitskaia et al. (2008); [7] Rossel et al. (2008); [8] Thiel et al. (2001a); [9] Elvert et al. (2003); [10] Krüger et al. (2008); [11] Hinrichs et al. (2000); [12] Elvert et al. (2005); [13] Pancost et al. (2001a); [14] Thiel et al. (2001b); [15] Blumenberg et al. (2006); [16] Werne et al. (2002)

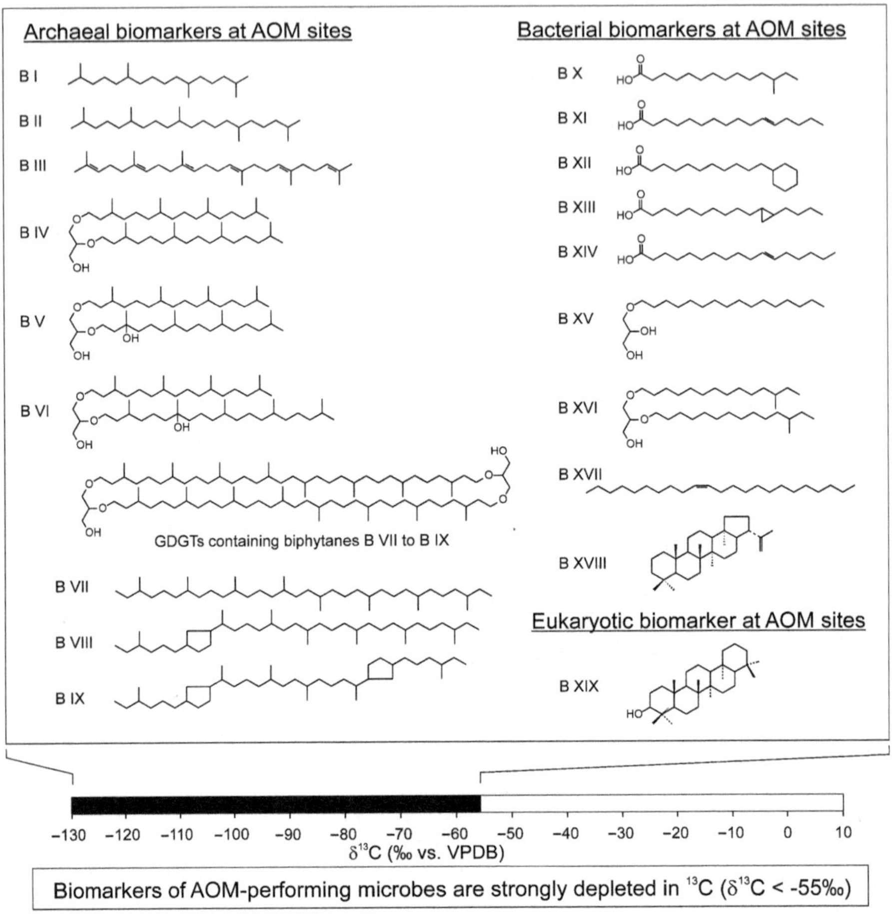

Fig. 4.10 Lipids of archaea and bacteria at AOM-settings (AOM = anaerobic oxidation of methane). Note that lipids are generally depleted in ^{13}C due to the use of methane carbon for biosynthesis. Names of the structures and further information are given in Table 4.2

either by molecular microbiology or lipid biomarkers, may be proper methods in future studies to give further insights into recent as well as fossil biogeochemical settings.

Biomarkers of bacteria. Due to the close association of ANME-archaea with sulfate-reducing bacteria (SRB), most bacterial biomarkers in Black Sea mats are suggested to originate from SRB. Like archaeal lipids, those of AOM-related SRB exhibit ^{13}C-depletion, because the AOM-derived carbon is also used by the bacterial partner (Orphan et al. 2002; Blumenberg et al. 2005; Wegener et al. 2008). SRB are suggested as sources for terminally branched anteiso-phospholipids fatty acids (e.g. *ai*-15 PLFA; B X; Table 4.2 and Fig. 4.10), but other PLFA with specific lengths and positions of the double bond are also indicative for certain SRB-groups, as well as for the bacterial/ANME association (see Table 4.2 and references cited therein; B XI, B XIII, B XIV).

Interesting and not yet well explained is the fact that specific SRB at AOM sites produce high amounts of glycerol ethers (MAGE and DAGE; e.g., Pancost et al. 2001a). Despite a few exceptions (Ollivier et al. 1991; Rütters et al. 2001; Ring et al. 2006), MAGE and DAGE have so far mainly been found in cultures of thermophilic bacteria (see above). But, temperatures are low at most AOM sites, so microbial adaptations to high temperature are not essential. It has been speculated that AOM-dependent sulfate-reducing bacteria are phylogenetically related to bacteria that branch deeply in the phylogenetic tree of life (e.g., Orphan et al. 2001). These ancestral bacteria were mostly thermophilic and their lipid properties may be conserved in AOM-related sulfate-reducing bacteria.

Other bacteria, with as yet unknown roles in the AOM process, are indicated by the presence of most likely non-SRB derived biomarkers. An example is a PLFA with a cyclohexylic ring (Table 4.2; B XII). Structural homologues have been described from *Bacilli* (De Rosa et al. 1972), but the role of these bacteria in AOM settings is unclear. Also unclear are the sources of hopanoids (e.g., B XVIII), components that have been thought to be produced by aerobic bacteria such as methanotrophs. Various hopanoids are abundant in Black Sea mats and carbonates (Thiel et al. 2003), and the ^{13}C-depletion suggests an autochthonous production in the anoxic mats. Although the production of hopanoids by strictly anaerobic bacteria has been recently shown (Sinninghe Damsté et al. 2004; Fischer et al. 2005; Blumenberg et al. 2006), the sources of hopanoids in Black Sea mats and other AOM settings have yet to be identified. Unfortunately, key-players in the AOM-process, as well as other prokaryotes, are still uncultured, preventing more detailed identifications of the sources of numerous bacterial (and archaeal) biomarkers present at AOM sites.

Biomarkers of eukaryotes. The GHOSTDABS-field is located in permanently anoxic waters, about 100 m below the oxic-anoxic transition zone in the water column. Consequently, the growth of macrofauna or other higher life is restricted, in contrast to other recent AOM settings such as the Gas Hydrate Ridge at the Cascadia Continental Margin (e.g., Treude et al. 2003). But the high abundance of tetrahymanol (Krüger et al. 2008), a pentacyclic triterpenoid (Table 4.2, Fig. 4.10; B XIX), suggests the growth of ciliated protozoans in the anoxic Black Sea mats. ^{13}C-depleted tetrahymanol is also known from other anoxic habitats in AOM settings (Werne et al. 2002) and has been proposed to be sourced by protozoans grazing on ^{13}C-depleted biomass. Numerous groups of ciliated protozoans are known to be capable of an anaerobic lifestyle (e.g., Van Hoek et al. 2000). Other biomarkers of eukaryotes, except for minor signals from allochthonous primary production in the overlying water column (e.g., algae), are absent in the Black Sea towers in the GHOSTDABS-field.

4.3.2.2 Chemofossils in AOM Carbonates

Numerous fossil carbonates that precipitated through the activity of AOM-performing microbes have been identified using structures and isotope signatures of chemofossils. This includes examples from the Neogene of Italy (Thiel et al. 1999;

Peckmann et al. 1999), the Paleogene of western Washington (Peckmann et al. 2002; Goedert et al. 2003; see also Chapter 14); the Late Cretaceous of the Western Interior Seaway of the USA (Birgel et al. 2006a) and Japan (Jenkins et al. 2008), the Early Cretaceous of California (Birgel et al. 2006b) and the Ukraine (Peckmann et al. 2009), and the oldest record so far is from the Carboniferous of Namibia (Birgel et al. 2008).

The chemofossil quality is high for hydrocarbons and low for functionalized structures (see Tables 4.1 and 4.2). Moreover, double bonds are less likely to be preserved, but if the structural characteristic is conserved after losses of double bonds, the resulting structures may still be applicable as chemofossils (e.g., high tricos-10-ene in recent and high tricosane in fossil carbonates; Thiel et al. 2001b). Saturated isoprenoid hydrocarbons (PMI and crocetane) are excellent biomarkers at recent AOM-sites and were also the main chemofossils discovered in Cenozoic AOM-carbonates (Peckmann et al. 1999; Thiel et al. 1999).

Functionalized chemofossils of methanotrophic archaea and sulfate-reducing bacteria have been observed also in carbonates with a low maturity. One good example is the Miocene Marmorito limestone, where ^{13}C-depleted archaeol (B IV) and terminally branched pentadecanoic alcohols are well preserved (Thiel et al. 1999). A very nice study was published by Peckmann et al. (2004) who demonstrated the preservation of chemofossils from a complex microbial community. They found chemofossils indicating, at the base of the food chain, anaerobic methanotrophic consortia producing sulfide for aerobic *Beggiatoa*-like sulfide-oxidizing bacteria and, in addition, biosignatures of grazing protozoans. A survey of recent literature that includes chemofossils from fossil AOM sites is presented by Peckmann and Thiel (2004).

It should be noted that the ratios of ^{13}C-depleted biomarkers, as has been demonstrated for recent AOM-sites (Blumenberg et al. 2004; Niemann and Elvert 2008), might be also useful for characterizing fossil ANME-associations and the paleoenvironment (Blumenberg et al. 2004; Peckmann et al. 2009). However, this must be done with caution, since diagenesis may have selectively influenced the distribution of individual components.

4.4 Conclusions and Outlook

Lipid biomarkers, the remnants of cell membranes, in combination with their stable isotope signatures are very helpful to learn about the distributions and functioning of microorganisms in extreme environments. Excellent examples include (a) specific black smoker settings, where the fossil remains of thermophilic and mesophilic, iron-sulfide leaching microbes are preserved in massive hydrothermal rocks and (b) the anoxic Black Sea, where microbial reefs are formed by strictly anaerobic consortia performing the anaerobic oxidation of methane (AOM). It has been demonstrated that biomarkers and stable isotope signatures allow for a differentiation between archaeal and bacterial signals and, moreover, provide further insights into phylogenetic groups and metabolisms (methanogenesis, methanotrophy, sulfide- and iron-oxidation, sulfate-reduction, etc.). In contrast to molecular microbiological techniques, many lipid biomarkers are also useful for reconstructing fossil environments because these chemofossils may resist diagenetic destruction for billions of years.

However, for future studies other innovative techniques must add to the existing knowledge of uncultured microbial associations. Promising strategies include techniques visualizing the distribution of biomarkers in recent mats and on rock surfaces. This was elegantly demonstrated in a recent study, showing the spatial distribution of microbes in AOM-mats using Time-of-Flight Secondary Ion-MS (ToF-SIMS; Thiel et al. 2007). Another method to get data on living microbial communities, the functioning of associations, and element flow is to perform in vitro experiments using stable isotope labeling (e.g., Boschker et al. 1998; Blumenberg et al. 2005; Wegener et al. 2008). Future studies should combine this technique with DNA-stable isotope probing (DNA-SIP) to enhance our knowledge of the role and taxonomy of prokaryotes (see Friedrich 2006). SIP combined with Nano-SIMS would be also helpful to scale down isotope labeling approaches.

Our knowledge of microbial life entrapped in fossil hydrothermal sulfides and AOM-carbonates should also be enhanced. Improvements in DNA extraction techniques from geological samples and in existing PCR methods are promising. Combining the lipid biomarker and the DNA extraction approaches works well to characterize environmental changes recorded in the growth and decease of microbial communities (e.g., Coolen et al. 2008). However, the stability of DNA is limited, so the study of geological samples with an age of millions to billions years old can only be accomplished through biomarker approaches. Nevertheless, the future of geomicrobiological studies is to combine molecular microbiological and organic-geochemical techniques, and biomarker analyses, including structural and isotopic investigations, will form a substantial part of these attempts.

Acknowledgements I am very thankful to Walter Michaelis, Richard Seifert, Dietmar Keyser (all University of Hamburg), Karen Hissmann (IfM-GEOMAR Kiel), and MARUM (University of Bremen) for the permission to show photos. I thank Helge Niemann and Steffen Kiel for their careful reviews, considereably improving the manuscript. I thank the Deutsche Forschungsgemeinschaft for financial support (DFG BL 971/1-1). Richard Seifert, Walter Michaelis and Volker Thiel (University of Göttingen) are also acknowledged for fruitful discussions.

References

Barnes RO, Goldberg ED (1976) Methane production and consumption in anaerobic marine sediments. Geology 4:297–300

Birgel D, Peckmann J, Klautzsch S, Thiel V, Reitner J (2006a) Anaerobic and aerobic oxidation of methane at Late Cretaceous seeps in the Western Interior Seaway, USA. Geomircobiol J 23:565–577

Birgel D, Thiel V, Hinrichs K-U, Elvert M, Campbell KA, Reitner J, Farmer JD, Peckmann J (2006b) Lipid biomarker patterns of methane-seep microbialites from the Mesozoic convergent margin of California. Org Geochem 37:1289–1302

Birgel D, Himmler T, Freiwald A, Peckmann J (2008) A new constraint on the antiquity of anaerobic oxidation of methane: late Pennsylvanian seep limestones from southern Namibia. Geology 36:543–546

Blumenberg M, Seifert R, Reitner J, Pape T, Michaelis W (2004) Membrane lipid patterns typify distinct anaerobic methanotrophic consortia. Proc Natl Acad Sci USA 101:11111–11116

Blumenberg M, Seifert R, Nauhaus K, Pape T, Michaelis W (2005) *In vitro* study of lipid biosynthesis in an anaerobically methane oxidizing microbial mat. Appl Environ Microbiol 71:4345–4351

Blumenberg M, Krüger M, Nauhaus K, Talbot HM, Oppermann B, Seifert R, Pape T, Michaelis W (2006) Biosynthesis of hopanoids by sulfate-reducing bacteria (genus *Desulfovibrio*). Environ Microbiol 8:1220–1227

Blumenberg M, Seifert R, Petersen S, Michaelis W (2007) Biosignatures present in a hydrothermal massive sulphide from the Mid-Atlantic Ridge. Geobiology 5:435–450

Boetius A, Ravenschlag K, Schubert CJ, Rickert D, Widdel F, Gieseke A, Amann R, Jørgensen BB, Witte U, Pfannkuche O (2000) A marine microbial consortium apparently mediating anaerobic oxidation of methane. Nature 407:623–626

Boschker HTS, Nold SC, Wellsbury P, Bos D, de Graaf W, Pel R, Parkes RJ, Cappenberg TE (1998) Direct linking of microbial populations to specific biogeochemical processes by ^{13}C-labelling of biomarkers. Nature 392:801–805

Bradley AS, Hayes JM, Summons RE (2009) Extraordinary ^{13}C enrichment of diether lipids at the Lost City Hydrothermal Field suggests a carbon-limited ecosystem. Geochim Cosmochim Acta 73:102–118

Brazelton WJ, Schrenk MO, Kelley DS, Baross JA (2006) Methane- and sulfur-metabolizing microbial communities dominate the Lost City hydrothermal field ecosystem. Appl Environ Microbiol 72:6257–6270

Brocks JJ, Pearson A (2005) Building the biomarker tree of life. Rev Miner Geochem 59: 233–258

Brocks JJ, Logan GA, Buick R, Summons RE (1999) Archean molecular fossils and the early rise of eukaryotes. Science 285:1033–1036

Burhan RYP, Trendel J-M, Adam P, Wehrung P, Albrecht P, Nissenbaum A (2002) Fossil bacterial ecosystem at methane seeps: origin of organic matter from Be'eri sulfur deposit, Israel. Geochim Cosmochim Acta 66:4085–4101

Cady SL, Farmer JD, Grotzinger JP, Schopf JW, Steele A (2003) Morphological biosignatures and the search for life on Mars. Astrobiology 3:351–368

Campbell BJ, Summers Engel A, Porter ML, Takai K (2006) The versatile e-proteobacteria: key players in sulphidic habitats. Nat Rev Microbiol 4:458–468

Chung A, Rainey F, Nobre M, Burghardt J, da Costa M (1997) *Meiothermus cerbereus* sp. nov., a new slightly thermophilic species with high levels of 3-hydroxy fatty acids. Int J Syst Bacteriol 47:1225–1230

Comita PB, Gagosian RB (1983) Membrane lipid from deep-sea hydrothermal vent methanogen: a new macrocyclic glycerol diether. Science 222:1329–1331

Coolen MJL, Overmann J (2007) 217000-year-old DNA sequences of green sulfur bacteria in Mediterranean sapropels and their implications for the reconstruction of the paleoenvironment. Environ Microbiol 9:238–249

Coolen MJL, Talbot HM, Abbas BA, Ward C, Schouten S, Volkman JK, Damste JSS (2008) Sources for sedimentary bacteriohopanepolyols as revealed by 16S rDNA stratigraphy. Environ Microbiol 10:1783–1803

Corliss JB, Dymond J, Gordon LI, Edmond JM, von Herzen RP, Ballard RD, Green K, Williams D, Bainbridge A, Crane K, van Andel TH (1979) Submarine thermal sprirngs on the Galapagos Rift. Science 203:1073–1083

De Rosa M, Gambacorta A, Minale L, BùLock JD (1972) The formation of ω-cyclohexyl fatty acids from shikimate in an acidophilic thermophilic *Bacillus*. Biochem J 128:751–754

DeNiro MJ, Epstein S (1976) You are what you eat (plus a few per mil): the carbon isotope cycle in food chains. Abstr Programs-Geol Soc Am Ann Meet 8:834–835

Eglinton G, Scott PM, Belsky T, Burlingame AL, Calvin M (1964) Hydrocarbons of biological origin from a one-billion-year-old sediment. Science 145:263–264

Eichorst SA, Breznak JA, Schmidt TM (2007) Isolation and characterization of soil bacteria that define *Terriglobus* gen. nov., in the Phylum Acidobacteria. Appl Environ Microbiol 73:2708–2717

Elvert M (1999) Microbial biomarkers in anaerobic marine environments dominated by methane. Dissertation, University of Kiel, Kiel, Germany

Elvert M, Suess E, Whiticar MJ (1999) Anaerobic methane oxidation associated with marine gas hydrates. Superlight C-isotopes from saturated and unsaturated C20 and C25 irregular isoprenoids. Naturwissenschaften 86:295–300

Elvert M, Suess E, Greinert J, Whiticar MJ (2000) Archaea mediating anaerobic methane oxidation in deep-sea sediments at cold seeps of the eastern Aleutian subduction zone. Org Geochem 31:1175–1187

Elvert M, Boetius A, Knittel K, Jørgensen BB (2003) Characterization of specific membrane fatty acids as chemotaxonomic markers for sulfate-reducing bacteria involved in anaerobic oxidation of methane. Geomircobiol J 20:403–419

Elvert M, Hopmans EC, Treude T, Boetius A, Suess E (2005) Spatial variations of methanotrophic consortia at cold methane seeps: implications from a high-resolution molecular and isotopic approach. Geobiology 3:195–209

Fautz E, Rosenfelder G, Grotjahn L (1979) Iso-branched 2- and 3-hydroxy fatty acids as characteristic lipid constituents of some gliding bacteria. J Bacteriol 140:852–858

Fischer WW, Summons RE, Pearson A (2005) Targeted genomic detection of biosynthetic pathways: anaerobic production of hopanoid biomarkers by a common sedimentary microbe. Geobiology 3:33–40

Fisher CR, Takai K, Le Bris N (2007) Hydrothermal vent ecosystems. Oceanography 20:14

Friedrich M (2006) Stable-isotope probing of DNA: insights into the function of uncultivated microorganisms from isotopically labeled metagenomes. Curr Opin Biotechnol 17:59–66

Furnes H, Banerjee NR, Muehlenbachs K, Staudigel H, de Wit M (2004) Early life recorded in Archean pillow lavas. Science 304:578–581

Galliker PK (1990) Zur Biosynthese der Etherlipide aus *Methanobacterium thermoautotrophicum*. Dissertation, University of Zurich, Zurich, Switzerland

Glamoclija M, Garrel L, Berthon J, Lopez-Garcia P (2004) Biosignatures and bacterial diversity in hydrothermal deposits of Solfatara Crater, Italy. Geomicrobiol J 21:529–541

Glynn S, Mills RA, Palmer MR, Pancost RD, Severmann S, Boyce AJ (2006) The role of prokaryotes in supergene alteration of submarine hydrothermal sulfides. Earth Planet Sci Lett 244:170–185

Goedert JL, Thiel V, Schmale O, Rau WW et al (2003) The late Eocene 'Whiskey Creek' methane-seep deposit (western Washington State) Part I: geology, palaeontology, and molecular geobiology. Facies 48:223–240

Golyshina OV, Timmis KN (2005) *Ferroplasma* and relatives, recently discovered cell wall-lacking archaea making a living in extremely acid, heavy metal-rich environments. Environ Microbiol 7:1277–1288

Haase KM, Petersen S, Koschinsky A, Seifert R, Devey CW, Dubilier N, Fretzdorff S, Garbe-Schönberg D, German CR, Giere O, Keir R, Kuever J, Lackschewitz KS, Mawick J, Marbler H, Melchert B, Mertens C, Ostertag-Henning C, Paulick H, Perner M, Peters M, Sander S, Schmale O, Shank TM, Stecher J, Stöber U, Strauss H, Süling J, Walter M, Warmuth M, Weber S, Westernströer U, Yoerger D, Zielinski F (2007) Young volcanism and related hydrothermal activity at 5°S on the slow-spreading southern Mid-Atlantic Ridge. Geochem Geophys Geosyst 8:Q11002

Hallam SJ, Putnam N, Preston CM, Detter JC, Rokhsar D, Richardson PM, DeLong EF (2004) Reverse methanogenesis: testing the hypothesis with environmental genomics. Science 305:1457–1462

Hayes JM (2001) Fractionation of the isotopes of carbon and hydrogen in biosynthetic processes. In: Valley JW, Cole DR (eds) Stable isotope geochemistry, reviews in mineralogy and geochemistry. Mineralogical Society of America, Washington, DC, pp 225–278

Hinrichs K-U, Hayes JM, Sylva SP, Brewer PG, DeLong EF (1999) Methane-consuming archaebacteria in marine sediments. Nature 398:802–805

Hinrichs K-U, Summons RE, Orphan VJ, Sylva SP, Hayes JM (2000) Molecular and isotopic analysis of anaerobic methane-oxidizing communities in marine sediments. Org Geochem 31:1685–1701

Hoefs J (2004) Stable isotope geochemistry. Springer, Berlin/New York

Holzer GU, Kelly PJ, Jones WJ (1988) Analysis of lipids from a hydrothermal vent methanogen and associated vent sediment by supercritical fluid chromatography. J Microbiol Meth 8:161–173

House CH, Schopf JW, Stetter KO (2003) Carbon isotopic fractionation by Archaeans and other thermophilic prokaryotes. Org Geochem 34:345–356

Huber R, Wilharm T, Huber D, Trincone A, Burggraf S, König H, Rachel R, Rockinger I, Fricke H, Stetter KO (1992) *Aquifex pyrophilus* gen. nov. sp. nov., represents a novel group of marine hyperthermophilic hydrogen-oxidizing bacteria. Syst Appl Microbiol 15:340–351

Huber R, Rossnagel P, Woese CR, Rachel R, Langworthy DE, Stetter KO (1996) Formation of ammonium from nitrate during chemolithoautotrophic growth of the extremely thermophilic bacterium *Ammonifex degensii* gen. nov. sp. nov. Syst Appl Microbiol 19:40–49

Inagaki F, Kuypers MMM, Tsunogai U, Ishibashi JI, Nakamura KI, Treude T, Ohkubo S, Nakaseama M, Gena K, Chiba H, Hirayama H, Nunoura T, Takai K, Jorgensen BB, Horikoshi K, Boetius A (2006) Microbial community in a sediment-hosted CO_2 lake of the southern Okinawa Trough hydrothermal system. Proc Natl Acad Sci USA 103:14164–14169

Ivanov MV, Polikarpov GG, Lein AY, Gal'chenko VF, Egorov VN, Gulin SB, Gulin MB, Rusanov II, Miller YM, Kuptsov VI (1991) Biogeochemistry of the carbon cycle in the region of methane seeps in the Black Sea. Dokl Akad Nauk SSSR 320:1235–1240 (in Russian)

Jahnke LL, Eder W, Huber R, Hope JM, Hinrichs K-U, Hayes JM, Des Marais DJ, Cady SL, Summons RE (2001) Signature lipids and stable carbon isotope analyses of Octopus Spring hyperthermophilic communities compared with those of *Aquificales* representatives. Appl Environ Microbiol 67:5179–5189

Jenkins RG, Hikida Y, Chikaraishi Y, Ohkouchi N, Tanabe K (2008) Microbially induced formation of ooid-like coated grains in the Late Cretaceous methane-seep deposits of the Nakagawa area, Hokkaido, northern Japan. Isl Arc 17:261–269

Jones GA, Gagnon AR (1994) Radiocarbon chronology of Black Sea sediments. Deep Sea Res I 41:531–557

Kashefi K, Lovley DR (2003) Extending the upper temperature limit of life. Science 301:934

Kelley DS, Karson JA, Blackman DK, Fruh-Green GL, Butterfield DA, Lilley MD, Olson EJ, Schrenk MO, Roe KK, Lebon GT, Rivizzigno P, AT3-60 Shipboard Party (2001) An off-axis hydrothermal vent field near the Mid-Atlantic Ridge at 30°N. Nature 412:145–149

Kelley DS, Karson JA, Fruh-Green GL, Yoerger DR, Shank TM, Butterfield DA, Hayes JM, Schrenk MO, Olson EJ, Proskurowski G, Jakuba M, Bradley A, Larson B, Ludwig K, Glickson D, Buckman K, Bradley AS, Brazelton WJ, Roe K, Elend MJ, Delacour A, Bernasconi SM, Lilley MD, Baross JA, Summons RE, Sylva SP (2005) A serpentinite-hosted ecosystem: the Lost City Hydrothermal Field. Science 307:1428–1434

Kerger BD, Nichols PD, Antworth CP, Sand W, Bock E, Cox JC, Langworthy TA, White DC (1986) Signature fatty-acids in the polar lipids of acid-producing *Thiobacillus* spp - methoxy, cyclopropyl, alpha-hydroxy-cyclopropyl and branched and normal monoenoic fatty-acids. FEMS Microbiol Ecol 38:67–77

Killops S, Killops V (2005) Introduction to organic geochemistry. Blackwell, Oxford, UK

Knittel K, Lösekann T, Boetius A, Kort R, Amann R (2005) Diversity and distribution of methanotrophic Archaea (ANME) at cold seeps. Appl Environ Microbiol 71:467–479

Kohring LL, Ringelberg DB, Devereux R, Stahl DA, Mittelman MW, White DC (1994) Comparison of phylogenetic relationships based on phospholipid fatty acid profiles and ribosomal RNA sequence similarities among dissimilatory sulfate-reducing bacteria. FEMS Microbiol Lett 119:303–308

Konhauser K (2006) Introduction to geomicrobiology. Blackwell, Malden, MA/Oxford, UK/ Carlton, Australia

Krüger M, Blumenberg M, Kasten S, Wieland A, Kanel L, Klock J-H, Michaelis W, Seifert R (2008) A novel, multi-layered methanotrophic microbial mat system growing on the sediment of the Black Sea. Environ Microbiol 10:1934–1947

Langworthy TA, Holzer G, Zeikus JG, Tornabene TG (1983) Iso- and anteiso-branched glycerol diethers of the thermophilic anaerobe *Thermodesulfotobacterium commune*. Syst Appl Microbiol 4:1–17

Lein AY, Galkin SV, Gebruk AV, Miller YM, Pimenov NV, Moskalev LI, Ivanov MV (1997) Contribution of the bacterial chemosynthesis to the feeding of animal community inhabiting vent Field Broken Spur (29°N, Mid-Atlantic Ridge). Dokl Akad Nauk 357:410–413

Lein AY, Glushchenko NN, Osipov GA, Ul'yanova NV, Ivanov MV (1998) Sulfide ore biomarkers of modern and ancient "black smokers". Dokl Akad Nauk 359:525–528 (in Russian)

Lein AY, Peresypkin VI, Simoneit BRT (2003) Origin of hydrocarbons in hydrothermal sulfide ores in the Mid-Atlantic Ridge. Lithol Miner Res V38:383–393

Little CTS, Vrijenhoek RC (2003) Are hydrothermal vent animals living fossils? Trends Ecol Evol 18:582–588

Lopez-Garcia P, Duperron S, Philippot P, Foriel J, Susini J, Moreira D (2003) Bacterial diversity in hydrothermal sediment and epsilonproteobacterial dominance in experimental microcolonizers at the Mid-Atlantic Ridge. Environ Microbiol 5:961–976

Lopez-Garcia P, Vereshchaka A, Moreira D (2007) Eukaryotic diversity associated with carbonates and fluid-seawater interface in Lost City hydrothermal field. Environ Microbiol 9:546–554

Madigan MT, Martinko JM, Parker J (2002) Brock biology of microorganisms. Prentice Hall, Upper Saddle River, NJ

Michaelis W, Seifert R, Nauhaus K, Treude T, Thiel V, Blumenberg M, Knittel K, Gieseke A, Peterknecht K, Pape T, Boetius A, Amann R, Jørgensen BB, Widdel F, Peckmann J, Pimenov NV, Gulin MB (2002) Microbial reefs in the Black Sea fueled by anaerobic oxidation of methane. Science 297:1013–1015

Mycke B, Michaelis W, Degens ET (1988) Biomarkers in sedimentary sulfides of Precambrian age. Org Geochem 13:619–625

Nakagawa S, Takai K (2008) Deep-sea vent chemoautotrophs: diversity, biochemistry and ecological significance. FEMS Microbiol Ecol 65:1–14

Nakagawa T, Takai K, Suzuki Y, Hirayama H, Konno U, Tsunogai U, Horikoshi K (2006) Geomicrobiological exploration and characterization of a novel deep-sea hydrothermal system at the TOTO caldera in the Mariana Volcanic Arc. Environ Microbiol 8:37–49

Nercessian O, Fouquet Y, Pierre C, Prieur D, Jeanthon C (2005) Diversity of Bacteria and Archaea associated with a carbonate-rich metalliferous sediment sample from the Rainbow vent field on the Mid-Atlantic Ridge. Environ Microbiol 7:698–714

Niemann H, Elvert M (2008) Diagnostic lipid biomarker and stable carbon isotope signatures of microbial communities mediating the anaerobic oxidation of methane with sulphate. Org Geochem 39:1668–1677

Niemann H, Losekann T, de Beer D, Elvert M, Nadalig T, Knittel K, Amann R, Sauter EJ, Schluter M, Klages M, Foucher JP, Boetius A (2006) Novel microbial communities of the Haakon Mosby mud volcano and their role as a methane sink. Nature 443:854–858

Nisbet EG, Sleep NH (2001) The habitat and nature of early life. Nature 409:1083–1091

Ollivier B, Hatchikian CE, Prensier G, Guezennec J, Garcia JL (1991) *Desulfohalobium retbaense* gen. nov., sp. nov., a halophilic sulfate-reducing bacterium from sediments of a hypersaline lake in Senegal. Int J Syst Bacteriol 41:74–81

Orphan VJ, House CH, Hinrichs K-U, McKeegan KD, DeLong EF (2001) Methane-consuming Archaea revealed by directly coupled isotopic and phylogenetic analysis. Science 293:484–487

Orphan VJ, House CH, Hinrichs K-U, McKeegan KD, DeLong EF (2002) Multiple archaeal groups mediate methane oxidation in anoxic cold seep sediments. Proc Natl Acad Sci USA 99:7663–7668

Pancost RD, Bouloubassi I, Aloisi G, Sinninghe Damsté JS, Party tMS (2001a) Three series of non-isoprenoidal dialkyl glycerol diethers in cold-seep carbonate crusts. Org Geochem 32:695–707

Pancost RD, Hopmans EC, Sinninghe Damsté JS, Party TMss (2001b) Archaeal lipids in Mediterranean cold seeps: molecular proxies for anaerobic methane oxidation. Geochim Cosmochim Acta 65:1611–1627

Pancost RD, Pressley S, Coleman JM, Talbot HM, Kelly K, Farrimond P, Schouten S, Benning L, Mountain BW (2006) Composition and implications of diverse lipids in New Zealand Geothermal sinters. Geobiology 4:71–92

Peckmann J, Thiel V (2004) Carbon cycling at ancient methane-seeps. Chem Geol 205:443–467
Peckmann J, Thiel V, Michaelis W, Clari P, Gaillard C, Martire L, Reitner J (1999) Cold seep deposits of Beauvoisin (Oxfordian; southeastern France) and Marmorito (Miocene; northern Italy): microbially induced authigenic carbonates. Int J Earth Sci 88:60–75
Peckmann J, Thiel V, Michaelis W, Reitner J (2002) A comprehensive approach to the study of methane seep deposits from the Lincoln Creek Formation, western Washington State, USA. Sedimentology 49:855–873
Peckmann J, Thiel V, Reitner J, Taviani M, Aharon P, Michaelis W (2004) A microbial mat of a large sulfur bacterium preserved in a Miocene methane-seep limestone. Geomicrobiol J 21:247–255
Peckmann J, Birgel D, Kiel S (2009) Molecular fossils reveal fluid composition and flow intensity at a Cretaceous seep. Geology 37:847–850
Peresypkin VI, Lein AY, Bogdanov YA, Bortnikov NS (1999) Lipids in the hydrothermal formations in the regions at 14°45'N and 29° N, the Mid-Atlantic Ridge. Oceanology (transl from Okeanologiya) 39:234–244
Perner M, Kuever J, Seifert R, Pape T, Koschinsky A, Schmidt K, Strauss H, Imhoff JF (2007) The influence of ultramafic rocks on microbial communities at the Logatchev Hydrothermal Field, located 15°N on the Mid-Atlantic Ridge. FEMS Microbiol Ecol 61:97–109
Pimenov NV, Rusanov II, Poglazova MN, Mityushina LL, Sorokin DY, Khmelenina VN, Trotsenko YA (1997) Bacterial mats on coral-like structures at methane seeps in the Black Sea. Mikrobiologiya 66:354–360
Proskurowski G, Lilley MD, Seewald JS, Fruh-Green GL, Olson EJ, Lupton JE, Sylva SP, Kelley DS (2008) Abiogenic hydrocarbon production at Lost City Hydrothermal Field. Science 319:604–607
Rasmussen B, Fletcher IR, Brocks JJ, Kilburn MR (2008) Reassessing the first appearance of eukaryotes and cyanobacteria. Nature 455:1101–1104
Reeburgh WS (1976) Methane consumption in Cariaco Trench waters and sediments. Earth Planet Sci Lett 28:337–344
Reeburgh WS (2007) Oceanic methane biogeochemistry. Chem Rev 107:486–513
Reitner J, Peckmann J, Reimer A, Schumann G, Thiel V (2005) Methane-derived carbonate build-ups and associated microbial communities at cold seeps on the lower Crimean shelf (Black Sea). Facies 51:66–79
Reysenbach A-L, Cady SL (2001) Microbiology of ancient and modern hydrothermal systems. Trends Microbiol 9:79–86
Ring MW, Schwar G, Thiel V, Dickschat JS, Kroppenstedt RM, Schulz S, Bode HB (2006) Novel Iso-branched ether lipids as specific markers of developmental sporulation in the myxobacterium *Myxococcus xanthus*. J Biol Chem 281:36691–36700
Ritger S, Carson B, Suess E (1987) Methane-derived authigenic carbonates formed by subduction-induced pore-water expulsion along the Oregon/Washington margin. Geol Soc Am Bull 98:147–156
Rossel PE, Lipp JS, Fredricks HF, Arnds J, Boetius A, Elvert M, Hinrichs K-U (2008) Intact polar lipids of anaerobic methanotrophic archaea and associated bacteria. Org Geochem 39:992–999
Russell NJ (1990) Cold adaptation of microorganisms. Philos Trans R Soc Lond B 326:595–611
Rütters H, Sass H, Cypionka H, Rullkötter J (2001) Monoalkylether phospholipids in the sulfate-reducing bacteria *Desulfosarcina variabilis* and *Desulforhabdus amnigenus*. Arch Microbiol 176:435–442
Schidlowski M (2001) Carbon isotopes as biogeochemical recorders of life over 3.8 Ga of Earth history: evolution of a concept. Precambrian Res 106:117–134
Schoell M (1988) Multiple origins of methane in the earth. Chem Geol 71:1–10
Schouten S, Wakeham SG, Hopmans EC, Sinninghe Damsté JS (2003) Biogeochemical evidence that thermophilic Archaea mediate the anaerobic oxidation of methane. Appl Environ Microbiol 69:1680–1686
Schouten S, Baas M, Hopmans E, Reysenbach A-L, Damsté J (2008) Tetraether membrane lipids of Candidatus "*Aciduliprofundum boonei*", a cultivated obligate thermoacidophilic euryarchaeote from deep-sea hydrothermal vents. Extremophiles 12:119–124

Schrenk MO, Kelley DS, Delaney JR, Baross JA (2003) Incidence and diversity of microorganisms within the walls of and active deep-sea sulfide chimney. Appl Environ Microbiol 69:3580–3592

Schroeder T, John B, Frost BR (2002) Geologic implicationsvof seawater circulation through peridotite exposed at slowspreading mid-ocean ridges. Geology 30:367–370

Simoneit BRT, Lein AY, Peresypkin VI, Osipov GA (2004) Composition and origin of hydrothermal petroleum and associated lipids in the sulfide deposits of the Rainbow field (Mid-Atlantic Ridge at 36°N). Geochim Cosmochim Acta 68:2275–2294

Sinninghe Damsté JS, Rijpstra IC, Schouten S, Fuerst JA, Jetten MSM, Strous M (2004) The occurrence of hopanoids in planctomycetes: implications for the sedimentary biomarker record. Org Geochem 35:561–566

Stadnitskaia A, Bouloubassi I, Elvert M, Hinrichs KU, Sinninghe Damsté JS (2008) Extended hydroxyarchaeol, a novel lipid biomarker for anaerobic methanotrophy in cold seepage habitats. Org Geochem 39:1007–1014

Stetter KO (1996) Hyperthermophilic procaryotes. FEMS Microbiol Rev 18:149–158

Stetter KO (2006) Hyperthermophiles in the history of life. Philos Trans R Soc Lond B 361: 1837–1843

Strous M, Jetten MSM (2004) Anaerobic oxidation of methane and ammonium. Annu Rev Microbiol 58:99–117

Summons RE, Brassell SC, Eglinton G, Evans E, Horodyski RJ, Robinson N, Ward DM (1988) Distinctive hydrocarbon biomarkers from fossiliferous sediment of the Late Proterozoic Walcott Member, Chuar Group, Grand Canyon, Arizona. Geochim Cosmochim Acta 52:2625–2637

Summons R, Albrecht P, McDonald G, Moldowan J (2008) Molecular biosignatures. Space Sci Rev 135:133–159

Takai K, Nakagawa S, Reysenbach A-L, Hoek J (2006) Microbial ecology of mid-ocean ridges and back-arc basins. In: Christie D, Fisher CR, Lee S-M, Givens S (eds) Interpretations among physical, chemical, biological, and geological processes in Back-arc spreading systems. American Geophysical Union, Washington, DC, pp 185–214

Takai K, Nakamura K, Toki T, Tsunogai U, Miyazaki M, Miyazaki J, Hirayama H, Nakagawa S, Nunoura T, Horikoshi K (2008) Cell proliferation at 122°C and isotopically heavy CH_4 production by a hyperthermophilic methanogen under high-pressure cultivation. Proc Natl Acad Sci USA 105:10949–10954

Teixidor P, Grimalt JO, Pueyo JJ, Rodriguez-Valera F (1993) Isopranylglycerol diethers in non-alkaline evaporitic environments. Geochim Cosmochim Acta 57:4479–4489

Teske A, Hinrichs K-U, Edgcomb V, de Vera GA, Kysela D, Sylva SP, Sogin ML, Jannasch HW (2002) Microbial diversity of hydrothermal sediments in the Guaymas Basin: evidence for anaerobic methanotrophic communities. Appl Environ Microbiol 68:1994–2007

Thiel V, Peckmann J, Seifert R, Wehrung P, Reitner J, Michaelis W (1999) Highly isotopically depleted isoprenoids: molecular markers for ancient methane venting. Geochim Cosmochim Acta 63:3959–3966

Thiel V, Peckmann J, Richnow HH, Luth U, Reitner J, Michaelis W (2001a) Molecular signals for anaerobic methane oxidation in Black Sea seep carbonates and a microbial mat. Mar Chem 73:97–112

Thiel V, Peckmann J, Schmale O, Reitner J, Michaelis W (2001b) A new straight-chain hydrocarbon biomarker associated with anaerobic methane cycling. Org Geochem 32:1019–1023

Thiel V, Blumenberg M, Pape T, Seifert R, Michaelis W (2003) Unexpected occurrence of hopanoids at gas seeps in the Black Sea. Org Geochem 34:81–87

Thiel V, Heim C, Arp G, Hahmann U, Sjövall P, Lausmaa J (2007) Biomarkers at the microscopic range: ToF-SIMS molecular imaging of Archaea-derived lipids in a microbial mat. Geobiology 5:413–421

Tornabene TG, Langworthy TA, Holzer G, Oró J (1979) Squalenes, phytanes and other isoprenoids as major neutral lipids of methanogenic and thermoacidophilic “archaebacteria”. J Mol Evol 13:73–83

Treude T, Boetius A, Knittel K, Wallmann K, Joergensen BB (2003) Anaerobic oxidation of methane above gas hydrates at Hydrate Ridge, NE Pacific Ocean. Mar Ecol Prog Ser 264:1–14

Valentine DL (2002) Biogeochemistry and microbial ecology of methane oxidation in anoxic environments: a review. Antonie Leeuwenhoek 81:271–282

Van Hoek AHAM, Van Alen TA, Sprakel VSI, Leunissen JAM, Brigge T, Vogels GD, Hackstein JHP (2000) Multiple acquisition of methanogenic archaeal symbionts by anaerobic ciliates. Mol Biol Evol 17:251–258

Vereshchaka AL, Vinogradov GM, Lein AY, Dalton S, Dehairs F (2000) Carbon and nitrogen isotopic composition of the fauna from the Broken Spur hydrothermal vent field. Mar Biol 136:11–17

Wegener G, Niemann H, Elvert M, Hinrichs K-U, Boetius A (2008) Assimilation of methane and inorganic carbon by microbial communities mediating the anaerobic oxidation of methane. Environ Microbiol 10:2287–2298

Werne JP, Baas M, Sinninghe Damsté JS (2002) Molecular isotopic tracing of carbon flow and trophic relationships in a methane-supported benthic microbial community. Limnol Oceanogr 47:1694–1701

Whiticar MJ (1999) Carbon and hydrogen isotope systematics of bacterial formation and oxidation of methane. Chem Geol 161:291–314

Woese CR, Kandler O, Wheelis ML (1990) Towards a natural system of organisms: proposal for the domains Archaea, Bacteria, and Eucarya. Proc Natl Acad Sci USA 87:4576–4579

Chapter 5
Chemosymbiotic Bivalves

John D. Taylor and Emily A. Glover

Contents

5.1 Introduction

Although the remarkable chemosymbiosis between bivalve molluscs and sulphide- and methane-oxidizing bacteria was originally recognized in a few species of spectacular, large mussels (*Bathymodiolus*) and clams (*Calyptogena*) from deep water hydrothermal vents it is now realized that there are hundreds of species of chemosymbiotic bivalves living from the intertidal zone to hadal depths. Far from being restricted to vent and seep habitats they are found in a wide range of reducing environments from sands and muds, mangrove sediments, seagrass beds, areas of sunken vegetation including wood pulp, offshore sewage sites and whale falls. Chemosymbiotic bivalves have been reported from six distinct families; Solemyidae, Nucinellidae, Mytilidae, Lucinidae, Thyasiridae and Vesicomyidae. The symbiosis has been identified in all species of Lucinidae, Solemyidae and Vesiocomyidae studied so far and is likely obligate, while in Thyasiridae many species possess symbionts but others lack them. In Mytilidae, chemosymbiosis is confined to members of the subfamily Bathymodiolinae, while other mytilids are asymbiotic. Little is known of the Nucinellidae with symbiosis inferred from internal morphology of two species (Reid 1998). Additionally, mention should also be made of the families Xylophagainae

J.D. Taylor (✉) and E.A. Glover
Department of Zoology, The Natural History Museum, London SW7 5BD, United Kingdom
e-mail: j.taylor@nhm.ac.uk

S. Kiel (ed.), *The Vent and Seep Biota*, Topics in Geobiology 33,
DOI 10.1007/978-90-481-9572-5_5,

and Teredinidae of wood-boring bivalves (Distel and Roberts 1997) that possess cellulolytic endosymbionts held in a similar position within the gills.

Phylogenetic trees of Bivalvia (Figs. 5.1 and 5.2) demonstrate that chemosymbiosis occurs in widely separated groups that differ considerably in basic anatomy and life habits. Amongst these, the Solemyidae and Nucinellidae are probably the most closely related. The Mytilidae are the only epifaunal byssate group of living bivalves to have acquired chemosymbiosis, although the habit has been claimed for some extinct fossils groups such as the Inoceramidae (Kauffman et al. 2007). By far the most diverse group of bivalves is the Heterodonta (Fig. 5.2) that includes most burrowing species and chemosymbiosis occurs in three distinct families, Thyasiridae, Lucinidae and Vesicomyidae. Previous morphology-based classifications placed Thyasiridae and Lucinidae together in the superfamily Lucinoidea but recent molecular analyses (Williams et al. 2004; Taylor et al. 2007a) have shown that the two families are not closely related and the morphological similarities a result of convergence. Molecular analyses indicate that the Vesicomyidae are related to a group of bivalves including *Glossus*, *Arctica* and shallow-water clams of the family Veneridae (Giribet and Distel 2004; Mikkelsen et al. 2006; Taylor et al. 2007b).

Burgeoning interest in chemosymbiosis and deep-sea exploration has stimulated taxonomic research and many new species and genera of host bivalves from deep and shallow waters have been described over the last 20 years. For example, 35 species of

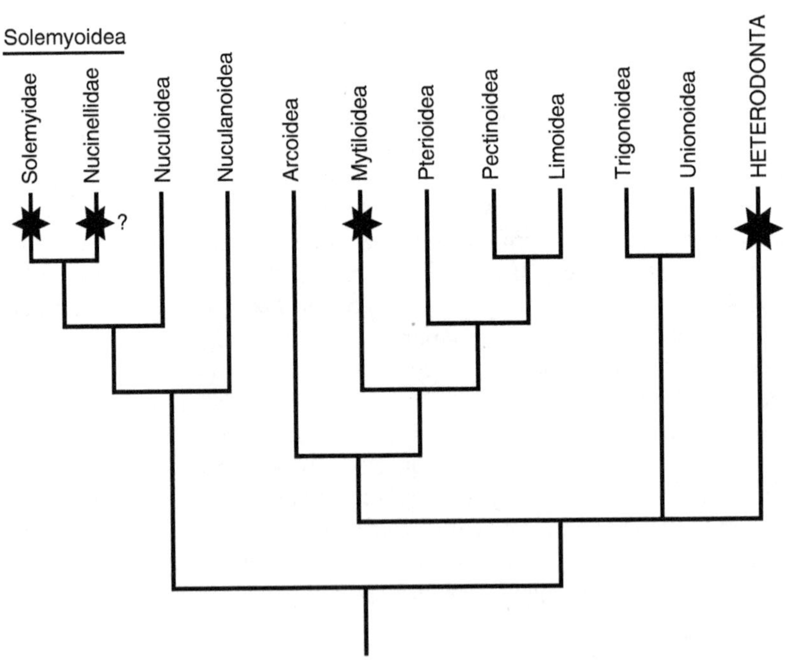

Fig. 5.1 Phylogenetic tree of bivalve superfamilies but with Heterodonta shown in more detail in Fig. 5.2. Families with chemosymbionts with star (Tree derived from Bieler and Mikkelsen 2006, Giribet 2008)

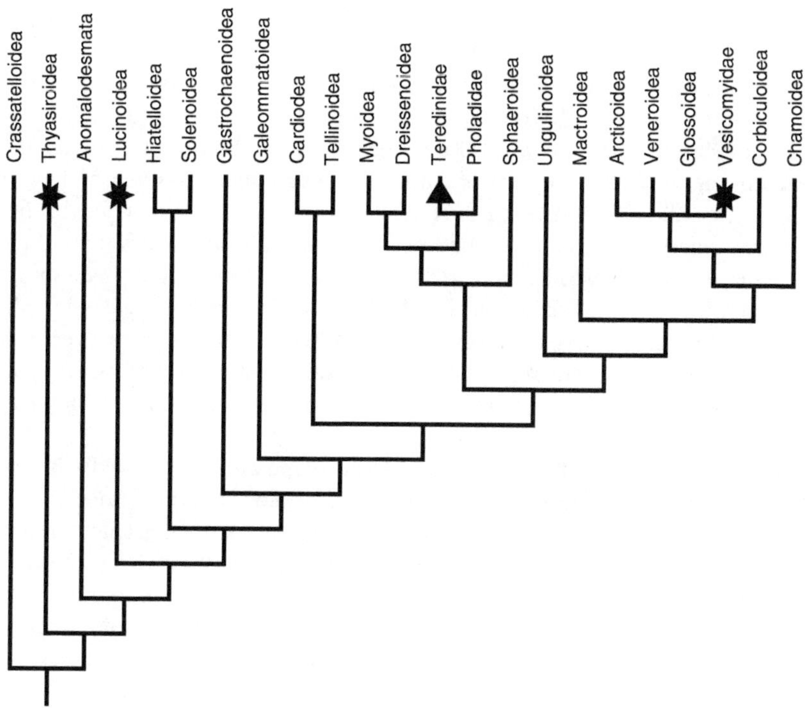

Fig. 5.2 Phylogenetic tree of Heterodonta families. Families with chemosymbionts with star, cellulolytic symbionts with triangle (Tree derived from Taylor et al. 2007a)

bathymodioline bivalves have been described since 1985 (Olu-Le Roy et al. 2007). Amongst Lucinidae, 32 new species and nine new genera from bathyal depths were recently described by von Cosel and Bouchet (2008), while previously, Glover and Taylor (2007) described 18 new species and nine genera from around New Caledonia and von Cosel (2006) introduced eight new species and six new genera from West Africa. Similarly, many new species of Thyasiridae have been described over the last 5 years (Holmes et al. 2005; Oliver and Levin 2006; Oliver and Holmes 2007; Rodrigues et al. 2008). Despite this activity, many species still remain undescribed and new discoveries continue apace.

Chemosymbiosis has been most intensively studied in Vesicomyidae, Mytilidae, Solemyidae and Lucinidae and some species, such as *Solemya velum*, '*Calpytogena*' *magnifica*, *Bathymodiolus azoricus*, and *Codakia orbicularis* have become foci of detailed investigations.

5.2 Solemyidae

Solemyidae are distinctive and unusual bivalves geographically widespread from the tropics to temperate latitudes and with a depth range from the intertidal to 5,350 m (Fujiwara 2003). They inhabit muds, silts and sands (Stanley 1970; Conway et al. 1992;

Krueger et al. 1996a) but are often associated with organically enriched habitats such as seagrass beds (Stanley 1970; Reid and Brand 1987), wood pulp accumulations (Reid 1980), human hair in sewage dumps (Coan et al. 2000), beneath whale carcasses (Fujiwara et al. 2009) and deep-sea cold seeps and vents (Métivier and von Cosel 1993; Neulinger et al. 2006; Kamenev 2009). *Acharax japonica* has been reported living in settling tanks of a seawater intake system at a marine laboratory (Yamanaka et al. 2008). Around 30 described species are known in the family, although it is certain that there are undescribed species from deep water cold seeps (see Neulinger et al. 2006; Kamenev 2009) as well as unregarded small species in tropical shallow waters (Taylor et al. 2008). Most shallow water solemyids are small, with lengths of less than 30 mm, but *Acharax* species from deeper water (Fig. 5.3, Plate 9) reach much larger sizes for example *Acharax bartschii* grows to over 200 mm, while a fossil *Acharax* from the Miocene measured a massive 300 mm (Kanie and Kuramochi 1995).

Solemyids have an elongate, near cylindrical shape and are covered with glossy, thick, brown periostracum extended as a broad, pleated, flexible fringe (Fig. 5.3). They have thin, lightly calcified, organic-rich, flexible, toothless shells, with extensive ventral mantle fusion forming a large anterior aperture, through which protrudes the large, papillae fringed foot. Large ctenidia occupy most of posterior mantle cavity and the gut is either much reduced, or even absent in some species. Solemyids live in "U" or "Y" shaped burrows (Stanley 1970; Reid 1998; Stewart and Cavanaugh 2006) and are capable of swimming by jet propulsion through flexure of the shell margins with a pulsing expulsion of water through the posterior aperture (Reid 1980).

All species investigated in any detail possess the endosymbionts (Stewart and Cavanaugh 2006); these are usually rod shaped (Fig. 5.3e–g) and have been identified as belonging to a group of Gammaproteobacteria that cluster with phylotypes of symbionts recognised from lucinid and thyasird bivalves. The symbionts of *Acharax* species, however, form a distinct clade (Imhoff et al. 2003). The bacteria are housed in the large posterior ctenidia where they pack bacteriocytes occupying the abfrontal zone of the individual ctenidial filaments (Fig. 5.3d–g). The biology of the chemosymbiosis of solemyids has been widely studied particularly, the North American *Solemya velum* and *Solemya reidi* (Reid 1980; Powell and Somero 1985; Fisher and Childress 1986; Gustafson and Reid 1986; Conway et al. 1989; Stewart and Cavanaugh 2006). The systematics of North Pacific species were reviewed by Kamenev (2009) who concluded that *S. pervernicosa* is the correct name for *S. reidi*. While a number of solemyids characteristically have small or reduced alimentary systems (Stempell 1899; Owen 1961; Kuznetsov et al. 1990; Reid 1998; Taylor et al. 2008), the gut is absent in *Solemya pervernicosa, S. tagiri, Acharax alinae, Petrasma atacama*, *Acharax eremita*, and *Petrasma borealis* (Reid 1980; Kuznetsov and Schileyko 1984; Conway et al. 1992; Métivier and von Cosel 1993; Kamenev 2009). Despite the dependence on symbionts a weak suspension feeding capability is retained in *Solemya velum* (Krueger et al. 1992).

Little is known of the biology of the generally deeper water *Acharax* species. They have been reported from sites in Atlantic, Indian and Pacific Oceans, including

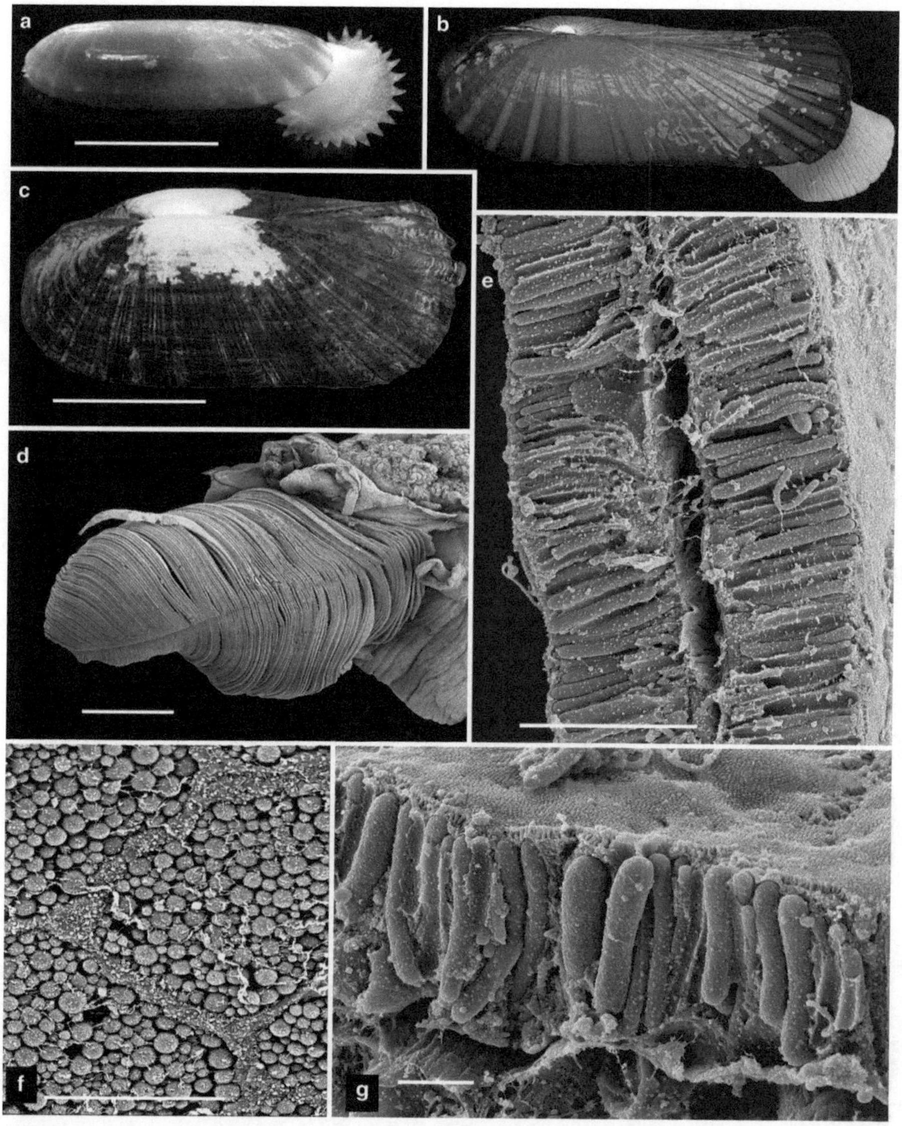

Fig. 5.3 Solemyidae and symbionts. (**a**) *Solemyarina velesiana* Iredale, 1931. Moreton Bay, Queensland, Australia Scale bar = 10 mm. (Image courtesy of Gonzalo Giribet) (**b**) *Acharax* sp. off Costa Rica 2,274 m Scale bar = 50 mm (Image by J Turner, National Museum of Wales) (**c**) *Acharax* sp. off Costa Rica 1013 m Scale bar = 50 mm. (Image by J Turner, National Museum of Wales). (**d**) Right ctenidiium of *Solemyarina velesiana.* Scale bar = 1.0 mm. (**e**) Section through ctenidial filament with bacteriocytes and central blood space Scale bar = 20 μm. (**f**) Surface view of bacteriocytes with apical cell wall removed showing tips of symbionts. Scale bar = 10 μm. (**g**) Detail of symbionts oriented beneath microvilli-covered outer wall of bacteriocytes. Scale bar = 2 μm. Colour versions of some of these images can be found in Appendix I (Plate 9)

species from methane seeps off Oregon, Java, Pakistan, Aleutian Islands, Costa Rica and Peru (Neulinger et al. 2006). *Acharax johnsoni* is recorded from hydrocarbon seeps of the north and eastern Pacific from the Sea of Okhotsk to Chile (Kamenev 2009), while *Acharax alinae* was described from 1,890 m at a cold seep on the Valufa Ridge, Lau Basin, SE of Fiji where it lived buried 20–30 cm deep (Métivier and von Cosel 1993). An *Acharax* community was recorded around a 770 m deep gas hydrate deposit on the Hydrate Ridge of the Cascadia convergent margin, NE Pacific (Sahling et al. 2002). Here the gas hydrate mounds are capped by *Beggiatoa* and surrounded by beds of *Calyptogena pacifica* and *C. kilmerae* with *Acharax* burrowing around the perimeter.

Geologically, Solemyidae are the oldest chemosymbiotic bivalves with a fossil record extending back into the Lower Palaeozoic. The earliest known is *Ovatoconcha* from the early Ordovician (ca 475 mya) (Cope 1996), while other solemyids are known from the middle Ordovician and younger rocks (Pojeta 1988; Liljedahl 1984). The general similarity in morphology of these fossils to living species suggests that they possessed similar life habits including chemosymbiosis. In view of their great antiquity, the evolutionary relationships of solemyids to other bivalves are of considerable interest. Most classifications of bivalves based on morphological analyses, place the Solemyidae along with the Nucinellidae (=Manzanellidae) in the superfamily Solemyoidea, and together with the Nuculoidea and Nuculanoidea in the subclass Protobranchia, occupying a basal position relative to all other bivalves (Zardus 2002). Alternatively, some classifications suggest placement of Solemyoidea in a separate subclass. More recently, molecular analyses indicate that Solemyidae and Nuculidae group as sister families but are distinct from Nuculanoidea and the rest of the bivalves (Giribet and Distel 2004; Taylor et al. 2008). This combined Solemyidae-Nuculidae grouping was recently named Opponobranchia by Giribet (2008).

5.3 Nucinellidae

The Nucinellidae are small, mainly deep-water bivalves with around 20 living species classified into two genera *Nucinella* and *Huxleyia*. Although most living species are small (less than 5 mm), La Perna (2005) described the 25 mm *Nucinella boucheti* from 1,600 m in the Philippines. They have oval shells with a few hinge teeth, a single adductor muscle and a large divided foot fringed with papillae (Fig. 5.4, Plate 9). From anatomical evidence, such as the similar foot structure, the Nucinellidae are usually classified within the Solemyoidea but this has yet to be confirmed by molecular analysis.

Knowledge of nucinellids is very limited, with anatomical studies confined to three species. The deep-water Atlantic species, *Nucinella serrei,* has fairly large ctenidia, well-developed labial palps and a relatively large stomach, oesophagus and hind-gut (Allen and Sanders 1969). From the northern Pacific (Kuznetsov and Schileyko 1984), *Nucinella viridula* has large ctenidia, very small labial palps and a small stomach but lacks an oesophagus and hind gut, while *Nucinella maxima*

Fig. 5.4 Nucinellidae. *Nucinella* sp. Panglao Island, Philippines, 0–3 m, soft bottom with seagrass Scale bar = 1.0 mm (Image courtesy of Pierre Lozouet, Museum National d'Histoire Naturelle, Paris. Panglao Marine Biodiversity Project 2004). A colour version of this image can be found in Appendix I (Plate 9)

possesses large ctenidia but the digestive system and labial palps are absent. Possible chemosymbiosis has been inferred from the reduced or absent gut and large ctenidia of *N. viridula* and *N. maxima* (Reid 1998) but not yet confirmed by electron microscopy of the ctenidia or molecular identification of symbionts. From off Japan, *Nucinella viridis* has been recorded from 2,630 to 3,581 m deep cold seep sites (Sasaki et al. 2005) but the possible chemosymbiosis has not been investigated.

Nucinella-like bivalves appeared in the early Jurassic with a patchy but continuous record to the Recent (Vokes 1956; Pojeta 1988). A large species has been recorded from a Cretaceous (Campanian) bathyal, putative cold-seep site in Japan (Amano et al. 2007).

5.4 Mytilidae

Chemosymbiotic mussels are the subject of a separate chapter in this book (Chapter 6) and are only briefly mentioned here.

Among the most spectacular early discoveries at hydrothermal vents were the populations of large mussels resembling shallow water *Modiolus* but now described under the new name, *Bathymodiolus*. Since then many *Bathymodiolus* species (Fig. 5.5),

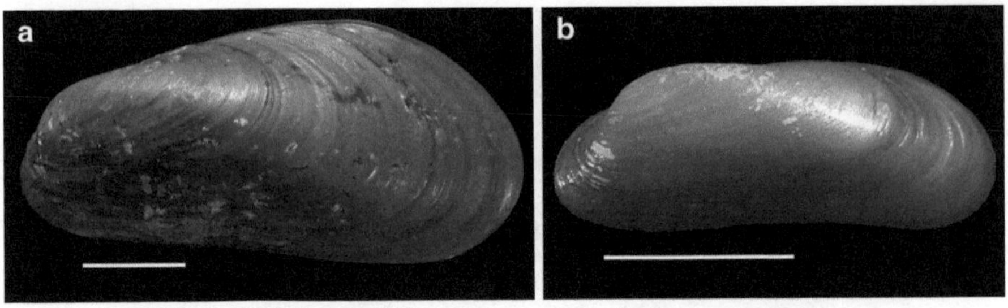

Fig. 5.5 Mytilidae. (**a**) *Bathymodiolus azoricus* von Cosel and Comtet, 1998. Menez – Gwen hydrothermal vent, Azores (BMNH). Scale bar = 20 mm. (**b**) *Idas simpsoni* (Marshall, 1900). From sunken whale bone, Shetland Islands, Scotland (BMNH). Scale bar = 10 mm

classified into Bathymodiolinae a subfamily of Mytilidae have been recognized at vents and seeps around the world as well as on sunken whale bones. Amongst *Bathymodiolus* species several distinct clades have been identified by molecular and morphological analyses, with some classified into the separate genera *Tamu* and *Gigantidas* (Jones et al. 2006; von Cosel and Janssen 2008). Some bathymodiolines are very large, for example *Bathymodiolus boomerang* reaches shell lengths of 370 mm and *Gigantidas gladius* to 316 mm (von Cosel and Marshall 2003). Symbionts have been reported for all studied species of bathymodiolines. These are housed in bacteriocytes within the thickened abfrontal zones of the filibranch ctenidia (Fiala-Médioni et al. 1986; Distel et al. 1995). In addition to sulphide-oxidizing symbionts, some species (e.g. *B. azoricus* and *B. heckarae*) have dual symbioses with both sulphide and methane-oxidising symbionts, both occupying the same bacteriocytes (Fisher et al. 1993; Duperron et al. 2005).

A group of smaller mussels, variously called *Idas, Idasola, Adipicola* or *Myrina* occur in deeper water associated with wood falls and other sunken vegetation, decaying whale carcases and bones, oil impregnated muds from oil-drilling platforms, cold seeps and hydrothermal vents (Smith and Baco 2003; Duperron et al. 2005; Southward 2008). Bacterial symbionts have been recognized in several *Idas* and *Adipicola* species; in some hosts these occur extracellularly amongst the microvillli of bacteriocytes while in other host species they are located intracellularly within the bacteriocytes (Duperron et al. 2005; Southward 2008). It should be stressed that not all *Idas* species are chemosymbiotic, for, surprisingly, the deep-water, wood-associated, *Idas argenteus* (the type species of the genus) has been discovered to be a predator on small wood-boring bivalves, *Xyloteredo* (Ockelmann and Dinesen in press).

5.5 Thyasiridae

Amongst chemosymbiotic bivalves Thyasiridae are second only to the Lucinidae in species diversity, morphological disparity and habitat range (Fig. 5.6a–e, Plate 9). With around 90 described living species (but many more yet unnamed) they are

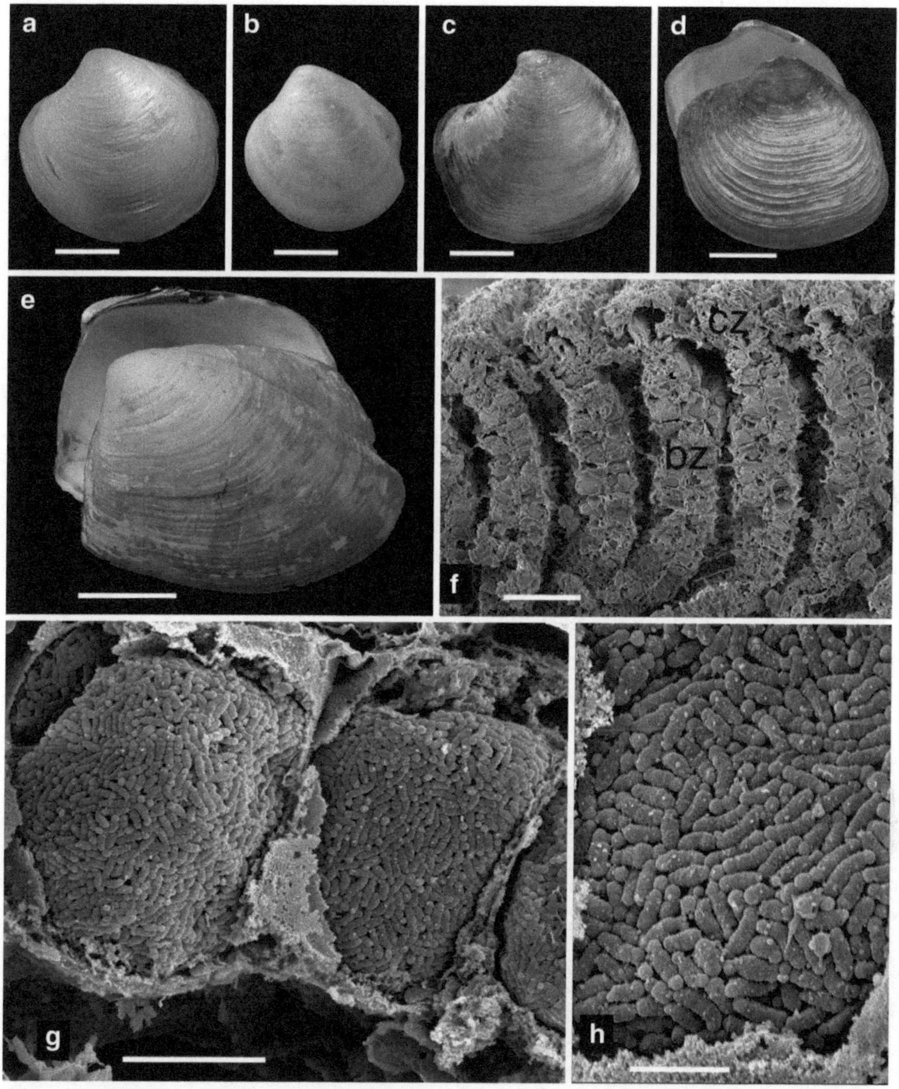

Fig. 5.6 Thyasiridae. (**a**) *Thyasira sarsi* (Philippi, 1845). Bukken, Norway (BMNH). Scale bar = 5 mm. (**b**) *Thyasiria flexuosa* (Montagu, 1803), Loch Fyne, Scotland, (BMNH). Scale bar = 5 mm. (**c**) *Axinus cascadiensis* Oliver and Holmes 2007. Baby Bare Seamount, NE Pacific 2592 m. Holotype (FMNHC 280989), Scale bar = 10 mm (Image Courtesy of Graham Oliver). (**d**) '*Maorithyas*' *hadalis* Okutani, Fujikura and Kojima, 1999, paratype NSMT, Japan Trench, 7,326 m. Scale bar = 10 mm. (**e**) *Conchocele bisecta* (Conrad, 1849) Off Vancouver Island, Canada, 366 m (BMNH) Scale bar = 20 mm. A colour version of this image can be found in Appendix I (Plate 9). (**f**) *Thyasira* sp. West Pacific vent species. Section with ctenidial filaments showing thickened abfrontal bacteriocyte zone. Scale bar = 50 μm. cz ciliated zone; bz bacteriocyte zone. (**g**) Two adjacent bacteriocytes packed with symbionts. Scale bar = 5 μm. (**h**) Detail of symbionts. Scale bar = 2 μm

globally distributed from polar to equatorial latitudes and live at all depths from the intertidal to hadal zones, although most species occur offshore (Payne and Allen 1991). Thyasirids live in a wide range of habitats including muds, silts and sands but are often associated with sediments enriched in hydrocarbons (Oliver and Killeen 2002; Oliver and Holmes 2006a), oxygen minimum zones (Oliver and Levin 2006), cold seeps (Clarke 1989; Fujikura et al. 1999; Kamenev et al. 2001; Holmes et al. 2005), mud volcanoes (Gebruk et al. 2003; Olu-Le Roy et al. 2004; Rodrigues et al. 2008) and hydrothermal vents (Oliver and Holmes 2006a, 2007). Although thyasirids range in size from less than 1 mm to around 110 mm (*Conchocele*), most are small, usually with lengths less than 10 mm. Externally, the smooth, relatively featureless shells of Thyasiridae appear rather uniform but their internal anatomy shows considerable variation. For example, some species have two ctenidial demibranchs while others have just a single inner demibranch and Dufour (2005) has recognized three structurally different types of ctenidial filament based on the amount of abfrontal thickening. Moreover, not all thyasirids are chemosymbiotic, for over half of the 26 species examined by Dufour (2005) lacked symbionts and there is varying nutritional dependence on the symbiosis amongst the symbiotic species (Dando and Spiro 1993). Some of these differences have been recognized in classification systems (Payne and Allen 1991) and the Thyasiridae are sometimes into two subfamilies, the Thyasirinae and Axinopsidinae (Coan et al. 2000), the latter containing some genera with single gill demibranchs but this division is not wholly supported by molecular analysis (Taylor et al. 2007a). Living thyasirids have been divided into around 12 genera largely based on rather poorly defined shell characters and this has led to confusion with species rather arbitrarily assigned to them or simply called *Thyasira* (see Oliver and Killeen 2002).

On current evidence, symbionts mainly occur in those thyasirid species with two ctenidial demibranchs, an exception is *Axinulus croulinensis* with only a single demibranch in which Dufour (2005) records a few symbionts amongst the microvilli. For those Thyasiridae with symbionts, the ctenidial filaments are thickened abfrontally (Fig. 5.6f) and the symbionts associated with bacteriocytes. Between species, however, there are marked differences in the location of the symbionts possibly reflecting progressive stages in the integration of the symbiosis. Extracellular bacteria can occur amongst the microvilli of bacteriocytes as in *Thyasira equalis* (Dufour 2005) or *T. falklandica* (Passos et al. 2007). In other species, the symbionts are contained within extracellular apical vesicles covered by a cuticle of modified microvilli as in *T. flexuosa* (Southward 1986; Herry and Le Pennec 1987; Dufour 2005) and *T. sarsi* (Southward 1986). In the most integrated state, the bacteria pack probable intracellular vesicles within the apical areas of the cells (Fig. 5.6g, h) as in *T. methanophila* (Oliver and Sellanes 2005), *Axinus cascadiensis* (Oliver and Holmes 2007) or '*Maorithyas*' *hadalis* (Fujiwara et al. 2001).

The thyasirid symbionts (Fig. 5.6g, h) are usually small (ca 0.6–1.0 μm), short, ovoid or rod-shaped forms (Southward 1986; Herry and Le Pennec 1987; Dufour 2005; Passos et al. 2007) and have been molecularly characterized from three host species – the shallow water *Thyasira flexuosa,* the cold seep-associated *Conchocele* and the hadal '*Maorithyas*' *hadalis.* They classify within the Gammaproteobacteria

(Imhoff et al. 2003; Dubilier et al. 2008). The symbiont from *T. flexuosa* groups with those from Lucinidae and Solemyidae hosts, while the bacteria from the deeper water *Conchocele* and '*Maorithyas*' *hadalis* fall within a group of symbionts from bathymodioline and vesicomyid bivalve hosts.

All thyasirids burrow in sediment, with some using the cylindrical vermiform foot to construct a network of radiating tunnels, extending up to 30 times the length of the shell into deeper anoxic sediment layers (Oliver and Killeen 2002; Dufour and Felbeck 2003). The bivalves are effectively mining sulphide from interstitial pore water. The mechanism by which sulphide is transferred to the symbionts in the ctenidia is unknown. This might be via diffusion into the extended foot as suggested by Dufour and Felbeck (2003), by hydraulic pumping of water into the mantle cavity by movements of the foot, or, simply by diffusion through tunnel water to the ctenidia. By contrast, some small and probably asymbiotic thyasirids such as *Adontorhina keegani* frequently have hydroids attached to the posterior shell indicating very shallow burial in the sediment (Barry and McCormack 2007).

Although Thyasiridae are widespread and often abundant in offshore reducing environments (Oliver and Killeen 2002), relatively few species have been recorded from vents and seeps. From the North Sea *Thyasira sarsi* has been recorded from methane seeps and also occurs in lubricant-enriched tailing mud around oil drilling sites (Dando et al. 1994; Oliver and Killeen 2002). *Thyasira oleophila* is recorded from hydrocarbon seeps on the Louisiana Slope at depths of around 500 m (Clarke 1989; Cordes et al. 2009) where it occurs in association with *Bathymodious childressi, Calyptogena ponderosa, Vesicomya cordata* and the lucinids *Lucinoma atlantis* and *Jorgenia louisiana*. From a methane seep at 780 m off Conception, Chile, *Thyasira methanophila* is found with *Calyptogena gallardoi* (Oliver and Sellanes 2005). The largest living thyasirid, *Conchocele bisecta,* is known from hydrocarbon seeps at 130–800 m in the Sea of Okhotsk and elsewhere in the North Pacific (Kamenev et al. 2001; Sasaki et al. 2005) and *Conchocele novaeguinensis* was described from a likely cold seep off New Guinea (Okutani 2002). In the Gulf of Cadiz, *Thyasira vulcolutre* is restricted to active seeps of mud volcanoes at depths of 1,300–2,200 m (Rodrigues et al. 2008). An unusual habitat of sisal, beans and sunflower seeds in the hold of a sunken ship is recorded for *Spinaxinus sentosus* recovered together with *Idas* spp. from 1,160 m off Vigo, Spain (Oliver and Holmes 2006a). Additionally, some thyasirids have been recorded associated with whale falls but not described in detail (Smith and Baco 2003, A. Glover personal communication).

From hydrothermal vents, *Thyasira southwardae* occurs at the Atlantic Logatchev hydrothermal site at 3,038 m in association with *Calyptogena* and *Bathymodiolus* (Oliver and Holmes 2006a). In the northeastern Pacific, *Axinus cascadensis* (Fig 5.6c) is recorded from 2,591 m at warm hydrothermal springs at the Baby Bare Seamount, Cascadia Basin (Oliver and Holmes 2007). Other thyasirids have been collected from Pacific vents but not yet formally described (Taylor et al. 2007a).

'*Maorithyas*' *hadalis* recovered from hadal depths of 7,326–7,434 m is amongst the deepest recorded chemosymbiotic bivalves (Fujikura et al. 1999; Fujiwara et al. 2001). The classification of this species (Fig. 5.6d) is uncertain but it does not belong to *Maorithyas* as originally described (see Oliver and Selanes 2005) or *Axinulus* as classified by Matsukuma (2000). Bacteria are held in large aggregations within apical vacuoles of bacteriocytes and resemble apical aggregations of symbionts described from other thyasirids (Southward 1986; Dufour 2005).

Thyasiridae have a fossil record extending back to the early Cretaceous; the earliest confirmed records being *Thyasira tanabei* from Albian cold-seep deposits in Japan (Kiel et al. 2008) and the *Axinus*-like '*Lucina*' *sculpta* from the Albian of England (Taylor et al. 2007a). However, the phylogenetic position of Thyasiridae in molecular analyses would suggest a longer fossil record (see discussion in Taylor et al. 2007a).

5.6 Lucinidae

Lucinidae are by far the most diverse group of chemosymbiotic bivalves with around 400 living species distributed from latitudes 60°N to 55°S at water depths from the intertidal zone down to around 2,500 m. Although lucinids are often considered to be a predominantly shallow-water family, new research is revealing a surprising diversity in tropical bathyal habitats (von Cosel and Bouchet 2008). Lucinids are the most easily and frequently encountered chemosymbiotic bivalves occurring in a broad range of habitats from mud and sands, mangrove sediments, seagrass beds, sites of high organic input such as sewage disposal sites and locations where sunken vegetation accumulates. Additionally, lucinids are associated with oxygen minimum zones, hydrocarbon seeps, mud volcanoes and vents. Shallow water lucinids can reach population densities of several thousand per square metre (Meyer et al. 2008). Living lucinids vary in size from a few millimetres to lengths of 150 mm but an Eocene fossil species reached over 300 mm (Taylor and Glover 2009a). Lucinidae have a long fossil record extending back into the Lower Palaeozoic with increasing diversity and association with fossil seeps sites and dysaerobic habitats through the Mesozoic (Taylor and Glover 2000; 2006). By the early Cenozoic (Eocene), lucinid diversity levels were comparable with those of the present day.

The symbiosis between lucinid bivalves and sulphide-oxidising bacteria is now well-known through numerous investigations (e.g. Giere 1985; Southward 1986; Reid and Brand 1986; Distel and Felbeck 1987; Frenkiel and Mouëza 1995; Frenkiel et al. 1996; Gros et al. 1996, 2000; Ball et al. 2009). All studied species, representing a broad spectrum of taxa and sampled from a wide range of habitats, possess symbionts and the chemosymbiosis is presumed obligate for the family. Nevertheless, lucinids are functionally capable of particulate feeding (Duplessis et al. 2004). Experiments by Gros et al. (1998a, 2003) have shown how the symbionts are acquired by environmental transmission from the sediment.

The ctenidia of lucinids are typically large and thick, comprising single inner demibranchs. Individual ctenidal filaments consist of a ciliated frontal zone, and a thick abrontal zone consisting of sheets of bacteriocytes separated by a central blood space. Narrow intercalary cells with spiky apices are located between the bacteriocytes (Fig. 5.7c). In many lucinids sections of adjacent ctenidial filaments are often fused to form cylindrical channels that are lined with bacteriocytes (Figs. 5.7a, b). Bacteriocytes are packed with intracellular symbionts usually housed in single vacuoles (Fig, 5.7e, f). Compared with thyasirids and vesicomyids the symbionts of lucinids are relatively large, up to 20 μm length in *Anodontia* species (Fig. 5.7d). Molecular analyses of lucinid symbionts places them in the Gammaproteobacteria, in a clade including symbionts from *Solemya* and vestimentiferan tube worms (Cavanaugh et al. 2006; Dubilier et al. 2008).

All lucinids are burrowers, living at the interface of oxic and suboxic zones in the sediment, sometimes up to 50 cm deep, and usually vertically aligned with the hinge and umbones uppermost. In many species, the foot is worm-like, capable of considerable extension and is responsible for the formation of ventral tunnels to acquire porewater sulphide (x-rays of burrowing lucinids in Stanley 1970) as well as constructing an anterior mucus-lined tunnel to the sediment surface in order to provide an oxygenating current. The bizarre looking lucinids, *Rasta thiophila* and *R. lamyi*, have several long ventral tubes constructed from fused periostracum that probably function as permanent conduits for transport of interstitial water to the mantle cavity (Taylor and Glover 1997; Taylor et al. 2005).

To date, only a single species of Lucinidae has been recorded from a hydrothermal vent. This is *Bathyaustriella thionipta* (Fig. 5.8a, Plate 10) described from depths of 480–504 m on the active Macauley Cone situated within the Macauley Caldera on the Kermadec Ridge, north of New Zealand (Glover et al. 2004). Here it lives alongside populations of the bathymodioline mussel *Gigantidas gladius*. The lucinid is also recorded from shallower depths of 144–175 m on the nearby Giggenbach Cone. *Bathyaustriella thionipta,* usually about 50 mm long, resembles the shallow water, mangrove-associated *Austriella corrugata* in general shell morphology and molecular analysis places it in a clade of shallow water Lucinidae close to *Austriella* and *Indoaustriella* (Glover et al. 2004, 2008).

At shallower hydrocarbon seeps on the Louisiana Slope, lucinids as dead shells, have been recorded in sufficient abundance to define a lucinid biofacies (Callender and Powell 1997, 2000). The latter authors record *Lucinoma atlantis* (Fig. 5.8d, Plate 10) and *Lucinoma* sp. occurring together with other chemosymbiotic bivalve species *Thyasira oleophila, Vesicomya cordata* and *Bathymodiolus.* The '*Lucinoma* sp.' is not closely related to *Lucinoma* and has recently been described as a new genus and species, *Jorgenia louisiana* (Fig. 5.8b, Plate 10), along with two other *Jorgenia* species from other likely seep sites in other parts of the Gulf of Mexico and Caribbean (Taylor and Glover 2009a). Nevertheless, *Lucinoma* species are the lucinids most widely reported from deeper water habitats and several have been reported in association with seeps and mud volcanoes (summarized in Fig. 5.9). *Lucinoma atlantis* occurs at seeps of the Louisiana slope (Cordes et al. 2009); *Lucinoma anemiophila* at a 780 m methane seep off Chile (Holmes et al. 2005); *Lucinoma myriamae* occurs at seep sites from 360–425 m off West Africa (von Cosel 2006); *Lucinoma taiwanensis* was described from an area of known hydrothermal activity off N Taiwan 205–650 m (von Cosel and Bouchet 2008), and *Lucinoma yoshidai* is reported at Japanese seeps

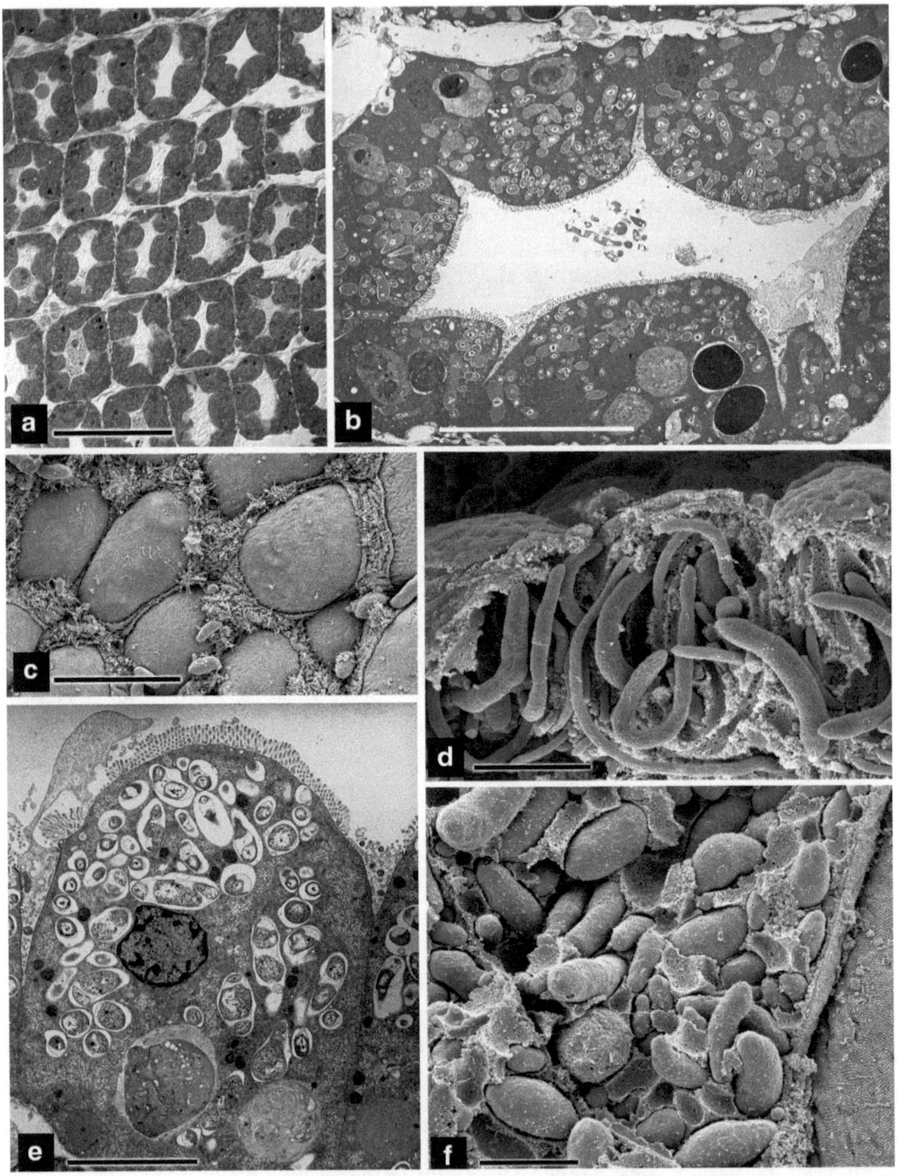

Fig. 5.7 Lucinidae – ctenidia and symbionts. (**a**) Semi-thin section through ctenidium of *Anodontia ovum* (Reeve 1850) showing cylindrical bacteriocyte lined channels. Scale bar = 100 μm. (**b**) TEM section of *A. ovum* through single bacteriocyte channel. Scale bar = 20 μm. (**c**) *Pseudolucinisca lacteola* (Tate 1897), surface SEM view of domed apices of bacteriocytes separated by spiky intercalary cells. Scale bar = 20 μm. (**d**) SEM section through bacteriocyte of *Anodontia philippiana* (Reeve 1850) with elongate symbionts. Scale bar = 5 μm. (**e**) TEM section of single bacteriocyte of *Anodontia ovum* with on left side and intercalary cell with overlapping apical extension. Scale bar = 5 μm. (**f**) SEM section through apex of bacteriocyte of *Pseudolucinisca lacteola* with short rod symbionts. Scale bar = 5 μm

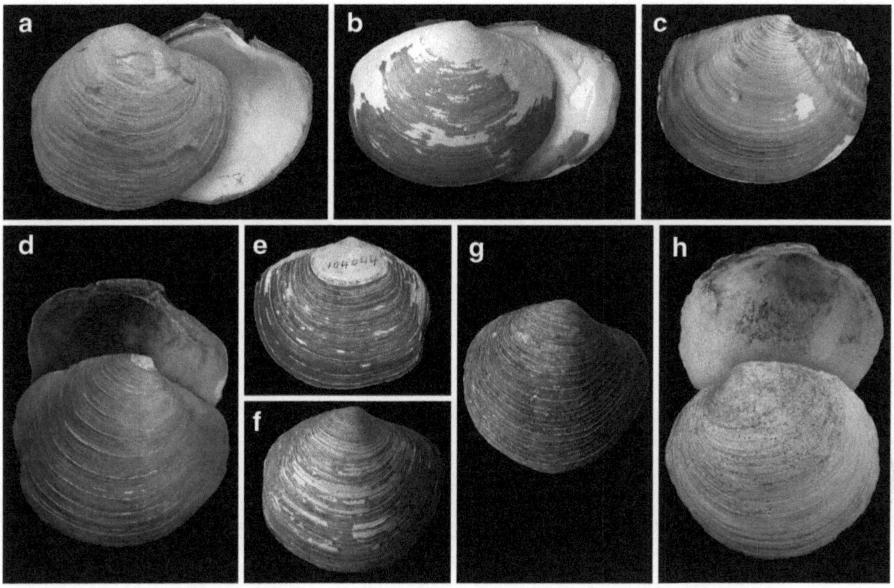

Fig. 5.8 Some Lucinidae from vent and seeps. (**a**) *Bathyaustriella thionipta* Glover, Taylor and Rowden 2004, Kermadec Ridge, New Zealand (holotype NIWA3277 H-838). (**b**) *Jorgenia louisiana* Taylor and Glover 2009a, Louisiana Slope, USA (holotype USNM 1116114). (**c**) *Jorgenia luteophila* Taylor and Glover 2009a, Louisiana Slope, 610–850 m (paratype Emilio Garcia collection EFG 27398). (**d**) *Lucinoma atlantis* (McLean 1936) Louisiana Slope, USA (Emilio Garcia collection 24039). (**e**) *Lucinoma aequizonata* (Stearns 1890), Santa Barbara Channel 505 m (syntype USNM 104044). (**f**) *Lucinoma gagei*, Oliver and Holmes 2006a, off S. India, 786 m (BMNH). (**g**) *Lucinoma myriamae* von Cosel, 2005 off Angola 360 m (BMNH). (**h**) *Meganodontia acetabulum* Bouchet and von Cosel, 2004, off North Taiwan 370 m (paratype MNHN). A colour version of this figure can be found in Appendix I (Plate 10)

between 100 and 1,000 m (Okutani and Hashimoto 1997). From the eastern Mediterranean *Lucinoma kazani* is described from 1,700 m deep mud volcanoes (Salas and Woodside 2002, Olu-Le Roy et al. 2004). Additionally two species of *Lucinoma* are known from sediments in oxygen minimum zones, these are: *L. aequizonata* from 400 to 650 m off southern California (Cary et al. 1989) and *Lucinoma gagei* from southern Oman at 675–967 m (Oliver and Holmes 2006a). Some 38% of *Lucinoma* species have been described in the last 8 years suggesting that the diversity has not yet been fully sampled.

Other lucinids recorded as associated with hydrocarbon seeps are: *Graecina karinae* from 360–425 m from off Nigeria and northern Angola (von Cosel 2006); the large *Meganodontia acetabulum* (Fig. 5.8h, Plate 10) from an area of hydrothermal activity at 256–472 m off northeastern Taiwan (Bouchet and von Cosel 2004); *Mesolinga soliditesta* from a seep at 363 m on Kanesu-no-Se Bank, Japan (Hashimoto et al. 1995; Okutani and Hashimoto 1997) and *Myrtea amorpha* associated with eastern Mediterranean mud volcanoes at 1,700–2,000 m (Olu-LeRoy et al. 2004). It often appears that Lucinidae are more abundant at fossil than Recent

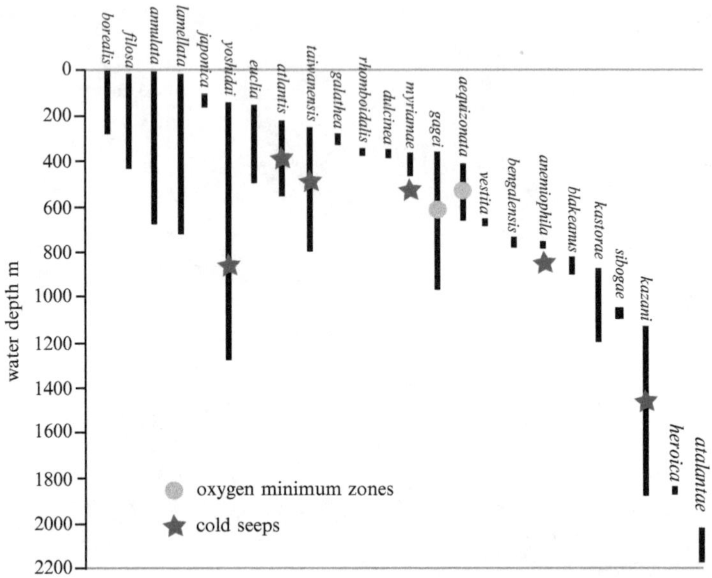

Fig. 5.9 Depth distribution of *Lucinoma* species. *Stars* indicate species associated with cold seeps and circles indicate species recorded from oxygen minimum zones (Compiled from various sources including von Cosel 2006; von Cosel and Bouchet 2008; Holmes et al. 2005; Oliver and Holmes 2006b; Britton 1970; Coan et al. 2006)

seeps; for example, the massive accumulations of *Lucinoma* shells described from some Japanese Pliocene deposits (Majima et al. 2003). This may reflect a sampling deficiency, for lucinids are often deeply buried in the sediment and not frequently captured with gear from ROVs.

It seems likely that significant populations of deeper water lucinids are present only at locations with some organic enrichment, including sites of accumulation of sunken vegetation, for example around Indonesia and Philippines (see von Cosel and Bouchet 2008), oxygen minimum zones (Cary et al. 1989; Oliver and Holmes 2006b), hydrocarbon seeps and mud volcanoes.

5.7 Vesicomyidae

Prior to the 1980s Vesicomyidae were considered one of the most obscure and little studied groups of bivalves, known from a few specimens of mainly dead shells, recovered by deep-sea expeditions. This situation was transformed by the discoveries of spectacular populations of large living vesicomyids, such as '*Calyptogena*' *magnifica,* around hydrothermal vents but later found to be associated with hydrocarbon seeps and whale falls. The discoveries of these clams at vents and seeps, most of them previously unknown to science, has catalyzed taxonomic effort and it is now estimated that there are around 100 living species with others yet to be described (von Cosel and Olu 2009, Krylova et al. 2010). The small '*Vesicomya*' *sergeevi*, remarkably abundant at hadal depths (6,090–9,530 m) in the Kuril-Kamchatka Trench, is the deepest known chemosymbiotic bivalve (Krylova et al. (2000). Vesicomyids embrace a wide size range from less than 5 mm to the extremely large species from vents and seeps that reach shell lengths up to 280 mm.

Most vesicomyids live shallowly burrowed in sediment usually with the posterior half of the shell protruding (Krylova et al. 2010). '*Calyptogena*' *magnifica* (Fig. 5.10a, Plate 9) occupies fissures and crevices in basaltic pillow-lava flows of the East Pacific Rise (Kennish and Lutz 1992) nestled with anterior ends downwards into waters of slightly elevated temperatures. Vesicomyids often occur in spectacularly high densities around vents and seeps, for example on the Edison seamount off Papua New Guinea, Stecher et al. (2003) record densities of up to 240 clams/m^2 of bivalves up to 23 cm in length but averaging 16.5 cm. In addition to seep and vents, vesicomyids are regularly associated with whale falls, living both on bones and the sediment in close proximity to the carcase (Bennett et al. 1994; Smith and Baco 2003). Species associated with whale falls are also found at vents and seeps (Baco et al. 1999).

Previously, Vesicomyidae have been classified in the superfamily Glossoidea but more recent molecular phylogenies of bivalves place '*Calyptogena*' species close to the families Arcticidae, Veneridae and Glossidae and often as a sister clade to Veneridae (Giribet and Distel 2004; Mikkelsen et al. 2006; 2010 Taylor et al. 2007a). Some 16 generic names are in use for Recent and fossil species of Vesicomyidae, with the main characters reviewed by Krylova and Sahling (2006, Table 12, Krylova and Sahling 2010) but most seep and vent species have been loosely classified as '*Calyptogena*'. Ongoing taxonomic studies (Krylova and Sahling 2006; Amano and Kiel 2007) coupled with molecular analyses (Peek et al. 1997; Goffredi et al. 2003; Kojima et al. 2004) demonstrate that several distinct lineages are represented. The concept of *Calyptogena* is now restricted to ten living species mainly associated with hydrocarbon seeps (Krylova and Sahling 2006). One result of this taxonomic revision is that the best-known species '*Calyptogena*' *magnifica* is now excluded from *Calyptogena.* A notable anatomical difference is that it has both inner and outer ctenidial demibranchs compared with only inner demibranchs in true *Calyptogena.* A new genus, *Laubiericoncha*, was introduced for another group of species with two demibranchs, including taxa from seeps off Barbados and West Africa (von Cosel and Olu 2008) and two new genera *Wareniconcha* and *Elenaconcha* introduced for species off West Africa (von Cosel and Olu 2009). '*Calyptogena*' species with very elongate shells and single demibranchs have recently been included in a new genus, *Abyssogena* (Krylova et al. 2010). Many of the smaller vesicomyids from the eastern Atlantic were reviewed by von Cosel and Salas (2001) who divided them into several distinct genera, *Vesicomya, Waisiuconcha, Isorropodon* and *Callogonia. Isorropodon perplexum* is associated with mud volcanoes in the eastern Mediterranean and *I. bigoti* recorded close to an oil-drilling platform off the mouth of the Congo River (von Cosel and Salas 2001). From this last study it becomes clear that many bivalves referred to as '*Vesicomya*' should be classified

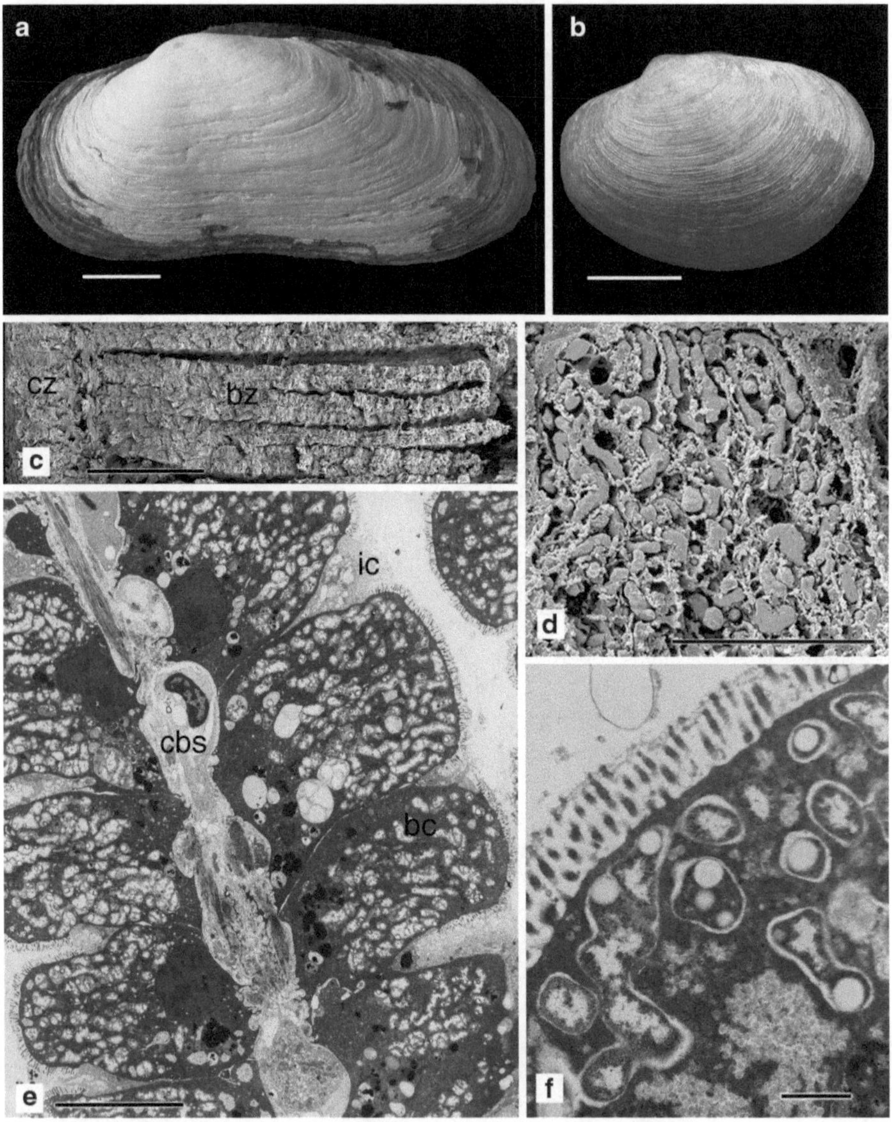

Fig. 5.10 Vesicomyidae. (**a**) '*Calyptogena*' *magnifica* (Boss and Turner 1980), Galapagos Ridge, 2,450 m (BMNH), Scale bar = 20 mm. (**b**) '*Calyptogena*' *lepta* (Dall 1896), Gulf of California, 1558 m (BMNH). Scale bar = 20 mm. (**c**) SEM section through ctenidial filaments of *Archivesica gigas* (Dall 1896) showing thickened abfrontal lateral zone containing bacteriocytes. Scale bar = 200 μm. cz, ciliated zone; bz bacteriocyte zone. (**d**) SEM of section through bacteriocyte of *Archivesica gigas* showing abundant bacteria. Scale bar = 10 μm. (**e**) TEM section through ctenidial filament of "*Calyptogena*" *kilmeri* (Bernard 1974). Monterey Bay, California. Scale bar = 20 μm. bc bacteriocyte; cbs central blood space; ic intercalary cell. Image courtesy of Shana Goffredi. (**f**) TEM section of apical portion of bacteriocyte of *Archivesica gigas* Monterey Bay, California with symbionts occupying individual vesicles. Empty circular patches in symbionts were occupied by elemental sulphur removed by solvents during processing. Scale bar = 1 μm (Image courtesy of Shana Goffredi)

into other genera. This is because the type species of the genus, *Vesicomya atlantica,* is a small (4 mm) rounded form from the north Atlantic, very different in shape and size from the larger species commonly classified under this name (e.g. Coan et al. 2000). Recently, Krylova and Sahling (2010) divided the Vesicomyidae into two subfamilies Vesicomyinae and Pliocardiinae; the latter including most of the larger vent and seep species.

Vesicomyids have large thickened ctenidia with single demibranchs only in *Calyptogena sensu stricto* (Krylova and Sahling 2006), *Isorropodon* species (von Cosel and Salas 2001) and other species loosely called *Calyptogena,* but with two demibranchs in other taxa including '*Calyptogena*' *magnifica.* Labial palps are reduced to plicated ridges around the mouth and the gut is simple with a small stomach (Morton 1986). Symbionts have been recognised in all species studied in any detail and the symbiosis is probably obligate for the family. Electron microscopy of ctenidial filaments (Fiala-Médioni and Le Pennec 1988; Goffredi et al. 2004) shows a frontal ciliated zone with an elongate abfrontal zone consisting of sheets of bacteriocytes separated by a central blood space (Fig. 5.10c, e). Bacteriocytes have domed, microvilli-covered apices (Fig. 5.10d, e) and are separated by narrow intercalary cells. Intracellular symbionts pack the bacteriocytes and in some species these are elongate and 'knobbly' in appearance (Figs. 5.10d–f). Structural differences in bacteriocytes and symbiont density were observed between seven vesicomyid species from Monterey Bay (Goffredi et al. 2004).

Molecular analyses of symbionts from various species of vesicomyids (originally classified as '*Calyptogena*', '*Ectenagena*' and '*Vesicomya*') from widely separated localities shows that they belong to one group of Gammaproteobacteria (Cavanaugh et al. 2006; Dubilier et al. 2008). These form a sister clade to sulphide-oxidising symbionts indentified from *Bathymodiolus* species and distinct from the symbiont clade associated with shallow water Lucinidae, Thyasiridae and Solemyidae.

It is widely reported that sulphide uptake in vesicomyids is through the foot that protrudes down into the sulphidic water in rock crevices or sediment (Arp et al. 1984; Fisher 1990; Goffredi and Barry 2002), while oxygen uptake is through the ctenidia. Sulphides are then transported via the blood from the foot to the ctenidia. Nevertheless, the apices of bacteriocytes are microvilli covered and the symbionts tend to be concentrated near the apices (Fig. 5.10e, f) and away from the central blood space. Moreover, some vesicomyid species have ctenidial filaments organized into cylindrical channels lined with bacteriocytes (Fiala-Médioni and Le Pennec 1988; Krylova et al. 2010), similar to those of Lucinidae (see Fig. 5.7a, b) suggesting that uptake of sulphides from water passing over the ctenidia is also highly likely.

5.8 Bivalves and Evolution of Chemosymbiosis

It is remarkable that chemosymbiosis has evolved in such phylogenetically divergent bivalve families ranging from the ancient Solemyidae, the bathymodioline mussels to the three heterodont families Thyasiridae, Lucinidae, and Vesicomyidae.

These bivalves have different anatomical organization particularly in relation to the structure of the ctenidia and very different life habits ranging from epifaunal byssate species to deep sediment burrowers. All evidence suggests that chemosymbiosis has evolved independently and at least five times in these different bivalve groups. Previous evolutionary scenarios of chemosymbiosis in Lucinoidea (Reid and Brand 1986) have been refuted by molecular analyses that demonstrate that the Thyasiridae are not closely related to the Lucinidae, with the conclusion that the symbiosis has been acquired independently in the two families (Williams et al. 2004; Taylor et al. 2007a).

Evidence from Solemyidae, Lucinidae and Vesicomyidae, where the symbionts have been recorded in all studied species, suggests that the symbiosis is obligate; most hosts have simplified or reduced alimentary systems although only in the Solemyidae and Mazanellidae are there species entirely lacking a gut. Structural modifications to accommodate the symbiosis are reflected in shell characters that can be recognized in the fossil record, for example, the elongate anterior adductor muscle and pallial blood vessel of lucinids (Taylor and Glover 2000; Amano et al. 2007). For the Solemyidae and Lucinidae both families have a long fossil record into the Lower Palaeozoic and morphological and palaeohabitat evidence suggests that the symbiosis is geologically ancient (Taylor and Glover 2000, 2006; Taylor et al. 2008). The course of evolution and habitat occupation patterns of these bivalve families is thus likely closely coupled to the symbiosis. The other three families are much more recent in origin. Thyasiridae are recognized from the Early Cretaceous with a contemporaneous association with cold seeps (Kiel et al. 2008; Taylor et al. 2007a). For the Vesicomyidae, confirmed fossils indicate an early Cenozoic origin and following adaptive radiation to seeps, vents and whale falls (Amano and Kiel 2007). Similarly the fossil record of *Bathmodiolus* and chemosymbiotic Mytilidae dates from the early Cenozoic (Kiel and Little 2006), although it would be difficult to recognize a chemosymbiotic mytilid from shell morphology alone.

Chemosymbionts occur in bivalves with three structurally different types of gill – the protobranch ctenidia with simple leaflets of Solemyidae, the filibranch ctenidia of Bathymodiolinae and the eulamellibranch form of Lucinidae, Thyasiridae and Vesicomyidae. In all these families the ctenidial filaments are modified in an essentially similar way to house the symbionts. The frontal parts of the filaments have ciliary tracts similar to non-symbiotic bivalves but the abfrontal portions are greatly thickened by proliferation of sheets of bacteriocytes either side of a central blood space. In Lucinidae, adjacent ctenidial filaments are often fused to form cylindrical channels lined with bacteriocytes (Fig. 5.6) and remarkably similar channels have been reported in some Vesicomyidae (Fiala-Médioni and Le Pennec 1988; Krylova et al. 2010) and maybe in *Conchocele* (Thyasiridae). The cylindrical channels allow a greater number of bacterioctyes to be accommodated compared to lamellar filaments and the resulting three-dimensional network of blood space may provide a more efficient circulatory system. The cylinders could also provide greater structural rigidity of the much thickened filaments whilst still facilitating water flow. The bacteriocytes are also remarkably similar in the different bivalves with domed, microvilli-covered apices and separated by narrow intercalary cells. Variation is seen in the way in which symbionts are housed, for example each symbiont is usually enclosed in single vacuole in Lucinidae but masses of symbionts are held in single vacuoles of some thyasirid species.

Interestingly, remarkably similar ctenidial bacteriocytes housing bacterial symbionts occur in species of the wood-boring bivalves Xylophagainae and Teredinidae (Distel and Roberts 1997; Luyten et al. 2006). Here the symbionts, identified as cellulolytic, nitrogen-fixing Gammaproteobacteria, are thought to produce enzymes enabling the bivalves to exploit cellulose from the wood habitat.

The family Thyasiridae is an outstanding subject for future research concerning the evolution of chemosymbiosis since there are both symbiotic and asymbiotic species. The symbiotic species show possible progressive grades of integration of symbionts with the host, so there are species with a low abundance of extracellular symbionts held amongst microvilli, while in other thyasirids the symbionts are contained within extracellular 'vacuoles' beneath a microvilli cuticle and finally, there are species where the bacteria are tightly packed within intracellular vacuoles.

Major questions concern the acquisition of the symbionts (see also Chapter 8); for a few species of Lucinidae these are known to be acquired by environmental transmission from the sediment (Gros et al. 2003), as also occurs in the Bathymodiolinae (Won et al. 2003, 2008). In Solemyidae evidence suggests vertical transmission through the embryo (Krueger et al. 1996b) as is also claimed for Vesicomyidae (Hurtado et al. 2003) but where lateral transmission is also suspected (Stewart et al. 2008). The mechanisms and signals by which the bivalve host recognizes and allows entry to the symbiont and how this triggers modifications of gill structure to form bacteriocytes are poorly known. Gros et al. (1998b) showed in post-larval, asymbiotic *Codakia orbicularis* that the bacteria, free-living in the sediment (Gros et al. 2003) are endocytosed into the apical areas of a few undifferentiated abfrontal cells. This is rapidly followed by proliferation of bacteriocytes. Significantly, most symbionts of shallow water hosts (Lucinidae, Thyasiridae, *Solemya*) belong to the same clade of Gammaproteobacteria and differ from those from deep-water bivalve hosts ('*Calyptogena*', '*Maorithyas*' '*Bathymodiolus*') which group in another major symbiont clade (Cavanaugh et al. 2006; Dubilier et al. 2008). In addition to the dual symbiosis of some bathymodioline mussels with sulphide-oxidising and methane-oxidising symbionts, multiple bacterial symbionts are now being increasingly recognised in a variety of bivalve hosts (Duperron et al. 2008; Dubilier et al. 2008; Ball et al. 2009) but their significance is as yet uninvestigated.

Why has bivalve chemosymbiosis evolved? A hypothesis is that the symbiosis enables bivalves to live in adaptation to environments where normal food sources for suspension feeding animals are poor or non-existent. This is particularly obvious in the deep sea where chemosymbiosis has equipped bivalves to exploit alternative energy sources, either in normal deep-sea sediments (e.g. Thyasiridae) or at cold seeps, mud volcanoes and hydrothermal vents, as well as patchily ephemeral nutritional substrates such as sunken wood and whale falls. In shallower environments, by far the most widespread sulphidic habitat is the subsurface anoxic zone of marine sediments (Ott et al. 2004; Stewart et al. 2005). Lucinidae commonly live at the interface of oxic/anoxic zones of sediments but are often surprisingly diverse and abundant in the clear oligotrophic waters of tropical seagrass beds and coral reef lagoons where suspension-feeding bivalves are low in numbers (Glover and Taylor 2007). The lucinids exploit sulphides from decomposing organic detritus, including coral mucus, within the sediment. Similarly, small tropical Solemyidae are often abundant in seagrass beds (Taylor et al. 2008) at similar locations. In some coral lagoons the dominant bivalves are often those possessing nutritional inputs from either photosymbionts (cardiids including *Tridacna*) or chemosymbionts.

Research over the last 30 years has revealed an extraordinary diversity of chemosymbiotic bivalves not only from deep-sea exploration but also resulting from new discoveries and reassessment of shallow water faunas. Description of new species and genera in all the families progresses at an accelerating rate, indicating a far greater diversity than was ever imagined (Krylova and Sahling 2006, 2010; Glover and Taylor 2007; Oliver and Holmes 2006a, 2007; von Cosel and Bouchet, 2008; von Cosel and Olu 2008, 2009; von Cosel and Janssen 2008; Taylor and Glover 2009a). More sophisticated taxonomic analyses using both molecular and morphological techniques are providing new insights into the evolutionary history of the host bivalve families and improved understanding of the adaptations to the chemosymbiosis. This progress is exemplified amongst the Vesicomyidae where the many species loosely called '*Calyptogen*a' are being are being divided into clades possessing different internal anatomies and with distinct ecological preferences and separate evolutionary histories.

Acknowledgements We are grateful to Shana Goffredi for generously providing electron micrographs of *Calyptogena* ctenidia, to Graham Oliver for the image of *Axinus cascadensis*, James Turner for images of *Acharax* species and Pierre Lozouet for the image of *Nucinella*. Philippe Bouchet and Rudo von Cosel kindly arranged a loan of preserved Vesicomyidae.

References

Allen JA, Sanders HL (1969) *Nucinella serrei* (Bivalvia: Protobranchia), a monomyarian solemyid and possible living actinodont. Malacologia 7:381–396

Amano K, Kiel S (2007) Fossil vesicomyid bivalves from the North Pacific region. Veliger 49:270–293

Amano K, Jenkins RG, Kurihara Y, Kiel S (2007) A new genus for *Vesicomya inflata* Kanie & Nishida, a lucinid shell convergent with that of vesicomyids, from Cretaceous strata of Hokkaido, Japan. Veliger 50:255–262

Arp AJ, Childress JJ, Fisher CR (1984) Metabolic and blood gas transport characteristics of the hydrothermal vent bivalve *Calyptogena magnifica*. Physiol Zool 57:648–662

Baco AR, Smith CR, Peek AS, Roderick GK, Vrijenhoek RC (1999) The phylogenetic relationships of whale-fall vesicomyid clams based on mitochondrial COI DNA sequences. Mar Ecol Prog Ser 182:137–147

Ball AD, Purdy KJ, Glover EA, Taylor JD (2009) Ctenidial structure and three bacterial symbiont morphotypes in *Anodontia (Euanodontia) ovum* (Reeve, 1850) from the Great Barrier Reef, Australia (Bivalvia: Lucinidae). J Moll Stud 75:175–185

Barry PJ, McCormack GP (2007) Two new species of *Adontorhina* Berry, 1947 (Bivalvia: Thyasiridae) from the Porcupine Bank, off the west coast of Ireland. Zootaxa 1526:37–49

Bennett BA, Smith CR, Glaser B, Maybaum HL (1994) Faunal community structure of a chemoautotrophic assemblage on whale bones in the deep northeast Pacific Ocean. Mar Ecol Prog Ser 108:205–223

Bouchet P, von Cosel R (2004) The world's largest lucinid is an undescribed species from Taiwan (Mollusca: Bivalvia). Zool Stud 43:704–711

Britton JC (1970) The Lucinidae (Mollusca: Bivalvia) of the Western Atlantic Ocean. Ph.D. dissertation, George Washington University, University Microfilms, pp 71–12, 288

Brooks JM, Kennicutt MC, Fisher CR, Macko SA, Cole K, Childress JJ, Bidigare RR, Vetter RD (1987) Deep-sea hydrocarbon seep communities: evidence for energy and nutritional carbon sources. Science 238:1138–1142

Callender WR, Powell EN (1997) Autochthonous death assemblages from chemautotrophic communities at petroleum seeps: palaeoproduction, energy flow and implications from the fossil record. Hist Biol 12:165–198

Callender WR, Powell EN (2000) Long-term history of chemautotrophic clam-dominated faunas of petroleum seeps in the northwestern Gulf of Mexico. Facies 43:177–204

Cary SC, Fry B, Felbeck H, Vetter RD (1989) Habitat characterization and nutritional strategies of the endosymbiont-bearing bivalve *Lucinoma aequizonata*. Mar Ecol Prog Ser 55:31–45

Cavanaugh CM, McKiness ZP, Newton ILG, Stewart FJ (2006) Marine chemosynthetic symbioses. In: Dworkin M, Falkow SI, Rosenberg E, Schlelfer KH, Stackebrandt E (eds) The Prokaryotes 1. Springer, New York, pp 475–507

Clarke AH (1989) New mollusks from undersea oil seep sites off Louisiana. Malacology Data Net 2:123–134

Coan EV, Scott PV, Bernard FR (2000) Bivalve seashells of western North America. Santa Barbara Museum of Natural History Monographs 2, Santa Barbara, 764 p

Conway NM, Capuzzo JM, Fry B (1989) The role of endosymbiotic bacteria in the nutrition of *Solemya velum*: evidence from a stable isotope analysis of endosymbionts and hosts. Limnol Oceanogr 34:249–255

Conway NM, Howes BL, McDowell Capuzzo JE, Turner RD (1992) Characterization and site description of *Solemya borealis* (Bivalvia; Solemyidae), another bivalve-bacteria symbiosis. Mar Biol 112:601–613

Cope JCW (1996) Early Ordovician (Arenig) bivalves from the Llangynog Inlier, South Wales. Palaeontology 39:979–1025

Cordes EE, Bergquist DC, Fisher CR (2009) Macro-ecology of Gulf of Mexico cold seeps. Ann Rev Mar Sci 1:143–168

von Cosel R (2006) Taxonomy of West African bivalves VIII. Remarks on Lucinidae, with descriptions of five new genera and nine new species. Zoosystema 28:805–851

von Cosel R, Bouchet P (2008) Tropical deep-water lucinids (Mollusca: Bivalvia) from the Indo-Pacific: essentially unknown, but diverse and occasionally gigantic. In: Héros V, Cowie R, Bouchet P (eds) Tropical deep sea benthos, vol 25. Mém Mus nat Hist natur, Paris 196:115–213

von Cosel R, Janssen R (2008) Bathymodioline mussels of the *Bathymodiolus* (s.l.) *childressi* clade from methane seeps near Edison Seamount, New Ireland, Papua New Guinea. Arch Molluskenkunde 137:195–224

von Cosel R, Marshall BA (2003) Two new species of large mussels (Bivalvia Mytilidae) from active submarine volcanoes and a cold seep off the eastern North Island of New Zealand, with descriptions of a new genus. Nautilus 117:31–46

von Cosel R, Olu K (2008) A new genus and new species of Vesicomyidae (Mollusca, Bivalvia) from cold seeps on the Barbados accretionary prism, with comments on other species. Zoosystema 30:929–944

von Cosel R, Olu K (2009) Large Vesicomyidae (Mollusca: Bivalvia) from cold seeps in the Gulf of Guinea off the coasts of Gabon, Congo and northern Angola. Deep Sea Res II 56:2350–2379

von Cosel R, Salas C (2001) Vesicomyidae (Mollusca: Bivalvia) of the genera *Vesicomya*, *Waisiuconcha*, *Isorropodon* and *Callogonia* in the eastern Atlantic and the Mediterranean. Sarsia 86:333–366

Dando PR, Bussmann I, Niven SJ, O'Hara SCM, Schaljohann R, Taylor LJ (1994) A methane seep in the Skagerrak, the habitat of the pogonophore *Siboglinum poseidoni* and the bivalve mollusc *Thyasira sarsi*. Mar Ecol Prog Ser 107:157–167

Dando PR, Southward AJ, Southward EC (1986) Chemoautotrophic symbionts in the gills of the bivalve mollusc *Lucinoma borealis* and the sediment chemistry of its habitat. Proc Roy Soc Lond B 227:227–247

Dando PR, Spiro B (1993) Varying nutritional dependence of the thyasirid bivalves *Thyasira sarsi* and *T. equalis* on chemoautotrophic symbiotic bacteria, demonstrated by isotopic ratios of tissues carbon and shell carbonate. Mar Ecol Prog Ser 92:131–158

Distel DL (1998) Evolution of chemoautotrophic endosymbioses in bivalves. Bioscience 48: 277–286

Distel DL, Felbeck H (1987) Endosymbiosis in the lucinid clams *Lucinoma aequizonata*, *Lucinoma annulata* and *Lucina floridana*: a reexamination of the functional morphology of the gills as bacteria-bearing organs. Mar Biol 96:79–86

Distel DL, Lee HKW, Cavanaugh CM (1995) Intracellular coexistence of methano- and thioautotrophic bacteria in a hydrothermal vent mussel. Proc Nat Acad Sci USA 92:9598–9602

Distel DL, Roberts SJ (1997) Bacterial endosymbionts in the gills of the deep-sea wood-boring bivalves *Xylophaga atlantica* and *Xylophaga washingtona*. Biol Bull 192:253–261

Dubilier N, Bergin C, Lott C (2008) Symbiotic diversity in marine animals: the art of harnessing chemosynthesis. Nat Rev Microbiol 6:725–740

Dufour SC (2005) Gill anatomy and evolution of symbiosis in the bivalve family Thyasiridae. Biol Bull 208:200–212

Dufour SC, Felbeck H (2003) Sulphide mining by the superextensible foot of symbiotic thyasirids. Nature 426:65–67

Duperron S, Halary S, Lorion J, Sibuet M, Gaill F (2008) Unexpected co-occurrence of six bacterial symbionts in the gills of the cold seep mussel *Idas* sp. (Bivalvia: Mytilidae). Environ Microbiol 10:433–445

Duperron S, Nadalig T, Caprais JC, Sibuet M, Fiala-Medioni A, Amann R, Dubilier N (2005) Dual symbiosis in a *Bathymodiolus* mussel from a methane seep on the Gabon continental margin (South East Atlantic): 16S rRNA phylogeny and distribution of the symbionts in the gills. Appl Environ Microbiol 71:1694–1700

Duplessis MR, Dufour SC, Blankenship LE, Felbeck H, Yayanos AA (2004) Anatomical and experimental evidence for particulate feeding in *Lucinoma aequizonata* and *Parvilucina tenuisculpta* (Bivalvia: Lucinidae) from the Santa Barbara Basin. Mar Biol 145:551–561

Fiala-Médioni A, Metivier C, Herry A, Le Pennec M (1986) Ultrastructure of the gill of the hydrothermal-vent mytilids *Bathymodiolus* sp. Mar Biol 92:65–72

Fiala-Médioni A, Le Pennec M (1988) Structural adaptations in the gill of the Japanese subduction zone bivalves (Vesicomydae) *Calyptogena phaseoliformis* and *Calyptogena laubieri*. Ocean Acta 11:185–192

Fisher CR (1990) Chemoautotrophic and methanotrophic symbioses in marine invertebrates. Revs Aqua Sci 2:399–436

Fisher CR, Childress JJ (1986) Translocation of fixed carbon from symbiotic bacteria to host tissues in the gutless bivalve *Solemya reidi*. Mar Biol 93:59–68

Fisher CR, Brooks JM, Vodenichar JS, Zande JM, Childress JJ, Burke RA (1993) The co-occurrence of methanotrophic and chemoautotrophic sulfur oxidizing bacterial symbionts in a deep-sea mussel. Mar Ecol 14:277–289

Frenkiel L, Mouëza M (1995) Gill ultrastructure and symbiotic bacteria in *Codakia orbicularis* (Bivalvia, Lucinidae). Zoomorph 115:51–61

Frenkiel L, Gros O, Mouëza M (1996) Gill structure in *Lucina pectinata* (Bivalvia: Lucinidae) with reference to hemoglobin in bivalves with symbiotic sulphur-oxidising bacteria. Mar Biol 125:511–524

Fujikura K, Kojima S, Tamaki K, Maki Y, Hunt OT (1999) The deepest chemosynthesis-based community yet discovered from the hadal zone, 7326 m deep, in the Japan Trench. Mar Ecol Prog Ser 190:17–26

Fujiwara Y (2003) Endosymbioses between invertebrates and chemosymbiotic bacteria. Jap J Benthol 58:26–33
Fujiwara Y, Kato C, Masui N, Fujikura K, Kojima S (2001) Dual symbiosis in the cold-seep thyasirid clam *Maorithyas hadalis* from the hadal zone in the Japan Trench, western Pacific. Mar Ecol Prog Ser 214:151–159
Fujiwara Y, Okutani T, Yamanaka T, Kawato M, Mizota C, Fujikura K, Yamamoto T, Okoshi K (2009) *Solemya pervernicosa* lives in sediment underneath submerged whale carcasses: its biological significance. Venus 68:27–37
Gebruk AV, Krylova EM, Lein AY, Vinogradov GM, Anderson E, Pimenov NV, Cherkasev GA, Crane K (2003) Methane seep community of the Håkon Mosby mud volcano (the Norwegian Sea): composition and trophic aspects. Sarsia 88:394–403
Giere O (1985) Structure and position of bacterial endosymbionts in the gill filaments of Lucinidae from Bermuda (Mollusca, Bivalvia). Zoomorph 105:296–301
Giribet G (2008) Bivalvia. In: Ponder WF, Lindberg DR (eds) Phylogeny and evolution of the Mollusca. University of California Press, Berkeley, CA, pp 105–141
Giribet G, Distel DL (2004) Bivalve phylogeny and molecular data. In: Lydeard C, Lindberg D (eds) Molecular Systematics and Phylogeography of Mollusks. Smithsonian Institution Press, Washington, DC, pp 45–90
Giribet G, Wheeler W (2002) On bivalve phylogeny: a high-level analysis of the Bivalvia (Mollusca) based on combined morphology and DNA sequence data. Invert Biol 121:271–324
Glover EA, Taylor JD (2007) Diversity of chemosymbiotic bivalves on coral reefs: Lucinidae of New Caledonia and Lifou (Mollusca, Bivalvia). Zoosystema 29:109–181
Glover EA, Taylor JD, Rowden AA (2004) *Bathyaustriella thionipta,* a new lucinid bivalve from a hydrothermal vent on the Kermadec Ridge, New Zealand and its relationship to shallow water taxa (Mollusca: Bivalvia: Lucinidae). J Moll Stud 70:283–295
Glover EA, Taylor JD, Williams ST (2008) Mangrove associated lucinid bivalves of the central Indo-West Pacific: review of the "*Austriella*" group with a new genus and species (Mollusca: Bivalvia: Lucinidae). Raffles Mus Bull Zool Suppl 18:25–40
Goffredi SK, Barry JP (2002) Species-specific variation in sulfide physiology between closely related vesicomyid clams. Mar Ecol Prog Ser 225:227–238
Goffredi SK, Barry JP, Buck KR (2004) Vesicomyid symbioses from Monterey Bay (central California) cold seeps. Symbiosis 36:1–27
Goffredi SK, Hurtado LA, Hallam S, Vrijenhoek RC (2003) Evolutionary relationships of deep-sea vent and cold seep clams (Mollusca: Vesicomyidae) of the '*pacifica/lepta*' species complex. Mar Biol 142:311–320
Gros O, Durand P, Frenkiel L, Mouëza M (1998a) Putative environmental transmission of sulfur-oxidising gill endosymbionts in four tropical lucinid bivalves, inhabiting various environments. FEMS Microbiol Letts 160:257–262
Gros O, Frenkiel L, Mouëza M (1996) Gill ultrastructure and symbiotic bacteria in the tropical lucinid, *Linga pensylvanica* (Linné). Symbiosis 20:259–280
Gros O, Frenkiel L, Mouëza M (1998b) Gill filament differentiation and experimental colonization by symbiotic bacteria in aposymbiotic juveniles of *Codakia orbicularis* (Bivalvia: Lucinidae). Invert Reprod Develop 34:219–231
Gros O, Frenkiel L, Felbeck H (2000) Sulfur-oxidising endosymbioisis in *Divaricella quadrisulcata* (Bivalvia: Lucinidae): morphological, ultrastructural and phylogenetic analysis. Symbiosis 29:293–317
Gros O, Liberge M, Heddi A, Khatchadourian C, Felbeck H (2003) Detection of free-living forms of sulfide-oxidising gill endosymbionts in the lucinid habitat (*Thalassia testudinum* environment). Appl Environ Microbiol 69:6264–6267
Gustafson RG, Reid RGB (1986) Association of bacteria with larvae of the gutless protobranch bivalve *Solemya reidi* (Cryptodonta: Solemyidae). Mar Biol 97:389–401
Hashimoto J, Fujikura K, Fujiwara Y, Tanishima M, Ohta S, Kojima S, Yieh SM (1995) Observations of a deep-sea biological community co-dominated by lucinid bivalve, *Lucimoma spectabilis* (Yokoyama, 1920) and vestimentiferans at the Kanesu-no-Se Bank, Enshu-Nada, Central Japan. JAMSTEC J Deep Sea Res 11:211–217

Herry A, Le Pennec M (1987) Endosymbiotic bacteria in the gills of the littoral bivalve molluscs *Thyasira flexuosa* (Thyasiridae) and *Lucinella divaricata* (Lucinidae). Symbiosis 4:25–36

Holmes AM, Oliver PG, Sellanes J (2005) A new species of *Lucinoma* (Bivalvia: Lucinoidea) from a methane gas seep off the southwest coast of Chile. J Conch 38:673–682

Hurtado LA, Mateos M, Lutz RA, Vrijenhoek RC (2003) Coupling of bacterial endosymbiont and host mitochondrial genomes in the hydrothermal vent clam *Calyptogena magnifica.* Appl Environ Microbiol 69:2058–2064

Imhoff JF, Sahling H, Sühling J, Kath T (2003) 16S rDNA-based phylogeny of sulphur-oxidising bacterial endosymbionts in marine bivalves from cold-seep habitats. Mar Ecol Prog Ser 249:39–51

Jones WJ, Won YJ, Maas PAY, Smith PJ, Lutz RA, Vrijenhoek RC (2006) Evolution of habitat use by deep-sea mussels. Mar Biol 148:841–851

Kamenev GM (2009) North Pacific species of the genus *Solemya* Lamarck, 1818 (Bivalvia: Solemyidae), with notes on *Acharax johnsoni* (Dall, 1891). Malacologia 51:233–261

Kamenev MK, Nadtochy VA, Kuznetsov AP (2001) *Conchocele bisecta* (Conrad, 1849) (Bivalvia: Thyasiridae) from cold water methane-rich areas of the Sea of Okhotsk. Veliger 44:84–94

Kanie Y, Kuramochi T (1995) *Acharax yokoyskensis* n.sp. (giant bivalve) from the Miocene Hayama Formation of the Miura Peninsula, south-central Japan. Sci Rept Yokosuka City Mus 43:51–57

Kauffman EG, Harries PJ, Meyer C, Villamil T, Arango C, Jaecks G (2007) Paleoecology of giant Inoceramidae (*Platyceramus*) on a Santonian (Cretaceous) seafloor in Colorado. J Paleont 81:64–81

Kennish MJ, Lutz RA (1992) The hydrothermal vent clam, *Calyptogena magnifica* (Boss and Turner, 1980): a review of existing literature. Revs Aquatic Sci 6:29–66

Kiel S, Little CTS (2006) Cold seep mollusks are older than the general marine mollusk fauna. Science 313:1429–1431

Kiel S, Amano K, Jenkins RG (2008) Bivalves from Cretaceous cold-seep deposits on Hokkaido, Japan. Acta Palaeont Polon 53:525–537

Kojima S, Fujikura K, Okutani T (2004) Multiple trans-Pacific migrations of deep-sea vent/seep-endemic bivalves in the family Vesicomyidae. Mol Phylog Evol 32:396–406

Krueger DM, Cavanaugh CM (1997) Phylogenetic diversity of bacterial symbionts of *Solemya* hosts based on comparative sequence analysis of 16S rRNA genes. Appl Environ Microbiol 63:91–98

Krueger DM, Dubilier N, Cavanaugh CM (1996a) Chemoautotrophic symbiosis in the tropical solemyid *Solemya occidentalis* (Bivalvia: Protobranchia): ultrastructural and phylogenetic analysis. Mar Biol 126:55–64

Krueger DM, Gallager SM, Cavanaugh CM (1992) Suspension feeding on phytoplankton by Solemya velum, a symbiont-containing clam. Mar Ecol Prog Ser 86:145–151

Krueger DM, Gustafson RG, Cavanaugh CM (1996b) Vertical transmission of chemoautotrophic symbionts in the bivalve *Solemya velum* (Bivalvia: Protobranchia). Biol Bull 190:195–202

Krylova EM, Drozdov AL, Mironov AN (2000) Presence of bacteria in gills of hadal bivalve '*Vesicomya*' *sergeevi* Filatova, 1971. Ruthenica 10:76–79

Krylova EM, Sahling H (2006) Recent bivalve molluscs of the genus *Calyptogena* (Vesicomyidae). J Moll Stud 72:359–395

Krylova EM, Sahling H (2010) Vesicomyidae (Bivalvia): current taxonomy and distribution. PLoS ONE 5(4) e9957:1–9

Krylova EM, Sahling H, Janssen R (2010) *Abyssogena*: a new genus of the family Vesicomyidae from deep water vents and seeps. J Moll Stud 76:107–132

Kuznetsov AP, Schileyko AA (1984) On gutless Protobranchia (Bivalvia). Biogicheskie Nauki 1984 2:39–49 (In Russian)

Kuznetsov AP, Ota C, Endow K (1990) Morphofunctional consequences of bacterial symbiotrophy in *Solemya* (*Petrasma*) *pusilla* (Protobranchia, Bivalvia) from Sagami Bay (central Japan). Izvest Akad Nauk SSSR, Ser Biol 1991(6):895–903

LaPerna R (2005) A gigantic deep-sea Nucinellidae from the tropical West Pacific (Bivalvia: Protobranchia). Zootaxa 881:1–10

Liljedahl L (1984) *Janeia silurica*, a link between nuculoids and solemyoids (Bivalvia). Palaeontology 27:693–698

Luyten YA, Thompson JR, Morrill W, Polz MF, Distel DL (2006) Extensive variation in intracellular symbiont community composition among members of a single population of the wood-boring bivalve *Lyrodus pedicellatus* (Bivalvia: Teredinidae). Appl Environ Microbiol 72:412–417

Majima R, Ikeda K, Wada H, Kato K (2003) An outer-shelf cold-seep assemblage in forearc basin fill, Pliocene Takanabe Formation, Kyushu Island, Japan. Paleont Res 7:297–311

Matsukuma A (2000) Family Thyasiridae. In: Okutani T (ed) Marine Mollusks in Japan. Tokai University Press, Tokyo, pp 933–935

Métivier B, von Cosel R (1993) *Acharax alinae* n.sp., Solemyidae (Mollusca: Bivalvia) géante du bassin de Lau. Comptes Rend l'Acad Sci ser III Sci Vie 316(3):229–237

Meyer E, Nilkerd B, Glover EA, Taylor JD (2008) Ecological importance of chemoautotrophic lucinid bivalves in a peri – mangrove community in Eastern Thailand. Raffles Mus Bull Zool Suppl 18:41–55

Mikkelsen PM, Bieler R, Kappner I, Rawlings TA (2006) Phylogeny of Veneroidea (Mollusca: Bivalvia) based on morphology and molecules. Zool J Linn Soc 148:439–521

Morton B (1986) The functional morphology of the organs of feeding and digestion of the hydrothermal vent bivalve *Calyptogena magnifica* (Vesicomyidae). J Zool 208:83–98

Neulinger SC, Sahling H, Sühling J, Imhoff JF (2006) Presence of two phylogenetically distinct groups in the deep-sea mussel *Acharax* (Mollusca: Bivalvia: Solemyidae). Mar Ecol Prog Ser 312:161–168

Ockelmann KW, Dinesen GE (in press) Life on wood – the carnivorous deep-sea mussel *Idas argenteus* (Bathymodiolinae, Mytilidae, Bivalvia). Mar Biol Res

Okutani T (2002) A new thyasirid *Conchocele novaeguinensis* n.sp. from a thanatocoenosis associated with a possible cold seep activity off New Guinea. Venus 61:141–145

Okutani T, Hashimoto J (1997) A new species of lucinid bivalve (Heterodonta: Lucinidae) from Kanesu-no-Se bank near the mouth of Suruga Bay, with a review of the Recent species of the chemosynthetic genus *Lucinoma* from Japan. Venus 56:271–280

Oliver PG, Holmes AM (2006a) New species of Thyasiridae (Bivalvia) from chemosynthetic communities in the Atlantic Ocean. J Conch 39:175–183

Oliver PG, Holmes AM (2006b) A new species of *Lucinoma* (Bivalvia: Lucinoidea) from the oxygen minimum zone of the Oman margin, Arabian Sea. J Conch 39:63–77

Oliver PG, Holmes AM (2007) A new species of *Axinus* (Thyasiridae: Bivalvia) from the Baby Bare Seamount, Cascadia Basin, NE Pacific with a description of the anatomy. J Conch 39:363–375

Oliver PG, Killeen IJ (2002) The Thyasiridae (Mollusca: Bivalvia) of the British Continental Shelf and North Sea Oil fields. Studies in Marine Biodiversity and Systematics from the National Museum of Wales, BIOMOR Reports, 3: 73p

Oliver PG, Levin L (2006) A new species of the family Thyasiridae (Mollusca: Bivalvia) from the oxygen minimum zone of the Pakistan Margin. J Mar Biol Assn UK 86:411–416

Oliver PG, Sellanes J (2005) New species of Thyasiridae from a methane seepage area off Concepcion, Chile. Zootaxa 1092:1–20

Olu-Le Roy K, Sibuet M, Fiala-Médioni A, Gofas S, Salas C, Mariotti A, Foucher JP, Woodside J (2004) Cold seep communities in the deep eastern Mediterranean Sea: composition, symbiosis and spatial distribution on mud volcanoes. Deep Sea Res I 51:1915–1936

Olu-Le Roy K, von Cosel R, Houdez S, Carney SL, Jollivet D (2007) Amphi-Atlantic cold-seep *Bathymodiolus* species complexes across the equatorial belt. Deep-Sea Res I 54:1890–1911

Ott J, Bright M, Bulgheresi S (2004) Symbioses between marine nematodes and sulfur-oxidising bacteria. Symbiosis 36:103–126

Owen G (1961) A note on the habits and nutrition of *Solemya parkinsoni* (Protobranchia: Bivalvia). Quart J Micros Sci 102:15–21

Passos FD, Meserani GDC, Gros O (2007) Structural and ultrastrucural analysis of the gills in the bacterial-bearing species *Thyasira falklandica* (Bivalvia, Mollusca). Zoomorph 126:153–162

Payne CM, Allen AJ (1991) The morphology of deep-sea Thyasiridae (Mollusca: Bivalvia) from the Atlantic Ocean. Phil Trans Roy Soc Lond B334:481–562

Peek AS, Gustafson R, Lutz RA, Vrijenhoek RC (1997) Evolutionary relationships of deep-sea hydrothermal vent and cold-water seep clams (Bivalvia: Vesicomyidae): results from mitochondrial cytochrome oxidase subunit 1. Mar Biol 130:151–161

Pojeta J (1988) The origin and Paleozoic diversification of solemyoid pelecypods. Mem New Mexico Bur Mines Min Res 44:201–271

Powell MA, Somero GN (1985) Sulfide oxidation occurs in the animal tissue of the gutless clam, *Solemya reidi*. Biol Bull 169:164

Reid RGB (1980) Aspects of the biology of a gutless species of *Solemya* (Bivalvia: Protobranchia). Can J Zool 58:386–393

Reid RGB (1998) Order Solemyoida. In: Beesley PL, Ross GJB, Wells A (eds) Mollusca: the Southern Synthesis. Fauna of Australia, vol 5, Part A. CSIRO Publishing, Melbourne, pp 2241–247

Reid RGB, Brand DG (1986) Sulfide-oxidising symbiosis in lucinaceans: implications for bivalve evolution. Veliger 29:3–24

Reid RGB, Brand DG (1987) Observations on Australian Solemyidae. J Malac Soc Austral 8:41–50

Rodrigues CF, Oliver PG, Cunha MR (2008) Thyasiroidea (Mollusca: Bivalvia) from mud volcanoes of the Gulf of Cadiz. Zootaxa 1752:41–56

Sahling H, Rickert D, Lee RW, Linke P, Suess E (2002) Macrofaunal community structure and sulfide flux at gas hydrate deposits from the Cascadia convergent margin, NE Pacific. Mar Ecol Prog Ser 231:121–138

Salas C, Woodside J (2002) *Lucinoma kazani* n.sp. (Mollusca, Bivalvia): evidence of a living community associated with a cold seep in the Eastern Mediterranean Sea. Deep Sea Res I 49:991–1005

Sasaki T, Okutani Y, Fujikura K (2005) Molluscs from hydrothermal vents and cold seeps in Japan: a review of taxa recorded in twenty recent years (1984–2004). Venus 64:87–133

Smith CR, Baco AR (2003) Ecology of whale falls at the deep-sea floor. Oceanogr Mar Biol Ann Rev 41:311–354

Southward EC (1986) Gill symbionts in thyasirids and other bivalve molluscs. J Mar Biol Assn UK 66:889–914

Southward EC (2008) The morphology of bacterial symbioses in the gills of mussels of the genera *Adipicola* and *Idas* (Bivalvia: Mytlilidae). J Shellfish Res 27:139–146

Stanley SM (1970) Relation of shell form to life habits of the Bivalvia (Mollusca). Geol Soc America Mem 125:1–296

Stecher J, Tunnicliffe V, Türkay M (2003) Population characteristics of abundant bivalves (Mollusca, Vesicomyidae) at a sulphide-rich seafloor site near Lihir Island, Papua New Guinea. Can J Zool 81:1815–1824

Stempell W (1899) Zur Anatomie von *Solemya togata* Poli. Zoologische Jahrbücher, Abtheilung für Anatomie und Ontogenie der Thiere 13:89–170

Stewart FJ, Cavanaugh CM (2006) Bacterial endosymbioses in *Solemya* (Mollusca: Bivalvia) – model systems for studies of symbiont-host adaptation. Antonie van Leeuwenhoeck 90:343–360

Stewart FJ, Newton ILG, Cavanaugh CM (2005) Chemosymbiotic endosymbioises: adaptations to oxic-anoxic interfaces. Trends Microbiol 13:439–448

Stewart FJ, Young CR, Cavanaugh CM (2008) Lateral symbiont acquisition in a maternally transmitted chemosynthetic clam endosymbiosis. Mol Biol Evol 25:673–687

Taylor JD, Glover EA (1997) A chemosymbiotic lucinid bivalve (Bivalvia: Lucinoidea) with periostracal pipes; functional morphology and description of a new genus and species. In: Wells FE (ed) The marine flora and fauna of the Houtman Abrolhos Islands, Western Australia. Western Australian Museum, Perth, pp 335–361

Taylor JD, Glover EA (2000) Functional anatomy, chemosymbiosis and evolution of the Lucinidae. In: Harper, E, Taylor JD, Crame, JA (eds) The evolutionary biology of the bivalvia. Spec Pub Geol Soc Lond 177:207–225

Taylor JD, Glover EA (2005) Cryptic diversity of chemosymbiotic bivalves: a systematic revision of worldwide *Anodontia* (Mollusca: Bivalvia: Lucinidae). Systemat Biodiv 3:281–338

Taylor JD, Glover EA (2006) Lucinidae (Bivalvia) – the most diverse group of chemosymbiotic molluscs. Zool J Linn Soc 148:421–438

Taylor JD, Glover EA (2009a) A giant lucinid bivalve from the Eocene of Jamaica – systematics, life habits and chemosymbiosis (Mollusca: Bivalvia: Lucinidae). Palaeontol 52:95–109

Taylor JD, Glover EA (2009b) New lucinid bivalves from hydrocarbon seeps of the Western Atlantic (Mollusca: Bivalvia: Lucinidae). Steenstrupia 30:127–140

Taylor JD, Glover EA, Williams ST (2008) Ancient shallow water chemosymbiotic bivalves: systematics of Solemyidae of eastern and southern Australia. Mem Qld Mus – Nature 54:75–104

Taylor JD, Glover EA, Zuschin M, Dworschak PC, Waitzbauer W (2005) Another bivalve with dreadlocks: living *Rasta lamyi* from Aqaba, Red Sea (Bivalvia: Lucinidae). J Conch 38:489–497

Taylor JD, Williams ST, Glover EA (2007a) Evolutionary relationships of the bivalve family Thyasiridae (Mollusca: Bivalvia), monophyly and superfamily status. J Mar Biol Assn UK 87:565–574

Taylor JD, Williams ST, Glover EA, Dyal P (2007b) A molecular phylogeny of heterodont bivalves (Mollusca: Bivalvia: Heterodonta): new analyses of 18S rRNA and 28S rRNA genes. Zool Scripta 36:587–606

Vokes H (1956) Notes on the Nucinellidae (Pelecypoda) with description of a new species from the Eocene of Oregon. J Paleont 30:652–671

Williams ST, Taylor JD, Glover EA (2004) Molecular phylogeny of the Lucinoidea (Bivalvia): non-monophyly and separate acquisition of bacterial chemosymbiosis. J Moll Stud 70:187–202

Won YJ, Hallam SJ, O'Mullan GD, Pan IL, Buck KR, Vrijenhoeck RC (2003) Environmental acquisition of thiotrophic endosymbionts by deep-sea mussels of the genus *Bathymodiolus*. Appl Environ Microbiol 69:6785–6792

Won YJ, Jones WJ, Vrijenhoeck RC (2008) Absence of cospeciation between deep-sea mytilids and their thiotrophic endosymbioints. J Shellfish Res 27:129–138

Yamanaka T, Shimizota C, Matsuyama-Serisawa K, Kakegawa T, Miyazaki J, Mampuku M, Tsutsumi H, Fujiwara Y (2008) Stable isotopic characterization of carbon, nitrogen and sulfur uptake of *Acharax japonica* from central Japan. Plankton Benthos Res 3:36–41

Zardus JD (2002) Protobranch bivalves. Adv Mar Biol 42:2–65

Chapter 6
The Diversity of Deep-Sea Mussels and Their Bacterial Symbioses

Sébastien Duperron

Contents

6.1 Introduction

Deep-sea chemosynthesis-based ecosystems are inhabited by diverse groups of metazoans. Although remote from the euphotic layer, and thus from photosynthetic primary producers, these ecosystems harbour high animal biomasses, orders of magnitude above biomasses usually reported in the deep-sea (Sibuet and Olu 1998; Van Dover 2000). The key to this high productivity is chemoautotrophy, a type of

S. Duperron (✉)
UMR 7138 (UPMC CNRS IRD MNHN), Systématique Adaptation Evolution – Adaptation aux Milieux Extrêmes, Université Pierre et Marie Curie, Bât. A, 7 quai St Bernard, 75005 Paris, France
e-mail: sebastien.duperron@snv.jussieu.fr

S. Kiel (ed.), *The Vent and Seep Biota*, Topics in Geobiology 33,
DOI 10.1007/978-90-481-9572-5_6, © Springer Science+Business Media B.V. 2010

metabolism by which many prokaryotes fix inorganic carbon into organic molecules using the chemical energy resulting from the oxidation of reduced compounds present in their environment (Madigan et al. 2002). Indeed, hydrothermal vents and cold seeps are characterized by the occurrence of fluid emissions originating from the subsurface, which bring reduced compounds into mixing with bottom oxygenated seawater (reviewed in Von Damm 1995; Sibuet and Olu 1998; Van Dover 2000; Tunnicliffe et al. 2003). To summarize, hydrothermal vents occur mostly on oceanic ridges, where bottom seawater circulates into the newly formed crust. The high geothermic gradient linked with the presence of a magma chamber few kilometres below the ridge causes water to heat and to be enriched in reduced compounds (metals, sulphide…). Heated fluids, displaying lower densities, then reach back the seafloor along cracks and are emitted. Fluid interaction with cold seawater provokes the precipitation of dissolved metals and minerals in the form of complexes with sulphide, yielding typical (and often spectacular) chimneys. At cold seeps, which are mostly located along continental margins, fluids originate from the subsurface. The thermogenic or biogenic reduction of buried organic matter produces methane and other hydrocarbons which seep to the seafloor, and processes such as the anaerobic oxidation of methane coupled with sulphate-reduction can account for local enrichments in sulphide. Co-occurrence of effective electron donors in the form of reduced compounds such as sulphide or methane, and of good acceptors such as oxygen or nitrate, provides large amounts of energy to prokaryotes able to use them, and triggers significant primary production (Jannasch and Mottl 1985). Although free-living chemoautotrophs are abundant, the most remarkable feature of hydrothermal vents and cold seeps is the occurrence of symbiotic associations between chemoautotrophic bacteria and dominant invertebrates. Although no definition of 'symbiosis' is universally accepted, the term is used herein to describe a close association between a metazoan host and bacteria allowing the host to gain novel metabolic capabilities (Douglas 1994). Many metazoan taxa, such as siboglinid tubeworms, molluscs and arthropods indeed live in symbiosis with chemoautotrophic bacteria, mostly sulphur-oxidizers (this volume, Dubilier et al. 2008).

Among symbiotic metazoans found at deep-sea hydrothermal vents and cold seeps, members of a clade within the family Mytilidae are the group which has evolved interactions with the most diversified bacteria. Although related to coastal mytilids, they represent a highly specialized group with a restricted habitat range. Deep-sea mytilids occur worldwide at hydrothermal vents and cold seeps at depths from 200 to 3,600 m (Plate 11, Table 6.1). They are also reported from large sunken organic debris such as wood falls and carcasses of large vertebrates, types of habitats which present analogies with vents and seeps, in particular due to the presence of reduced compounds (Distel et al. 2000; Baco and Smith 2003; Pailleret et al. 2007). To date, no such mytilid was reported from the background deep-sea fauna. The most remarkable biological feature distinguishing deep-sea mussels from their coastal relatives is the presence of chemosynthetic bacterial symbionts associated with their gill epithelial cells. Understanding this symbiotic association

Table 6.1 List of mussel species from which molecular data regarding symbiosis are available. The name, when given, is displayed

Area	Name	HAB	Sites	Reported depth range (m)	Symbionts	Ref symbionts	Accession number symbionts	Ref host	Accession number host	I, E	δ^{13}C range (‰)	Ref isotopes
Atlantic	*B. azoricus*	H	Menez Gwen, Lucky Strike, Rainbow, Broken Spur	840–3,350	M, S	23	AM083950 & 74	24	AY649795	I	−21.3 to −36	7, 10, 11
	B. aff. *azoricus*	H	Lost City	800	M,S	22	DQ899960–900018					
	B. puteoserpentis	H	Broken Spur, Snake Pit, Logatchev, TAG	3,000–3,510	M, S	23	AM083950 & 74	24	AY649796	I	−21.4 to −37.3	5, 10 '6, 11
	B. cf. *boomerang*	S	Regab, Barbados accretionary prism	3,150	M, S	18	AJ745717–8	26	DQ513451	I	−62.4 to −67.0	27
Gulf of Mexico	*B. heckerae*	S	Florida Escarpment, Blake Ridge	2,155–3,314	M, 2S, My	25	AM236325–8	24	AY649794	I	−56 to −93.8	4, 12, 21, 25
	B. brooksi	S	Alaminos Canyon, Atwater Canyon (Lower Louisiana Slope), Florida Escarpment	1,893–3,314	M, S	25	AM236330–1	24	AY649797	I	−44.4 to −67.3	4, 25
	I. macdonaldi	S	Louisiana Slope (Garden Banks)	650	S	31	EU326225	24	ay649804	?		

(continued)

Table 6.1 (continued)

Area	Name	HAB	Sites	Reported depth range (m)	Symbionts	Ref symbionts	Accession number symbionts	Ref host	Accession number host	I, E	$\delta^{13}C$ range (‰)	Ref isotopes
	B. childressi	S	Alaminos canyon (Lower Louisiana Slope), upper Louisiana Slope	540–2,226	M	3	AM236329	24	AY649800	I	−37.5 to −67.1	4
Eastern mediterranean	*Idas* sp. Med	S	Central Province	1,150–3,000	M, 2S, My, G, CFB	28	AM402955–60	28	EF210072	?	−38.3	28
Eastern Pacific	*B.* sp. JDF	H	Juan de Fuca Ridge	2,200–2,500	S	19	DQ077893	19	DQ077892	?	−26.6	19
	B. thermophilus	H	East Pacific Rise 13°N to 22°S, 32°S, Galapagos Ridge	2,000–2,747	S	1	M99445	18	AF456285	I	−29.2 to −37.3	2, 16
	B. aff. *thermophilus*	H	East Pacific Rise (32°S)	2,331	S	31	DQ321717	14	AF456317	?		
Western Pacific	*B. japonicus*	H, S	Sagami Bay, Minami Ensei Knoll Iheya (Okinawa Through)	550–1,180	M	8	AB036711	15	AB101423	I		
	B. platifrons	H, S	Sagami Bay, Iheya Ridge and Izena Caldron (Okinawa Trough)	1,028–1,523	M	8	AB036710	15	AB101421	I	−67.5 to −68.1	9

B. septemdierum	H	Suiyo Seamount, Myojin Knoll (Izo Ogasawara Island Arc), Okinawa trough, Mariana back Arc	1,288–3,600	S	8	AB036709	15	AB101430	I	–37	7
B. brevior	H	Lau Basin, Mariana Through, north Fiji	1,750–3,289	S	31	DQ321714	24	AY649799	I	–30.8 to –35.8	5, 30
MOTU 16	W	Bohol sea	357–372	S	33	AM931532	33	EU350070	E		
A. longissima (BC 279)	W	Bohol sea	450–1,764	S	29	AM851094	33	EU350072	E		
B. sp. Manus	H	Manus Basin	1,629	S	31	EU326223	15	AB101431	?		
B. tangaroa	S	KERMadec Arc – Cape Turnagain – Cape Kidnappers	920–1,205	S	31	EU326222	17	AY608439	?		
G. gladius	H	KERM Rumble 3 and 5 seamount	216–755	S	31	EU326224	24	AY649802	?		
"*B.*" sp. NZ3	H	KERM – Maucauley Cone	200	S	31	DQ321718	17	AY608440	?		
Idas sp. C	W, B	Near Vanuatu	290–802	S	32	EU683305	32	EU702374	E		
Idas sp. D (= MOTU5, BC288 and 294)	W	Bohol Sea	762–1,764	S	29	AM503926, AM503922	32, 33	EU702359, EU350071	E		

(continued)

Table 6.1 (continued)

Area	Name	HAB	Sites	Reported depth range (m)	Symbionts	Ref symbionts	Accession number symbionts	Ref host	Accession number host	I, E	δ13C range (‰)	Ref isotopes
	BC1007	W	Bohol Sea, Maribojoc island	387–453	S	29	AM503923			E		
	Adipicola crypta	W, B	Philippines, Vanuatu, Japan (off Kagoshima)	440	S	32	EU683302; EU683308	32	EU702317	I		
Indian ocean	*B. marisindicus* (= *B.* aff. *brevior*)	H	Central Indian Ridge (Kairei and Edmond)	2,450–3,300	S	20	DQ077891	20, Smith unp.	AY275543	I	−20 to −30.8	20

HABitats: H: hydrothermal; S: cold seeps; W: wood falls; B: bone falls. Symbionts: M: methanotroph; S: thiotroph; My: Methylotroph; G: unknown Gammaproteobacterium; CFB: Bacteroidetes. Accession numbers are given for bacterial 16 rRNA symbionts and COI hosts. I, E: Intra- or Extracellular localization of bacteria

1 Distel et al. 1988; 2 Fisher et al. 1988; 3 Distel and Cavanaugh 1994; 4 Nelson and Fisher 1995; 5 Dubilier et al. 1998; 6 Robinson et al. 1998b; 7 Trask and Van Dover 1999; 8 Fujiwara et al. 2000; 9 Barry et al. 2002; 10 Colaco et al. 2002; 11 Fiala-Médioni et al. 2002; 12 Van Dover et al. 2003; 13 Won et al. 2003b; 14 Won et al. 2003a; 15 Miyazaki et al. 2004; 16 Raulfs et al. 2004; 17 Smith et al. 2004; 18 Duperron et al. 2005; 19 McKiness et al. 2005; 20 McKiness and Cavanaugh 2005; 21 Salerno et al. 2005; 22 DeChaine et al. 2006; 23 Duperron et al. 2006; 24 Jones et al. 2006; 25 Duperron et al. 2007; 26 Olu-LeRoy et al. 2007a; 27 Olu-LeRoy et al. 2007b; 28 Duperron et al. 2008b; 29 Duperron et al. 2008a; 30 Henry et al. 2008; 31 Won et al. 2008; 32 Lorion et al. 2009; 33 Duperron et al. 2009

has stimulated the interest of many investigators since its discovery in the early 1980s. Many species harbour sulphur-oxidizing symbionts comparable to those described from other bivalves families (Lucinidae, Thyasiridae, Solemyidae and Vesicomyidae), but some species harbour methane-oxidizing symbionts, a very rare feature among bivalves. Multiple symbioses involving two to six distinct bacterial types are another originality of these deep-sea mussels, and appear as a good way to deal with highly variable environments they are experiencing (Fisher et al. 1993; Distel et al. 1995).

Although no symbiont has yet been isolated in pure culture, the diversity, role, physiology, ecology and evolution of mussel symbioses have been documented. The development of tools applied to deep-sea biology, such as molecular techniques or experimental devices, and the input from other disciplines such as deep-sea imagery or in situ chemistry has allowed a better understanding of the ecology of chemosynthesis-based ecosystems. The present contribution gives a general overview of the diversity of mussels associated with deep-sea chemosynthesis-based ecosystems and their bacterial symbionts, with an emphasis on recent advances in the understanding of interactions between hosts, bacteria, and their environment. Finally, some new questions and directions for future research are presented.

6.2 Mussel Distribution and Habitats

Mussels have been collected over the years from many deep-sea chemosynthesis-based environments spread over the world including hydrothermal vents, cold seeps and organic falls such as whale bone or wood deposits (Plate 11). Mussels are mostly epibenthic and live attached by byssal threads to hard substrates, sometimes forming tight aggregates (Fig. 6.1, Plate 12). Notable exceptions are *Bathymodiolus boomerang* from the Barbados accretionary prism and *Gigantidas horikoshi*, which live two-thirds buried into the sediment (Hashimoto et al. 1995; von Cosel and Olu 1998; Desbruyères et al. 2006). The mussels' reliance on chemosynthetic symbionts located in their gills implies that reduced compounds are needed in the surrounding fluid, as mussels do not have access to the anoxic sediment. This does not normally occur in the deep-sea despite the anoxy of the sediment because fluxes are not sufficient to emit reduced compounds into the bottom seawater. On the contrary, fluids emitted at cold seeps and hydrothermal vents contain huge amounts of reduced compounds such as sulphide and methane (Van Dover 2000; Le Bris et al. 2006b; Sarrazin et al. 2006). Similar compounds, for example sulphide and methane, are also produced during the microbially-mediated decay of large debris of organic matter such as whale or wood falls (Leschine 1995; Baco and Smith 2003; Goffredi et al. 2008; Treude et al. 2009).

Around vents, mussels usually occur on chimneys or in the vicinity of cracks where diffuse flow occurs (Fig. 6.1b) (Johnson et al. 1994; Fisher 1995; Cuvelier et al. 2009; Le Bris and Duperron in press). At seeps mussels usually live attached

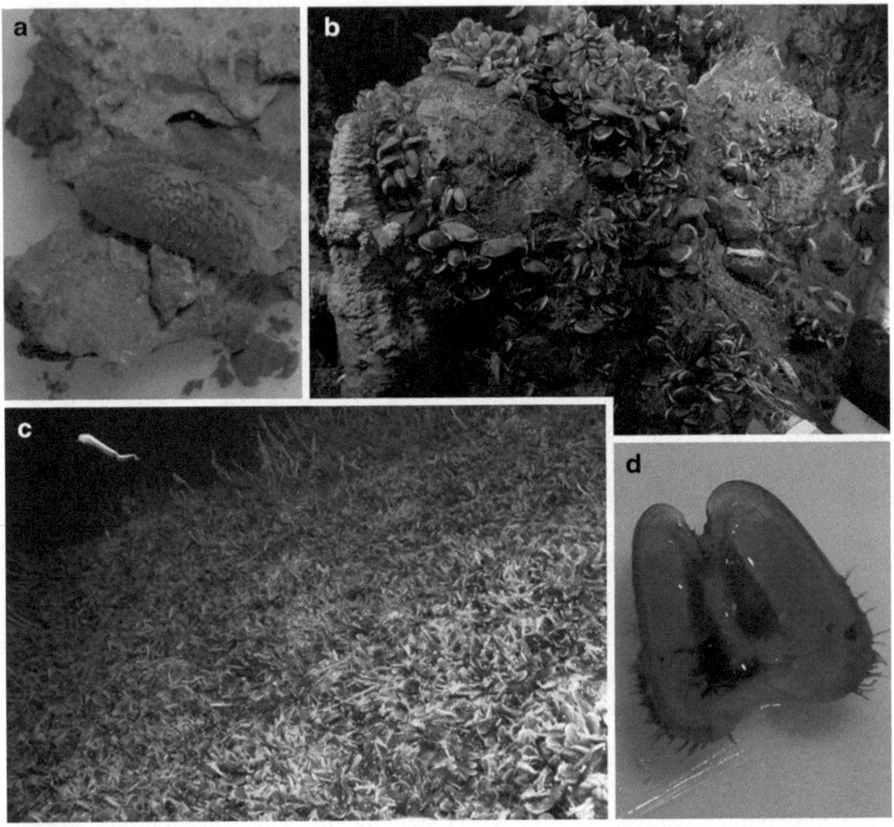

Fig. 6.1 Morphology and habitats of bathymodiolin mussels. (**a**) *Idas* sp. Med attached to a carbonate crust recovered from cold seeps in the eastern Mediterranean (MEDECO cruise, 2007, photo by O. Gros). Note the two whitish siphons in the posterior part. Shell length 1 cm. (**b**) *Bathymodiolus azoricus* specimens attached to a fluid-emitting edifice at the Rainbow hydrothermal vent site (Mid Atlantic Ridge, 2,277 m depth). Notice the iron deposits on chimneys and shells, and the presence of *Rimicaris exoculata* shrimps on the right of the image. Image © Ifremer, MOMARDREAM cruise, 2008 (chief scientist: J. Dyment). (**c**) Dense bed of the mussel *Bathymodiolus* sp. at a cold seep site situated in a depression on top of a diapir structure in the lower Congo Basin (2,700 m depth). The area displayed gas bubble sources. Other metazoans include tubeworms (*Lamellibrachia* sp., in the back), shrimps, and limpets (small white dots visible on mussel shells). Image © Marum, University of Bremen, GUINECO M76/3a cruise, 2008 (chief scientist: A. Boetius). (**d**) *Idas* sp. Med, opened valves. Note the white foot, brownish visceral mass below and gills in the left and right valves. A colour version of this figure can be found in Appendix (Plate 12)

to authigenic carbonates, sometimes covering large areas (Fig. 6.1c), or other hard substrates such as tubes of vestimentiferans (Olu-LeRoy et al. 2007b). Living at the oxic-anoxic interface can be risky. In a documented example, specimens of *B. childressi* form a ring, a few meters wide, surrounding a 190 m^2 methane-rich anoxic hypersaline brine pool in the northern Gulf of Mexico (GoM). However, some mussels located in the inner rim were dead after being entirely submerged by an elevation of the brine level (MacDonald et al. 1990). Similarly, small specimens of *B. azoricus* reach densities above 10,000 individuals m^{-2} on the walls of active venting chimneys at Lucky Strike despite the risk of exposure to deadly hot fluids (von Cosel et al. 1999). Specimens of *B. brevior* were shown to survive in acidic waters with pH values as low as 5.36 at the Eifuku volcano (Mariana Arc), although with shell thickness and daily growth rates only half those measured in mussels occurring in neutral waters (Tunnicliffe et al. 2009). Aside of such occasional extremes, mussels have to tolerate only a relatively moderate temperature range. *B. brevior* has an upper temperature limit of 35°C and tolerates 2 h exposure at 19°C (Henry et al. 2008). Heat shock experiments for 2 h at 25°C and 30°C were shown to result in global depression in gene expression in *B. azoricus*, suggesting that mussels are not adapted to 'hot' temperatures (Boutet et al. 2009). At vents, mussels are indeed usually out-competed at the hottest habitats by specialists such as the gastropod *Ifremeria nautilei* at Lau Basin, the annelid polychaete *Alvinella pompejana* at sites of the East Pacific Rise (EPR), or the shrimp *Rimicaris exoculata* at sites of the Mid Atlantic Ridge (MAR). They rather occupy more peripheral areas with more diffuse venting and temperatures slightly above those of ambient seawater, up to 15°C (Sarradin et al. 1999). Although in situ measurement of physico-chemical parameters is difficult, and some compounds such as methane can not be monitored directly, several studies have described local conditions around mussels. An important point is that physico-chemical characteristics are highly variable in both space and time, particularly at vents, resulting in varying sharp gradients to which the organisms must adapt (Johnson et al. 1994; Le Bris et al. 2006b). Mussels thrive in environments with moderate concentrations of reduced compounds (compared to end-member fluids), in the range of micromolars for sulphide and methane. For example, sulphide concentrations between 0.5 and 62 μM were reported in the vicinity of mussel beds at the Lucky Strike vent site (<0.5 to 26 μM at Menez Gwen), with larger mussels found at sulphide-rich locations (Sarradin et al. 1999), and up to 325 μM sulphide was recorded at Rose Garden (Galapagos Rift) (Johnson et al. 1986, 1994; Fisher et al. 1988). Methane concentrations up to 33.7 μM were reported from mussel beds at the Régab cold seep (Duperron et al. 2005; Olu-LeRoy et al. 2007b). The presence of additional chemical compounds can influence the habitat. For example, end member fluids at the Rainbow vent site display low sulphide concentrations, and the abundance of dissolved iron (which forms deposits on shells, see Fig. 6.1b) further limits the bioavailability of sulphide for mussels (Le Bris et al. 2006a). At vents, toxic heavy metals can also be abundant, even in diffuse flow, and mussels use a variety of systems to protect themselves (Geret et al. 1998; Rousse et al. 1998). Mussels 'engineer' their vent habitat by redirecting fluid flow and dispersing it throughout the mussel bed (Johnson et al. 1994). They also display a limited mobility, and can move a bit to optimize access to their substrates (Childress 1988; Johnson et al. 1994). Compared to vent and seep habitats, precise characteristics of organic falls have rarely been documented, although some data indicate enrichment in sulphide in the sediment close to whale bones, and the occurrence of methane (Smith et al. 1998; Fujiwara et al. 2007; Treude et al. 2009). Recent results confirm the direct emission of sulphide from wood deployed in a shallow mangrove habitat (Laurent et al. 2009). In summary, observations and results indicate that mussels require habitats where oxygen is available as well as electron donors and energy sources for their bacterial symbionts.

6.3 The Diversity of Mussels

6.3.1 General Description

Around 37 mussel species reported from chemosynthesis-based ecosystems have been assigned a name (22 large mussels within the genera *Bathymodiolus* and *Gigantidas*, and around 16 small species within the genera *Idas* and relatives including *Adipicola*, *Benthomodiolus*, or *Tamu*) (Dell 1987; von Cosel 2002; von Cosel and Marshall 2003; von Cosel and Janssen 2008; Pelorce and Poutiers 2009). Many more possible 'species' have been identified based on molecular studies (Smith et al. 2004; McKiness and Cavanaugh 2005; McKiness et al. 2005; Jones et al. 2006; Olu-LeRoy et al. 2007a; Samadi et al. 2007; Lorion et al. 2009). They display several common features. Shell is modioliform and usually brownish (Fig. 6.1). Adult shell length varies from ~10 to 360 mm, larger species (>40 mm) being restricted to hydrothermal vents and cold seeps. The mantle delimiting the pallial cavity displays different degrees of fusion. In some species (such as *Idas* sp. from the eastern Mediterranean, see Fig. 6.1a), siphons comparable to those of infaunal bivalve species have been described on living specimens despite their epibenthic lifestyle (von Cosel 2002; Okutani et al. 2003; Fujiwara et al. 2007). Compared to their shallow, non-symbiotic relatives, mussels display reduced labial palps and guts, the latter varying among species from curved or S-shaped to straight (Page et al. 1991; von Cosel 2002). Gills are typically enlarged, occupying for example 77% of shell length in *B. boomerang*, and very often dark coloured, a common trend in symbiotic bivalves (Fisher 1990; von Cosel and Olu 1998). The gills are filibranch. They display inner and outer demibranchs composed of ascending and descending lamellae, with few-to-no inter-lamellar junctions, and ciliary junctions between consecutive filaments (von Cosel et al. 1999). Gill epithelium is constituted of one layer of cells, of which some harbour symbionts (see below and Fig. 6.2) while others are almost symbiont-free and harbour cilia (intercalary cells) (Fiala-Médioni 1984). Gills retains the ability to filter feed at rates comparable to shallow mussels, as shown in *B. azoricus* and *"B." childressi* (the use of quotation marks is explained in Section 6.3.2), and the digestive tract is functional (Page et al. 1991; Pile and Young 1999). To colonize scattered habitats efficiently, mussels have developed specific strategies.

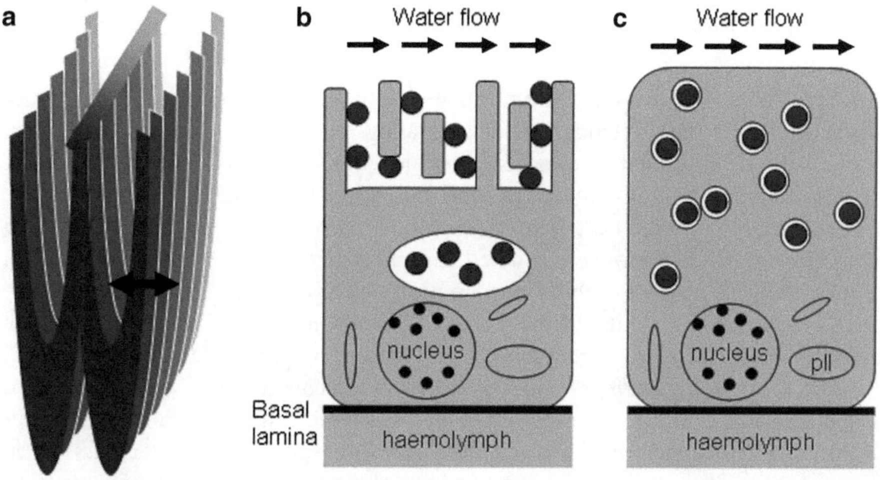

Fig. 6.2 (**a**) Frontal view of a single mussel gill. Arrow indicates how transverse sections through filaments (used for electron and fluorescence microscopy) are obtained. (**b–c**) Detail of two mussel gill epithelial cells as seen in transverse sections. Notice the basal lamina underlying the gill epithelium and separating cells from the hemolymph, and the nucleus and phagolysosome-like bodies (pll) in the basal part of cells. Arrows indicate the direction of water flow. Sulphur-oxidizing bacteria are displayed in dark grey. (**b**) Extracellular bacteria located between microvilli, or engulfed in a large vacuole, as typically observed in some small bathymodiolines from organic falls; (**c**) Intracellular bacteria inside of host cells within vacuoles, as observed in most large species from hydrothermal vents and cold seeps

Bathymodiolinae are dioecious or successive hermaphrodites, but simultaneous hermaphrodites rarely occurred in the species investigated (Le Pennec and Beninger 2000). For example *Idas washingtonia* displayed evidence of protandric hermaphroditism (Tyler et al. 2009). Males represented 74% of individuals investigated and included the smallest mussels. Female and hermaphrodites represented 10% and 14% of specimens, respectively. Although not well-documented, the development of mussels appears to involve small and numerous gametes and eggs (Tyler et al. 2007, 2009). The large size of the larval shell indicative of a long larval life, planktotrophic larvae probably have large dispersal abilities (Lutz et al. 1980, 1984; Comtet et al. 2000; Arellano and Young 2009, reviewed in Le Pennec and Beninger 2000 and Tyler and Young 2003). Mussel lifespan is hard to estimate compared to that of many coastal bivalves, because no comparable typical cycles (seasonal, lunar, circadian, tidal) are proven to be linked with shell increments. Rough estimates of an age around 18 years for a 14 cm-long specimen of *B. brevior* were suggested, and *B. thermophilus* specimens may get older than 25 years (Schöne and Giere 2005). This is not an unusual lifespan among mytilids, as *Modiolus modiolus*, a non symbiotic mussel, was reported to live 50 years (von Cosel and Olu 1998).

6.3.2 *Taxonomy and Classification*

The taxonomy of mussels from chemosynthesis-based ecosystems is under discussion. Not many distinctive features allow comparisons among species, and some could be attributed to allometric growth (von Cosel and Olu 1998). Most large species found at hydrothermal vents and cold seeps are classified within the genus *Bathymodiolus* of which the type species is *B. thermophilus* from the EPR (Kenk and Wilson 1985). However, detailed morphological investigations indicate that *Bathymodiolus* can be subdivided in three groups: the *"B." childressi*, the *B. thermophilus*, and the *B. aduloides* group (von Cosel 2002; von Cosel and Janssen 2008). A second genus of large mussels is *Gigantidas*, morphologically distinct from *Bathymodiolus* and described from vents around New Zealand (von Cosel and Marshall 2003). Although known since the end of the nineteenth century from trawls (example in Jeffreys 1876), smaller species (shell lengths <3 cm, Fig. 6.1a) have been less investigated and morphological descriptions are sometimes ancient and rarely backed by molecular data. Members of the genera *Idas* and *Adipicola* (classified within the sub-family Modiolinae), or the more recently erected genus *Tamu*, have been reported from cold seeps as well as organic falls from various locations. Except for their smaller size, their morphology resembles that of large *Bathymodiolus*. Soft parts have however not been studied in great detail.

Molecular phylogenies based on host nuclear and mitochondrial genes support the monophyly of symbiont-associated mussels within the family Mytilidae. The subfamily name 'Bathymodiolinae' has been proposed. The monophyly of *Bathymodiolus* is not supported. Analyses of COI, ND4 and 28S rRNA-encoding genes distinguish three *Bathymodiolus* clades, not contradictory to the groups suggested by von Cosel (Miyazaki et al. 2004; Jones et al. 2006; Samadi et al. 2007; Lorion et al. 2009). The three groups emerge from a multifurcation which includes smaller mussels classified as *Idas*, *Tamu* or *Adipicola*. To distinguish between 'true' *Bathymodiolus* from the *B. thermophilus* group and others, Gustafson suggested the use of quotation marks for representatives of other groups until the taxonomy is fully clarified (Gustafson et al. 1998). Molecular studies also point to the existence of cryptic species (Moraga et al. 1994; Won et al. 2003a; Smith et al. 2004; Olu-LeRoy et al. 2007a). A re-assessment of the nomenclature is thus needed which should combine morphological as well as molecular data, as used for example by Maas and co-workers to distinguish between mid-Atlantic ridge (MAR) vent species (Maas et al. 1999).

6.4 The Diversity of Bacterial Symbionts

Bacteria have been shown to occur in all Bathymodiolinae species investigated to date. As for other chemosynthetic symbioses, no symbiont has been isolated in pure culture despite attempts (Nelson et al. 1995). Bacteria are associated with gill epithelial cells located in the lateral zone of gill filaments (Fig. 6.2). This allows bacteria to gain access to compounds present in circulating fluid, including energy and carbon sources. Comparable gill symbioses, though involving only sulphur-oxidizers (thiotrophs), have also been reported from the bivalve families Thyasiridae, Lucinidae, Solemyidae and Vesicomyidae (Fisher 1990; Stewart et al. 2005; Stewart and Cavanaugh 2006; Taylor and Glover 2006).

6.4.1 Sulphur-Oxidizing Bacteria

Of the 27 mussel 'species' for which molecular information about symbiosis is available, 24 harbour bacteria related to chemoautotrophic sulphur-oxidizing (or thiotrophic) Gammaproteobacteria, based on comparative analysis of their16S rRNA-encoding gene (Table 6.1). Initially discovered in the EPR vent species *B. thermophilus*, sulphur oxidizing symbionts are small (0.5 μm diameter on electron micrographs) Gram negative bacteria (Rau and Hedges 1979; Felbeck et al. 1981; Cavanaugh 1983; Fiala-Médioni 1984). They fix carbon autotrophically using the Calvin Benson Bassham (CBB) cycle as demonstrated by the presence of the functional form I of Ribulose 1,5 bisphosphate carboxylase/oxygenase (RubisCO) and its encoding gene in several species (Fisher et al. 1988; Robinson et al. 1998a; Pimenov et al. 2002; Spiridonova et al. 2006). Measurements of enzyme activities and gene sequencing also demonstrated the presence of adenosine phosphosulfate (APS) reductase and ATP sulfurylase, two enzymes involved in sulphur metabolism (Nelson and Fisher 1995). Thiosulphate and sulphide stimulate the fixation of inorganic carbon based on incubation experiments of animal tissue or living specimens of *B. thermophilus* and *B. azoricus* (Belkin et al. 1986; Nelson et al. 1995; Riou et al. 2008). The use of alternate energy sources, such as hydrogen, has been suspected but has not been proven to date. Contrary to what is observed for example in lucinid clams, symbionts do not deposit sulphur granules (Dando et al. 1986).

In most species, thiotrophic symbionts occur within vacuoles containing one to several bacteria each that are located inside of host gill epithelial cells named 'bacteriocytes' (Fig. 6.2) (Fiala-Médioni 1984; Dubilier et al. 1998). Recently however, thiotrophs were shown to occur extracellularly between microvilli of epithelial cells in several small species such as MOTU 16, *A. longissima*, *Idas* sp. C and *Idas* sp. D from organic falls of the western Pacific (Fig. 6.2) (Gros and Gaill 2007; Gros et al. 2007; Duperron et al. 2008a; Lorion et al. 2009). *Idas simpsoni*, *I. washingtoniana* and *Adipicola* sp. JDF (for Juan de Fuca ridge) also display extracellular bacteria as evident from microscopic investigation (Southward 2008). In many species, symbiont localization was not investigated, or ambiguous such as in *Bathymodiolus* sp. from the Juan de Fuca ridge (McKiness et al. 2005; Duperron et al. 2008b; Won et al. 2008). Either intra- or extracellular, symbiont localization appears to be consistent and conserved within a given host species (Lorion et al. 2009).

All sulphur-oxidizing symbionts of mussels cluster within a clade of Gammaproteobacteria that also includes sequences from a few free-living uncultivated bacteria, and a clade of maternally-inherited endosymbionts associated with vesicomyid clams (Distel et al. 1988; Peek et al. 1998; Duperron et al. 2008a). Grouping free-living bacteria as well as extracellular, intracellular, and maternally-inherited (with reduced genome) symbionts, this clade is a good target group to study the processes leading from free-living to strictly endosymbiotic bacteria. One 16S rRNA symbiont phylotype (a phylotype is a version of the sequence) is usually associated specifically with a single mussel species, although exceptions occur. The two MAR vent species *B. azoricus* and *B. puteoserpentis* for example share a single 16S rRNA symbiont phylotype (Duperron et al. 2006), and the co-occurrence of two distinct thiotroph-related 16S rRNA phylotypes representing two distinct symbiont strains was reported in *B. heckerae* and in *Idas* sp. Med (Duperron et al. 2007, 2008b).

6.4.2 Methane-Oxidizing Bacteria

To date, at least ten species from cold seeps and hydrothermal vents are reported to harbour bacteria related to Type I methane-oxidizing (or methanotrophic) Gammaproteobacteria, of which three have only methanotrophs, five also have sulphur-oxidizers, and two – *Bathymodiolus heckerae* and *Idas* sp. Med – have four and six symbiont types in total, respectively (Table 6.1). Methanotrophic symbiosis is rare among metazoans, and was suggested only for some sponges (Vacelet et al. 1996), a few siboglinid annelids (Schmaljohann et al. 1990), the gastropod *Ifremeria nautilei* (Borowski et al. 2002), the thyasirid bivalve *Conchocele bisecta* (Kamenev et al. 2001), and epibionts of the shrimp *Rimicaris exoculata* (Zbinden et al. 2008). Methanotrophic symbionts were initially discovered in mussels from the Gulf of Mexico cold seeps. *B. childressi* grows using methane as a carbon and energy source (Childress et al. 1986; Cary et al. 1988). *Bathymodiolus heckerae* harbours large bacteria with a diameter of 1.5 μm and stacked internal membranes typical of free-living Type I methanotrophs. This species also displays unexpectedly negative $\delta^{13}C$ tissue values, very close to those reported for local methane (Cavanaugh et al. 1987). Methanol dehydrogenase activity, the second enzyme of the aerobic oxidation of methane, was shown (Fisher et al. 1987; Robinson et al. 1998b; Pimenov et al. 2002). The presence of methane mono-oxygenase, the diagnostic enzyme for aerobic methane oxidation, was confirmed based on gene sequence analysis as well as in situ hybdridization of its mRNA. However, its activity is difficult to measure because of its unstability (Pernthaler and Amann 2004; Spiridonova et al. 2006; Duperron et al. 2007). Besides methane utilization, RubisCO appears to be present in methane-oxidizing symbionts of a *Bathymodiolus* from Okinawa Through, indicating the possible presence of a functional CBB cycle (Elsaied et al. 2006). When present, methanotrophs are always intracellular, located within vacuoles inside bacteriocytes.

Symbiotic methanotrophs form a clade within the Gammaproteobacteria which is the sister-group of free-living Type I methanotrophs (Distel and Cavanaugh 1994; Duperron et al. 2005). To date, very few free-living bacteria cluster within this clade questioning the existence of active free-living forms of symbionts. Similar to thiotrophs, one 16S rRNA symbiont phylotype is usually associated with a single mussel species. Exceptions again are *B. azoricus* and *B. puteoserpentis* that share a single methanotroph-related phylotype (Duperron et al. 2006).

6.4.3 Additional Symbionts and Multiple Symbiosis

Besides the presence of methanotrophic symbionts, the main originality of bathymodiolines is the occurrence of multiple symbioses. 'Dual symbiosis' involving sulphur- and methane-oxidizers co-occurring within a single bacteriocyte was first suspected in *B. heckerae* (Cavanaugh et al. 1987), hypothesized in *B. brooksi* from the Gulf of Mexico based on microscopy (Fisher et al. 1993), and demonstrated formally using 16S rRNA gene sequencing and fluorescence in situ hybridization (FISH) in the vent species *B. puteoserpentis* (Distel et al. 1995). Dual symbiosis has subsequently been documented in *B. azoricus*, *B.* aff. *azoricus* from Lost City, *B. boomerang* from the Barbados accretionary prism, *B.* cf. *boomerang* from Régab (Plate 13a–b) and re-investigated in *B. brooksi* (Olu et al. 1996; Fiala-Médioni et al. 2002; Duperron et al. 2005, 2007; DeChaine et al. 2006).

Recently, thanks to the sequencing of larger bacterial 16S rRNA clone libraries from gill tissues, new types of bacterial symbionts have been identified in the gills of two cold seep species, *B. heckerae* and *Idas* sp. Med, and confirmed using FISH (Duperron et al. 2007, 2008b). Both mussel species harbour one methanotroph- and two distinct thiotroph-related 16S rRNA symbiont phylotypes, labelled M, S1 and S2 in Table 6.1, respectively (Plate 13d). Both mussels also harbour phylotypes related to free-living methylotrophs of the genus *Methylophaga* (Plate 13c). Although the metabolism of these symbionts is unknown, it can be hypothesized that they are actually methylotrophs, i.e. bacteria utilizing mono carbon compounds such as methanol, but not methane. Use of methane is restricted to the subgroup of methanotrophs (Anthony 1982). Finally, *Idas* sp. has two additional symbionts compared with *B. heckerae*. One is a Gammaproteobacterium unrelated to any cultivated bacterium, making it impossible to propose any metabolic scheme. The other is a member of the Bacteroidetes group, related to members of the genus *Tenacibaculum* and cellulose-degrading bacteria such as *Polaribacter dokdonensis*. Its role is unknown, and could be linked either to the degradation of organic compounds or pathogenicity. Intra- or extracellular localization of these recently discovered symbionts has not yet been ascertained.

Bacteria suspected to be deleterious have also been identified in mussel gill tissues. "Candidatus *Endonucleobacter bathymodioli*", a Gammaproteobacterium related to sequences from a bacterium causing mass mortalities in Pacific razor clams, has recently been identified within nuclei of gill intercalary cells in several species of *Bathymodiolus* from various geographical origins (Zielinski et al. 2009). This filament-shaped bacterium apparently establishes itself within host cells nuclei, multiplies and destroys the content of the nucleus. It then causes the nucleus to break open, releasing offspring bacteria to the environment, similar to the way that is used by some viruses. "Ca. *E. bathymodioli*" is apparently closely related to several bacterial species based on sequences recovered from marine animal samples, including two which also have been hypothesized to originate from endonuclear bacteria (Kerk et al. 1992). This led Zielinski et al. (2009) to propose that endonuclear bacteria could be widespread among marine invertebrates.

Clearly, recent discoveries of new bacterial partners and of the occurrence of up to six symbionts within a single mytilid host indicate that the diversity of bacteria associated with mussel tissues has been underestimated. Cheaper high-throughput sequencing technologies should allow in the future to gain a more exhaustive and systematic overview of bacteria associated with bathymodioline mussels.

6.5 The Symbiotic System: How Does It Work?

Mussels and their sometimes diverse associated bacteria form complex 'symbiotic systems' in which each partner interacts with others and with the environment. Although the nature of these interactions is rarely well-understood, the following chapter aims at summarizing current knowledge concerning aspects of these interactions.

6.5.1 Interactions Between Mussels and Their Bacterial Symbionts

How bacteria are acquired by host tissue has not been fully resolved. Although early developmental stages of dual symbiotic species display sulphur- and methane-oxidizing bacteria (Salerno et al. 2005), these were not observed from gonad tissue or eggs, suggesting an early environmental acquisition. The extracellular localization of thiotrophs in several small mussel species, and the observations of open vacuoles engulfing sulphide-oxidizing bacteria in the most apical part of bacteriocytes in many species with intracellular symbionts support this hypothesis (Dubilier et al. 1998; Duperron et al. 2008a; Southward 2008; Lorion et al. 2009). Detailed genetic investigations have shown the co-occurrence of several 16S rRNA or ITS (Internal Transcribed Spacer, a more variable marker gene allowing to distinguish between closely related bacterial strains) phylotypes of sulphide-oxidizers in the gills of mussels from the Mid-Atlantic Ridge, Gulf of Mexico, eastern Mediterranean and western Pacific (Won et al. 2003b; Duperron et al. 2007, 2008b; Lorion et al. 2009). Because some of these strains were usually found in mussel species occurring nearby, these observations are interpreted as further support for environmental acquisition. The transmission of methanotrophs and other types of symbionts has not been studied in detail. To date only one 16S rRNA methanotroph-related phylotype was reported from a given mussel species. Genomic investigations will help documenting whether bacteria can live outside of their hosts, and how hosts and bacteria recognize each other and establish the association.

Contrary to tubeworms such as *Riftia pachyptila*, or clams such as *Calyptogena* spp., no specific H_2S transporter has been identified in mussels. In tubeworms, bacteria are located inside a special organ, the trophosome, remote from the gill where H_2S is taken up. In *Calyptogena* (vesicomyid) clams the foot mines for sulphide in the sediment or emitted waters whereas bacteria sit in the gills (Childress et al. 1993; Scott and Fisher 1995; Flores et al. 2005). Both are highly-integrated symbioses in which reduced sulphur needs to be transported to the symbionts. In mussels, symbionts occurring in (or on) gill epithelial cells are very close to the circulating fluid that contains both their electron donors, including sulphide, and acceptors (Fig. 6.2). Transporters are therefore not needed.

The nutritional role of symbionts has been well-documented. Stable isotope analyses are a classical tool used to trace carbon sources in ecology (Lajhta and Michener 1994). Tissue carbon isotope signatures can be compared with signatures of carbon sources to estimate their quantitative importance in the nutrition of animals. In hydrothermal vents and cold seeps, the role of sulphur- or methane-oxidizing symbionts in mussel nutrition can be, to a certain extent, evaluated using this approach (Fisher 1995). Along with direct incubations of gill material or whole living mussels in the presence of radio-labelled inorganic carbon and methane, these methods have demonstrated the transfer of bacterially-acquired carbon to the host tissue from both sulphur- and methane-oxidizing symbionts (Table 6.1) (Childress et al. 1986; Fisher et al. 1987; Fisher and Childress 1992; Kochevar et al. 1992; Jahnke et al. 1995; Nelson et al. 1995; Riou et al. 2008). Estimates for *B. heckerae* mussels from Blake Ridge show 60% of their organic carbon from methane- and 40% from sulphur-oxidizers based on stable isotopes investigation (Van Dover et al. 2003). How carbon transfer is achieved is however still debated. Early work has shown the digestion of bacteria within phagolysosome-like bodies located in the basal part of bacteriocytes (Fig. 6.2) (Fisher et al. 1987; Fiala-Médioni et al. 1994). No such phagolysosome-like bodies were convincingly observed in extracellular sulphur-oxidizers, questioning their role as carbon providers (Duperron et al. 2008a). Other ways of transfer are known from other symbioses, for example the release of organic compounds by photosynthetic symbionts in corals. Another possibility was suggested following the observation of bacteria-filled bacteriocytes, detached from the gill, in the food groove. Mussels could possibly perform a type of autophagy. Direct evidence is lacking, but these alternate pathways can not be ruled out and could play a role. Besides their role in carbon nutrition, thiotrophs were also suggested to detoxify sulphide, which is toxic for host cytochrome oxidase, protecting mussels against occasional bursts of sulphide-rich fluids (Powell and Somero 1986; Dubilier et al. 1998; Duperron et al. 2005). Methanotrophs were suggested to contribute nitrogen uptake (Lee and Childress 1994; Lee et al. 1999; Dattagupta et al. 2004).

6.5.2 Interactions Among Bacteria

Dividing stages of sulphur- and methane-oxidizers have been observed in many mussel species, confirming that symbionts benefit from their cellular environment (examples in Fiala-Médioni et al. 1986; Duperron et al. 2005). In the limited volume available for growth within bacteriocytes, competition is likely. In multiple symbioses, direct competition between symbionts with distinct metabolisms is possibly prevented by their use of distinct carbon and energy sources. Relative densities appear to depend on environmental parameters that favour either metabolism (see below). Competition between two symbionts with the same presumed metabolism was suggested based on the observation of mutual exclusion between the two thiotrophs S1 and S2 from *B. heckerae*. They very rarely co-occur within a single bacteriocyte, but instead tend to occupy distinct areas of the gill filaments (Plate 13d) (Duperron et al. 2007). A similar co-occurrence of two thiotrophs has been reported in *Idas* sp. Preliminary observations indicate that one phylotype occurs mostly in the periphery of filaments while the other is spread over the whole lateral zone, but mutual exclusion has not been observed (Duperron et al. 2008b; Halary et al. personal observation). In both cases, it is not clear if the presence of two thiotrophs leads to true functional redundancy, or whether they display slight differences in their metabolisms.

Besides competition, syntrophic relationships have been suggested between methanotrophs and methylotroph-related bacteria in *B. heckerae* (Duperron et al. 2007). Methylotrophs were suggested to use methanol, a C1 intermediate compound in the methane oxidation pathway used by methanotrophs. This hypothesis was based on the close proximity between the two symbiont types observed in hybridized sections (Plate 13c). Syntrophy among symbionts was previously hypothesized between the sulphur-oxidizing and sulphate-reducing bacteria associated with gutless oligochetes (Dubilier et al. 2005; Woyke et al. 2006).

The nature of inter-bacteria interactions is very hard to study. Furthermore, the host likely controls what occurs within bacteriocytes, maybe not only through digestion. In vesicomyid clams for example, recent investigation of the genome content of the thiotrophic symbiont *Vesicomyosocius okutanii* indicated the absence of *ftsZ*, an important gene involved in bacterial division. Authors suggested a close control of bacterial division directly by the host (Kuwahara et al. 2007).

6.5.3 Interaction Between the Symbiotic System and Its Environment

Generally, even mussels associated with symbionts can filter-feed. Filter feeding can significantly contribute to mussel nutrition, as suggested by stable isotope-based evidence from "*B.*" *childressi* (Page et al. 1991; Pile and Young 1999). Recently, modelling approaches have indicated a shift from reliance on photosynthesis-derived carbon towards total reliance on bacterial-derived carbon during the growth of *B. azoricus* (Martins et al. 2008). However, transplant experiments in which *B. thermophilus* specimens were transferred from their vent habitat to bare rock or sediment devoid of fluid emissions have shown that mussels were quickly losing their symbionts and dying (Raulfs et al. 2004). This fits with the observation that specimens from a waning vent site were in bad condition, displaying low condition indexes and thin bacteriocytes with almost no symbionts. Altogether, this makes clear that organic matter available in the deep-sea is not sufficient to sustain symbiotic mussels without the help of their symbionts (Raulfs et al. 2004).

Deep-sea mussels can not survive in the absence of reduced compounds. However, detailed studies of the interaction between mussel symbioses and environmental parameters are still in their infancy due to our lack of precise knowledge about the physico-chemical environment of mussels and to the difficulties of experimentation on deep-sea animals. Deep-sea environments, in particular vent habitats, are highly variable in space and time even at the scale of centimeters. These steep gradients are caused by multiple distinct sources of fluids, biological activity, convective processes and fluid circulation, currents, and so on (see section 6.2). Determination of local physico-chemical parameters in the deep-sea requires the use of adapted in situ sensors, if they exist, and necessitates time-series measurements that are expensive to perform. Despite these limits, transplant experiments in which specimens of *"B." childressi* were transferred from a low-methane to a high-methane site (and vice versa) have clearly demonstrated increased mussel growth caused by increased abundance of methane, confirming the direct link between environmental parameters and mussels health (Nix et al. 1995; Bergquist et al. 2004). Experimentation is another difficulty. Certain mussel species from shallower sites, such as *B. azoricus* and *"B". childressi*, can be kept alive in aquaria at ambient pressure (Kadar et al. 2005; Riou et al. 2008). For deeper-dwelling species, pressurized aquaria in which specific compounds can be injected and monitored, such as the IPOCAMP™ device, are becoming available (Shillito et al. 2001). However, without detailed knowledge of habitat characteristics, it is not possible to properly simulate a realistic deep-sea vent or seep environment in an experimental setting. Even though observed trends can usually be extrapolated to the 'real' environment, exact physiological rates must therefore be taken with caution (Cary et al. 1988; Kochevar et al. 1992).

Mussels from different sites can display markedly different carbon stable isotope signatures that indicate different relative contributions of sulphur- and methane-oxidizers, even within a single species. (Trask and Van Dover 1999). This was shown for mussels from the Mid-Atlantic ridge associated with both sulphur- and methane-oxidizing symbionts, where isotope signatures were correlated with sulphide concentration in the end-member fluid (Colaco et al. 2002). To test whether these differences in the nutritional role of symbionts arise from differences in bacterial abundances, accurate quantification methods are required. However quantifying bacterial densities is not straightforward. Using bacteria-specific indicators such as lipids or quinones, densities of 1.7×10^{10} to 2×10^{11} bacteria per gram of gill tissue were estimated (Powell and Somero 1986; Yamamoto et al. 2002). Methanotrophic symbionts of *"B." childressi* were estimated at 5.6×10^{9} bacteria per gram tissue (Kochevar et al. 1992). Quantifying variations of symbiont densities is also difficult. Several methods have been employed, including direct counts on TEM pictures (benefiting from the distinct morphologies of sulphur- and methane-oxidizers), quantification of 16S rRNA, and quantification of volumes occupied by distinct symbionts in bacteriocyte sections (Fiala-Médioni 1984; Trask and Van Dover 1999; Fiala-Médioni et al. 2002; Duperron et al. 2007; Halary et al. 2008; Riou et al. 2008). Using such methods on mussels collected from several MAR hydrothermal vent sites, the authors demonstrated that symbiont populations were flexible. Relative abundances of methane- and sulphur-oxidizers were correlated with the composition of the end-member vent fluids, with the former becoming predominant at methane-rich sites and vice versa. Symbiont densities in *B. azoricus* specimens that were experimentally exposed to either sulphide or methane responded within hours to changes in their environment, with an increased abundance of sulphur- or methane-oxidizers, respectively (Kadar et al. 2005; Halary et al. 2008; Riou et al. 2008). Dynamic multiple symbioses thus appear to allow mussels to quickly optimize their access to available resources in variable environments, and cope with more diverse types of habitats, as suggested shortly after their discovery (Distel et al. 1995). Recently, we suggested that the specific bacterial densities observed in specimens of a single mussel species found on wood or whale falls are a consequence of sulphide availability (Lorion et al. 2009). Tools are becoming available to study the interaction between mussels, symbiont populations, and the environment in detail. However, more interdisciplinary studies coupling biological, chemical and modelling approaches are required (Martins et al. 2008).

6.6 The Symbiotic System: How Did It Evolve?

6.6.1 The Evolution of Mussels

Mussels occurring at chemosynthesis-based ecosystems and organic falls group in a single clade within the Mytilidae. The question of how the group evolved, what its common ancestor looked like, what the environment it lived in was or whether symbiotic bacteria were present, is hard to answer. Bathymodiolines are known from fossil seeps since the late Eocene, but the occurrence of symbionts is not possible to establish in the fossil record (Kiel and Goedert 2006). Molecular studies suggest an age between 23.7 and 74.3 MY for the origin of the group (Little and Vrijenhoek 2003; Jones et al. 2006). Early work suggested the evolution of vent mussel species from seep ancestors (Craddock et al. 1995). The later addition of small species from organic falls in phylogenetic reconstructions lead Distel et al. to propose the hypothesis that vent and seep mussels derived from ancestors associated with wood and whale fall ecosystems (Distel et al. 2000). However, results from recent bathymodioline phylogenies draw a more complex picture. Some species associated with organic falls such as *Benthomodiolus lignicola* or *Idas* sp. Sal 3 appear to be basal to seep and vent species (Samadi et al. 2007) while at least one vent species, "*Bathymodiolus*" NZ3, possibly a relict from an early vent colonization event, is also quite basal (Jones et al. 2006). Apart from these deep-branching species, the bathymodioline tree today is a large multifurcation which includes at least three vent/seep clades (named the *B. thermophilus*, *"B." childressi* and *"B." aduloides* clades) as well as numerous small species from whale and wood falls (Iwasaki et al. 2006; Jones et al. 2006; Lorion et al. 2009). Authors identified a recurrent trend from shallow (<1,000 m) to deep habitats in the evolution of the three groups of *Bathymodiolus*-like mussels. This would advocate multiple events of seep and vent colonization from a pool of small species associated with organic falls (Fig. 6.3, Plate 12). Habitat shifts during the evolution of mussels are plausible. Indeed, although most species today are specialists of a single type of environment, several ubiquist species were recently identified using reliable genetic approaches. *B. platifrons* and *B. japonicus* occur both at seeps and vents around Japan, and *Adipicola crypta* and *Idas* sp. C were both associated with wood- and whale-falls in the western Pacific (Miyazaki et al. 2004; Lorion et al. 2009).

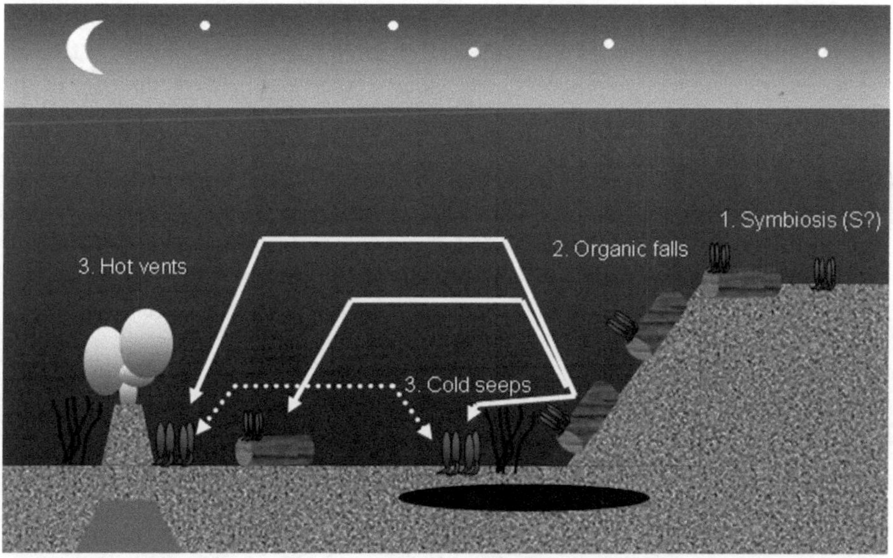

Fig. 6.3 The 'stepping stone' hypothesis. Ancestors of bathymodiolines were shallow species (on the right), which might have acquired the ability to associate with bacteria (1), most likely sulphur oxidizers. Mussels able to derive nutrition from both filter-feeding and bacterial chemoautotrophs colonized organic habitats such as wood and whale falls occurring at various depths (2). From such habitats, mussels then colonized hydrothermal vents at ridges and cold seeps on margins (3), giving rise to large *Bathymodiolus*-like species. This event probably occurred several times during the evolution of the group. Shifts between vent and seep probably occur, as evidenced by the existence of species colonizing both habitats (dotted arrow). A color version of this figure can be found in Appendix (Plate 12)

6.6.2 The Evolution of Mussel Symbiosis

Bacterial symbionts do not co-speciate with their mussel hosts (Won et al. 2008). However, very closely related mussel species, for example *B. azoricus* and *B. puteoserpentis* from the MAR, *B. boomerang* and *B.* aff. *boomerang* from the Gulf of Mexico (GoM) and Gulf of Guinea (GoG), or *Idas* sp. C and *Idas* sp. D from the Vanuatu and Philippine waters, harbour very closely related bacteria (Duperron et al. 2006, 2007; Lorion et al. 2009). Overall, symbiont phylogenies are not well resolved and sometimes ambiguous due to the low variation levels observed in 16S rRNA gene sequences (Duperron et al. 2005, 2008b; DeChaine et al. 2006; Won et al. 2008). Clearly, the understanding of symbiont biogeography and evolution would greatly benefit from the use of multi-marker approaches besides 16S rRNA. Reconstructions indicate that the occurrence of thiotrophic symbionts in the latest common ancestor of bathymodiolines is likely (Won et al. 2008; Duperron et al. 2009). Symbiont localization in the ancestor is however not clear. Thiotrophs are consistently intra- or extracellular depending on the species and mostly intracellular in seep and vent mussels and often extracellular in small *Idas*-like species from organic falls. However, symbiont localization is not documented or ambiguous for the deepest-branching species such as *Bathymodiolus* from the Juan de Fuca Ridge, *Benthomodiolus lignicola* or *"Bathymodiolus"* NZ3, thus not allowing a prediction (McKiness et al. 2005; Lorion et al. 2009). Whatever the localization, the acquisition of a flexible sulphur-oxidizing symbiosis in the common ancestor of bathymodiolines must have constituted an advantage for the mussels. Such an adaptation opened new potential ecological niches, and could explain the subsequent specialization of the group towards reduced ecosystems. It probably triggered a rapid diversification of the group which translated into the large multifurcation observed in host phylogenetic trees. Based on our current knowledge, methanotrophic symbiosis seems to appear several times in host trees, and could be more opportunistic from the animal's point of view. It apparently occurred only at methane-rich sites. However, host – symbiont specificity appears to be high as evidenced by the consistent occurrence of two distinct methanotrophs in the gills of *B. brooksi* and *B. childressi*, two mussels which co-occur within the same mussel beds at Alaminos Canyon (Fujiwara et al. 2000; Duperron et al. 2007). Other symbioses involving methylotrophs, Bacteroidetes or other Gammaproteobacteria have been discovered recently and it is too early to propose any hypothesis. Nevertheless, it is intriguing that dual and multiple symbioses are to date reported only from the Atlantic area, including MAR, the GoM, GoG and Mediterranean (Table 6.1). Noteworthy, at least three mussel species from this area, two *Bathymodiolus* and one *Idas*, display amphi-Atlantic distributions (Olu-LeRoy et al. 2007a; Duperron et al. 2008b; Génio et al. 2008). No mussel with multiple symbionts is reported to date from the Pacific, despite 16 mussel 'species' investigated in the area. Clearly, although hypotheses can be proposed based on current knowledge, much information is needed to clarify our understanding of symbiosis evolution in mytilids.

6.7 Conclusion

Mytilids associated with deep-sea chemosynthesis-based ecosystems provide fascinating examples of symbiosis. How the diversity of symbionts is accommodated within a single bacteriocyte in some species is not understood, but the worldwide occurrence of mytilids at seeps, vents and organic falls and the diversity of their associated bacteria illustrate the success of this group. The ability to associate with metabolically diverse bacteria and to adapt symbiont densities to environmental parameters is probably a key to this success, allowing mytilids to cope with diverse and variable habitats.

Many questions remain. Our inability to cultivate symbionts, and difficulties of both in situ and ex situ experimentation on living mussels are major issues hampering research on these ‘symbiotic systems’. The existence of free-living forms of symbionts remains to be proven, and how associations establish is not known yet. The metabolic potential and diversity of symbionts are probably underestimated. Genomic approaches should provide indications, yield testable hypotheses, and allow comparisons between mytilid symbioses and other well-documented associations for which genomic information is available, be they chemosynthetic (*Riftia*, *Calyptogena*…) or not (insects).

Finally, geographical trends observed and evolutionary hypotheses proposed to date result from the study of a limited number of species from regions where oceanographic research has focused over the last decades (GoM, GoG, MAR…), leaving many blank spots on the map (Indian Ocean, southern Atlantic, Antarctic…). There is no doubt that new species discovered in newly explored areas will in the future considerably change our views, and yield a more complete picture of the biogeography and evolution of symbiosis in deep-sea mytilids.

Acknowledgments I thank S. Kiel for his invitation to contribute this chapter, and the two reviewers for their helpful comments. S. Gaudron is gratefully acknowledged for providing helpful comments on the manuscript. Thanks to S. Halary and J. Lorion for fruitful discussions. The author’s work is funded by Université Pierre et Marie Curie, French ANR Deep Oases, and EU programs HERMES and CHEMECO.

References

Anthony C (1982) The biochemistry of methylotrophs. Academic, London

Arellano SM, Young CM (2009) Spawning development and the duration of larval life in a deep-sea cold-seep mussel. Biol Bull 216:149–162

Baco AR, Smith CR (2003) High species richness in deep-sea-chemoautotrophic whale skeleton communities. Mar Ecol Prog Ser 260:109–114

Barry JP, Buck KR, Kochevar RK, Nelson DC, Fujiwara Y Goffredi, SK HJ (2002) Methane-based symbiosis in a mussel *Bathymodiolus platifrons* from cold seeps in Sagami Bay, Japan. Invertebr Biol 121:47–54

Belkin S, Nelson DC, Jannash HW (1986) Symbiotic assimilation of CO2 in two hydrothermal vent animals the mussel *Bathymodiolus thermophilus* and the tube worm *Riftia pachyptila*. Biol Bull 170:110–121

Bergquist DC, Fleckenstein C, Szalai EB, Knisel J, Fisher CR (2004) Environment drives physiological variability in the cold seep mussel *Bathymodiolus childressi*. Limnol Oceanogr 49:706–715

Borowski C, Giere O, Krieger J, Amann R, Dubilier N (2002) New aspects of the symbiosis in the provannid snail *Ifremeria nautilei* from the North Fiji Nack Arc Basin. Cah Biol Mar 43:321–324

Boutet I, Tanguy A, Le Guen D, Piccino P, Hourdez S, Legendre P, Jollivet D (2009) Global depression in gene expression as a response to rapid thermal changes in vent mussels. Proc R Soc Lond B 276:3071–3079

Cary SC, Fisher CR, Felbeck H (1988) Mussel growth supported by methane as sole carbon and energy source. Science 240:78–80

Cavanaugh CM (1983) Symbiotic chemoautotrophic bacteria in marine invertebrates from sulphide-rich habitats. Nature 302:58–61

Cavanaugh CM, Levering PR, Maki JS, Mitchell R, Lidstrom ME (1987) Symbiosis of methylotrophic bacteria and deep-sea mussels. Nature 325:346–347

Childress JJ (1988) Biology and chemistry of a deep-sea hydrothermal vent on the Galapagos Rift; the Rose Garden in 1985 an introduction. Deep Sea Res 35:1677–1680

Childress JJ, Fisher CR, Brooks JM, Kennicutt MC II, Bidigare R, Anderson AE (1986) A methanotrophic marine molluscan (Bivalvia Mytilidae) symbiosis: mussels fueled by gas. Science 233:1306–1308

Childress JJ, Fisher CR, Favuzzi JA, Arp AJ, Oros DR (1993) The role of a zinc-based serum-borne sulphide-binding component in the uptake and transport of dissolved sulphide by the chemoautotrophic symbiont containing clam *Calyptogena elongata*. J Exp Biol 179:131–158

Colaco A, Dehairs F, Desbruyères D, Le Bris N, Sarradin PM (2002) d^{13}C signature of hydrothermal mussels is related with the end-member fluid concentration of H_2S and CH_4 at the Mid-Atlantic Ridge hydrothermal vent fields. Cah Biol Mar 43:259–262

Comtet T, Jollivet D, Khripounoff A, Segonzac M (2000) Molecular and morphological identification of settlement-stage vent mussel larvae *Bathymodiolus azoricus* (Bivalvia Mytilidae) preserved in situ at active vent fields on the Mid-Atlantic Ridge. Limnol Oceanogr 45:1655–1661

Craddock C, Hoeh WR, Gustafson RG, Lutz RA, Hashimoto J, Vrijenhoek RJ (1995) Evolutionary relationships among deep-sea mytilids (Bivalvia: Mytilidae) from hydrothermal vents and cold water methane/sulfide seeps. Mar Biol 121:477–485

Cuvelier D, Sarrazin J, Colaço A, Copley J, Desbruyères D, Glover AG, Tyler P, Santos RS (2009) Distribution and spatial variation of hydrothermal faunal assemblages at Lucky Strike (Mid-Atlantic Ridge) revealed by high-resolution video image analysis. Deep Sea Res I 56:2026–2040

Dando PR, Southward AJ, Southward EC (1986) Chemoautotrophic symbionts in the gills of the bivalve mollusc *Lucinoma borealis* and the sediment chemistry of its habitat. Proc R Soc Lond B 227:227–247

Dattagupta S, Bergquist DC, Szalai EB, Macko SA, Fisher CR (2004) Tissue carbon nitrogen and sulfur stable isotope turnover in transplanted *Bathymodiolus childressii* mussels: relation to growth and physiological condition. Limnol Oceanogr 49:1144–1151

DeChaine EG, Bates AE, Shank T, Cavanaugh CM (2006) Off-axis symbiosis found: characterization and biogeography of bacterial symbionts of *Bathymodiolus* mussels from Lost City hydrothermal fields. Environ Microbiol 8:1902–1912

Dell RK (1987) Mollusca of the family Mytilidae (Bivalvia) associated with organic remains from deep water off New Zealand with revisions of the genera *Adipicola* Dautzenberg 1927 and *Idasola* Iredale 1915. Natl Mus N Z Rec 3:17–36

Desbruyères D, Segonzac M, Bright M (2006) Handbook of deep-sea hydrothermal vent fauna, Denisia 18. Linz Museum, Austria, p 544

Distel DL, Cavanaugh CM (1994) Independent phylogenetic origins of methanotrophic and chemoautotrophic bacterial endosymbioses in marine bivalves. J Bacteriol 176:1932–1938

Distel D, Lane D, Olsen G, Giovannoni S, Pace B, Pace NR, Stahl DA, Felbeck H (1988) Sulfur-oxidizing bacterial endosymbionts - Analysis of phylogeny and specificity by 16S ribosomal RNA sequences. J Bacteriol 170:2506–2510

Distel DL, Lee HKW, Cavanaugh CM (1995) Intracellular coexistence of methano- and thioautotrophic bacteria in a hydrothermal vent mussel. Proc Natl Acad Sci USA 92:9598–9602

Distel D, Baco A, Chuang E, Morrill W, Cavanaugh C, Smith C (2000) Do mussels take wooden steps to deep-sea vents? Nature 403:725–726

Douglas AE (1994) Symbiotic interactions. Oxford University Press, Oxford

Dubilier N, Windoffer R, Giere O (1998) Ultrastructure and stable carbon isotope composition of the hydrothermal vent mussel *Bathymodiolus brevior* and *B.* sp affinis *brevior* from the North Fiji Basin western Pacific. Mar Ecol Prog Ser 165:187–193

Dubilier N, Blazejak A, Ruehland C (2005) Symbioses between bacteria and gutless marine oligochaetes. In: Overmann J (ed) Molecular basis of symbiosis. Springer, Berlin, pp 251–275

Dubilier N, Bergin C, Lott C (2008) Symbiotic diversity in marine animals: the art of harnessing chemosynthesis. Nat Rev Microbiol 6:725–740

Duperron S, Nadalig T, Caprais JC, Sibuet M, Fiala-Médioni A, Amann R, Dubilier N (2005) Dual symbiosis in a *Bathymodiolus* mussel from a methane seep on the Gabon continental margin (South East Atlantic): 16S rRNA phylogeny and distribution of the symbionts in the gills. Appl Environ Microbiol 71:1694–1700

Duperron S, Bergin C, Zielinski F, McKiness ZP, DeChaine EG, Cavanaugh CM, Dubilier N (2006) A dual symbiosis shared by two mussel species *Bathymodiolus azoricus* and *B. puteoserpentis* (Bivalvia: Mytilidae) from hydrothermal vents along the northern Mid-Atlantic Ridge. Environ Microbiol 8:1441–1447

Duperron S, Sibuet M, MacGregor BJ, Kuypers MM, Fisher CR, Dubilier N (2007) Diversity, relative abundance and metabolic potential of bacterial endosymbionts in three *Bathymodiolus* mussels (Bivalvia: Mytilidae) from cold seeps in the Gulf of Mexico. Environ Microbiol 9:1423–1438

Duperron S, Laurent MCZ, Gaill F, Gros O (2008a) Sulphur-oxidizing extracellular bacteria in the gills of Mytilidae associated with wood falls. FEMS Microbiol Ecol 63:338–349

Duperron S, Halary S, Lorion J, Sibuet M, Gaill F (2008b) Unexpected co occurrence of 6 bacterial symbionts in the gill of the cold seep mussel *Idas* sp (Bivalvia:Mytilidae). Environ Microbiol 10:433–445

Duperron S, Lorion J, Samadi S, Gros O, Gaill F (2009) Symbioses between deep-sea mussels (Mytilidae: Bathymodiolinae) and chemosynthetic bacteria: diversity, function and evolution. C R Biol 332:298–310

Elsaied HE, Kaneko R, Naganuma T (2006) Molecular characterization of a deep-sea methanotrophic mussel symbiont that carries a RuBisCO gene. Mar Biotechnol 8:511–520

Felbeck H, Childress JJ, Somero GN (1981) Calvin-Benson cycle and sulphide oxidation enzymes in animals from sulphide-rich habitats. Nature 293:291

Fiala-Médioni A (1984) Mise en évidence par microscopie électronique à transmission de l'abondance de bactéries symbiotiques dans la branchie de Mollusques bivalves de sources hydrothermales profondes. C R Acad Sci 298:487–492

Fiala-Médioni A, Métivier C, Herry A, Le Pennec M (1986) Ultrastructure of the gill filament of an hydrothermal vent mytilid *Bathymodiolus* sp. Mar Biol 92:65–72

Fiala-Médioni A, Michalski JC, Jollès J, Alonso C, Montreuil J (1994) Lysosomic and lysozyme activities in the gill of bivalves from deep hydrothermal vents. C R Acad Sci 317:239–244

Fiala-Médioni A, McKiness ZP, Dando P, Boulègue J, Mariotti A, Alayse-Danet AM, Robinson JJ, Cavanaugh CM (2002) Ultrastructural biochemical and immunological characterisation of two populations of he Mytilid mussel *Bathymodiolus azoricus* from the Mid Atlantic Ridge: evidence for a dual symbiosis. Mar Biol 141:1035–1043

Fisher CR (1990) Chemoautotrophic and methanotrophic symbioses in marine invertebrates. Rev Aquat Sci 2:399–613

Fisher CR (1995) Towards an appreciation of hydrothermal-vent animals: their environment physiological ecology and tissue stable isotope values. In: Humphris SE, Zierenberg RA, Mullineaux LS, Thomson RE (eds) Seafloor hydrothermal systems: physical, chemical, biological, and geochemical interactions. American Geophysical Union, Washington, DC, pp 297–316

Fisher CR, Childress JJ (1992) Organic carbon transfer from methanotrophic symbionts to the host hydrocarbon-seep mussel. Symbiosis 12:221–235

Fisher CR, Childress JJ, Oremland RS, Bidigare RR (1987) The importance of methane and thiosulphate in the metabolism of the bacterial symbionts of two deep-sea mussels. Mar Biol 96:59–71

Fisher CR, Childress JJ, Arp AJ, Brooks JM, Distel DL, Favuzzi JA, Felbeck H, Hessler R, Johnson KS, Kennicutt MC, Macko SA, Newton A, Powell MA, Somero GN, Soto T (1988) Microhabitat variation in the hydrothermal vent mussel *Bathymodiolus thermophilus* at the Rose Garden vent on the Galapagos Rift. Deep Sea Res 35:1769–1791

Fisher CR, Brooks JM, Vodenichar JS, Zande JM, Childress JJ, Burke RA Jr (1993) The co-occurrence of methanotrophic and chemoautotrophic sulfur oxidizing bacterial symbionts in a deep-sea mussel. Mar Ecol 14:277–289

Flores JF, Fisher CR, Carney SL, Green BN, Freytag JK, Schaeffer SW, Royer WEJ (2005) Sulfide binding is mediated by zinc ions discovered in the crystal structure of a hydrothermal vent tubeworm hemoglobin. Proc Natl Acad Sci USA 102:2713–2718

Fujiwara Y, Takai K, Uematsu K, Tsuchida S, Hunt JC, Hashimoto J (2000) Phylogenetic characterization of endosymbionts in three hydrothermal vent mussels: influence on host distribution. Mar Ecol Prog Ser 208:147–155

Fujiwara Y, Kawato M, Yamamoto T, Yamanaka T, Sato Okoshi W, Noda S et al (2007) Three-year investigations into sperm whale-fall ecosystems in Japan. Mar Ecol 28:219–232

Génio L, Johnson SB, Vrijenhoek RC, Cunha MR, Tyler PA, Kiel S, Little CTS (2008) New record of "*Bathymodiolus*" *mauritanicus* Cosel 2002 from the Gulf of Cadiz (NE Atlantic) mud volcanoes. J Shellfish Res 27:53–61

Geret F, Rousse N, Riso R, Sarradin PM, Cosson R (1998) Metal compartmentalization and metallothionein isoforms in mussels from the Mid-Atlantic Ridge; preliminary approach to the fluid-organism relationship. Cah Biol Mar 39:291–293

Goffredi SK, Wilpiszeski R, Lee RW, Orphan VJ (2008) Temporal evolution of methane cycling and phylogenetic diversity of archaea in sediments form a deep-sea whale fall in Monterey Canyon California. ISME J 2:204–220

Gros O, Gaill F (2007) Extracellular bacterial association in gills of "wood mussels". Cah Biol Mar 48:103–109

Gros O, Guibert J, Gaill F (2007) Gill-symbiosis in Mytilidae associated with wood-fall environments. Zoomorphology 126:163–172

Gustafson R, Turner R, Lutz R, Vrijenhoek R (1998) A new genus and five new species of mussels (Bivavlia Mytilidae) from deep-sea sulfide/hydrocarbon seeps in the Gulf of Mexico. Malacologia 40:63–112

Halary S, Riou V, Gaill F, Boudier T, Duperron S (2008) 3D FISH for the quantification of methane- and sulphur-oxidising endosymbionts in bacteriocytes of the hydrothermal vent mussel *Bathymodiolus azoricus*. ISME J 2:284–292

Hashimoto J, Ohta S, Fujikura K, Miura T (1995) Microdistribution pattern and biogeography of the hydrothermal vent communities of the Minami-Ensei Knoll in the mid-Okinawa trough Western Pacific. Deep Sea Res I 42:577–598

Henry MS, Childress JJ, Figueroa D (2008) Metabolic rates and thermal tolerances of chemoautotrophic symbioses from Lau Basin hydrothermal vents and their implication for species distribution. Deep Sea Res I 55:679–695

Iwasaki H, Kyuno A, Shintaki M, Fujita Y, Fujiwara Y, Fujikura K, Hashimoto J, Martins LD, Gebruk A, Miyazaki JI (2006) Evolutionary relationships of deep-sea mussels inferred by mitochondrial DNA sequences. Mar Biol 149:1111–1122

Jahnke LL, Summons RE, Dowling LM, Zahiralis KD (1995) Identification of methanotrophic lipid biomarkers in cold seep mussel gills: chemical and isotopic analysis. Appl Environ Microbiol 61:576–582

Jannasch HW, Mottl MJ (1985) Geomicrobiology of deep-sea hydrothermal vents. Science 229:717–725

Jeffreys JG (1876) New and peculiar Mollusca of the Pecten Mytilus and Arca families procured in the 'Valorous' expedition. Ann Mag Nat Hist 4:424–436

Johnson KS, Beehler CL, Sakamoto-Arnold CM, Childress JJ (1986) In situ measurements of chemical distributions in a deep-sea hydrothermal vent field. Science 231:1139–1141

Johnson KS, Childress JJ, Beehler CL, Sakamoto CM (1994) Biogeochemistry of hydrothermal vent mussel communities – the deep-sea analog to the intertidal zone. Deep Sea Res I 41:993–1011

Jones WJ, Won YJ, Maas PAY, Smith PJ, Lutz RA, Vrijenhoek RC (2006) Evolution of habitat use by deep-sea mussels. Mar Biol 148:841–851

Kadar E, Bettencourt R, Costa V, Serrão Santos R, Lobo-da-Cunha A, Dando PR (2005) Experimentally induced endosymbiont loss and re-acquirement in the hydrothermal vent bivalve *Bathymodiolus azoricus*. J Exp Mar Biol Ecol 318:99–110

Kamenev GM, Nadtochy VA, Kuznetsov AP (2001) *Conchocoele bisecta* (Conrad, 1849) (Bivalvia: Thyasiridae) from cold-water methane-rich areas of the sea of Okhotsk. Veliger 44:84–94

Kenk VC, Wilson BR (1985) A new mussel (Bivlavia: Mytilidae) from hydrothermal vents in the Galapagos Rift zone. Malacologia 26:253–271

Kerk D, Gee A, Dewhirst FE, Drum AS, Elston RA (1992) Phylogenetic placement of "Nuclear Inclusion X (NIX)" into the gamma subclass of proteobacteria on the basis of 16S ribosomal RNA sequence comparisons. Syst Appl Microbiol 15:191–196

Kiel S, Goedert JL (2006) A wood-fall association from late Eocene deep-water sediments of Washington State, USA. Palaios 21:548–556

Kochevar RE, Childress JJ, Fisher CR, Minnich E (1992) The methane mussel: roles of symbiont and host in the metabolic utilization of methane. Mar Biol 112:389–401

Kuwahara H, Yoshida T, Takaki Y, Shimamura S, Nishi S, Harada M, Matsuyama K, Takishita K, Kawato M, Uematsu K, Fujiwara Y, Sato T, Kato C, Kitagawa M, Kato I, Maruyama T (2007) Reduced genome of the thioautotrophic intracellular symbiont in a deep-sea clam *Calyptogena okutanii*. Curr Biol 17:881–886

Lajhta K, Michener RH (1994) Stable isotopes in ecology and environmental sciences. Oxford University Press, Oxford

Laurent MCZ, Gros O, Brulport JP, Gaill F, Le Bris N (2009) Sunken wood habitat for thiotrophic symbiosis in mangrove swamps. Mar Environ Res 67:83–88

Le Bris N, Duperron S (2010) Chemosynthetic communities and biogeochemical energy pathways along the MAR: the case of *Bathymodiolus azoricus*. In: Rona PA, Devey CW, Dyment J, Murton BJ (eds): Diversity of hydrothermal systems on slow spreading ocean ridges. Geophysical Monograph Series, American Geophysical Union 188:350

Le Bris N, Rodier P, Sarradin PM, Le Gall C (2006a) Is temperature a good proxy for sulfide in hydrothermal vent habitats? Cah Biol Mar 47:465–470

Le Bris N, Govenar B, Le Gall C, Fisher CR (2006b) Variability of physico-chemical conditions in 9°50N EPR diffuse flow vent habitats. Mar Chem 98:167–182

Le Pennec M, Beninger PG (2000) Reproductive characteristics and strategies of reducing-system bivalves. Comp Biochem Physiol A 126:1–16

Lee RW, Childress JJ (1994) Assimilation of inorganic nitrogen by marine invertebrates and their chemoautotrophic symbionts. Appl Environ Microbiol 60:1852–1858

Lee RW, Robinson JJ, Cavanaugh CM (1999) Pathways of inorganic nitrogen assimilation in chemoautotrophic bacteria-marine invertebrate symbioses: expression of host and symbiont glutamine synthetase. J Exp Biol 202:289–300

Leschine SB (1995) Cellulose degradation in anaerobic environments. Ann Rev Microbiol 49:399–426

Little CTS, Vrijenhoek RC (2003) Are hydrothermal vent animals living fossils? Trends Ecol Evol 18:582–588

Lorion J, Duperron S, Gros O, Cruaud C, Couloux A, Samadi S (2009) Several deep-sea mussels and their associated symbionts are able to live both on wood and on whale falls. Proc R Soc Lond B 276:177–185

Lutz RA, Jablonski D, Rhoads DC, Turner R (1980) Larval dispersal of a deep-sea hydrothermal vent bivalve from the Galapagos rift. Mar Biol 57:127–133

Lutz RA, Jablonski D, Turner R (1984) Larval development and dispersal at deep-sea hydrothermal vents. Science 226:1451–1454

Maas PAY, O'Mullan GD, Lutz RA, Vrijenhoek RC (1999) Genetic and morphometric characterization of mussels (Bivalvia: Mytilidae) from Mid-Atlantic hydrothermal vents. Biol Bull 196:265–272

MacDonald IR, Reilly JF II, Guinasso NL Jr, Brooks JM, Carney RS, Bryant WA, Bright TJ (1990) Chemosynthetic mussels at a brine-filled pockmark in the northern Gulf of Mexico. Science 248:1096–1099

Madigan MT, Martinko JM, Parker J (2002) Brock biology of microorganisms. Pearson Education, Upper Saddle River, NJ

Martins I, Colaço A, Dando PR, Martins I, Desbruyères D, Sarradin PM, Marques JC, Serrao Santos R (2008) Size-dependant variations on the nutritional pathway of *Bathymodiolus azoricus* demonstrated by a C-flux model. Ecol Model 217:59–71

McKiness ZP, Cavanaugh CM (2005) The ubiquitous mussel: *Bathymodiolus* aff *brevior* symbiosis at the Central Indian Ridge hydrothermal vents. Mar Ecol Prog Ser 295:183–190

McKiness ZP, McMullin ER, Fisher CR, Cavanaugh CM (2005) A new bathymodioline mussel symbiosis at the Juan de Fuca hydrothermal vents. Mar Biol 148:109–116

Miyazaki JI, Shintaku M, Kyuno A, Fujiwara Y, Hashimoto J, Iwasaki H (2004) Phylogenetic relationships of deep-sea mussels of the genus *Bathymodiolus* (Bivalvia: Mytilidae). Mar Biol 144:527–535

Moraga D, Jollivet D, Denis F (1994) Genetic differentiation across the western Pacific populations of the hydrothermal vent bivalve *Bathymodiolus* spp and the eastern Pacific (13°N) population of *Bathymodiolus thermophilus*. Deep Sea Res I 41:1551–1567

Nelson DC, Fisher CR (1995) Chemoautotrophic and methanotrophic endosymbiotic bacteria at deep-sea vents and seeps. In: Karl DM (ed) Microbiology of deep-sea hydrothermal vents. CRC Press, Boca Raton, FL, pp 125–167

Nelson DC, Hagen KD, Edwards DB (1995) The gill symbiont of the hydrothermal vent mussel *Bathymodiolus thermophilus* is a psychrophilic chemoautotrophic sulfur bacterium. Mar Biol 121:487–495

Nix ER, Fisher CR, Vodenichar JS, Scott KM (1995) Physiological ecology of a mussel with methanotrophic endosymbionts at three hydrocarbon seep sites in the Gulf of Mexico. Mar Biol 122:605–617

Okutani T, Fujiwara Y, Fujikura K, Miyake H, Kawato M (2003) A mass aggregation of the mussel *Adipicola pacifica* (Bivalvia:Mytilidae) on submerged whale bones. Venus 63:61–64

Olu K, Sibuet M, Hermeignies F, Foucher J-P, Fiala-Médioni A (1996) Spatial distribution of diverse cold seep communities living on various diapiric structures of the southern Barbados prism. Prog Oceanogr 38:347–376

Olu-LeRoy K, von Cosel R, Hourdez S, Jollivet D (2007a) Amphi-Atlantic cold-seep *Bathymodiolus* species complexes across the equatorial belt. Deep Sea Res 54:1890–1911

Olu-LeRoy K, Caprais JC, Fifis A, Fabri MC, Galéron J, Budzinski H, Le Menach K, Khripounoff A, Ondreas H, Sibuet M (2007b) Cold seep assemblages on a giant pockmark off West Africa: spatial patterns and environmental control. Mar Ecol 28:1–16

Page H, Fiala-Médioni A, Fisher C, Childress J (1991) Experimental evidence for filter-feeding by the hydrothermal vent mussel *Bathymodiolus thermophilus*. Deep Sea Res I 38:1455–1461

Pailleret M, Haga T, Petit P, Privé-Gill C, Saedlou N, Gaill F, Zbinden M (2007) Sunken woods from the Vanuatu islands: identification of wood substrates and preliminary description of associated fauna. Mar Ecol 27:1–9

Peek AS, Feldman RA, Lutz RA, Vrijenhoek RC (1998) Cospeciation of chemoautotrophic bacteria and deep sea clams. Proc Natl Acad Sci USA 95:9962–9966

Pelorce J, Poutiers JM (2009) Une nouvelle espèce de Bathymodiolinae (Mollusca Bivalvia Mytilidae) associée à des os de baleine coulés en Méditerranée. Zoosystema 31:975–985

Pernthaler A, Amann R (2004) Simultaneous fluorescence in situ hybridization of mRNA and rRNA in environmental bacteria. Appl Environ Microbiol 70:5426–5433

Pile AJ, Young CM (1999) Plankton availability and retention efficiencies of cold-seep symbiotic mussels. Limnol Oceanogr 44:1833–1839

Pimenov NV, Kalyuzhnaya MG, Khmelenina VN, Mityushina LL, Trotsenko YA (2002) Utilization of methane and carbon dioxide by symbiotrophic bacteria in gills of Mytilidae (*Bathymodiolus*) from the Rainbow and Logatchev hydrothermal fields on the Mid-Atlantic Ridge. Microbiology 71:587–594

Powell MA, Somero GN (1986) Adaptations to sulfide by hydrothermal vent animals: sites and mechanisms of detoxification and metabolism. Biol Bull 171:274–290

Rau G, Hedges J (1979) C-13 depletion in a hydrothermal vent mussel – Suggestion of a chemosynthetic food source. Science 203:648–649

Raulfs EC, Macko SA, Van Dover CL (2004) Tissue and symbiont condition of mussels (*Bathymodiolus thermophilus*) exposed to varying levels of hydrothermal activity. J Mar Biol Assoc UK 84: 229–234

Riou V, Halary S, Duperron S, Bouillon S, Elskens M, Bettencourt R, Santos RS, Dehairs F, Colaço A (2008) Influence of CH4 and H2S availability on symbiont distribution carbon assimilation and transfer in the dual symbiotic vent mussel *Bathymodiolus azoricus*. Biogeosciences 5:1681–1691

Robinson JJ, Stein JL, Cavanaugh CM (1998a) Cloning and sequencing of a form II Ribulose-15-bisphosphate carboxylase/oxygénase from the bacterial symbiont of the hydrothermal vent tubeworm *Riftia pachyptila*. J Bacteriol 180:1596–1599

Robinson JJ, Polz MF, Fiala-Médioni A, Cavanaugh CM (1998b) Physiological and immunological evidence for two distinct C1-utilizing pathways in *Bathymodiolus puteoserpentis* (Bivalvia:Mytilidae) a dual endosymbiotic mussel from the Mid-Atlantic Ridge. Mar Biol 132:625–633

Rousse N, Boulègue J, Cosson R, Fiala-Médioni A (1998) Bioaccumulation of metal within the hydrothermal mytilidae *Bathymodiolus* sp from the Mid-Atlantic Ridge. Oceanol Acta 21:597–607

Salerno JL, Macko SA, Hallam SJ, Bright M, Won Y, McKiness ZP, Van Dover CL (2005) Characterization of symbiont populations in life-history stages of mussels from chemosynthetic environments. Biol Bull 208:145–155

Samadi S, Quéméré E, Lorion J, Tillier A, von Cosel R, Lopez P, Cruad C, Couloux A, Boisselier Dubayle MC (2007) Molecular phylogeny in mytilids supports the wooden steps to deep-sea vents hypothesis. C R Acad Sci 330:446–456

Sarradin PM, Caprais JC, Riso R, Kerouel R, Aminot A (1999) Chemical environment of the hydrothermal mussel communities in the Lucky Strike and Menez Gwen vent fields Mid Atlantic Ridge. Cah Biol Mar 40:93–104

Sarrazin J, Walter C, Sarradin PM, Brind'amour A, Desbruyères D, Briand P, Fabri MC, Van Gaever S, Vanreusel A, Bachraty C, Thiébaut E (2006) Community structure and temperature dynamics within a mussel assemblage on the southern East Pacific Rise. Cah Biol Mar 47:483–490

Schmaljohann R, Faber E, Whiticar MJ, Dando PR (1990) Co-existence of methane- and sulphur-based endosymbioses between bacteria and invertebrates at a site in the Skagerrak. Mar Ecol Prog Ser 61:119–124

Schöne BR, Giere O (2005) Growth increment and stable isotope variation in shells of the deep-sea hydrothermal vent bivalve mollusk *Bathymodiolus brevior* from the North Fiji Basin Pacific Ocean. Deep Sea Res I 52:1896–1910

Scott KM, Fisher CR (1995) Physiological ecology of sulfide metabolism in hydrothermal vent and cold seep vesicomyid clams and vestimentiferan tubeworms. Am Zool 35:102–111

Shillito B, Jollivet D, Sarradin PM, Rodier P, Lallier FH, Desbruyères D, Gaill F (2001) Temperature resistance of *Hesiolyra bergi* a polychaetous annelid living on vent smoker walls. Mar Ecol Prog Ser 216:141–149

Sibuet M, Olu K (1998) Biogeography biodiversity and fluid dependence of deep sea cold seep communities at active and passive margins. Deep Sea Res II 45:517–567

Smith CR, Maybaum HL, Baco AR, Pope RH, Carpenter SD, Yager PL, Macko SA, Deming JW (1998) Sediment community structure around a whale skeleton in the deep Northeast Pacific: macrofaunal microbial and bioturbation effects. Deep Sea Res II 45:335–364

Smith PJ, McVeagh SM, Won Y, Vrijenhoek RC (2004) Genetic heterogeneity among New Zealand species of hydrothermal vent mussels (Mytilidae: *Bathymodiolus*). Mar Biol 144:537

Southward EC (2008) The morphology of bacterial symbioses in the gills of mussels of the genera *Adipicola* and *Idas* (Bivalvia: Mytilidae). J Shellfish Res 27:139–146

Spiridonova EM, Kuznetsov BB, Pimenov NV, Tourova TP (2006) Phylogenetic characterization of endosymbionts of the hydrothermal vents mussel *Bathymodiolus azoricus* by analysis of the 16S rRNA cbbL and pmoA genes. Microbiology 75:798–806

Stewart FJ, Cavanaugh CM (2006) Bacterial endosymbioses in *Solemya* (Mollusca:Bivalvia) - model systems for studies of symbiont-host adaptation. Antonie Leeuwenhoek 90:343–360

Stewart FJ, Newton LG, Cavanaugh CM (2005) Chemosynthetic endosymbioses: adaptations to oxic-anoxic interfaces. Trends Microbiol 13:439–448

Taylor JD, Glover E (2006) Lucinidae (Bivalvia) – the most diverse group of chemosymbioic molluscs. Zool J Linn Soc 148:421–438

Trask JL, Van Dover CL (1999) Site-specific and ontogenetic variations in nutrition of mussels (*Bathymodiolus* sp) from the Lucky Strike hydrothermal vent field Mid-Atlantic Ridge. Limnol Oceanogr 44:334–343

Treude T, Smith CR, Wenzhöfer F, Carney E, Bernardino AF, Hannides AK, Krüger M, Boetius A (2009) Biogeochemistry of a deep-sea whale fall: sulfate reduction sulfide efflux and methanogenesis. Mar Ecol Prog Ser 382:1–21

Tunnicliffe V, Juniper SK, Sibuet M (2003) Reducing environments of the deep-sea floor. In: Tyler PA (ed) Ecosystems of the world: the deep-sea. Elsevier, Amsterdam, The Netherlands, pp 81–110

Tunnicliffe V, Davies KTA, Butterfield DA, Embley RW, Rose JM, Chadwick WW (2009) Survival of mussels in extremely acidic waters on a submarine volcano. Nat Geosci 2:344–348

Tyler PA, Young CM (2003) Dispersal at hydrothermal vents: a summary of recent progress. Hydrobiologia 503:9–19

Tyler PA, Young CM, Dolan E, Arellano SM, Brooke SD, Baker M (2007) Gametogenic periodicity in the chemosynthetic cold-seep mussel "*Bathymodiolus childressi*". Mar Biol 150:829–840

Tyler PA, Marsh L, Baco-Taylor A, Smith CR (2009) Protandric hermaphroditism in the whale-fall bivalve mollusc *Idas washingtonia*. Deep Sea Res II 56:1689–1699

Vacelet J, Fiala-Médioni A, Fisher CR, Boury-Esnault N (1996) Symbiosis between methane-oxidizing bacteria and a deep-sea carnivorous cladorhizid sponge. Mar Ecol Prog Ser 145:77–85

Van Dover CL (2000) The ecology of deep-sea hydrothermal vents. Princeton University Press, Princeton, NJ

Van Dover SL, Aharon P, Bernhard JM et al (2003) Blake Ridge methane seeps: characterization of a soft-sediment chemosynthetically based ecosystem. Deep Sea Res I 50:281–300

von Cosel R (2002) A new species of bathymodioline mussel (Mollusca Bivalvia Mytilidae) from Mauritania (West Africa) with comments on the genus *Bathymodiolus* Kenk & Wilson 1985. Zoosystema 24:259–271

von Cosel R, Janssen R (2008) Bathymodioline mussels of the *Bathymodiolus* (sl) *childressi* clade from methane seeps near Edison Seamount New Ireland Papua New Guinea. Arch Molluskenkunde 137:195–224

von Cosel R, Marshall BA (2003) Two new species of large mussels (Bivalvia: Mytilidae) from active submarine volcanoes and a cold seep off the eastern North Island of New Zealand with description of a new genus. Nautilus 117:31–46

von Cosel R, Olu K (1998) Gigantism in Mytilidae A new *Bathymodiolus* from cold seep areas on the Barbados accretionary Prism. C R Biol 321:655–663

von Cosel R, Comtet T, Krylova EM (1999) *Bathymodiolus* (Bivalvia:Mytilidae) from hydrothermal vents on the Azores Triple junction and the Logatchev hydrothermal field Mid-Atlantic Ridge. Veliger 42:218–248

Von Damm KL (1995) Controls on the chemistry and temporal variability of seafloor hydrothermal vents. In: Humphris S, Zierenberg RA, Mullineaux LS, Thomson RE (eds) Seafloor hydrothermal systems: physical chemical biological and geological interactions, Geophysical Monographs. American Geophysical Union, Washington, DC, pp 222–247

Won Y, Young CR, Lutz RA, Vrijenhoek RC (2003a) Dispersal barriers and isolation among deep-sea mussel populations (Mytilidae: *Bathymodiolus*) from Eastern Pacific hydrothermal vents. Mar Ecol 12:169–184

Won YJ, Hallam SJ, O'Mullan D, Pan IL, Buck KR, Vrijenhoek RC (2003b) Environmental acquisition of thiotrophic endosymbionts by deep-sea mussels of the genus *Bathymodiolus*. Appl Environ Microbiol 69:6785–6792

Won YJ, Jones WJ, Vrijenhoek RC (2008) Absence of cospeciation between deep-sea mytilids and their thiotrophic symbionts. J Shellfish Res 27:129–138

Woyke T, Teeling H, Ivanova N, Hunteman M, Richter M, Glockner FO, Boffelli D, Anderson IJ, Barry KW, Shapiro HJ, Szeto E, Kyrpides NC, Musmann M, Amann R, Bergin C, Ruehland C, Rubin EM, Dubilier N (2006) Symbiosis insights through metagenomic analysis of a microbial consortium. Nature 443:950–955

Yamamoto H, Fujikura K, Hiraishi A, Kato K, Maki Y (2002) Phylogenetic characterization and biomass estimation of bacterial endosymbionts associated with invertebrates dwelling in chemosynthetic communities of hydrothermal vents and cold seeps. Mar Ecol Prog Ser 245:61–67

Zbinden M, Shillito B, Le Bris N, de Villardi de Montaur C, Roussel E, Guyot F, Gaill F, Cambon-Bonavita MA (2008) New insights on the metabolic diversity among the epibiotic microbial community of the hydrothermal vent shrimp *Rimicaris exoculata*. J Exp Mar Biol Ecol 359:131–140

Zielinski FU, Pernthaler A, Duperron S, Raggi L, Giere O, Borowski C, Dubilier N (2009) Widespread occurrence of an intranuclear bacterial parasite in vent and seep bathymodiolin mussels. Environ Microbiol 11:1150–1167

Chapter 7
Gastropods from Recent Hot Vents and Cold Seeps: Systematics, Diversity and Life Strategies

Takenori Sasaki, Anders Warén, Yasunori Kano, Takashi Okutani, and Katsunori Fujikura

Contents

T. Sasaki (✉)
The University Museum, The University of Tokyo, 7-3-1 Hongo, Bunkyo-ku, Tokyo, 113-0033, Japan
e-mail: sasaki@um.u-tokyo.ac.jp

A. Warén
Swedish Museum of Natural History, Box 50007, SE-10405 Stockholm, Sweden

Y. Kano
Atmosphere and Ocean Research Institute, The University of Tokyo, 5-1-5 Kashiwanoha, Kashiwa, Chiba 277-8564, Japan

T. Okutani and K. Fujikura
Japan Agency for Marine-Earth Science and Technology, 2-15 Natsushima, Yokosuka 237-0061, Japan

S. Kiel (ed.), *The Vent and Seep Biota*, Topics in Geobiology 33,
DOI 10.1007/978-90-481-9572-5_7,

7.1 Introduction

Since the first discovery of hydrothermal vents at the Galapagos Spreading Center in 1977, gastropods have gained considerable attention as a major constituent of the chemosynthesis-based biological communities, especially the colonies of large species like *Alviniconcha*, *Ifremeria* and *Lepetodrilus*, or morphologically strange ones like the scaly-foot snail. Various types of symbiotic relations to bacteria have broadened the interest in them. During more than 30 years numerous vent and seep biotopes have been found mainly in temperate seas, but recently hydrothermal vents off Jan Mayen in the North Atlantic at 80°N and cold seeps off Norway between 67°N and 70°N (Haakon Mosby Mud Volcano) with chemosynthesis-based fauna (see Warén and Bouchet 2001; Desbruyères et al. 2006: 516–517 for map). Biology of these vent/seep organisms is still in an early state of exploration; much of what is known is summarized in the monographs by Van Dover (2000) and Desbruyères et al. (2006). Vent/seep taxa have also been a major target in research on higher phylogeny and systematics of gastropods. Several new families were established in the 1980s, based on seemingly "great differences" in morphology.

In this review, we attempt to provide a brief overview of the phylogeny, anatomical organization (Appendix 7.1), protoconch morphology (Appendix 7.2), and biology for taxonomic groups, supplemented by an exhaustive bibliography. We also compiled a list of species described until 2009 (Appendix 7.3), which is an updated version from Warén and Bouchet (2001). Still new taxa are regularly found at new sites or by refined methods of investigation, and a set of more than 50 species is under description. At this point the great surprises are rare and many of the former new "vent families" have been recognized as parts of others. Among gastropods the greatest enigma is the Neomphalina, a group of three families that defies all attempts to classify it with or in other subgroups of Gastropoda, especially molecular attempts. The survey is arranged systematically following Bouchet et al. (2005) except for some part. The systematics of Trochoidea and Seguenzioidea were updated, following recent phylogenetic analyses (Kano 2008; Williams et al. 2008; Kano et al. 2009). Neomphalina is treated as a seventh clade in addition to Patellogastropoda, Vetigastropoda, Cocculiniformia (mainly Cocculinidae), Neritimorpha, Caenogastropoda and Heterobranchia.

The following abbreviations are used for institutions. IFREMER – Institut français de recherche pour l'exploitation de la mer; JAMSTEC – Japan Agency for Marine-Earth Science and Technology; NSMT – National Museum of Nature and Science, Tokyo (formerly National Science Museum, Tokyo); SMNH – Swedish Museum of Natural History; UMUT – The University Museum, The University of Tokyo.

7.2 Groups of Vent/Seep-Associated Gastropods

It is not always obvious which species are seep and vent-associated. Often a large majority of the organisms in a seep or vent belong to a small number of species which do not occur in the periphery but congregate on chimneys or in seepages.

Many of these are endemic to vents, even at genus and occasionally at higher taxonomic level. Other species are regularly found also in other biotopes, but they occur in much larger densities in chemosynthetic environments. Such species are common in cold seeps, rare in hot vents but may also be favoured by food falls. We term them vent- or seep favoured. However, sometimes species do not want to follow human classifications or categorizations; *Lepetodrilus corrugatus* from the Endeavour Ridge, probably the best known locality for vent faunas, is obviously a vent-seep species like its 26 congeners (Johnson et al. 2008) but is known from a single specimen only. No more specimens have been found among hundreds of thousands of specimens of *Lepetodrilus* from all kinds of environments at the Ridge (V. Tunnicliffe, personal communication, 2007).

7.2.1 Clade Patellogastropoda

This clade is currently composed of seven families, Patellidae, Nacellidae, Neolepetopsidae, Lepetidae, Lottiidae, and Eoacmaeidae (Fig. 7.1) (see Nakano and Ozawa 2007: fig. 3; Lindberg 2008). Of these, Neolepetopsidae and Pectinodontidae (= 'Acmaeidae') have independently adapted to vents and seeps. Both feed by grazing with a very long radula. Fertilization is external, and larval development is lecithotrophic. All species lack a copulatory organ, a prostate gland and a receptaculum seminis. The protoconch is symmetrical and consists of an almost complete but poorly visible coil. It is difficult to find the protoconch, since it is rapidly corroded and specimens smaller than a millimeter are needed.

Family Neolepetopsidae (Fig. 7.1a)

This family consists of ten described species in three genera (see Appendix 7.3; McLean 2008). They are mostly vent/seep-endemic, but *Neolepetopsis nicolasensis* and *Paralepetopsis clementensis* were described from whale carcasses off California (McLean 2008). The Neolepetopsidae was originally proposed as an independent suborder Lepetopsina based on two characters, the articulation of radular teeth and non-mineralized radula (McLean 1990b). In a molecular phylogenetic analysis by Harasewych and McArthur (2000), *Eulepetopsis* and *Paralepetopsis* form a clade which is closest to *Acmaea mitra*, a shallow water species of the family Lottiidae. The shells are characterized by foliated structure (McLean 1990b, pl. VIII, fig. h; Kiel 2004). The base of the shell is flat in most species, but in some species it adjusts to the substratum. For example, the holotype of *Neolepetopsis gordensis* has a low apex and flat apertural margin in lateral view, but one of paratypes is laterally compressed, and its apertural margin is convex in lateral view (McLean 1990b: pl. 1, 2008: fig. 1a–b). These types occur on bivalves and tubes of Vestimentifera respectively and may give an impression of distinct species. Occasionally one can find specimens that have changed habitat, which is reflected in the shell.

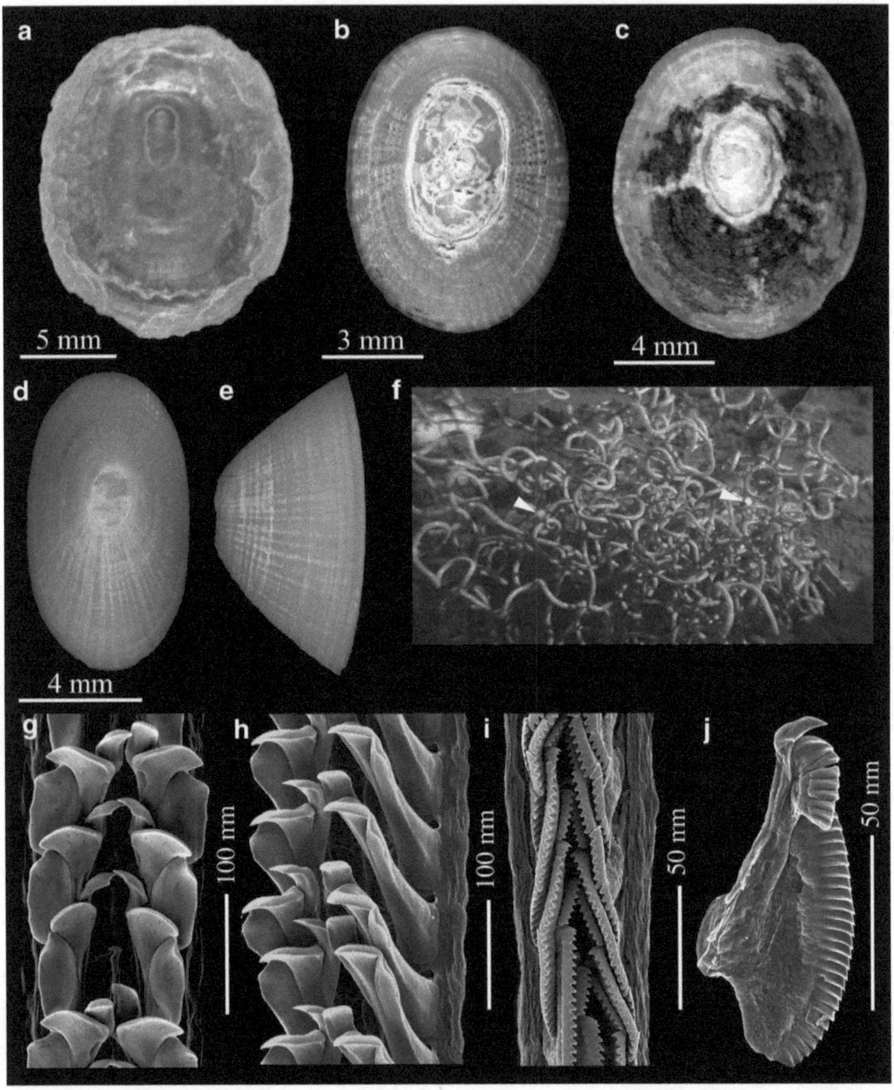

Fig. 7.1 Patellogastropoda. (**a**) *Eulepetopsis vitrea* McLean 1990b. SMNH 43437, East Pacific Rise, 9°N, 2,517 m, vent. (**b**) *Bathyacmaea subnipponica* Sasaki et al. 2003. Holotype UMUT RM27959. Ryuyo Canyon, Nankai Trough, Japan, 1,100 m, seep. (**c**) H. *Bathyacmaea tertia* Sasaki et al. 2003. Holotype, UMUT RM27955. North Knoll of Iheya Ridge, Okinawa Trough, Japan, 969 m, vent. (**d–f**) *Serradonta vestimentifericola* Okutani et al. 1992. (**d–e**) Holotype NSMT-Mo 69643. Off Hatsushima, Sagami Bay, Japan, 1,110–1,200 m, seep. (**f**) Habitat on vestimentiferan tubes. *Hyper-Dolphin* Dive 528, off Hatsushima, 1,095 m. (**g**) *Bathyacmaea secunda* Okutani et al. 1993. Paratype UMUT RM 29199. (**i–j**) Radula of *Serradonta kanesunosensis* Sasaki et al. 2003. Holotype, UMUT RM27961. Kanesunose Bank, Nankai Trough, Japan, 322 m, seep (a: Photo taken by T. Sasaki at SMNH; b–c, h–j: Sasaki et al. 2003; f: Sasaki et al. 2007; g: Sasaki et al. 2006a)

Neolepetopsids lack a gill and the gas exchange is restricted to the surface of the pallial cavity. The radula is formulated as 2-3-1-3-2 and characterized by a well-developed rachidian tooth, pluricuspid outermost lateral teeth, and two broad and scale like marginal teeth (Warén and Bouchet 2001: fig. 2). The radular teeth were formerly believed to be non-minelarized (McLean 1990b) but revealed to be a mis-interpretation. Cruz and Farina (2005) reported that the radula of *P. ferrugivora* has a similar elemental composition to those of other patellogastropods having dark-colored, well-mineralized radular teeth. However, the iron oxide in *P. ferrugivora* is amorphous unlike shallow-water patellogastropods showing crystlline phases (Cruz and Farina 2005). Our experience is that these radulae are even more sensitive to chemical cleaning than other patellogastropods and it is only the anterior part that is fully mineralised. McLean's SEM photos show the immature radula. Other striking characters of the family include the absence of osphradia, the absence of basal plates in radula, and posteriorly torted esophagus. Oral lappets which might be related to feeding are present in *Eulepetopsis vitrea* (Fretter 1990: fig. 2), *Paralepetopsis sasakii* (Warén and Bouchet 2009: fig. 3) and *Paralepetopsis ferrugivora* (Warén and Bouchet 2001: fig. 4a).

Warén and Bouchet (2001) reported that the gut is usually filled with ferruginous crusts (Warén and Bouchet 2001: 123 for *Eulepetopsis vitrea*; 125 for *Paralepetopsis ferrugivora*). Nematodes and ciliate fragments are also found in the intestine of *Paralepetopsis ferrugivora* (Warén et al. in Desbruyères et al. 2006: 85). *Paralepetopsis sasakii* had its gut full of calcareous fragments and diatom tests (Warén and Bouchet 2009). *Eulepetopsis vitrea* and *Paralepetopsis floridensis* are gonochoristic and their eggs are heavily yolked and large (Gustafson and Lutz 1994: 88). Larval development of neolepetopsids is lecithotrophic with a planktonic dispersal stage (Warén et al. in Desbruyères et al. 2006: 83–85) on the basis of egg size and protoconch morphology. The protoconch is symmetrical in all patellogastropods including *E. vitrea* (McLean 1990b: pl. VII) and *P. floridensis* (McLean 1990b: pl. X).

Family Pectinodontidae ("Acmaeidae")(Fig. 7.1b–j)

The Pectinodontidae has frequently been classified in the Acmaeidae (e.g. Sasaki et al. 2003), but the type species of genus *Acmaea*, *A. mitra,* formed a clade with *Niveotectura pallida* within the Lottiidae (Nakano and Ozawa 2007), and much of this classification is in a state of limbo. To avoid misunderstandings and unnecessary changes forth and back we use Pectinodontidae. The family contains three genera, *Bathyacmaea* and *Serradonta* from vents or seeps and *Pectinodonta* from sunken wood (Fig. 7.1b–j).

1. Genus *Bathyacmaea* (Figs. 7.1b–c, g–h): Five species (see Appendix 7.3) are known from the western Pacific (and an unnamed species from seeps off Costa Rica). The habitat is either vents or seeps and attached to rocks and shells of

other molluscs. The shell of *Bathyacmaea* is characterized by three layers of semi-foliated structure of the shell (Fuchigami and Sasaki 2005). Radula morphology is strange among gastropods: the teeth are long and spoon-like with cusps at the top, and robust basal plates (e.g. Okutani et al. 1992: figs. 10–11, 1993: figs. 18–19; Sasaki et al. 2003: figs. 5–7). Sexes are separate. Their mode of reproduction is unknown, but external fertilization is plausible as in other patellogastropods. The protoconch has not been found.

2. Genus *Serradonta* (Fig. 7.1d–f, i–j): This genus is represented by two species from seeps in Japan. The shell is elongate, laterally compressed, and its apertural margin is convex in lateral view. This morphology allows the shell to fit a tube-shaped substrate. A similar shell morphology is also known in Late Cretaceous fossil species (Jenkins et al. 2007a, b), but also in the genera *Paralepetopsis* and *Neolepetopsis*. Species of *Serradonta* are similar to those of *Bathyacmaea* in pale shells with a finely reticulate sculpture, but their distinction is unmistakable in radular morphology. The radular teeth of *Serradonta* consist of two longitudinal rows of composite teeth, paralleling the wood-eating genus *Pectinodonta* (Okutani et al. 1992: figs. 5–6; Sasaki et al. 2003: fig. 8). The habitat is restricted to the surface of vestimentiferan tubes in *S. vestimentifericola* (Sasaki et al. 2007), but unknown for *S. kanesunosensis*. Protoconch and eggs have not been observed.

7.2.2 Clade Vetigastropoda

The phylogeny of Vetigastropoda remains contentious (Williams and Ozawa 2006; Williams et al. 2008; Kano 2008, Kano et al., 2009) (Figs. 7.2, 7.3, 7.4, and 7.5). Geiger and Thacker (2006) and Geiger et al. (2008) define Vetigastropoda to encompass Cocculiniformia and Neomphalina, based on the shared presence of bursicles in the gills of the latter hot-vent taxa with Vetigastropoda in a strict sense (Sasaki 1998), although conflicting phylogenetic relationships have been suggested for the three groups (e.g. Ponder and Lindberg 1997; McArthur and Harasewych 2003; Aktipis et al. 2008). On the other hand, the monophyly of Vetigastropoda *s.s.* (including Pleurotomarioidea, Haliotoidea, Scissurelloidea, Fissurelloidea, Lepetodriloidea, Lepetelloidea, Seguenzioidea, Phasianelloidea, Angarioidea and Trochoidea) is well supported by both morphological and molecular characters, except perhaps the inclusion of Pleurotomaroidea which lack sensory bursicles in the gill leaflets and sensory papillae on any part of the head-foot, as well as having an enigmatic radula. Molecular evidence, however, supports this concept. The internal relationship of the clade is far from resolved. Recent phylogenetic analyses demonstrate that fundamental revision is needed for the classification of superfamilies and families currently assigned to Trochoidea (Williams and Ozawa 2006; Williams et al. 2008) and Seguenzioidea (Kano 2008; Kano et al. 2009). Some species from

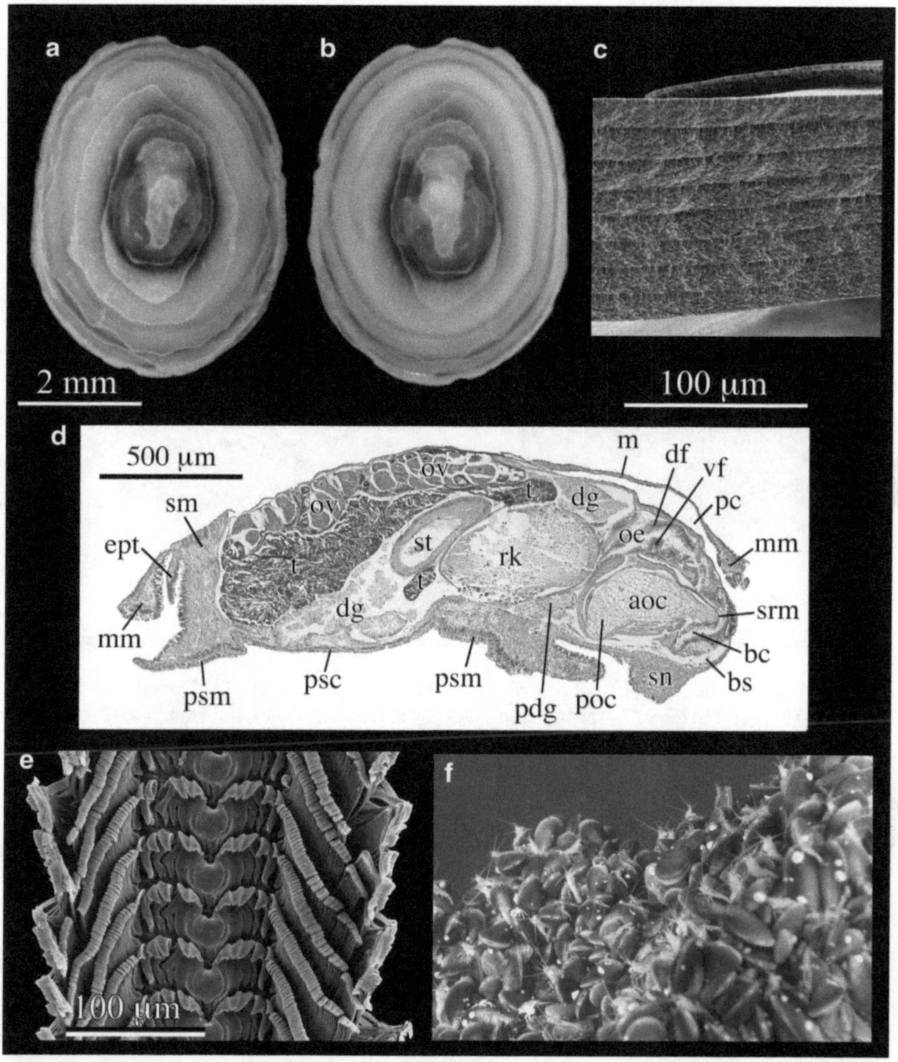

Fig. 7.2 Pyropeltidae. *Pyropelta ryukyuensis* Sasaki et al. 2008. (**a**) Exterior of shell. (**b**) Interior of shell (a–b, Holotype UMUT RM29412). (**c**) Vertical section of shell. (**d**) Longitudinal section of animal. (**e**) Radula. Hatoma Knoll, Okinawa Trough, 1,451 m, vent. (**f**) Habitat on *Bathymodiolus platifrons*. Hatoma Knoll, ca. 1,480 m. *Shinkai 2000* Dive 1361. Abbreviations: aoc, anterior odontophoral cartilage; bc, buccal cavity; bs, buccal sinus; df, dorsal fold of oesophagus; dg, digestive glands; e, oesophagus; ept, epipodial tentacle; i, intestine; m, mantle; mm, mantle margin; ov, ovary; pc, pallial cavity; pdg, pedal ganglion; poc, posterior odontophoral cartilage; psc, central part of pedal sole; psm, marginal part of pedal sole; sm, shell muscle; sn, snout; srm, subradular membrane; st, stomach; t, testis; vf, ventral fold of oesophagus. A color plate of this figure can be found in Appendix I (Plate 14) (a–f: Sasaki et al. 2008)

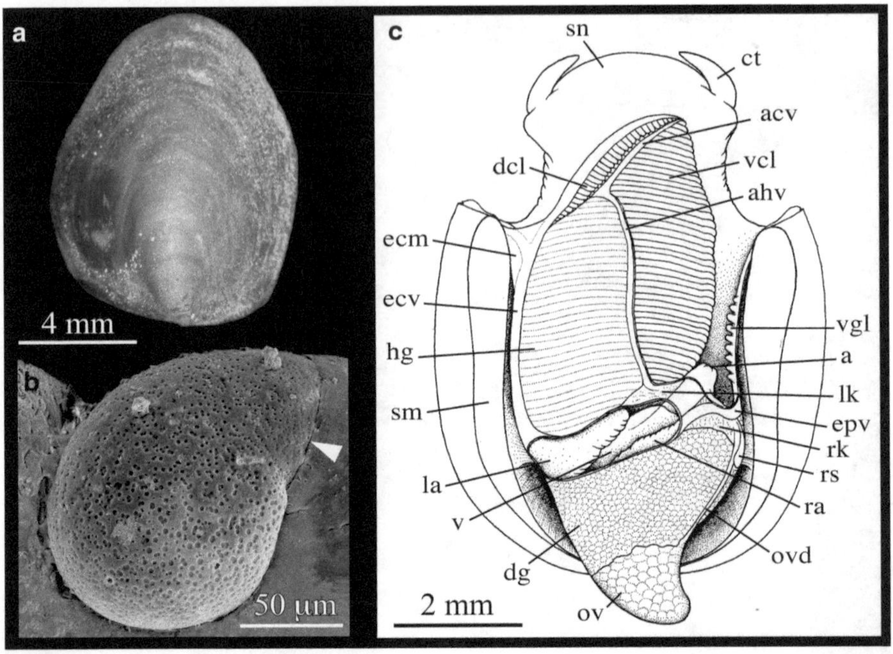

Fig. 7.3 Lepetodrilidae. (**a–c**) *Lepetodrilus nux* (Okutani et al. 1993). (**a**) Dorsal view of shell. Holotype, NSMT-Mo 69972, Izena Hole, Okinawa Trough, 1,340 m, vent. (**b**) Protoconch. *Arrowhead* indicates the boundary between the protoconch and the teleoconch. UMUT RM 27967. North Knoll of Iheya Ridge, Okinawa Trough, 1,390 m, vent. (**c**) Animal with most part of mantle removed. Abbreviations: a, anus; acv, afferent ctenidial sinus; ahv, afferent hypobranchial sinus; ct, cephalic tentacle; dg, digestive gland; ecm, efferent ctenidial membrane; ecv, efferent ctenidial sinus; epv, efferent pallial sinus; hg, hypobranchial gland; la, left auricle; lk, left kidney; ov, ovary; ovd, oviduct; ra, right auricle; rk, right kidney; rs, seminal receptacle; sm, shell muscle; sn, snout; v, ventricle; vcl, ventral ctenidial lamellae; vgl, vestigial gill leaflets (a: Sasaki et al. 2005; b: Sasaki et al. 2003; c. Sasaki 1998)

vent/seep environments are in indeterminate systematic positions among other vetigastropods (e.g. *Sahlingia xandaros*).

Superfamily Fissurelloidea

Family Fissurellidae

The family is highly diversified in shallow-water environment (Geiger et al. 2008), but only seven species are known from vents/seeps in the genera *Clathrosepta*, *Cornisepta* and *Puncturella* (see Appendix 7.3). The biology is poorly known for deep-sea species. This family is not a common member of vent/seep communities.

Fig. 7.4 Collonidae (a–b) and Turbinidae (c–d). (**a**) *Cantrainea macleani* Warén and Bouchet 1993. SMNH21179. Gulf of Mexico, off Louisiana, 546 m, hydrocarbon seep. (**b**) *Cantrainea jamsteci* (Okutani and Fujikura 1990). Holotype, NSMT-Mo 66235. Minami Ensei Knoll, Okinawa Trough, 720 m, vent. (**c**) *Margarites shinkai* Okutani et al. 1992. Holotype, NSMT-Mo 69635. Off Hatsushima, Sagami Bay, 1,110–1,200 m, seep. (**d**) *Margarites ryukyuensis* Okutani et al. 2000. Holotype, UMUT RM27884. North Knoll of Iheya Ridge, Okinawa Trough, 968 m, vent (a: Photo taken by T. Sasaki at SMNH; c–d: Sasaki et al. 2005)

Superfamily Lepetelloidea

Family Pyropeltidae (Fig. 7.2, Plate 14)

The family was established for a single genus *Pyropelta* currently contains seven species from northeastern, northwestern and southwestern Pacific (see Appendix 7.3; also McLean 1992b). Four of seven species are vent-endemic, but two species, *P. musaica* and *P. corymba* were collected also from whale bone and/or seeps in addition to vents (McLean 1992b; Smith and Baco 2003: table 6). *Pyropelta corymba* and *P. ryukyuensis* were found to attach to the shells of *Provanna* spp. (Warén and Bouchet 2001: fig. 6) and *Bathymodiolus* (Fig. 7.2f), respectively. *Pyropelta*

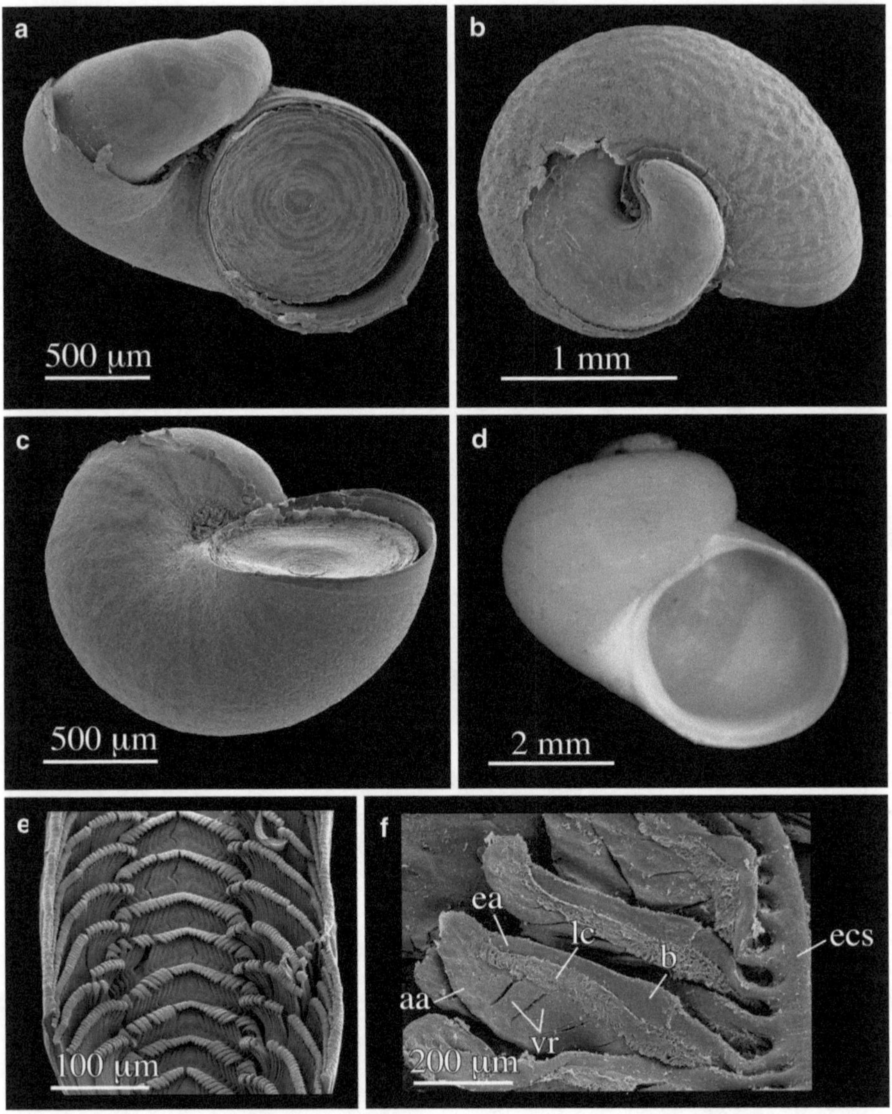

Fig. 7.5 Skeneinae, Turbinidae. (**a–c**) *Bruciella wareni* Okutani et al. 2004. Kairei Field, Indian Ocean, 2,434 m, vent. (**a**) Apertural view of Holotype. UMUT RM28666. (**b**) Apical view. Paratype, UMUT RM28668. (**c**) Basal view. Paratype, UMUT RM28669. (**d–f**) *Iheyaspira lequios* Okutani et al. 2000. North Knoll of Iheya Ridge, Okinawa Trough, Japan, ca. 1,000 m, vent. (**d**) Shell of holotype. (**e**) Radula. (**f**) Ctenidium. Abbreviations: aa, afferent ctenidial axis; b, bursicle; ea, efferent ctenidial axis; ecs, efferent ctenidial sinus; lc, lateral cilia; vr, vertical ridge of ctenidial lamella. (d, f) Holotype, UMUT RM27887. (e) Paratype, UMUT RM27888. (a–c: Okutani et al. 2004; d: Sasaki et al. 2005; e–f: Okutani et al. 2000)

wakefieldi McLean 1992 is known only from whale bones. *Pyropelta* has several distinctive conchological characters among gastropods (Sasaki et al. 2008: 314), including irregular surface sculpture, several repeated layers of homogeneous and prismatic structures (Fig. 7.2c), the myostracum exposed exterior by corrosion, the microtubes penetrating the myostracum, the shell growth front deeply inside of the interior away from the shell margin. Sequential secretion of shell layers might be adaptive in the deep sea, where shell corrosion is more extensive than in shallow-water environments. The repetitive thin prismatic layers are a useful character to identify fossil pyropeltids (Kiel 2006: fig. 3).

The main food source of *Pyropelta* are bacteria (Warén et al. in Desbruyères et al. 2006: 93) which are scraped with a rhipidoglossate radula (Fig. 7.2d). All species examined are hermaphroditic with separate testis and ovary (Fig. 7.2d). The right cephalic tentacle of male is modified into a copulatory organ (McLean and Haszprunar 1987), and the spermatozoa are conveyed along the neck with ciliary tract (Warén and Bouchet 2009: fig. 7.4e). The eggs have the vitelline layer which is one of characters restricted to Vetigastropoda. Larval development is lecithotrophic with planktonic dispersal stage (Warén et al. Desbruyères et al. 2006: 93).

Family Pseudococculinidae

The Pseudococculinidae are rare in vents and seeps, and there are only a few records (see Appendix 7.3). Most species of the family live on sunken wood (e.g. Marshall 1986). Some species of the Pseudococculinidae have a concentric shell sculpture similar to that of *Pyropelta* (e.g. McLean 1988b, 1991; McLean and Harasewych 1995). However, the single known study on the shell microstructure of Pseudococculindae (Lindberg and Hedegaard 1996: fig. 12) did not show the repetition of microstructual units. The Pseudococculinidae and Pyropeltidae are clearly separated by radular morphology (e.g. Warén and Bouchet 2009: figs. 4–5).

Superfamily Lepetodriloidea

Family Lepetodrilidae (Fig. 7.3)

The family currently comprises four genera: *Lepetodrilus*, *Gorgoleptis*, *Clypeosectus*, and *Pseudorimula*. *Gorgoleptis* was originally given an independent family status (Gorgoleptidae: McLean 1988a), and the latter two genera were united as the Clypeosectidae (McLean 1989b). However, these were later transferred into this family by Warén and Bouchet (2001: 155). Molecular phylogenetic data support the closeness of *Clypeosectus* and *Pseudorimula* to *Lepetodrilus* (Warén et al. 2003; Johnson et al. 2008). *Lepetodrilus* and *Gorgoleptis* share many similar characters, as they were originally defined as Lepetodriloidea: Diagnostic characters of these are: non-papillate tentacles, a single left ctenidium which is bipectinate at the anterior free tip but monopectinate posteriorly, a single left hypobranchial gland, the gonad discharging through the right kidney, the vas deferens functioning as the seminal vesicle, the prostate, the receptaculum seminis, a ciliated genital groove in female (see Sasaki 1998: 215). On the other hand, *Lepetodrilus* and *Gorgoleptis* are distinguished by the configuration of shell muscles, the presence or absence of operculum, the elaboration of epipodial tentacles, and the structure and position of the penis. In *Gorgoleptis*, all species keep the operculum until just below the maximum size (Warén and Bouchet 2001: 154).

1. Genus *Pseudorimula*: The shell of the genus has an obliquely elongate slit which is finally closed at the margin. No operculum is retained after metamorphosis. In *P. midatlantica*, the shell is frequently encrusted with bacterially precipitated iron and manganese oxides (Warén and Bouchet 2001: 156). Male has the prostate gland (Haszprunar 1989a: 7, fig. 9C) but lack a penis and seminal groove. Femals lack a receptaculum seminis (Haszprunar 1989a: 8). Development is without planktotrophic stage (Warén et al. in Desbruyères et al. 2006: 91).
2. Genus *Clypeosectus:* This genus also has an oblique slit in the shell, but it is not closed at the margin. No operculum is retained after metamorhosis. Reproductive system differs from that of *Pseudorimula.* In male the proximal vas deferens functions as a seminal vesicle. The prostate gland is present around the urinogenital opening (Haszprunar 1989a: 7, fig. 5B). The oviduct is connected to the renopericardial duct and also to the right kidney; female has an isolated receptaculum seminis opening at the right side of the pallial cavity (Haszprunar 1989a: figs. 3C, 4B). The eggs are yolky and large (maximum diameter 200 μm: Haszpruar 1989a: 7; Gustafson and Lutz 1994: 87). *C. curvus* is a specialist grazer on folliculinid ciliates (Bergquist et al. 2007). The protoconch of *C. delectus* was figured by McLean (1989b: fig. 9C, F).
3. Genus *Gorgoleptis*: The genus (three species) is a rare group confined to vents of East Pacific Rise (Warén and Bouchet 2001: appendix 2). The shell of the genus has no slit. The operculum is small, multispiral and positioned at the posterior side of the foot. Anatomically the genus is diagnosed by a pair of separate shell muscle attachments, long epidpodial tentacles whose number is five on the left and four on the right, the penis developed from the left side of the snout. The male has a prominent prostate gland; female lacks a receptaculum seminis (Fretter 1988). Larval development is lecithotrophic with planktonic dispersal stage (Warén et al. in Desbruyères et al. 2006: 87).
4. Genus *Lepetodrilus* (Fig. 7.3): The genus exhibits the highest diversty among vent/seep-associated gastropods. There are 14 described species: 10 species are distributed in the East Pacific, four in the West Pacific, and a single species on the Mid-Atlantic Ridge and in the West African seeps. Johnson et al. (2008) barcoded 16 species or 19 DNA types, but recognised six more of which no material was available for barcoding. The shell is limpet-shaped with a posterior apex, and without a slit. At microscopic level the shell consists of three layers: outermost prismatic, middle crossed-lamellar and innermost prismatic layers, among which the middle layer is prominently thick (Hunt 1992: figs. 4–8). Diagnostic characters of *Lepetodrilus* include the horseshoe-shaped shell muscle attachment, the lack of an operculum in adults, and three pairs of short epipodial tentacles

(Sasaki 1998: 215). The penis has a deep seminal groove near the base of the right cephalic tentacles when present. The penis is prominent or absent, depending on species or individuals. In *L. shannonae* it there is no penial difference between male and female, but the male has what seems to be a newly evolved penis of pallial origin.

Lepetodrilus have two different ways of feeding: ciliary feeding and grazing. Morphologically, ciliary feeding is correlated with enlargement of frontal and abfrontal ciliary pads at the tip of ctenidial lamellae (Bates 2007a: fig. 10). Limén et al. (2007: fig. 2) confirmed that food particles were collected by cilia at the same time as chemoautotrophic bacteria are cultivated on the gill. Rejected food particles are expelled as pseudofeces (Johnson et al. 2006: fig. 4G). Grazing is suggested by wear of radular teeth (Bates 2007a: 94) and inclusion of mineral particles in the gut. Among species of *Lepetodrilus*, there is variability in the relative size of the radula and stomach: the differences probably represent the degree of dependence on grazing and suspension feeding. Filamentous episymbiotic bacteria are farmed on the ctenidium, transported along the neck region, and ingested, as was found in food particles (Beck 1993: pl. 5, fig. 5, *L. schrolli*; Johnson et al. 2006: fig. 4H, L. *L. gordensis*; Bates 2007a: fig. 9A, 2007b: fig. 2, *L. fucensis*). Stable carbon and nitrogen isotopic composition also indicated that epibiotic bacteria on the ctenidium are used for nutrition (Levesque et al. 2006).

Species from the East Pacific Rise have been most intensively analyzed at population level with molecular characters (especailly COI gene of mtDNA). Johnson et al. (2006) revealed that northern and southern lineages of formerly called *L. fucensis* are distinct species. A new species *L. gordensis* was proposed for a lineage from Gorda Ridge and Escanaba Trough, and *L. fucensis* sensu stricto is confined to the Explorer and Juan de Fuca ridge systems. Morphologically *L. gordensis* lacks a sensory neck papilla and has a more tightly coiled teleoconch. Similarly Matabos et al. (2007) demonstrated the presence of distinct lineages in *L. elevatus* from 13°N to 9°50′N and from 9°50′N to 17°S (Matabos et al. 2008: fig. 5). Divergence time of these southern and northern lineages was estimated to be 11.6 Ma by Plouviez et al. (2009). Johnson et al. (2008) barcoded *Lepetodrilus* species, using 1,000 base pairs of COI gene, and revealed five cryptic species along the EPR, of which up to three occurred sympatrically.

Two cases of predation have been reported for *Lepetodrilus*. *Buccinum thermophilum* is a predator of *L. elevatus* (Voight and Sigwart 2007) [possibly *L. fucensis*, not *L. elevatus*], and the octopod *Graneledone boreopacifica* for *L. fucensis* (Voight 2000a). The shell is damaged in the former species, and the shell fragments were found in the gut of the predator in the latter species.

Lepetodrilus are generally gonochoristic and exert internal fertilization. The female have an isolated receptaculum seminis on the right side; the male has the prostate (Fretter 1988; Sasaki 1998). The male has a penis with a seminal furrow on the right side of the head in *L. fucensis* (Johnson et al. 2006: fig. 4B; Bates 2007a: fig. 8D; Kelly and Metaxas 2007: fig. 3e) and in *L. gordensis* (Johnson et al. 2006: fig. 4H). McLean (1993) reported that a penis was absent in *L. elevatus* from Mariana Trough. Warén and Bouchet (2001: 154) also noted that most males of

Lepetodrilus from North Fiji Basin lack a penis, although some individuals will develop it. *Lepetodrilus fucensis* was confirmed to form euspermatozoa only (Hodgson et al. 1997). Elongate head (nucleus) and flagellum (Hodgson et al. 1997: fig. 4, ant-aquasperm type) are suggestive of fertilization within the pallial cavity rather than of ect-aquasperm type which is typical of species conducting external fertilization (Hodgson et al. 1997: 94). Development is probably through free-swimming lecithotrophic larvae with a planktonic period (Warén et al. in Desbruyères et al. 2006: 88). The protoconch has been illustrated for several species (Table 7.2). Growth rate of *L. fucensis* from Juan de Fuca Ridge is 0.4 cm per year (Kelly and Metaxas 2008: table 7). Reproduction starts at 3.8 (male) and 3.9 (female) mm in shell length in *L. fucensis* (Kelly and Metaxas 2007). Histological condition of the gonad indicates gametogenesis is continuous, and reproduction is asynchronous in *L. fucensis* (Kelly and Metaxas 2007).

Family Sutilizonidae

The family was originally classified as a part of Scissurellidae: *Temnocinclis* and *Temnozaga* in Temnocinclinae and *Sutilizona* in Sutilizoninae (McLean 1989b). Warén and Bouchet (2001) recognized the family Sutilizonidae for these groups, but they are likely to end up in the Lepetodrilidae as indicated by similarities in the protoconch. The family has a limpet-shaped shell, asymmetrical in *Sutilizona* or nearly symmetrical in *Temnocinclis* and *Temnozaga*, with a long slit and selenizone, a multispiral operculum, paired monopectinate ctenidia, non-papillate cephalic and epipodial tentacles. The number of epipodial tentacles varies among genera. The anatomy of *Temnocinclis euripes*, *Temnozaga parilis*, and *Sutilizona theca* was described by Haszprunar (1989a: 3–5). All species are gonochoristic. The ovary contains large yolky eggs, and the oviduct is ciliated and non-glandular. The male lacks a penis. In *S. theca* all stages of sperm development except ripe spermatozoa were observed (Haszprunar 1989a: 5). The gonoduct has a common opening with the right kidney. Eggs are large (200 μm in maximum) and exhibit all stages of development in *S. theca* (Haszprunar 1989a: 5). The above anatomical features indicate that spawning is continuous, and fertilization is likely to occur in the female mantle cavity (Gustafson and Lutz 1994: 82). The protoconch of *S. theca* (McLean 1989b: fig. 4F) has dense pits and spiral ridges, very similar or identical to that in Lepetodrilidae, an indication of close relations. The larval development is lecithotrophic with planktonic dispersal stage (Warén et al. in Desbruyères et al. 2006: 98–100).

Superfamily Seguenzioidea

Family uncertain

The Seguenzioidea was recently redefined to include the Cataegidae, Chilodontidae, Calliotropidae, Seguenziidae, and skeneimorph genera (e.g. *Adeuomphaus*, *Xyloskenea* and *Ventsia*) of uncertain familial assignment (Kano 2008; Kano et al. 2009).

The genus *Adeuomphalus* is characterized by a planispiral shell with axial lamellae, monopectinate gill, and anteriorly bifurcated foot with a pair of epipodial sense organs and epipodial tentacles (Warén and Bouchet 2001: 132; Kano et al. 2009). Among seven recent and a single fossil species of the genus, three species were collected near hydrothermal vents (see Appendix 7.3) together with carnivorous sponges. A parasitic mode of life was suggested by the absence of radulae as in other parasitic gastropods and a tubular snout (Kano et al. 2009). *Ventsia tricarinata* has been assigned to the Skeneidae, but anatomical characters are similar to those of *Adeuomphalus* (Kano et al. 2009: 414). Especially, the possession of the seminal receptacle suggests the inclusion of the genus in the Seguenzioidea (Kunze et al. 2008; Kano et al. 2009), and DNA also suggests its affinity to the Seguenzioidea (Kano 2008).

Family Seguenziidae

Bathymargarites symplector is similar to *Margarites* in shell form, but strikingly different in having "penis" which is modified from the right neck lobe (Warén and Bouchet 1989: 87). The reproduction is therefore not considered to be external. The eyes are present at bases of cephalic tentacles with pigmented part of variable shape (Warén and Bouchet 1989). Feeding is by grazing on a surface layer of detritus. Stomach content consists of mineral particles and fragments of various organisms (Warén et al. in Desbruyères et al. 2006: 101). Larval development is lecithotrophic with planktonic dispersal stage (Warén et al. in Desbruyères et al. 2006: 101).

Superfamily Trochoidea

Traditional classification recognized only three families, namely Trochidae, Turbinidae and Skeneidae, in this superfamily (Hickman and McLean 1990). However, recent phylogenetic studies have revealed the traditional Trochoidea to be polyphyletic (Williams and Ozawa 2006; Kano 2008; Williams et al. 2008). Some of the traditional trochoideans have been transferred to Seguenzioidea (Kano 2008, Kano et al. 2009) as well as to the newly established Angarioidea and Phasianelloidea. The redefined Trochoidea comprises five families including Trochidae, Turbinidae, Solariellidae, Calliostomatidae and Liotiidae (Williams et al. 2008). This is probably not the final edition, since some basal groups show a tendency of jumping back and forth depending on the selection of species.

Family Turbinidae (Fig. 7.4c, d)

Williams et al. (2008) classified this trochoid family into five subfamilies, namely Turbininae, Prisogasterinae, Margaritinae, Tegulinae and Skeneinae. Vent/seep endemic turbinids belong to either Margaritinae or Skeneinae. Members of the genus *Margarites* (Fig. 7.4c–d) are common in shallow subtidal zones, but also known from vents and seeps. *Margarites shinkai* lives on muddy bottom in the seep-associated *Calyptogena* bed, together with *Provanna* species (Sasaki et al. 2007).

The subfamily Skeneinae (Fig. 7.5) had been treated as an independent family in Trochoidea (e.g. Hickman and McLean 1990). However, recent anatomical and molecular studies have revealed that 'Skeneidae' was a miscellany of little-known minute vetigastropods, often mixed up even with neomphalines, caenogastropods and heterobranchs (Heß et al. 2008; Kano 2008; Kunze et al. 2008; Kano et al. 2009; see below). The type genus *Skenea* and several similar genera have been classified in the family Turbinidae based on the phylogenetic position of a morphologically similar genus, *Protolira* (Warén et al. 2003) and *Dillwynella* (Kano 2008; Williams et al. 2008). The anatomy of the type species of *Skenea*, *S. serpuloides*, was described by Kunze et al. (2008). Probable members of Skeneinae from vent/seep environments include *Protolira* and *Bruceiella* (Fig. 7.5a–c), possibly also *Fucaria* and *Iheyaspira* (Fig. 7.5d).

In the genus *Fucaria*, food consists of detritus from the sea bottom as is suggested by gut contents. Larval development lecithotrophic with planktonic dispersal stage (Warén et al. in Desbruyères et al. 2006: 95). *Protolira thorvaldssoni* lives at vents and also on whale bones from which the species was originally described. Food consists of the detrital surface layer of the sea bottom. Species of *Skenea* and the closely related vent genus *Protolira* have a propodial penis, assumed internal fertilization and are hermaphrodites (Warén 1992; Warén and Bouchet 1993). Larval development lecithotrophic with planktonic dispersal stage (Warén et al. in Desbruyères et al. 2006: 97).

Family Colloniidae (Fig. 7.4a, b)

Phasianelloidea was recently established by Williams et al. (2008) for Phasianellidae and Colloniidae solely based on the topology of their multi-gene phylogeny. The two families had been classified in Turbinidae of Trochoidea at subfamilial rank (Phasianellinae and Colloniinae), based on opercular, shell and radular characters (e.g. Hickman and McLean 1990; Bouchet et al. 2005). However, the traditional 'Turbinidae' classification or species with calcareous opercula has been rejected by nuclear rRNA gene data but the conclusions are not consistent and some details seem unlikely (Williams and Ozawa 2006; Kano 2008; Williams et al. 2008). Future anatomical studies are needed to morphologically explore the clade. Vent/seep collonids are represented by the members of the genus *Cantrainea* only. Their shells (Fig. 7.4a–b) are thick for deep-sea gastropods, and opercula are also heavily calcified. There are four known species from vents and seeps (see also Appendix 7.3): *C. macleani* from off Louisiana (Warén and Bouchet 1993: 8; 2001: 132); *C. jamsteci* and *C. nuda* are both from off Ryukyu Islands, Japan; *C. panamense* off Chile.

7.2.3 *Clade Neomphalina*

The term Neomphalida (e.g. Heß et al. 2008) has also been used for this group, but it is a younger name (Fig. 7.6, Plate 15). The phylogenetic position of this clade is still uncertain regarding the status within or outside Vetigastropoda (Heß et al. 2008: 399). For example, anatomical characters indicate the inclusion of the Neomphalina ("hot-vent taxa") in the Vetigastropoda (Aktipis et al. 2008: figs. 9.3), but a combined analysis of morphological and molecular characters was considered to support relations to the Cocculinidae (Aktipis et al. 2008: figs. 9.4–9.7). Molecular phylogeny by Geiger and Thacker (2006) was interpreted to show that the Neomphalina are basal to vetigastropods. The morphological characterization is difficult due to an unusually great variation in morphology, with a number of misleading apomorphies but the monophyly is supported by molecular phylogeny (Kano and Warén in prep) as well as a mosaic pattern of plesiomorphies. The presence of shell pores (Batten 1984, *Neomphalus fretterae*; Warén and Bouchet 1993: 35, fig. 27B–D, *Ctenopelta porifera*; Heß et al. 2008: fig. 8, *Leptogyra* and *Leptogyropsis*; Kiel and Goedert 2007: fossil *Leptogyra* from the Eocene) had been regarded as one of putative shared characters but most neomphalins lack pores. The Neomphalina is the last mollusc group above the family level, considered endemic to chemosynthetic environments but it is now known from food-falls, seeps and vents (Kano 2008, Warén and Bouchet 2009). All species have a protoconch with no trace of planktotrophy, and a planktonic dispersal phase can be assumed since larvae of several species were retrieved in plankton traps during the French ATOS and HOPE programs on the EPR (Warén unpubl., see also Mills et al. 2007). Two types of protoconch morphology occur: a plesiomorphic one with a fine net-sculpture, especially on the initial part (Melanodrymiidae and Neomphalidae) and a strongly bent type with prominent spiral ridges in the Peltospiridae.

Family Melanodrymiidae (Fig. 7.6e)

The genus *Melanodrymia* had been assigned either Peltospiridae or Neomphalidae, but redefined as the type genus of Melanodrymiidae based on both anatomy and molecular phylogeny (Salvini-Plawen and Steiner 1996). The genera *Leptogyra*, *Xyleptogyra* and *Leptogyropsis* were also transferred from the Skeneidae to this family (Heß et al. 2008). Molecular evidence (but not all morphology) supports the monophyly of the family (Heß et al. 2008). The anatomy of *Melanodrymia aurantiaca* was described by Haszprunar (1989b). The presence of spermatophore was mentioned for *Melanodrymia* sp. ("rust covered") by Warén and Bouchet (2001: 161). They found lanceolate objects attached to the posterior part of the pallial cavity. The food of *Melanodrymia* consists of the detrital surface layer of the bottom (Warén et al. in Desbruyères et al. 2006: 106)

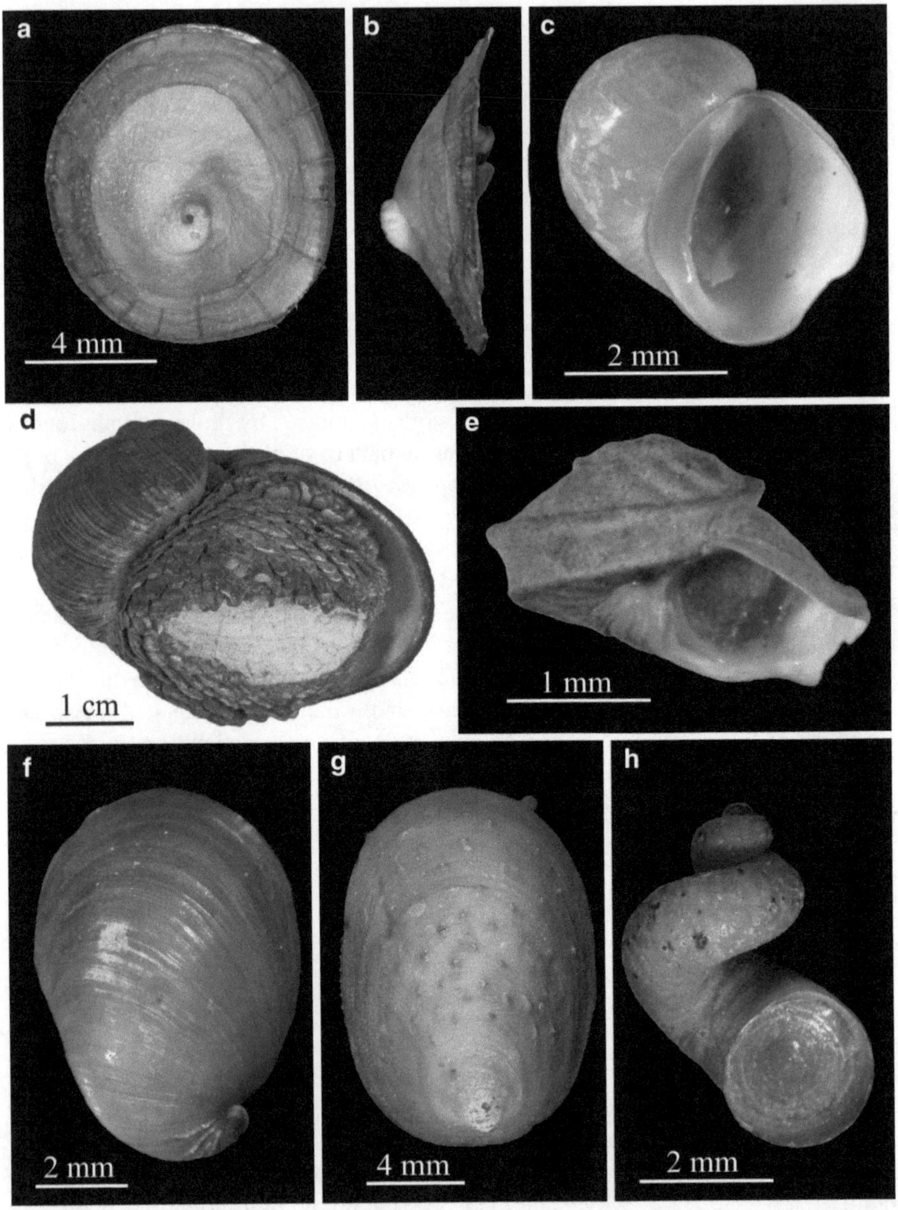

Fig. 7.6 Neomphalina. (**a–b**) *Neomphalus fretterae* McLean 1981. SMNH43346. East Pacific Rise, 13°N, 2,630 m, vent. (**c**) *Cyathermia naticoides* Warén and Bouchet 1989. SMNH21160. East Pacific Rise, 13°N, 2,630 m, vent. (**d**) Scaly-foot gastropod. Kairei Field, Indian Ocean, 2,422 m, vent. UMUT RM30208. (**e**) *Melanodrymia aurantiaca* Hickman 1984. SMNH43311, East Pacific Rise, 13°N, 2,632 m, vent. (**f**) *Peltospira operculata* McLean 1989. SMNH43168. East Pacific Rise, 13°N, 2,649 m, vent. (**g**) *Nodopelta heminoda* McLean 1989. SMNH43316. East Pacific Rise, 13°N, 2,630 m, vent. (**h**) *Pachydermia laevis* Warén and Bouchet 1989. SMNH43130. East Pacific Rise, 13°N, 2,650 m, vent. A color plate of this figure can be found in Appendix I (Plate 15). (a–b, e–h: Photo taken by T. Sasaki at SMNH)

Family Neomphalidae (Fig. 7.6a–c)

The family currently encompasses six genera: *Cyathermia, Lacunoides, Neomphalus, Planorbidella, Solutigyra* and *Symmetromphalus*. Familial assignment is not always consistent among past studies. For example, Cyathermiidae McLean 1990a was once proposed for *Cyathermia*; *Melanodrymia* currently in the Melanodrymiidae was contained in this family (Warén and Bouchet 2001) due to similarities in the protoconch. *Planorbidella* and *Pachydermia* were transferred from the Peltospiridae to the Neomphalidae, based on the similarities to *Neomphalus* in the protoconch (Warén and Bouchet 2001: 157), but *Pachydermia* was assigned to the Peltospiridae based on molecular phylogenetic tree (Heß et al. 2008). The family is with a couple of uncertain exceptions endemic to hydrothermal vent environments (e.g. *Retiskenea*).

The shell form is extremely variable among different genera: limpet-shaped (*Neomphalus fretterae*: Fig. 7.6f and *Symmetromphalus*), naticiform (*Cyathermia*: Fig. 7.6h), and high-spired with disjunct whorls (*Pachydermia*: Fig. 7.6d). The shells of other genera are *Velutina*-like or skeneimorph. Some species have characteristic conchological characters. *Cyathermia naticoides* has a deep notch at basal side of the outer lip (Fig. 7.6h). Its function is uncertain, but possibly used to increase the exposure of the gill to the sea water. The shell of *Neomphalus fretterae* (and some others) is penetrated by numerous pores (Batten 1984: fig. 12). Young stages of *Neomphalus fretterae* strikingly resemble *Cyathermia naticoides* (Warén and Bouchet 1989: 69). Sexual size dimorphism is known in *Pachydermia laevis* (Warén and Bouchet 1989: 80) and *Neomphalus fretterae*. McLean (1981) mentioned a 15% difference in maximum size of sexes in *Neomphalus*. The operculum is also variable among the family. In limpet like genera, a small operculum persists in subadult *Symmetromphalus* (Beck 1992a: pl. 4, fig. 2). In *Neomphalus*, the operculum is lost at a size of a few millimeter. Spirally coiled species retain a complete multisprial operculum.

The anatomy of the Neomphalidae is mainly described for *Neomphalus fretterae* by Fretter et al. (1981) and also for *Symmetromphalus* by Beck (1992a). Supposed general characters of the family are: a long neck, the epipodial tentacles along the posterior side of the foot, the male gonoduct with the prostate, and an open seminal groove in the left cephalic tentacle (see also Warén and Bouchet 2001 for *Lacunoides* and *Cyathermia*). Fertilization is likely to occur internally within the female oviduct (*Neomphalus*: Fretter et al. 1981). The anatomy of *Pachydermia laevis* was described by Israelsson (1998). The configuration of the shell muscle is often supposed to be of high value in taxonomic ranking, but the muscle is crescent-shaped in *Neomphalus* and symmetrically paired in *Symmetromphalus*. Filter feeding predominates in the limpet species while grazing may be the only way in species with a small gill like *Pachydermia*, judging from the relative size of the radula. *Cyathermia* uses the ctenidium for filter feeding and also grazes on bacterial growth on tube worms (Warén and Bouchet 1989: 71). Development is inferred to be non-planktotrophic from protoconch morphologies and dimensions (Gustafson and Lutz 1994: 83).

Family Peltospiridae (Fig. 7.6f–h)

The family is composed of nine described genera: *Ctenopelta, Depressigyra, Echinopelta, Hirtopelta, Lirapex, Nodopelta, Pachydermia, Peltospira* and *Rhynchopelta*. *Solutigyra* were originally assigned to this family (Warén and Bouchet 1989: 82) but later transferred to the Neomphalidae (Warén and Bouchet 2001). Current knowledge on anatomy is mainly based on *Rhynchopelta, Nodopelta, Echinopelta, Hirtopelta and Peltospira* (Fretter 1989). Shell form varies greatly in the family, from symmetrically limpet-like (*Echinopelta*, *Nodopelta* and *Rhynchopleta*), haliotiform limpet-like (*Ctenopelta* and *Hirtopelta*), to skeneiform (*Depressigyra*, *Lirapex*). In the genus *Peltospira*, the shell is variable from depressed haliotiform (*P. operculata*) to inflated *Vanikoro*-like shape (*P. smaragdina*). The shell sculpture is also variable: The surface is characteristically ornamented by spirally arranged small tubercles (*Ctenopelta*), sparsely scattered spines or tubercles in (*Echinopelta, Hirtopelta* and *Nodopelta*), commarginal elements (*Peltospira, Rhynchopelta*, and *Lirapex*), or nearly smooth (*Depressigyra*). The shell microstructure has not sufficiently been examined for the family. In *Ctenopelta porifera*, the shell has two different types of pores which consist of macropores (30–70 μm in diameter) and micropores (2–3 μm in diameter) (Warén and Bouchet 1993: 35, fig. 27B–D). The operculum is present in spirally coiled forms (*Depressigyra* and *Lirapex*). In limpet forms, the operculum is absent in most species, but present at the posterior end of the foot in *Hirtopelta* (Fretter 1989: fig. 16, *H. hirta*; Beck 2002: fig. 8, *H. tufari*). In the genus *Peltospira*, the operculum is present in *P. operculata* (McLean 1989a: figs. 4, 6; Fretter 1989: fig. 9), lost in *P. smaragdina* before reaching 1 mm (Warén and Bouchet 2001: 168) and also absent in adult in *P. delicata*.

Some species are confirmed or presumed to have ecological association with bacteria. *Hirtopelta tufari* possess bacteria in bacteriocytes of the ctenidium (Beck 2002: figs. 12–15). In *Ctenopelta porifera* tubular spiny shell sculpture and the setae of the foot might be related to bacterial symbioses (Warén and Bouchet 1993: 33). Sexes are separate, and females are larger than males in *Ctenopelta porifera* (Warén et al. in Desbruyères et al. 2006: 117). In *Rhyncopelta concentrica*, internal fertilization is suggested by filiform spermatozoa (introsperm) (Hodgson et al. 2009).

Family Peltospiridae: Scaly foot gastropod (Fig. 7.6d)

The 'Scaly foot gastropod' was discovered from the bases of black-smoker chimneys in the Kairei vent field at Rodriguez Triple Junction in the Indian Ocean (Warén et al. 2003). The species has not yet been named but will be taxonomically described by Anders Warén. It is nearly sedentary and firmly attached to rocks. The shell is rounded, swollen, and covered with thick black periostracum. The animal lacks the operculum, but the sclerites on the side of the foot function like the operculum of other gastropods. When the animal is retracted, the foot is folded in the midline and the aperture is closed by the sclerites of the epipodium. But, when in danger, another possibility is to pull the shell towards the substrate, and all tissues will be covered by the shell and the sclerites. The sclerites are mineralized with iron sulfide minerals containing pyrite (FeS_2) and greigite (Fe_3S_4) (Warén et al. 2003; Goffredi et al. 2004; Suzuki et al. 2006c). The presence of a conchiolin layer in the sclerites may suggest the homology between the sclerites and opercula of other gastropods (Warén et al. 2003).

An isolated sclerite is illustrated by Goffredi et al. (2004: fig. 3B). The surface of the sclerite is covered with filamentous and coccoid bacteria (Warén et al. 2003: fig. 1C, S2G; Goffredi et al. 2004: fig. 3E–F). The sclerites consist of three layers, *viz.* sulfide, mixed and conchiolin layers (Warén et al. 2003: fig. S2I, K; Suzuki et al. 2006c: fig. 2). Isotopic analyses suggest that sulfur and iron in the sclerites may be derived from hydrothermal fluids, not from bacterially precipitated substance (Suzuki et al. 2006c). The animal harbors thiotrophic γ-proteobacteria in an enormously enlarged oesophageal gland (Goffredi et al. 2004: fig. 2). Like in several gastropods suspected to have symbiotic bacteria the intestine is full of finely granular iron sulphides, particle size <0.2 μm. There are no obviously symbiotic bacteria in the ctenidum.

Apart from the possession of sclerites, overall organization conforms to that of peltospirids: the bipectinate ctenidium with sensory buriscles, the rhipidoglossate radula, an enlarged esophageal glands, hypoathroid nervous system with cross-connected pedal cords. The esophageal gland is 10–1,000 times as large as the relative volumes of that of other gastropods. Other parts of digestive system are reduced, the volume is 1/10–1/100 of that of other gastropods (Warén et al. 2003: supporting online material). The female has spermatophore producing organ (Warén et al. 2003: fig. S2M) and sperms are conveyed with spermatophores. Development lecithotrophic, presumably with a planktonic dispersal stage (Warén et al. in Desbruyères et al. 2006: 120). Scaly-foot is an extreme example of how adaptations to the chemosynthetic environment with very special selective pressures can modify the morphology.

7.2.4 Clade Neritimorpha

The members of this clade with a shallow-sea origin have invaded a wide range of habitats ranging from the deep-sea to freshwater and terrestrial environments (Kano et al. 2002; Sasaki and Ishikawa 2002) (Fig.7.7). They are grazers and feed on microalgae and bacteria with a rhipidoglossate radula. Fertilization is internal; eggs are deposited in lenticular capsules produced by the well-developed glandular gonoduct of the female. Larval development is either planktonic or benthic. In the Neritimorpha, the presence or absence of a planktotrophic larval period can be inferred from the morphology of the larval operculum, which is very often retained in adult individuals, as well as from the more vulnerable protoconch (Kano 2006).

Family Phenacolepadidae (Fig. 7.7b–g)

This family of small limpets is clearly divided into two groups: shallow-water genera (*Phenacolepas* and *Cinnalepeta*) and deep-water, vent-associated genera

Fig. 7.7 Neritimorpha. (**a**) *Bathynerita naticoidea* Clarke 1989. SMNH21177. Gulf of Mexico, off Louisiana, 546 m, hydrocarbon seep. (**b–g**) *Shinkailepas myojinensis* Sasaki et al. 2003. (**b**) Radula. r, rachidial tooth, L1–L4, first to fourth lateral teeth. (**c**) Dorsal view of shell, (**d**) Ventral view of shell, (**e**) Magnified view of shell pores, (**f**) Operculum, (**g**) Protoconch. Myoin Knoll, Japan, 1,260–1,290 m, vent. Holotype, UMUT RM27970 (a: Photo taken by T. Sasaki at SMNH; b–g: Sasaki et al. 2003)

(*Shinkailepas* and *Olgasolaris*) (Sasaki et al. 2006b). The shallow-water species inhabit reduced environments under half-buried boulders (Kano et al. 2002; Kano and Haga 2010). Animals of the family are vividly red when alive and contain erythrocytes in blood (Fretter 1984; Sasaki 1998: fig. 85d; Sasaki et al. 2006b: fig. 10b). Globular bacteria were found on the surface of ctenidial cilia of *Olgasolaris* (Beck 1992b: pl. 6, fig. 6), but their association has not yet been examined in detail. Species of the genus *Shinkailepas* (Fig. 7.7b–g) have been described from vents in the western Pacific (*S. kaikatensis, S. myojinensis,* and *S. tufari*) and from the Mid-Atlantic Ridge (*S. briandi*). At least seven undescribed species exist at vents in the western and eastern Pacific, Indian Ocean and Mid-Atlantic Ridge. In the monotypic genus *Olgasolaris*, the type species *O. tollmanni* is distributed in the Manus Back-Arc Basin. The shell of *O. tollmanni* differs from that of *Shinkailepas* in that the apex is positioned near the center, but not at the posterior end. The interior of the shell has a very shallow septum at the posterior region (Beck 1992a: pl. 4, fig. 4). The shells of some vent-associated phenacolepadids are penetrated by numerous microtubules from the mantle, and some of the microtubules reach the outer shell surface (Sasaki et al. 2003: fig. 12). Their opercula are partly calcified (Fig. 7.7f). Reproduction is via internal fertilization. The male has a grooved penis modified from the right cephalic lappet (Beck 1992b: pl. 5, fig. 6). Egg capsules are less than 1 mm in diameter and sometimes deposited on shells (Warén and Bouchet 2001: fig. 32b). The small veliger larvae ready to hatch from the egg capsule, the multispiral protoconch (Beck 1992b: pl. 1, fig. 5; Warén and Bouchet 2001: fig. 32f), and the initial part of the operculum consisting of small embryonic and large larval regions (type A of Kano 2006: fig. 4, table 2) all suggest a planktotrophic early development for the species of *Shinkailepas* and *Olgasolaris* (Kano 2006).

Family Neritidae (Fig. 7.7a)

The genus *Bathynerita* consists of a single species, *Bathynerita naticoidea* which is distributed at hydrocarbon seep sites in the Lousiana Slope of the Gulf of Mexico. The shell is mostly occupied by the body whorl; the aperture is wide and nearly semicircular. Inside of the whorls are resorbed (Warén and Bouchet 1993: 3). However, in contrast to shallow-water neritids, the shell is pale, without denticles in the inner lip. The shell wall has three layers (Kiel 2004: 176). The operculum is calcified only inside and lacks a peg (Warén and Bouchet 1993: fig. 1C). The right cephalic lappet in *Bathynerita* is replaced by a penis in males (Warén and Bouchet 1993: fig. 2A). The radula is of the typical neritid type (Warén and Bouchet 1993: fig. 3A–C). *Bathynerita naticoidea* further differs from shallow-water neritids in sperm microstructure; its axoneme is not bent at the junction of the mid-piece and tailpiece, and the tail is not terminally expanded, as they are in most shallow-water neritids (Hodgson et al. 1998). These features suggest that *Bathynerita* is phylogenetically closer to the deep-sea phenacolepadids than to neritids (Hodgson et al. 1998). Molecular phylogenies also favor this phenacolepadid affinity of *Bathynerita*

(McArthur and Koop 1999; McArthur and Harasewych 2003; see also Kano et al. 2002). *Bathynerita naticoidea* is commonly associated with beds of *Bathymodiolus childressi* and feeds on bacteria and decomposing periostracum of *Bathymodiolus* shells (Dattagupta et al. 2007). This species can tolerate a considerable degree of salinity fluctuations by closing the operculum tightly (Van Gaest et al. 2007).

7.2.5 *Clade Caenogastropoda*

Caenogastropods is a large group with more than 120 families (Bouchet et al. 2005), but members in vents and seeps mostly belong to Provannidae, Elachisinidae, Turridae and Buccinidae. A few species of Rissoidae have been found on food-falls, in vents and in cold seeps but their relations to each other and to the environment are poorly known (Figs. 7.8–7.12). Scattered species of other families (e.g. Capulidae, Ceritiopsidae, Vitrinellidae and Cancellariidae) have been found in vents and seeps (see Appendix 7.3), sometimes in considerable numbers.

Family Provannnidae (Figs. 7.8–7.9, Plate 16)

The members of the family belong to five genera, *Alviniconcha*, *Ifremeria*, *Provanna*, *Desbruyeresia,* and *Cordesia.* A sixth genus is so far known from two whale falls only is under description (Johnson et al. in press). The systematic position of the family has been uncertain, but is currently assumed to be a member of Abyssochrysoidea along with Abyssochrysidae and extinct Hokkaidoconchidae (Kaim et al. 2008a). Colgan et al. (2007) based on several genes arrived at a position in the vicinity of Littorinoidea. Anatomically the presence of pallial tentacle in *Provanna* is shared with *Abyssochrysos* (Abyssochrysidae) (Houbrick 1979). The shell form of this family is divided into two types. Genera *Provanna* and *Desbruyeresia* have small and slender shells, whereas in *Alviniconcha* and *Ifremeria*, the shells are much larger and swollen. These two shell forms seems to be totally different in adult stage, but the protoconch of *Desbryueresia* and *Alviniconcha* (see Kaim et al. 2008a: fig. 4) suggests that they have arisen from the common larval form. The shells of *Alviniconcha* and *Ifremeria* are probably enlarged to house a hypertrophied ctenidium harboring a large quantities colony of symbiotic bacteria. The protoconch is ornamented with a uniform reticulate sculpture formed by sharp axial and spiral ribs in *Desbruyeresia* (Warén and Bouchet 1993: fig. 44D; Kaim et al. 2008a: fig. 4A; but except *D. marisindica*) and *Alviniconcha* (Warén and Bouchet 1993: fig. 44C; Kaim et al. 2008a: fig. 4B). Before settling, the top whorls of the protoconch are lost by a characteristic mode of decollation (Kaim et al. 2008a: fig. 4). In *Provanna* the sculpture consists of scattered axial ribs (e.g. Warén and Ponder 1991: fig. 1). The protoconch has never been found in *Ifremeria nautilei* whose species brood the embryos in a brood chamber in the foot (Warén in Desbruyères et al. 2006: 130; Watanabe et al. unpublished).

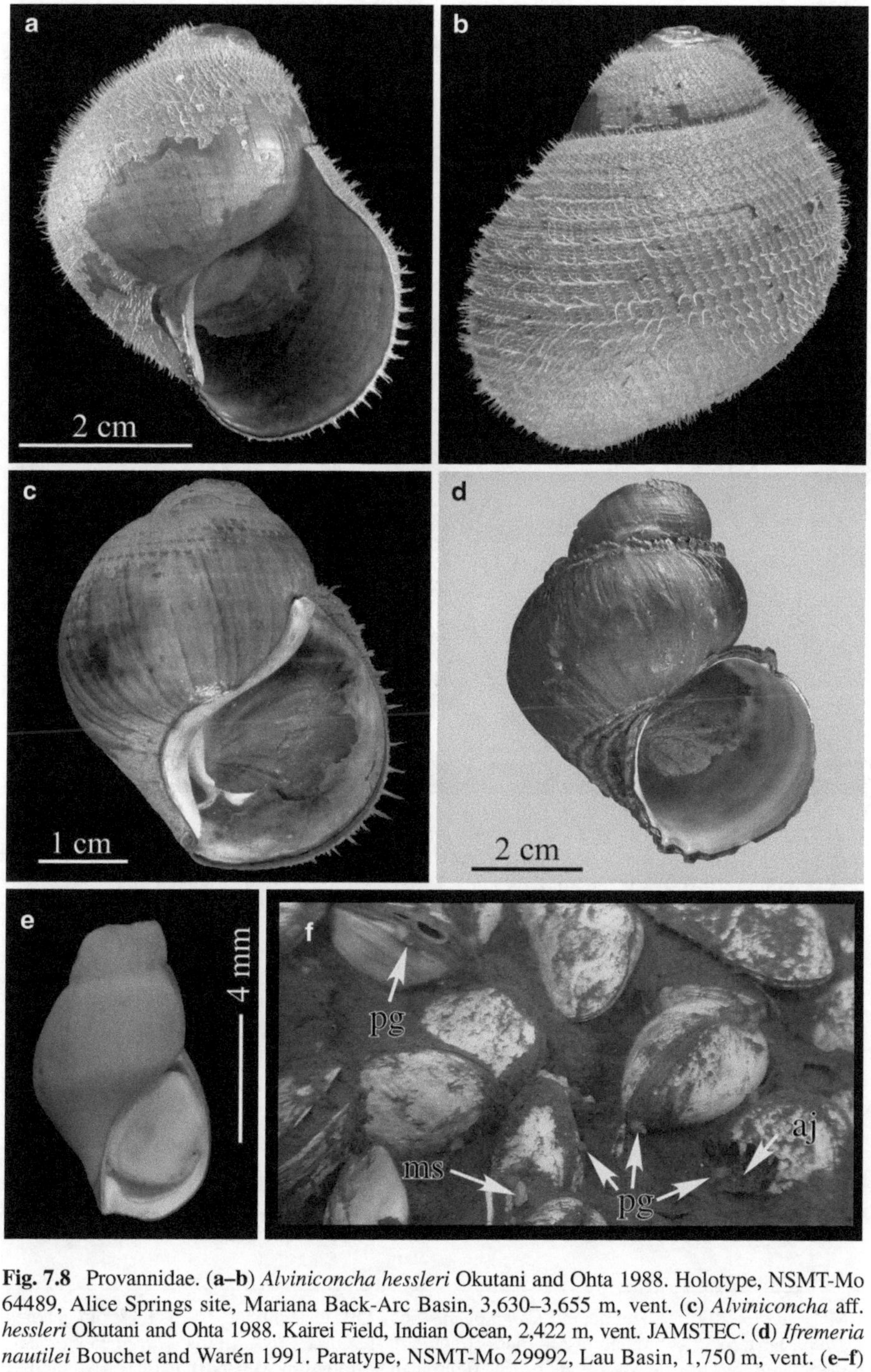

Fig. 7.8 Provannidae. (**a–b**) *Alviniconcha hessleri* Okutani and Ohta 1988. Holotype, NSMT-Mo 64489, Alice Springs site, Mariana Back-Arc Basin, 3,630–3,655 m, vent. (**c**) *Alviniconcha* aff. *hessleri* Okutani and Ohta 1988. Kairei Field, Indian Ocean, 2,422 m, vent. JAMSTEC. (**d**) *Ifremeria nautilei* Bouchet and Warén 1991. Paratype, NSMT-Mo 29992, Lau Basin, 1,750 m, vent. (**e–f**) *Provanna glabra* Okutani et al. 1992. Hatsushima seep site, Sagami Bay, Japan, seep. (**e**) Shell.

Fig. 7.9 Provannidae. (**a–d**) *Desbruyeresia marisindica* Okutani et al. 2004. Kairei Field, Indian Ocean, 2,434 m, vent. (**a**) Holotype UMUT RM28673. (**b**) Protoconch. UMUT RM28676. (**c**) Operculum. UMUT RM28674. (**d**) Radula. UMUT RM28675 (a–d: Okutani et al. 2004)

1. Genus *Alviniconcha* (Fig. 7.8a–c): *Alviniconcha* is one of most noticeable organisms of hydrothermal vents. The genus has been collected from vents of Mariana, Manus, North Fiji, Lau Back-Arc Basins and Kairei Field, but only a single species *Alviniconch hessleri* has been taxonomically described (type locality Mariana Back-arc Basin). Specimens from the southwestern Pacific and Indian Ocean are surprisingly similar in shell morphology in spite of their remotely separated habitats. Molecular phylogenetic studies have revealed several distinct genotypes (Denis et al. 1993; Kojima et al. 2001; Kojima 2004; Suzuki et al. 2005a, b, 2006a,

UMUT RM30209. (**f**) Habitat at *Calyptogena* bed. Off Hatsuhima, 1,173–1,175 m, *Hyper-Dolphin* Dive 525. Abbreviations: aj, periostracum of *Acharax johnsoni*; ms, *Margarites shinkai*; pg, *Provanna glabra*. A color plate of this figure can be found in Appendix I (Plate 16). (a–b, d: Photo taken by T. Sasaki at NSMT; c: Okutani et al. 2004; f: Sasaki et al. 2007)

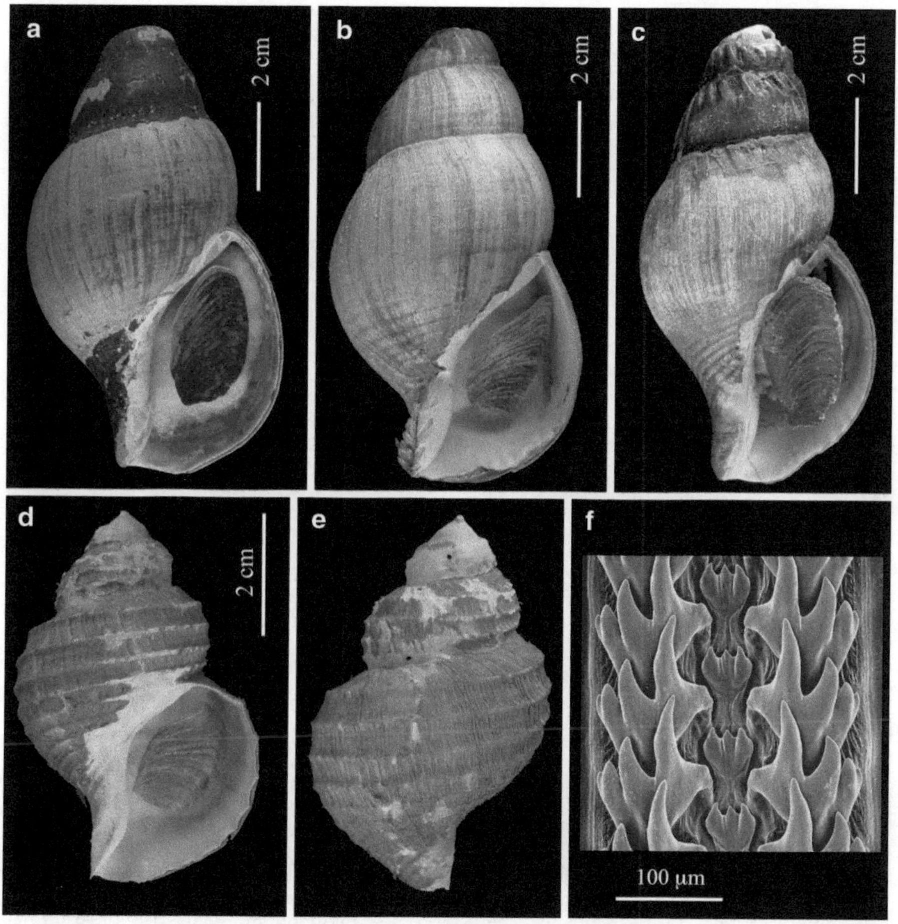

Fig. 7.10 Buccinidae. (**a**) *Eosipho desbruyeresi* Okutani and Ohta 1993. NSMT-Mo 69949. Lau Basin, 1,750 m, vent. (**b**) *Eosipho desbruyeresi nipponensis* Okutani and Fujiwara 2000. Holotype, NSMT-Mo 71689. North Knoll of Iheya Ridge, Okinawa Trough, Japan, 1,049 m, vent. (**c**) Paratype, NSMT-Mo 71691. Myojin Knoll, Japan, 1,268 m, vent. (**d–e**) *Calliloncha nankaiensis* Okutani and Iwasaki 2003. Holotype, NSMT-Mo 73525. Off Cape Muroto, Nankai Trough, Japan, 3,540 m, seep. (**f**) Radula. Paratype, NSMT-Mo 73526 (a, c: Photo taken by T. Sasaki at NSMT; b, d–e: Sasaki et al. 2005; f: Okutani and Iwasaki 2003)

Johnson and Vrijenhoek unpubl.), but so far the only general conclusion is that the Indian Ocean population is well separated and isolated from several others. The shell of *Alviniconcha* is swollen, fragile, and cracks when dried. The surface is covered with thick periostracum with spirally arranged bristles, formed by corresponding papillae on the pallial margin (Warén and Bouchet 1993: fig. 47). The bristles presumably have a protective function against fouling animals. The shell of *A. hessleri* consists of four layers of different microstructures from outer to inner sides: (1) periostracum, (2) simple prismatic, (3) complex crossed lamellar and (4) simple prismatic structures (Kiel 2004: 178).

Fig. 7.11 Conidae. (**a–b**) *Phymorhynchus starmeri* Okutani and Ohta 1993. Holotype, NSMT-Mo 69950. North Fiji Basin, 2,750 m, vent. (**c**) *Phymorhynchus turris* Okutani and Iwasaki 2003. Holotype, NSMT-Mo 73527. Off Cape Muroto, Nankai Trough, Japan, 3,540 m, seep. (**d**) *Oenopota sagamiana* Okutani and Fujikura 1992. Holotype, NSMT-Mo 69645. Off Hatsushima, Sagami Bay, Japan, 1,170 m, seep. (**e–g**) *Phymorhynchus buccinoides* Okutani et al. 1993. Holotype NSMT-Mo 73527. Off Hatsushima, Sagami Bay, Ca. 1,100m, seep. (a–b: Photo taken by T. Sasaki at NSMT; c–e: Sasaki et al. 2005; f–g: Fujikura et al. 2009)

There is no pallial tentacle. A pair of neck furrows is apparent on the right side of the head: one of these often contains a mucus string with mineral particles and fragments of other organisms (Warén and Bouchet 1993: 63) and possibly used to transport filtered food material from the ctenidium to the mouth. The ctenidial lamellae are attached to the mantle on the right side, but hung into the pallial cavity towards the left side (Warén and Bouchet 1993: fig. 48). The ctenidium of *Alviniconcha* is hypertrophied to cultivate symbiotic bacteria (Stein et al. 1988; Suzuki et al. 2005a: fig. 1). In *A.* aff. *hessleri* from the Kairei Field, spherical epsilon proteobractria are contained in elongate cells around the afferent ctenidial vessel (Suzuki et al. 2005a: fig. 2). In contrast, in *A. hessleri* from

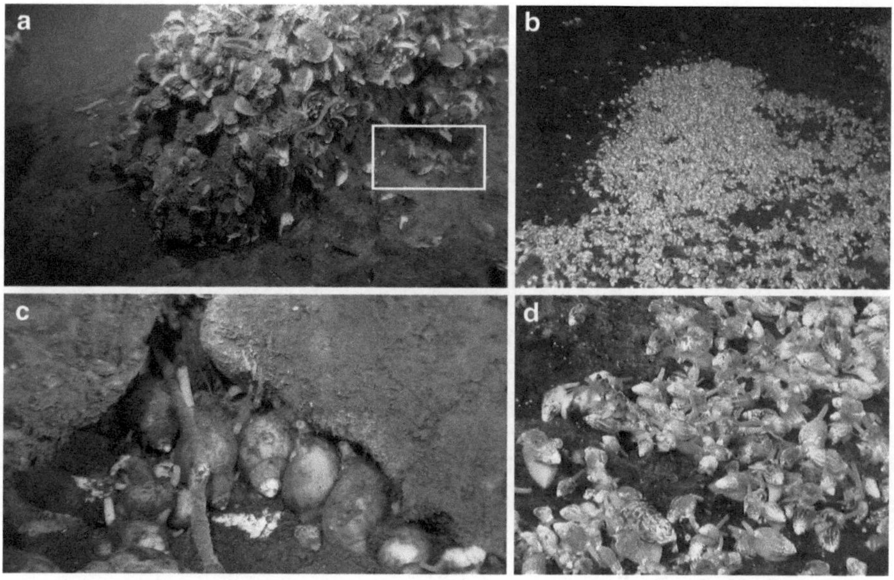

Fig. 7.12 Habitats of Conidae at the Hatsushima seep site in Sagami Bay, Japan. (**a–b**) *Phymorhynchus buccinoides* Okutani et al. 1993. *Hyper-Dolphin* Dive 524, 1,180 m. (**a**) Outcrop covered by *Bathymodiolus* spp. Rectangle is enlarged in B. (**b**) *P. buccinoides* extending siphon at base of outcrop. (**c–d**) *Oenopota sagamiana* Okutani and Fujikura 1992. *Hyper-Dolphin* Dive 525, 1,173 m. (**c**) Large colony of *O. sagamiana.* (**d**) Enlarged view of C. A color plate of this figure can be found in Appendix I (Plate 17) (a–d: Sasaki et al. 2007)

Mariana Back-Arc Basin, vermicular gamma proteobacteria are contained in round bacteriocytes on the surface of ctenidial lamellae (Suzuki et al. 2005b: fig. 1). It is interesting that different population of *Alviniconcha* have different lineages of endosymbionic bacteria. The digestive tract is reduced in volume. Species of *Provanna* have a stomach 100 times larger than *Alviniconcha* relative to whole animal volume (Warén and Bouchet 1993: 63).

2. Genus *Ifremeria* (Fig. 7.8d): *Ifremeria* is another large-sized gastropod forming massive aggregates in vents in southwestern Pacific. There is only a single species in the genus, *Ifremeria nautilei*. Contrary to *Alviniconcha*, genetic divergence is not remarkable in this species. Phylogenetic analysis using mitochondrial COI gene sequences suggests that populations from Manus, North Fiji and Lau Back-Arc Basins are the same biological species (Suzuki et al. 2006b; see also Kojima et al. 2000 for previous results). Similarly symbiotic bacteria from these populations belong to a single lineage of γ-proteobacteria (Suzuki et al. 2006b). The species is characterized by the black and shiny periostracum, the tall spire and nodular sculpture. The head-foot resembles *Alviniconcha* (Warén and Bouchet 1993: fig. 51) but numerous prominent warts cover the sides of the foot (Warén and Bouchet 1993: fig. 51–52). As in *Alviniconcha*, the ctenidium is enlarged (Windoffer and Giere 1997: figs. 1–2). Each ctenidial lamella is divided

into six distinct regions, including large area of bacteriocytes (Beck 1991: fig. 2; Windoffer and Giere 1997: figs. 3–4). The bacteriocytes are single-layered on either side and contains numerous elongate bacteria except basal areas (Windoffer and Giere 1997: fig. 9). The digestive system is reduced in size. Commensal polychaetes are known to occur in the pallial cavity (scale worms) and the umbilicus (*Amphisamytha* cf. *galapagensis*: Warén in Desbruyères et al. 2006: 130).

3. Genus *Provanna* (Fig. 7.9e–f): *Provanna* is one of most common members in vents and seeps (and more rarely on food-falls (Warén and Bouchet 2001: 178)). There are 16 described species up to 2009 (see Appendix 7.3), and there are more undescribed species (at least four in Japan). The types of habitat are vents, seeps, or food-falls depending on the species. The shell microstructure of *P. variabilis* was described by Kiel (2004: 178); the arrangement of shell layers is the same as that of *Alviniconcha hessleri*. The anatomy of *Provanna* was described by Warén and Ponder (1991). A striking feature is the presence an annulate pallial tentacle (Warén and Ponder 1991: figs. 3B–C, 4B). A similar tentacle is also possessed by *Abyssochrysos* sp. (Warén and Ponder 1991: fig. 19). The ctenidium is not hypertrophied (Warén and Ponder 1991: fig. 4A), and the digestive tract and stomach are not reduced in size (Warén and Ponder 1991: fig. 8). The male lacks a penis. The spermatozoa were described by Warén and Ponder (1991: fig. 13) and Healy (1990). *P. variabilis* is considered to be deposit feeder (Warén and Bouchet 1993; Levesque et al. 2006: table 1; Warén et al. in Desbruyères et al. 2006: 132) and grazes on bacteria and detritus. The possibility of symbiosis with autotrophic bacteria was proposed from stable carbon and nitrogen isotope composition (Bergquist et al. 2007), but may also originate from a diet of bacteria that have grown outside the body, a possibility rarely considered by microbiologists. From anatomical evidence, it is unlikely that *Provanna* have endosymbiotic bacteria, since the ctenidium of *Provanna* is not hypertrophied unlike in *Alviniconcha* and *Ifremeria*. Predation on *Provanna* is little known, but there are two records. Shell fragments of *P. variabilis* were found in the gut of the octopus *Graneledone* cf. *boreopacifica* (Voight 2000a), and a shell injury presumably caused by a crustacean was documented for a fossil (Oligocene) species (Kiel 2006).

   Adelphophagy has been reported for *P. lomana* but probably results from confusion with some other gastropods since no shell bearing veliger were observed (Warén in Kaim et al. 2008a: 430). Later observations at the seeps off Costa Rica revealed numerous egg-capsules with one, rarely up to three embryos, length 200 μm, which hatch as trochophore similar to those of *Ifremeria* (Warén unpubl.). This seems to be the second case of trochophore among the Caenogastropoda, *Ifremeria* being the other. The protoconch has been figured for several species (Table 7.2) and shell-bearing larvae can be regularly collected in sediment traps at the EPR. The lack of protoconch II and size of protoconch I suggest planktotrophic stage is absent.

4. Genus *Desbruyeresia* (Fig. 7.9): The distinction of this genus from *Provanna* was previously founded on the absence of a pallial tentacle (Warén and Bouchet 1993: 71), but later the tentacle was revealed to exist in better preserved specimens (Kaim et al. 2008a: 431). The diagnostic characters of the genus are slender

Melania-like teleoconch and the protoconch with rough cancellate sculpture (except *D. marisindica*). Other characters including the operculum and radula (Fig. 7.9) are similar to those of *Provanna*. The biology of the genus is little known. Feeding habit is grazing (Warén et al. Desbruyères et al. 2006: 129). Judging from protoconch morphology, development is probably planktotrophic in most species (e.g. *D. spinosa* and *D. cancellata*: Warén and Bouchet 1993: fig. 44D, E) but lecithotrophic in *D. marisindica* (Fig. 7.9b).

5. Genus *Cordesia*: This genus was recently described from a seep off West Africa. Its diagnostic characters are a cylindrical penis (Warén and Bouchet 2009: fig. 11B) and two right pallial tentacles (Warén and Bouchet 2009: fig. 11B–C). Its larva is inferred to have planktotrophic development from protoconch morphology (Warén and Bouchet 2009: fig. 11D–G). The larva ("Richter's larva") was collected 50–100 m below surface over the bottom of 4,570 m deep. This is obvious direct evidence for planktotrophic dispersal from deep-sea hydrocarbon seeps.

Subclade Neogastropoda (Figs. 7.10–7.12)

Neogastropods are generally predators and/or scavengers. In vents and seeps, Buccinidae and Turridae are two major groups; others are rare. Neogastropods are not known to harbor symbiotic bacteria.

Family Buccinidae (Fig. 7.10)

Vent/seep-associated buccinids include genera *Neptunea*, *Buccinum*, *Eosipho* (Fig. 7.10a–c), *Bayerius* and *Calliloncha* (Fig. 7.10d–f). *Eosipho* is a common predator-scavenger in vents and distributed from the West Pacific to the East Pacific Rise. Other groups have more limited occurrence. Species of *Neptunea* and *Buccinum* seem to favour seeps in the North Pacific. The species investigated most in detail biologically is *Buccinum thermophilum* Harasewych and Kantor 2002. The stomach contents indicate that the species is an active predator and opportunistic scavenger with a broad range of diet. Majid crabs are active predators of the species. Egg masses consist of more than 200 capsules and their morphology is similar to those of other buccinids (Martell et al. 2002). *Bayerius arnoldi* has the deepest habitat among vent/seep associated members of the family. The localities are widely spread at depths of 4,774–7,434 m along Japan Trench, Kurile-Kamchatka Trench and Aleutian Trench along with records on abyssal plain (Okutani and Fujikura 2005).

Family Conidae ("Turridae") (Figs. 7.11–7.12, Plate 17)

"Turrids" (conoidean gastropods) are common elements of predators in deep-sea environments in general. The family in a traditional sense will be split into several separate families (cf. Puillandre et al. 2008), but phylogenetic analysis of this large group is still in the beginning. The most well-known group in vents and seeps is *Phymorhynchus*. The biology of *Phymorhynchus* is relatively well examined in *P. buccinoides* which is endemic to seeps in Sagami Bay, Japan. The species inhabits exclusively large concretions covered by *Bathymodiolus* (Fig. 7.12a–b). The gut investigation could not reveal their prey, but isotopic data suggest the species utilizes *Bathymodiolus* as their food source (Fujikura et al. 2009). The egg capsules contain more than 1,000 eggs on average and are deposited on the shells of *Bathymodiolus* (Watanabe et al. 2009). The egg capsules of *Phymorhynchus* were also illustrated for an unidentified species from the Mid-Atlantic Ridge (Warén and Bouchet 2001: fig. 40b–c) and for *P. coseli* from off West Africa (Warén and Bouchet 2009: fig. 9K). *Phymorhynchus* has a well-developed funnel-shaped rhynchostome (Warén and Bouchet 2001: fig. 40d), and the true proboscis is small and everted internally near the posterior end of the rhynchostome (Warén and Bouchet 2001: 192). The radula (Fig. 7.11f–g) is small for the body size.

There are three records of *Phymorhynchus*' diet. The radula of *Neomphalus fretterae* was found in the gut of *Phymorphychus* sp. (Warén and Bouchet 1989: fig. 5), remains of *Bathymodiolus* were found in the anterior gut of *P. ovatus* (Warén and Bouchet 2001), and polychaete fragments are known from the oesophagus of *P. cingulatus* (Warén and Bouchet 2009: 2344). These are direct evidences of prey and predator (see also Voight 2000b).

Oenopota sagamiana is known exclusively from seeps. The species is associated with bacterial mats, and was found aggregated on a metachromatic area near and within the *Calyptogena* community in Sagami Bay (Okutani and Fujikura 1992: pl. 1, fig. 4). Later in 2006 a huge aggregate of extraordinarily high density were observed in a seep site in Sagami Bay (Sasaki et al. 2007; Fig. 7.12). The attractant of the colony was unidentified. One possible hypothesis is that they temporarily gather for reproduction. Seemingly there is not a large enough food source to sustain such a huge colony of predatory gastropods.

Family Cancellariidae

The family is rare in vent/seep environments, and there are only three records (see Appendix 7.3). The family is known to have a suctorial mode of feeding (O'Sullivan et al. 1987), and thus, the species from vents and seeps are assumed to parasitize other organisms (Warén and Bouchet 1993: 78).

7.2.6 Clade Heterobranchia

The diversity of heterobranchs is noticeably low compared to bathyal non-vent/seep communities (Appendix 7.3: species no. 208–218).

1. Family Pyramidellidae: The occurrence of pyramidellids is exceptional. A single species of *Odostomia* has been found at a few seeps off California and Oregon and a large species of *Eulimella* at similar biotopes from southern California to Costa Rica, among Vestimentifera and clams which are potential hosts for these parasitic species.
2. Cephalaspidea: Cephalaspids are not known from hydrothermal vents, although they are common in the surrounding deep-sea. A few specimens of *Parvamplustrum* sp. (Family Aplustridae) have been noticed in material from the seeps off Oregon.
3. Family Orbitestellidae: The genus *Lurifax* is a vent/seep-related group with species in Mediterranean and New Zealand seeps and Japanese and Mid-Atlantic vents. Nothing is known on its biology. Even its orthostrophic protoconch defies any attempt to infer the type of larval development.
4. Family Hyalogyrinidae: This family contains several species in the genera *Hyalogyra* and *Hyalogyrina*. *Hyalogyrina* may occur in large numbers on bacterial mats at sulphide-rich areas at seeps and whale falls as well as at vents (Braby et al. 2007; Warén unpublished).
5. Family Xylodisculidae: This family has several species known from food-falls, seeps and vents. Nothing is known about its biology, but their unusual radula type (Marshall 1988) may give a hint of unusual or specialized feeding biology. Both Hyalogyrinidae (*Hyalogyra* and *Hyalogyrina*) and Xylodisculidae (*Xylodiscula*) are originally known from sunken wood in the deep sea (Marshall 1988; Hasegawa 1997; Warén and Bouchet 2001).

7.3 Discussion

The following discussion of biological strategies of vent/seep gastropods concerns a wide range of topics. Interests in past publications were focused mostly on vent/seep-related life-strategies, feeding, reproduction and larval development, endemism to vents/seeps, the relation to fauna from biogenic substrates, the geographic distribution, and the correlation between type of larval development and distribution. Here the knowledge on these topics is summarized and evaluated.

7.3.1 Morphological Adaptation

Shells from deep-sea vents and seeps tend to exhibit common features (Warén and Bouchet 1993: 81) such as (1) simplification of shell shape, (2) development of periostracum especially in vents, (3) frequent damage by corrosion, and (4) crusts of mineral deposits. (1)–(3) are common features among deep-sea gastropods, and (4) is a phenomenon typical of vents/seeps or other reducing environments like on shells of species living in burrows or under rocks.

The shells of deep-sea gastropods are generally more simplified than those of their shallow-water counterparts. There are two possible factors: calcification and predation (Vermeij 1993). The shell of calcium carbonate is energetically more costly to produce and more easily corroded in deep-sea environments with colder temperatures and higher pressure. Therefore shells tend to be thin. Spines and teeth in the aperture are absent in gastropods from vents and seeps, contrary to shallow gastropods. But since the majority of vent/seep gastropods are limpets or have a very large aperture, spines and teeth should not be expected. Determinate growth forming thickened apertures (see Sasaki 2008: fig. 1) is also rare in deep-sea species. This is probably because shell-crushing or shell-peeling predation is less common in the deep sea (Vermeij 1993). Shell damage by corrosion is seen more frequently than predatory damages in the deep-sea. Especially when the periostracum is damaged by wear (or possibly bacterial activity), the corrosion starts from there and often forms a deep pit. Possibly other gastropods grazing on bacterial films on the shells contribute to the damage of the periostracum in vents and seeps. The problem with the periostracum is that it cannot be repaired, because the periostracum-secreting mantle of the animal cannot extend far away from the shell aperture. Mineral deposits are common on the shell surface at vents. The deposits are considered to be precipitated by bacterial activities (Warén and Bouchet 1989: 99, 1993: 81). They certainly contribute to the protection against corrosion and the last hope to obtain protoconchs with a preserved sculpture when very young specimens are not available is to remove crusts on the apex with bleach.

Eyes are often reduced or absent in deep-sea species. The reduction or loss of eyes has occurred independently in many gastropod lineages. Rudiments of eyes can be found as black spots in *Shinkailepas myojinensis* (Sasaki et al. 2006b), while in *S. briandi* from the MAR eyes are present only in the veliger larva (Warén and Bouchet 2001). In *Provanna*, the eyes contain the lens but lack a deeply pigmented layer of the retina (Warén and Ponder 1991). In *Bathymargarites symplector* the loss is evidently in progress since some specimens lack an eye and the pigment spots are often deformed or have split (Warén and Bouchet 1989). Other forms of specialization occur in the ctenidium and digestive tract. Among species having endosymbiotic bacteria, the ctenidium, especially the region of bacteriocytes, is hypertrophied. In filter feeding species, the ctenidial lamellae markedly develop to increase the surface area for filtering with cilia (see Declerck 1995 for general discussion) as are found in *Neomphalus* and *Lepetodrilus*. In addition, a groove is formed between the ctenidium and the mouth along the neck as in *Neomphalus, Lepetodrilus* and *Alviniconcha* (Fretter et al. 1981; Fretter 1988; Warén and Bouchet 1993). The digestive tract is reduced in relative size in the species depending largely on symbiotic bacteria (Warén and Bouchet 1993, 2001).

7.3.2 Feeding and Predation

Feeding of vent/seep gastropods is largely similar to shallow-water species or those in regular deep-sea environments. But there is one principal difference: It is not based on photosynthesis, but the community is largely sustained by chemosynthetic production by bacteria. The most common modes of feeding at vents and seeps can be categorized into four types.

Grazing: The surface layer of detritus and bacterial films is the food source for most species but not for most specimens, because the more specialized filter feeders and symbiont-dependent species usually occur in much larger numbers. Mechanical wear of radular teeth is regarded as the supportive evidence for grazing. Such wear can be recognized by the occurrence of loose radular teeth in the gut content and by SEM examination of the anterior end of the radula, where teeth are missing or broken.

Filter feeding: Neomphalus and *Lepetodrilus* employ filter feeding using ctenidial cilia in addition to grazing (Bates 2007a). *Cyathermia* and *Symmetromphalus* are also inferred to use filter feeding, based on the large size of ctenidial lamellae. This mode of feeding is accompanied by the modification of the ctenidial structure as mentioned above (Declerck 1995).

Predation and scavenging: Neogastropods are predator or scavenger in general. Houart and Sellanes (2006) reported possible borings by trophonine muricids on *Calyptogena gallardoi* and *Thyasira methanophoila* in a seep area off Chile. No naticid gastropods (specialized predators that drill holes in other molluscs) are known in vent/seep communities, but also in normal environments they seem to avoid prey with thick periostracum like mussels. When looking for stomach or gut content in predators or scavengers, it is common to find them empty. This is well known for cones (Conidae), and several *Neptunea amianta* (common in Californian seeps) were all reported to survive a year of starvation (Tamburri and Barry 1999).

Symbiotic bacteria: Alviniconcha and *Ifremeria* cultivate symbiotic bacteria intracellularly in the ctenidium (see above for references). *Lepetodrilus gordensis* and *L. fucensis* cultivate filamentous bacteria on the exterior of the gill (Johnson et al. 2006; Bates 2007a). The scaly foot gastropod cultivates bacteria in an enlarged esophageal gland. *Ctenopelta, Echinopelta* and *Hiropelta* are also to some extent probably dependent on bacteria, judging from the unusually small digestive system and that it is partly filled by extremely fine-grained iron sulphides, but low availability of specimens for investigation is hampering their study. The radula is also less important, when grazing is not a main method of feeding.

It is not always easy to specify actual food source in food web. The most direct and least complicated way for food investigations is by direct examination of the gut and stomach content, although it requires some basic knowledge on fragments of animals and sediment. The detection of bacteria in the gut also needs molecular characterization before the food is digested. Supplemented by stable isotope characterization of the tissues this gives a good picture; stable isotopes do not really indicate how the food is obtained. Predators on gastropod species have rarely been identified for species in vents. Shell repair is, however, common (*Buccinum thermophilum*: Martell et al. 2002; *Lepetodrilus fucensis*: Voight and Sigward 2007; *Lepetodrilus* spp., *Peltospira* spp. and *Cyathermia naticoides*: Warén unpublished) and predation has occasionally been more directly confirmed from the gut contents of predators (see Voight 2000b for review).

7.3.3 Reproductive Anatomy

Reproduction can be a relatively easily documented field in the studies of the ecology of deep-sea species. There is a wide variety of modes of reproduction throughout molluscs. The common questions concerning these are listed in Table 7.1. The reproductive strategy of various animals can be deduced from the anatomy of the reproductive organs.

Seasonality of reproduction: If the reproduction occurs during a limited period, the gonad development is not homogeneous throughout a year. The presence of various stages of gametes in the gonad suggests the reproduction is performed continuously. Seasonality has not been clearly documented in any gastropod in seeps and vents.

Hermaphroditism or gonochorism: Hermaphroditism is ubiquitous in heterobranchs and less common in non-heterobranch gastropods. In vents or seeps, *Protolira*, *Pyropelta* and *Adeuomphalus* are known to be simultaneously hermaphroditic. Sequential hermaphroditism (sex change) is not known in vent/seep gastropods. These features have however rarely been investigated, and vary greatly in several groups of gastropods.

Glands in pallial gonoduct: The development of glands in the gonoduct indicates internal fertilization in gastropods. The albumen gland secretes an albumen layer around eggs, and the capsule gland provides protective outer covers. In males the prostate provides mucous fluid to spermatozoa. These glands are absent in externally fertilizing species.

Table 7.1 Common questions concering reproduction and development of gastropods

Reproductive organ	Gametes in the gonad developing synchronously or continuously?
	Gonochoristic or hermaphroditic (simultaneous or sequential)?
	Prostate present or absent?
	Penis present or absent?
	Albumen and capsule glands present or absent?
	Receptaculum seminis present or absent?
	Receptaculum seminis connected to oviduct or isolated?
Sperm	Spermatozoa of aquasperm type or not?
	Spermatozoa dimorphic or euspermatoza only?
	Spermatophore formed or not?
	Fertilization external, internal or semi-internal (in pallial cavity)?
Egg	Eggs in ovary large and yolky or not?
	Eggs encapsulated, brooded or released freely?
	How many eggs per egg capsule?
	Embryos feeding on nurse eggs (adelphophagy) or not?
Larva	Larva with planktonic stage or entirely benthic?
	Larva hatching before or after metamorphosis?
	Larva planktotorohic or non-planktotrophic?
	How long planktonic larva can survive?
Protoconch	Protoconch symmetrical, paucispial or multispiral?
	Protoconch divided into PI and PII?
	Protoconch size within range of planktotorphic or non-prototrophic type?

Copulatory organ in male: The presence of a copulatory organ (penis) is another evidence for internal fertilization. The formation of the penis is inferred to have evolved several times in gastropods. For instance, the penis exists on the right side in most gastropods but on the left side in Neomphalidae. The position and innervation are morphological criteria to judge homology of various types of copulatory organs. Absence of a copulatory organ does not exclude internal fertilization.

Position of receptaculum seminis: The receptaculum seminis is connected to the pallial oviduct in the majority of caenogastropods and heterobranchs. However, it is located in an isolated position on the left side in Neomphalina and Seguenzoidea or isolated on the right side in *Lepetodilus*. In these groups, the spermatozoa transferred from the male are preserved in the receptaculum embedded in the pallial wall, and fertilization presumably occurs in the pallial cavity. This type of fertilization is termed "semi-internal," since the pallial cavity is not inside but not completely outside of the animal.

Sperm morphology: The shape of various parts of fertilizing sperms is known to be correlated with reproductive strategies. Most notably, the group exerting external fertilization has spermatozoa with a triangular head and long thin tail. Species with internal fertilization have a more elongate blunt head, and the difference in diameter from head to tail is less conspicuous than in externally fertilizing groups.

Sperm dimorphism: Many groups of internally fertilized caenogastropods have dimorphic sperms, eu- and paraspermatozoa, but other gastropods not. Paraspermatozoa possibly function as to convey euspermatozoa which are used for fertilization.

Spermatophore: Spermatophores occur in several internally or semi-internally fertilizing groups listed by Robertson (1989, 2007). Among the seep/vent-related gastropods, *Melanodrymia* sp. (Warén and Bouchet 2001) and the scaly foot gastropod were reported to produce spermatophores. The spermatophore from a male is deposited in the pallial cavity of a female and probably sperms are released and transferred to the receptaclum seminis from there. The number of sperm producing taxa is probably underestimated since their presence is probably very temporary after mating.

Egg size: The eggs tend to be larger in lecithotrophic species requiring a sufficient amount of yolk than in planktotrophic species. The dimensions of eggs are listed for a few vent/seep gastropods by Gustafson and Lutz (1994) and Tyler et al. (2008).

Egg capsule: Eggs are encased in an egg capsule and deposited on the substrate in internally fertilized species, but released into the seawater in externally fertilized species. The form of the egg capsules is variable, depending on taxonomic groups and sometimes also different between closely related species.

7.3.4 Larval Development

Larval ecology is basic information in the discussion of dispersal and evolution of the vent/seep fauna. Types of development can be categorized in various ways in marine invertebrates by nutritional modes, site of development, dispersal potential, and mode of morphogenesis (Levin and Bridges 1995). From an evolutionary perspective, different types of development have different selective advantages in the balance of maternal investment and larval survival (see McEdward 1995 for discussion).

Benthic or planktonic development: In species with external fertilization, the eggs are freely discharged, and larvae swim until metamorphosis. In species that deposit encapsulated eggs, larvae can be either benthic or planktonic. In benthic development, metamorphosed larvae hatch as crawling juveniles out of the egg capsules. In planktonic development, larvae can swim with the cilia and velum.

Planktotrophic or non-planktotrophic development: Some gastropods develop a plankton-feeding stage as the veliger larva (planktotrophy). During this phase the larva usually grows from a size of 0.1–0.2 mm to 0.4–2.0 mm and adds one or more whorls to the shell. This trait affects the duration of planktonic life, but little is known on the duration except in some common shallow water species. The ability of a species to shift between planktotrophy and lecithotrophy has been much discussed and occasionally reported but so far no cases can be considered well documented in deep-sea gastropods.

The type of larval development can usually be inferred from well preserved larval shells (Jablonski and Lutz 1983). The larva hatching from the egg capsule ideally has a shell of roughly one whorl with a certain sculpture; during the planktotrophic life 1–5 whorls, usually with a more prominent sculpture are added, by incremental growth at the pallial margin. These two parts are called protoconch 1 and protoconch 2, respectively. The term "non-planktotrophic" was applied originally from a paleobiological viewpoint to avoid the problem to infer the presence or absence of a planktonic dispersal phase in species with lecithotrophic development (Shuto 1974; Jablonski and Lutz 1983). This is called "Thorson's apex theory" (1950), but it is not always so easy or clearcut. To be sure about the applicability of these criteria it is necessary to have detailed knowledge of some closely related or very similar species, as pointed by Thorson (1950) but often conveniently forgotten. When there is different sculptures in the protoconchs 1 and 2, consisting of small and sharp elements, like in triphorids, cerithiids or turrids, planktotrophy can usually be safely inferred. If protoconch 2 is smooth adelphophagy cannot be excluded. For example, the world's largest gastropod *Syrinx auranus* has a protoconch of 4–5 smooth whorls, but the development is via adelphophagy. Many gastropod species at vents and seeps are presumed to be non-plankotrophic, but more interesting is the presence or absence of a planktonic dispersal phase (details of most species in Warén and Bouchet 1989, 1993, 2001). Multispiral protoconchs are limited to Neritimorpha and Caenogastropoda. Among the species in vents and seeps, almost all of the species in vents and seeps have lecithotrophic development as in Patellogastropoda, Neomphalina, and Vetigastropoda.

A special type of sculpture is formed in the protoconch of *Bathymargarites, Ventsia*, and *Bruciella* (Warén and Bouchet 1993: 82). The larval shell of these groups has two distinct sections, but this probably does not indicate planktotrophic stage as in caenogastropods. One possible hypothesis is that the section is formed when the larvae still use nutrients from eggs after metamorphosis (Warén and Bouchet 1993: 82). A similar sculptural change was also shown in the Calliostomatidae (teleoconch I: Marshall 1995: 385) and might be indicative of dietary change after metamorphosis.

7.3.5 Dispersal and Speciation

The capacity of dispersal presumably has an essential role in the evolution of vent/seep gastropods. Species with a feeding larval stage are assumed to extend their distributions to wider ranges. By contrast, the species without planktotrophic phase have little potential of long-distance dispersal, and therefore, have more restricted distributions (Table 7.2). However, this is not always the case in the deep sea. There seems to be no noticeable difference in distribution between species with a planktotrophic larva, a planktonic dispersal phase or those that crawl directly out of the egg capsule. This seems also to be the case for bivalves, where mussels (*Bathymodiolus*) with planktotrophic larvae have more geographic variation and smaller distributional areas than the clams (Vesicomyidae) with lecithotrophic development (Vrijenhoek pers. comm. 2006). Two details must be remembered here. We know nothing on how long the lecithotrophic larvae can survive. Pradillon et al. (2001) showed that some vestimentiferan larvae can hibernate and wake up when slightly warmer water is encountered. It is not only an advantage with a long larval life that may also carry the larvae far away from seeps and vents. The ideal would be to have the choice. And that may be the choice for many of the lecithotrophic ones. Most larvae will probably end up far away from any suitable environment. One possible dispersal strategy is rafting. Individuals attached to drifting object can stay afloat and drift a long distance. The only likely candidate known to us is *Buccinum* sp. They produce egg masses consisting of hundreds of egg capsules, some of which may die and the capsule fill with gas that make the whole mass buoyant. It can then drift along the bottom like a salt shaker regularly dropping young specimens (Warén and Bouchet 2001). The dispersal capabilities might be inferred from genetic distance among populations. For example in vent/seep gastropods, genetic divergence is notably different between *Alviniconcha* and *Ifremeria* (Suzuki et al. 2005a, b, 2006a, b; Kojima et al. 2001, 2004), and larval ecology of these snails may also be different. If a species is genetically homogeneous over a wide area, it probably has high dispersal capability. If a species is genetically differentiated among distant sites, its dispersal capability may be restricted. However, Johnson et al. (2008) showed that some species of *Lepetodrilus* showed virtually no variation in the COI gene from 20°N at the EPR to 38°S, while other species of the genus

Table 7.2 Assumptions on development and dispersal of gastropods. Capacity of dipersal increases to the right

Amount of yolk	Large	←→	Small
Planktotrophic stage	Absent	←→	Long
Velum	Weakly developed	←→	Well-developed
Protoconch whorls	Paucispiral	←→	Multispiral
Dispersal ability	Low	←→	High
Geographic range	Narrow	←→	Wide
Gene flow	Low	←→	High

had quite restricted distributions or considerable genetic variations, despite having a larval shell of roughly the same size and morphology.

The dispersal is considered a bottleneck for vent/seep animals (Craddock et al. 1997; Van Dover 2002; Tyler and Young 2003), because their habitats are isolated by considerable distances often hundreds of kilometers. In such a situation, the invasion and colonization of new localities may be aided by occasional or unusual habitats like when Johnson et al. (2008) found *Lepetodrilus ovalis*, normally a vents species, on whalebone, at Monterey Bay, California, 1,000 km north of its nearest known occurrence (stepping stone hypothesis). It is not clear whether long-distance dispersal of long-life planktotrophic larvae is effective for vents or seeps. If larvae cannot find suitable vents or seeps for the settlement, most larvae are consumed with high mortality. In lecithotrophic development larvae with larger yolk might be more advantageous for longer dispersal (Kelly and Metaxas 2007). In food-poor environments, nutrition from yolk is important for survival. Thus, yolk amount and egg size have a major effect on survival, dispersal and isolation of species (see discussion by Craddock et al. 1997). Reduction of hydrothermal fluid or seepage causes the decline of local production and eventually the extinction of local life. This has been well observed by repeated visits to the same sites (Fujikura personal observation, 2008; Van Dover 2000).

7.3.6 Endemism to Vents and/or Seeps

It has been widely recognized that many species in chemosynthetic communities are either vent- or seep-endemic (Tunnicliffe 1991, 1992; McArthur and Tunnicliffe 1998; Martin and Haney 2005). At family level, some groups are strict endemics. Peltospiridae and Sutilizonidae are strictly endemic to vents; Neomphalidae and Lepetodrilidae are almost endemic to vents. Provannidae and Neolepetopsidae are mainly distributed in vents but also seeps and foodfalls. In summary in 2001, at genus level, 31 out of 57 genera (54.3%) are endemic to vents; 8 genera (14%) are restricted to seeps, thus nearly 70% are endemic to these environments (Warén and Bouchet 2001: 212). Based on updated information (Tables 7.3, 7.4) and a not fully comparable data set, the number of endemic genera is 36 out of 101 (35.6%) in vents, 8 (7.9%) in seeps, and 5 (5.0%) for vents and seep. Therefore, 48.5% are endemic to vents and/or seeps. At species level, the number of endemic species is 138 out of 218 (63.3%) in vents, 52 (23.9%) in seeps, and 5 (2.3%) for vents and seep. Therefore, 89.5% are endemic to vents and/or seeps. A limited number of species occur both in vents and seeps. For examples, *Neolepetopsis gordensis*, *Margarites shinkai*, *Provanna laevis* and *P. variabilis* are known from both vents and seeps (see Appendix 7.3). As a general tendency the degree of endemism is lower in the seep fauna than at vents (Warén and Bouchet 1993: 81). It is likely that the seep fauna is more frequently invaded from ambient non-chemosynthetic environments than the vent fauna.

Table 7.3 Number of described species and type of habitat in vent/seep-assocaited gastropods at genus level

Higher taxon	Family or Superfamily	Genus	Distribution IO	WP	EP	AO	MS	Habitat of genus
Patellogastropoda	Neolepetopsidae	*Eulepetopsis*	–	–	1 sp.	–	–	Vent
		Neolepetopsis	–	–	4 spp.	–	–	Vent, seep, whale bone
		Paralepetopsis	–	2 spp.	1 sp.	2 spp.	–	Vent, seep, whale bone
	Pectinodontidae	*Bathyacmaea*	–	5 spp.	–	–	–	Vent, seep
		Serradonta	–	2 spp.	–	–	–	Seep
Vetigastropoda		*Sahlingia*	–	–	1 sp.	–	–	Seep
	Fissurellidae	*Clathrosepta*		1 sp.	1 sp.	–	–	Vent[a]
		Cornisepta	–	–	2 spp.	–	–	Vent[a]
		Puncturella	–	3 spp.	–	1 sp.	–	Vent, seep[a]
	Pseudococculinidae	*Amphiplica*	–	–	1 sp.	–	–	Vent
		Tentaoculus	–	–	–	1 sp.	–	Seep, anomuran carapaces, sunken seaweed
	Pyropeltidae	*Pyropelta*	–	3 spp.	2 sp.	2 spp.	–	Vent, seep, whale bone
	Lepetodrillidae	*Clypeosectus*	–	–	2 spp.	–	–	Vent
		Gorgoleptis	–	–	3 spp.	–	–	Vent
		Lepetodrilus	–	4 spp.	10 spp.	2 spp.	–	Vent (mostly), seep
		Pseudorimula	–	1 sp.	–	1 sp.	–	Vent
	Sutilizonidae	*Sutilizona*	–	–	2 spp.	1 sp.	–	Vent
		Temnocinclis	–	–	1 sp.	–	–	Vent
		Temnozaga	–	–	1 sp.	–	–	Vent
	Family uncertain	*Adeuomphalus*	–	2 spp.	1 sp.	–	–	Vent
		Akritogyra	–	–	1 sp.	–	–	Seep[a]
		Ventsia	–	1 sp.	–	–	–	Vent
	Seguenziidae	*Bathymargarites*	–	–	–	1 sp.	–	Vent
	Cataegidae	*Cataegis*	–	–	–	1 sp.	–	Seep
	Chilodontidae	*Bathybembix*	–	–	1 sp.	–	–	Seep[a]
		Putzeysia	–	–	–	–	1 sp.	Seep[a]

(continued)

Table 7.3 (continued)

Higher taxon	Family or Superfamily	Genus	Distribution IO	WP	EP	AO	MS	Habitat of genus
	Calliostomatidae	*Calliostoma*	–	–	1 sp.	–	–	Seep[a]
		Falsimargarita	–	–	1 sp.	–	–	Vent[a]
		Otsukaia	–	–	1 sp.	–	–	Seep[a]
	Trochidae	*Clelandella*	–	–	–	–	1 sp.	Seep[a]
	Turbinidae	*Gaza*	–	–	1 sp.	–	–	Seep[a]
		Margarites	–	2 spp.	–	1 sp.	–	Vent, seep[a]
		Bruceiella	1 sp.	1 sp.	1 sp.	–	–	Vent, seep
		Akritogyra	–	–	1 sp.	–	–	Seep[a]
		Fucaria	–	1 sp.	1 sp.	–	–	Vent, seep
		Iheyaspira	–	1 sp.	–	–	–	Vent
		Protolira	–	–	–	1 sp.	–	Vent, whale bone
		Solutigyra	–	–	1 sp.	–	–	Vent
	Collonidae	*Cantrainea*	–	2 spp.	1 sp.	1 sp.	–	Vent, seep[a]
	Solariellidae	*Zetela*	–	–	1 sp.	–	–	Seep[a]
Neomphalina	Melanodrymiidae	*Leptogyra*	–	1 sp.	–	1 sp.	–	Vent, seep, sunken wood
		Melanodrymia	–	–	3 spp.	–	–	Vent
	Family uncertain	*Helicrenion*	–	1 sp.	–	–	–	Vent
		Retiskenea	–	1 sp.	1 sp.	–	–	Seep
		Vetulonia	–	1 sp.	–	–	–	Vent[a]
	Neomphalidae	*Cyathermia*	–	–	1 sp.	–	–	Vent
		Lacunoides	–	–	2 spp.	–	–	Vent
		Neomphalus	–	–	1 sp.	–	–	Vent
		Planorbidella	–	1 sp.	1 sp.	–	–	Vent
		Symmetromphalus	–	2 spp.	–	–	–	Vent
	Peltospiridae	*Ctenopelta*	–	–	1 sp.	–	–	Vent
		Depressigyra	–	–	2 spp.	–	–	Vent
		Echinopelta	–	–	1 sp.	–	–	Vent

		Hirtopelta	–	–	2 spp.	–	–	Vent
		Lirapex	–	–	2 spp.	1 sp.	–	Vent
		Nodopelta	–	–	3 spp.	–	–	Vent
		Pachydermia	–	1 sp.	1 sp.	–	–	Vent
		Peltospira	–	–	3 spp.	1 sp.	–	Vent
		Rhynchopelta	–	–	1 sp.	–	–	Vent
Neritimorpha	Neritidae	*Bathynerita*	–	–	–	1 sp.	–	Seep
	Phenacolepadidae	*Olgasolaris*	–	1 sp.	–	–	–	Vent
		Shinkailepas	–	3 spp.	–	1 sp.	–	Vent
Caenogastropoda	Capulidae	*Hyalorisia*	–	–	–	1 sp.	–	Seep
	Cerithiopsidae	*Speculator*	–	–	1 sp.	–	–	Vent
	Elachisinidae	*Laeviphitus*	–	1 sp.	–	1 sp.	–	Vent, seep[a]
	Provannidae	*Alviniconcha*	–	1 sp.	–	–	–	Vent
		Cordesia	–	–	–	1 sp.	–	Seep
		Desbruyeresia	1 sp.	4 spp.	–	–	–	Vent
		Ifremeria	–	1 sp.	–	–	–	Vent
		Provanna	–	7 spp.	6 spp.	4 spp.	–	Vent, seep, sunken wood (rare)
	Ranellidae	*Fusitriton*	–	–	1 sp.	–	–	Seep[a]
	Rissoidae	*Alvania*	–	–	–	1 sp.	–	Vent[a]
		Benthonella	–	–	1 sp.	–	–	Seep[a]
		Pseudosetia	–	–	–	1 sp.	–	Vent[a]
	Vitrinelldiae	*Neusas*	–	–	–	1 sp.	–	Vent
	Buccinidae	*Bayerius*	–	1 sp.	2 spp.	–	–	Seep[a]
		Buccinum	–	1 sp.	1 sp.	1 sp.	–	Vent, seep[a]
		Calliloncha	–	1 sp.	–	–	–	Seep[a]
		Eosipho	–	1 sp. + 1 s. sp.	1 sp.	1 sp.	–	Vent, seep[a]

(continued)

Table 7.3 (continued)

Higher taxon	Family or Superfamily	Genus	Distribution IO	WP	EP	AO	MS	Habitat of genus
		Neptunea	–	2 spp.	2 sp.	–	–	Vent, seep[a]
		Kryptos	–	–	–	1 sp.	–	Seep[a]
	Cancellariidae	*Admete*	–	1 sp.	–	–	–	Seep[a]
		Cancellaria	–	–	1 sp.	–	–	Seep[a]
		Iphinopsis	1 sp.	–	–	–	–	Vent[a]
	Muricidae	*Trophon*	–	–	3 spp.	–	–	Seep[a]
	Conidae	*Bathybela*	–	–	–	1 sp.	–	Seep[a]
		Benthomangelia	–	–	–	–	1 sp.	Seep[a]
		Drilliola	–	–	–	–	1 sp.	Seep[a]
		Gymnobela	–	–	–	1 sp.	–	Seep[a]
		Oenopota	–	2 spp.	–	–	–	Vent, seep[a]
		Phymorhyunchus	–	5 spp.	1 sp.	5 spp.	–	Vent, seep[a]
		Taranis	–	–	–	–	1 sp.	Seep[a]
	Volutidae	*Tractolira*	–	–	1 sp.	–	–	Seep[a]
		Miomelon	–	–	–	1 sp.	–	Seep[a]
Heterobranchia	Hyalogyrinidae	*Hyalogyra*	–	1 sp.	–	–	–	Vent, sunken wood
		Hyalogyrina	–	–	3 spp.	1 sp.	–	Vent, seep, sunken wood
	Orbitestellidae	*Lurifax*	–	1 sp.	–	1 sp.	1 sp.	Vent, seep
	Pyramidellidae	*Eulimella*	–	–	1 sp.	–	–	Vent[a]
	Xylodisculidae	*Xylodiscula*	–	1 sp.	–	1 sp.	–	Vent, seep, sunken wood
	Dendronotidae	*Dendronotus*	–	–	–	1 sp.	–	Vent[a]

Data based on the species described by the end of 2009 (see Appendix 7.3). IO, Indian Ocean: Rodrigues Triple Junction; WP, Japan to New Zealand, EP, East Pacific: Aleutian to Chile; AO: Atlantic Ocean: Mid-Atlantic Ridge, Caribbean Sea and West Africa; MS, Mediterranean Sea

[a]Genera also known from 'normal' non-chemosynthetic enviroments (including possible records). sp., spp. = species, s. sp. = subspecies

Table 7.4 Habitats of described taxa at genus and species level

Habitat	Number of genera	Percentage	Number of species	Percentage
Seeps only	8	7.9%	52[a]	23.9%
Seeps and biogenic substrata	1	1.0%	3	1.4%
Seeps and regular environments	24	23.8%	18	8.2%
Vents only	36	35.6%	138[a]	63.3%
Vents and biogenic environments	2	2.0%	1	0.5%
Vents and regular environments	9	8.9%	0	0%
Vents and seeps	5	5.0%	5	2.3%
Vents, seeps and biogenic substrata	7	6.9%	1	0.5%
Vents, seeps and regular environments	9	8.9%	0	0%
Total	101	100%	218	100%

Data from Appendix 7.3
[a] Including possible records

7.3.7 Connection with Sunken Wood and Bone

Some species, genera and families are associated with vents/seeps and food falls (Smith et al. 1989; Warén and Bouchet 1989; Smith and Baco 2003), which may indicate an evolutionary link between them (Distel et al. 2000). At species level, overlapping habitats have been reported in four species (see Appendix 7.1 for references). (1) In the genus *Pyropelta,* two species, *P. musaica* and *P. corymba* have been recorded from vents and whale-fall. (2) The turbinid *Protolira thorvaldssoni* is known both from vents, seeps and whale bones. (In neither case species identity has been tested by molecular analysis). (3) *Lepetodrilus ovalis* and *L. elevatus* are common in hydrothermal vents at the EPR but have also been found and sequenced from sunken driftwood, off California (Johnson et al. 2008).

At genus or family level, connections between vent/seep and sunken wood/bone are obvious in several cases. (1) Species of Neolepetopsidae are mostly vent/seep-endemic, but two species, *Neolepetopsis nicolasensis* and *Paralepetopsis clementensis*, are known from whale skeletons off California (McLean 2008: 16, 18). (2) In Pectinodontidae, the genus *Bathyacmaea* is closely related to the wood-grazing genus *Pectinodonta* (Nakano and Ozawa 2007). Furthermore, *Serradonta* on vestimentiferan tubes has a radular morphology similar to that of *Pectinodonta.* (3) In the Melanodrymiidae, Kunze et al. (2008) and Heß et al. (2008) considered *Melanodrymia* is closely related to *Leptogyra, Xyleptogyra* and *Leptogyropsis* known from sunken wood (Marshall 1988). (4) *Pyropelta wakefieldi* was described from whale bone off California. In addition, *Pyropelta* is morphologically closely related to *Pseudococculina* which is common to sunken wood. (5) The family Provannidae is a mainly vent/seep group, but *P. macleani and P. pacifica* were collected from sunken driftwood, not from vents or seeps (Warén and Bouchet 2001: p. 179). A couple of species of *Provanna* and a new provannid genus were found on a whale skeleton off California (Braby et al. 2007; Johnson et al. in press). (6) The

Hyalogyrinidae (*Hyalogyra* and *Hyalogyrina*) and Xylodisculidae (*Xylodiscula*) also use sunken wood as habitat (Marshall 1988; Warén 1992; Haseagawa 1997).

This connection to the wood/bone-associated fauna is also evident in the fossil record. For example, *Leptogyra* from the Eocene, *Xylodiscula* from the Oligocene (Kiel and Goedert 2006, 2007), *Pyropelta*, *Retiskenea*, *Depressigyra*, *Provanna* and *Lurifax* from the Eocene to Oligocene (Kiel 2006) and *Provanna* from the Late Cretaceous (Kaim et al. 2008a) represent ancestors of modern vent/seep taxa. Before ocean-going whales radiated in the Eocene (Kiel and Little 2006), chemosysnthesis-associated gastropods utilized marine reptile bones (e.g. Kaim et al. 2008b). Abyssochrysoid Hokkaidoconchidae probably utilized a similar ecological niche as the Provannidae (Kaim et al. 2008a, b; Kiel et al. 2008).

While seep and food-fall related species seem to be recognized back to the Jurassic – Cretaceous boundary (Kiel et al. 2008), the few older records must be considered uncertain. No fossil vent faunas with identifiable gastropods are known except in Cretaceous ophiolites from Cyprus (Little and Cann 1999). This is probably due to a combination of deep sea-environment with high pressure and low temperature typical of the deep sea and low pH of the vent fluids, factors that increase the solubility of calcium carbonate. It is actually rare to find any accumulations of old shells in vents and usually the calcareous layer disappears before the periostracum.

Among the two major groups in vents, Neomphalina and Lepetodriloidea, only a few safe records, back to the Eocene, are known and only of species related to *Leptogyra* (Kiel and Goedert 2006) which now is classified in Neomphaloidea. Not a single of the larger and modified neomphalins or anything similar has been found. The same is true for the Lepetodriloidea, although they do occur also in seeps where conditions for fossilization are better. Records of *Retiskenea* from the early Cretaceous (Campbell et al. 2008) may represent the earliest known Neomphalina, but those fossils are more likely to belong to one of other skeneimorph groups.

7.3.8 Biogeography

The summary of the geographic distribution of each species (Appendix 7.3) shows that most species are endemic to a relatively narrow region in a single trench, ridge, trough, or basin. Examples of the widest distribution like *Lepetodrilus elevatus* which has been recorded the Galapagos Spreading Center, the EPR between 20°N and 17°S, to North Fiji and Lau Back-Arc Basins (Warén and Bouchet 2001) have turned out to be based on misidentifications (Johnson et al. 2008), but *L. ovalis* occurs along the EPR from 21°N to 38°S with virtually no variation in the COI gene. Although vents and seeps are widely distributed in the world oceans (see Desbruyères in Desbruyères et al. 2006: 513–517 for map), there is no wide-ranged species that occurs in two oceans, instead they follow geologic formations, like the EPR (21 – ca 25°S occasionally to 38°S), MAR (38°N to 09°S) or the seeps along the American West Coast. The EPR has been explored to 38°S (R. Vrijenhoek pers. comm.) and the MAR to 09°S (O. Giere pers. comm.). At 38°S of the EPR the fauna is largely similar to the more northern localities although often with different species; at 09°S of the MAR it is largely the same species as further north. Apparently the diversity within the vent-endemic families is the highest along the EPR, as is shown in the Neolepetopsidae (*Eulepetopsis* and *Neolepetopsis*), Lepetodriloidea (*Clypeosectus, Gorgoleptis*, *Temnocinclis* and *Temnozaga*), Neomphalidae (*Cyathermia*, *Lacunoides* and *Neomphalus*), Melanodrymidae (*Melanodrymia*) and Peltospiridae (*Ctenopelta*, *Depressigyra*, *Ecihnopelta*, *Hirtopelta*, *Nodopelta*, and *Rhynchopelta*).

The faunal link between the East Pacific and the Atlantic are suggested by a restricted occurrence of genera like *Sutilizona*, *Lirapex* and *Peltospira*, as well as a couple of species pairs, *Provanna ios – P. sculpta, P. pacifica – P. admetoides*. The northern connection of the East and West Pacific is supported by a few examples: similar species of Pyropeltidae and *Retiskenea diploura* and *Bayeryus arnoldi*, which occur in seeps along the trenches of the North Pacific. Supraspecific taxa endemic to West Pacific are represented only by *Iheyaspira*, *Serradonta*, *Symmetromphalus*, and *Ifremeria*. Similarities in the gastropod fauna of the seeps of the Central American west coast, the Gulf of Mexico and off West Africa were pointed out by Warén and Bouchet (2009). The vent fauna of the Indian Ocean is still poorly explored (Okutani et al. 2004; Johnson et al. 2008), but connections exist to the West Pacific as suggested by the presence of *Alviniconcha* and *Desbruyeresia*, and to the Atlantic by a species similar to *Lepetodrilus atlanticus*. On the other hand, two undescribed neolepetopsids, one very similar to *Eulepetopsis vitrea* and one to *Neolepetopsis densata* disturb these patterns (Warén, unpublished).

7.4 Future Topics

The fauna of especially the vents is now pretty well known and predictable, much better than the surrounding deep-sea. However, some geographical areas still remain unexplored, especially in the Indian Ocean and the Polar regions. Vent activity with vent related fauna has been reported from about 80°N at Jan Mayen in the North Atlantic but still little is known on the fauna, except that rissoid gastropods occur there (C. Schander pers. comm.). Hydrothermal activity has also been reported from the Gakkel Ridge in the Polar Basin, but only two shells of an unknown gastropod with possible vent affinity were retrieved from grabs (S. Kim pers. comm.). The area south of South America, the southern Mid-Atlantic Ridge and south of the Pacific vents are also unexplored. These areas will however be more difficult thanks to weather conditions with ice or reliably windy weather.

The faunistic knowledge is today fair, but still the gastropod fauna of some areas, like the Indian Ocean and the Manus Basin are incompletely described and explored. The seep areas are even less known, partly due to the problems with a larger and more diverse surrounding fauna which is mixed with the seep component. The evolution of the vent-seep fauna and its connection to food-falls and other temporary food sources is slowly getting documented but here the gastropod fauna of food-falls is the limiting factor. Systematic COI sequencing has usually turned out very useful to reveal cryptic species and should be routinely applied when exploring new faunas.

The feeding and symbiosis with bacteria of the gastropods is still poorly explored although there are more different types of symbioses than in any other group of animals and several probable cases remain to explore; *Ctenopelta*, *Echinopelta*, *Hirtopelta*, all species with reduced alimentary system. Also the mineralization of sulphides and in the alimentary canal of the host remains to be explored; it is usually filled with extremely finely granular sulphides or occasionally elementary sulphur. In addition to stable isotope studies of the feeding biology, also more direct examination and identification of gut content is needed to understand the ecology.

The evolution of the chemosynthetic faunas needs more exploration of fossil deposits to find better preserved material, especially from the first half of the Cretaceous. What little is known of Jurassic deposits seems to indicate a faunal shift in the upper Jurassic (Kaim 2004). The study of dispersal of vent organisms as well as other marine fauna needs more real data, not only assumptions, but studies of the transport of larvae with bottom currents, where do the planktotrophic species live and feed, and intensive study of settlement in new vents.

Acknowledgements We deeply thank Dr. Steffen Kiel (Christian-Albrechts-University Kiel) for his editorial work. We are also grateful to those who kindly reviewed this manuscript and supported our research on vent/seep molluscs since the 1980s. Many samples illustrated here were provided though kind help by crew and staff of IFREMER and JAMSTEC. Photographs of type specimens were taken at NSMT with help by Dr. Hiroshi Saito. Specimens, ecological and molecular information were given by Prof. Kazutaka Amano (Joetsu University of Education), Dr. Erik E. Cordes (Temple University), Dr. Daniel Desbruyères (IFREMER), Dr. Cindy Lee van Dover (Duke University), Dr. Yoshihiro Fujiwara (JAMSTEC), Dr. Shannon Johnson (Monterey Bay Aquarium Research Institute: MBARI), Dr. Andrzej Kaim (Polish Academy of Sciences), Dr. Steffen Kiel (Christian-Albrechts-University Kiel), Prof. Shigeru Kojima (Atmosphere and Ocean Research Institute, University of Tokyo: AORI), Dr. Lisa Levin (Scripps Institute of Oceanography), Dr. Tomoyuki Nakano (NSMT), and Dr. Heiko Sahling (Bremen University), Dr. Michel Segonzac (IFREMER), Dr. Verena Tunnicliffe (University of Victoria), Dr. Robert Vrijenhoek (MBARI) and Dr. Hiromi Watanabe (JAMSTEC). Dr. Robert G. Jenkins (AORI), Dr. Takami Nobuhara (Shizuoka University), and Dr. Ryuichi Majima (Yokohama National University) shared fruitful discussion on fossil vent/seep fauna with us. Part of this work was supported by a Grant-in-Aid from Japan Society of the Proton of Science (no. 18340165).

Appendix 7.1: References to Internal Anatomy of Vent/Seep-Associated Gastropods

Higher taxon	Family	Species	References
Patellogastropoda	Neolepetopsidae	*Eulepetopsis vitrea*	Fretter 1990: 531–544
		Neolepetopsis denstata	Fretter 1990: 545–546
		Neolepetopsis gordensis	Fretter 1990: 546–547
		Neolepetopsis occulta	Fretter 1990: 545
		Paralepetopsis floridensis	Fretter 1990: 547–548
		Paralepetopsis rosemariae	Beck 1996: 89–90
	Pectinodontidae	*Bathyacmaea jonassoni*	Beck 1996: 90–92
		Bathyacmaea secunda	Sasaki et al. 2006a: 295–303
Vetigastropoda	Pyropeltidae	*Pyropelta bohlei*	Beck 1996: 93
		Pyropelta corymba	McLean and Haszprunar 1987: 200–202
		Pyropelta musaica	McLean and Haszprunar 1987: 198–200
		Pyropelta ryukyuensis	Sasaki et al. 2008: 309–310
	Lepetodrilidae	*Clypeosectus curvus*	Haszprunar 1989a: 5–10
		Clypeosectus delectus	Haszprunar 1989a: 5–10
		Gorgoleptis emarginatus	Fretter 1988: 58–64
		Gorgoleptis patulus	Fretter 1988: 65–67
		Gorgoleptis spiralis	Fretter 1988: 64–65
		Lepetodrilus guaymasensis	Fretter 1988: 54–56
		Lepetodrilus cristatus	Fretter 1988: 53–54
		Lepetodrilus elevatus	Fretter 1988: 35–50
		Lepetodrilus fucensis	Fretter 1988: 56–58
		Lepetodrilus gordensis	Johnson et al. 2006: 155–156
		Lepetodrilus nux	Sasaki 1998: 94–100
		Lepetodrilus ovalis	Fretter 1988: 50–53
		Lepetodrilus pustulosus	Fretter 1988: 35–49
		Lepetodrilus schrolli	Beck 1993: 171–173
		Pseudorimula marianae	Haszprunar 1992: 5–10
	Sutilizonidae	*Sutilizona theca*	Haszprunar 1992: 3–5
		Temnocinclis euripes	Haszprunar 1992: 3
		Temnozaga parilis	Haszprunar 1992: 3
	family uncertain, Segenzioidea	*Ventsia tricarinata*	Kunze et al. 2008: 124
Neomphalina	Melanodrymiidae	*Melanodrymia aurantiaca*	Israelsson 1998: 105–108; Haszprunar 1989: 176–180
	Neomphalidae	*Neomphalus fretterae*	Fretter et al. 1981: 338–353
		Pachydermia laevis	Israelsson 1998: 95–105
		Symmetromphalus hageni	Beck 1992: 246–254
	Peltospiridae	*Echinopelta fistulosa*	Fretter 1989: 149–150
		Hirtopelta hirta	Fretter 1989: 150–152
		Hirtopelta tufari	Beck 2002: 252

(continued)

Appendix 7.1 (continued)

Higher taxon	Family	Species	References
		Nodopelta heminoda	Fretter 1989: 146–147
		Nodopelta subnoda	Fretter 1989: 148–149
		Peltospira delicata	Fretter 1989: 141–145
		Peltospira operculata	Fretter 1989: 141–145
		Rhynchopelta concentrica	Fretter 1989: 125–141
		scaly foot gastropod	Warén et al. 2003: supporting on-line material
Neritimorpha	Neritidae	*Bathynerita naticoidea*	Warén and Bouchet 1993: 3–6
	Phenacolepadidae	*Olgasolaris tollmanni*	Beck 1992: 266–269
		Shinakilepas tufari	Beck 1992: 262–265
		Shinkailepas myojinensis	Sasaki et al. 2006a: 2–14
Caenogastropoda	Provannidae	*Alviniconcha hessleri*	Warén and Bouchet 1993: 56–64
		Ifremeria nautilei	Beck 1991: 282–284 as *Olgaconcha tufari*; Warén and Bouchet 1993: 64–71
		Provanna sculpta, P. laevis, P. adametoides, P. variabilis and *P. lomana*	Warén and Ponder 1991: 28–48
	Buccinidae	*Buccinum thermophilum*	Harasewych and Kantor 2002: 40–43
	Conidae	*Phymorhynchus moskalevi*	Sysoev and Kantor 1995: 22–25

Appendix 7.2: References to Protoconch Morphology of Vent/Seep-Associated Gastropods

Higher taxon	Family	Species	References
Patellogastropoda	Neolepetopsidae	*Eulepetopsis vitrea*	McLean 1990a: pl. VII, fig. h
		Neolepetopsis densata	Gustafson and Lutz 1994: figs. 4.11–4.12; Warén and Bouchet 2001: fig. 15a–c
		Paralepetopsis floridensis	McLean 1990a: pl. X, fig. g
		Paralepetopsis sasakii	Warén and Bouchet 2009: fig. 3C, E, F
Vetigastropoda	Family uncertain, Vetigastropoda	*Sahlingia xandaros*	Warén and Bouchet 2001: fig. 15d
	Pseudococculinidae	*Amphiplica gordensis*	McLean 1991: figs. 29, 30
		Tentaoculus granulatus	Warén and Bouchet 2009: fig. 5D

(continued)

Appendix 7.2 (continued)

Higher taxon	Family	Species	References
	Lepetodrilidae	*Clypeosectus delectus*	McLean 1989a: fig. 9C, F; Mullineaux et al. 1996: fig. 2J–K
		Gorgoleptis emarginatus	Mullineaux et al. 1996: fig. 2H
		Gorgoleptis sp.	Mullineaux et al. 1996: fig. 2G
		Lepetodrilus atlanticus	Warén and Bouchet 2001: fig. 15i
		Lepetodrilus elevatus	Mullineaux et al. 1996: fig. 1E
		Lepetodrilus elevatus galriftensis	Mullineaux et al. 1996: fig. 1D
		Lepetodrilus fucensis	Johnson 2006: fig. 5A–B
		Lepetodrilus nux	Sasaki et al. 2003: fig. 11D
		Lepetodrilus ovalis	Mullineaux et al. 1996: fig. 1F
		Lepetodrilus schrolli	Beck 1993: pl. 5, fig. 1
		Lepetodrilus sp.	Mullineaux et al. 1996: fig. 1A–C
		Pseudorimula marianae	McLean 1989a: fig. 13B
		Pseudorimula midatlantica	Warén and Bouchet 2001: fig. 15h
	Sutilizonidae	*Sutilizona pterodon*	Warén and Bouchet 2001: fig. 15g
		Sutilizona theca	McLean 1989a: fig. 4F
		Sutilizona tunnicliffae	Warén and Bouchet 2001: fig. 15f
	Scissurellidae	*Sinezona* sp.	Warén and Bouchet 1989: fig. 4
	Family uncertain, Seguenzioidea	*Adeuomphalus collinsi*	Kano et al. 2009: fig. 4D, E
		Adeuomphalus elegans	Kano et al. 2009: fig. 3F, I
		Adeuomphalus trochanter	Warén and Bouchet 2001: fig. 15l; Kano et al. 2009: fig. 2F
		Ventsia tricarinata	Warén and Bouchet 1993: fig. 18I
	Seguenziidae	*Bathymargarites symplector*	Warén and Bouchet 1993: fig. 1A–C
	Turbinidae	*Bruceiella athlia*	Warén and Bouchet 2001: fig. 15e
		Bruceiella globulus	Warén and Bouchet 1993: fig. 18E–F
		Protolira valvatoides	Warén and Bouchet 1993: fig. 18D, G

(continued)

Appendix 7.2 (continued)

Higher taxon	Family	Species	References
Neomphalina	Family uncertain	*Helicrenion reticulatum*	Warén and Bouchet 1993: fig. 32C
		Vetulonia phalcata	Warén and Bouchet 1993: fig. 18C
	Melanodrymiidae	*Leptogyra costellata*	Warén and Bouchet 2009: fig. 8A
		Leptogyra inflata	Warén and Bouchet 1993: fig. 32A
		Melanodrymia aurantiaca	Warén and Bouchet 1989: fig. 82; Mullineaux et al. 1996: fig. 4A, C
		Melanodrymia galeronae	Warén and Bouchet 2001: fig. 15k
	Neomphalidae	*Lacunoides vitreus*	Warén and Bouchet 2001: fig. 29f
		Neomphalus fretterae	McLean 1981: fig. 10A; Warén and Bouchet 1989: fig. 79; Mullineaux et al. 1996: fig. 2D–E
		Planorbidella planispira	Warén and Bouchet 1989: fig. 84
		Retiskenea diploura	Warén and Bouchet 2001: fig. 15j; Okutani and Fujikura 2002: fig. 2B
		Solutigyra reticulata	Warén and Bouchet 1989: figs. 89–90
		Symmetromphalus hageni	Beck 1992: pl. 3, fig. 4
		Symmetromphalus regularis	McLean 1990b: fig. 14
	Peltospiridae	? Peltospiridae sp. A	Warén and Bouchet 1993: fig. 32F
		Ctenopelta porifera	Warén and Bouchet 1993: fig. 26F
		Depressigyra globulus	Warén and Bouchet 1989: fig. 83; Warén and Bouchet 2001: fig. 15o
		Echinopelta fistulosa	Gustafson and Lutz 1994: figs. 4.9
		Lirapex costellata	Warén and Bouchet 2001: fig. 15m
		Lirapex granularis	Warén and Bouchet 1989: fig. 88; Mullineaux et al. 1996: fig. 3G–H
		Lirapex humata	Warén and Bouchet 1989: fig. 87
		Nodopelta subnoda	McLean 1989b: fig. 29

(continued)

Appendix 7.2 (continued)

Higher taxon	Family	Species	References
		Pachydermia laevis	Warén and Bouchet 1989: figs. 85–86
		Pachydermia sculpta	Warén and Bouchet 1993: fig. 32E
		Peltospira operculata	Mullineaux et al. 1996: fig. 3E
		Peltospira operculata?	Mullineaux et al. 1996: fig. 3D
		Peltospira smaragdina	Warén and Bouchet 2001: fig. 15n
		Rhynchopelta concentrica	McLean 1989b: figs. 37–38; Mullineaux et al. 1996: fig. 3A–B
Neritimorpha	Neritidae	*Bathynerita naticoidea*	Warén and Bouchet 2001: fig. 30d–e
	Phenacolepadidae	*Olgasolaris tollmanni*	Beck 1992: pl. 5, figs. 1–2
		Shinkailepas briandi	Warén and Bouchet 2001: fig. 32f
		Shinkailepas myojinensis	Sasaki et al. 2003: fig. 13C
		Shinkailepas tufari	Beck 1992: pl. 1, fig. 5
Caenogastropoda	Elachisinidae	*Laeviphitus desbruyeresi*	Warén and Bouchet 2001: fig. 37b
		Laeviphitus japonicus	Okutani et al. 1993: figs. 39–40
	Provannidae	*Alviniconcha hessleri*	Warén and Bouchet 1993: fig. 44C
		Cordesia provannoides	Warén and Bouchet 2009: fig. 11D, G
		Desbruyeresia cancellata	Warén and Bouchet 1993: fig. 44E
		Desbruyeresia marisindica	Okutani et al. 2004: fig. 4B
		Desbruyeresia spinosa	Warén and Bouchet 1993: fig. 44D
		Provanna reticulata	Warén and Bouchet 2009: fig. 10J
		Provanna segonzaci	Warén and Bouchet 1993: fig. 57A–B
		Provanna spp.	Warén and Ponder 1991: fig. B, D, F
		Provanna variabilis	Gustafson and Lutz 1994: figs. 4.14–4.15
	Rissoidae	*Alvania stenolopha*	Bouchet and Warén 1993: fig. 1447
		Alvania cf. *stenolopha*	Warén and Bouchet 2001: fig. 37a
		Pseudosetia azorica	Bouchet and Warén 1993: figs. 1603–4

(continued)

Appendix 7.2 (continued)

Higher taxon	Family	Species	References
	Conidae	*Phymorhunchus* sp.	Gustafson and Lutz 1994: figs. 4.17
		Phymorhynchus alberti	Bouchet and Warén 1980: fig. 239
		Phymorhynchus ovatus	Warén and Bouchet 2001: fig. 42a
Heterobranchia	Hyalogyrinidae	*Hyalogyra vitrinelloides*	Warén and Bouchet 1993: fig. 42C
		Hyalogyrina globularis	Warén and Bouchet 2001: fig. 37f
		Hyalogyrina rissoella	Warén and Bouchet 2009: fig. 12K
		Hyalogyrina umbellifera	Warén and Bouchet 2001: fig. 37e, g
	Orbitestellidae	*Lurifax japonica*	Sasaki and Okutani 2005: fig. 1C
		Lurifax vitreus	Warén and Bouchet 2001: fig. 37c–d
	Xylodisculidae	*Xylodiscula major*	Warén and Bouchet 1993: fig. 42A–B

Appendix 7.3: A List of Gastropods Described from Recent Vents and/or Seeps Up to the End of 2009

Families, genera and species are arranged alphabetically within clades. Unavailable names that appeared in the past literature are excluded. Type localities are given for the species distributed in more than a single vent/seep area. Abbreviations: EPR, East Pacific Rise; JdF, Juan de Fuca Ridge; MAR, Mid-Atlantic Ridge; WP, West Pacific. *Undescribed or unidentified species. [ ] Probably not vent-seep species.

Clade Patellogastropoda
Family Neolepetopsidae

1. *Eulepetopsis vitrea* McLean 1990. McLean 1990b: 503, pls. VIIa–i, VIIIa–j, IXa–f; Warén and Bouchet 2001: 123, figs. 2b, g, 14c, appendix 2; Kiel 2004: 170, figs. 3–5; Warén et al. in Desbruyères et al. 2006: 83. EPR: 21°N–17°S (type locality: 21°N), Galapagos Spreading Center, 2,447–2,635 m, vent.
2. *Neolepetopsis densata* McLean 1990. McLean 1990b: 496, pls. IIIa–g, IVa–d; Gustafson and Lutz 1994: figs. 4.10–4.12; Warén and Bouchet 2001: 123, figs. 2a, 3a, 15a–c, appendix 2; Warén et al. in Desbruyères et al. 2006: 84. EPR: 13–12°N (type locality: 12°N), Galapagos Spreading Center, 2,630–2,735 m vent.
3. *Neolepetopsis gordensis* McLean 1990. McLean 1990b: 492, pls. Ia–h, IIa–g; Warén and Bouchet 2001: appendix 2; Warén et al. in Desbruyères et al. 2006: 84. EPR: Gorda Ridge, 41°N (type locality); South of Gulf of California, 20°N; possibly also off Peru, 5°S; 3,200–3,271 m, vent/seep.

**Neolepetopsis* cf. *gordensis* McLean 1990. Warén and Bouchet 2001: 120, figs. 2d, 3f–g; Kiel 2004: 170, figs. 1–2. Gorda Ridge, south of Gulf of California, off Peru, 5°S, 3,200–3,795 m.

4. *Neolepetopsis occulta* McLean 1990. McLean 1990b: 501, pl. VIa–g; Warén and Bouchet 2001: appendix 2; Warén et al. in Desbruyères et al. 2006: 84 EPR: 21°N, 1,990 m, vent.
5. *Neolepetopsis verruca* McLean 1990. McLean 1990b: 499, pl. Va–g; Warén and Bouchet 2001: appendix 2; Warén et al. in Desbruyères et al. 2006: 84. EPR: 21°N, 2,593 m, vent.
6. *Paralepetopsis ferrugivora* Warén and Bouchet 2001. Kiel 2004: 170, figs. 6–7; Warén and Bouchet 2001: 123, figs. 2e–f, 3b–e, 4a–b, e, 14a–b, appendix 2; Warén et al. in Desbruyères et al. 2006: 85. MAR: Lucky Strike, 37°N, ca. 1,700 m, vent.
7. *Paralepetopsis floridensis* McLean 1990 McLean 1990b: 510, pls. Xa–h, XIa–f; Warén and Bouchet 2001: appendix 2. Florida Escarpment, 3,270 m, seep.
8. *Paralepetopsis lepichoni* Warén and Bouchet 2001. Warén and Bouchet 2001: 125, figs. 2c, 5a–f, appendix 2; Okutani and Iwasaki 2003: 2, fig. 2B; Sasaki et al. 2005: 107. WP: Nankai Trough: south off Shizuoka Prefecture (type locality), south-southeast off Cape Muroto, Japan, 2,140–3,571 m, seep.
9. *Paralepetopsis rosemariae* Beck 1996. Beck 1996: text-fig. 2, pl. 1, figs. 1–7; Warén and Bouchet 2001: appendix 2 WP: Edison Seamount, 1,483 m, vent.
10. *Paralepetopsis sasakii* Warén and Bouchet 2009. Warén and Bouchet 2009: 2331, figs. 2A–E, 3A–I. Regab site, West Africa, 750–3,150 m, seep.

Family Pectinodontidae (“Acmaeidae”)

11. *Bathyacmaea jonassoni* Beck 1996. Beck 1996: text-fig. 3, pl. 2, figs. 1–7; Warén and Bouchet 2001: appendix 2. WP: Edison Seamount, 1,483 m, vent.
12. *Bathyacmaea nipponica* Okutani et al. 1992. Okutani et al. 1992: 140, figs. 7–11; Warén and Bouchet 2001: appendix 2; Sasaki et al. 2005: 109. WP: Off Hatsushima, Sagami Bay, 1,110–1,200 m, seep
13. *Bathyacmaea secunda* Okutani et al. 1993. Okutani et al. 1993: 130, figs. 13–19; Okutani and Fujiwara 2000: 124, figs. 2–3; Warén and Bouchet 2001: appendix 2; Sasaki et al. 2003, figs. 2, 3A; Fuchigami and Sasaki 2005: 156, fig. 10H; Sasaki et al. 2005: 109, fig. 12D. Okinawa Trough: Minami Ensei Knoll (type locality), Iheya Ridge, North Knoll of Iheya Ridge, Izena Hole, 700–1,049 m, vent.
14. *Bathyacmaea subnipponica* Sasaki et al. 2003. Sasaki et al. 2003: 193, fig. 6; Sasaki et al. 2005: 111, fig. 12E. WP: Nankai Trough: Ryuyo Canyon, 1,100 m, seep
15. *Bathyacmaea tertia* Sasaki et al. 2003. Sasaki et al. 2003: 190, figs. 3B, 4, 5; Sasaki et al. 2005: 111, fig. 12F. WP: Okinawa Trough: North Knoll of Iheya Ridge, 996–1,000 m, vent

**Bathyacmaea* sp. 1. Okutani and Iwasaki 2003: 2; Sasaki et al. 2005: 111. WP: Off Cape Muroto, Nankai Trough, Japan, 3,571 m, seep.

Bathyacmaea sp. 2. Sasaki et al. 2003: 195; Sasaki et al. 2005: 111. WP: Sumisu Caldera, Japan, 676 m, vent.

16. *Serradonta kanesunosensis* Sasaki et al. 2003. Sasaki et al. 2003: 195, fig. 8; Sasaki et al. 2005: 108, fig. 12A. WP: Kanesunose Bank, Nankai Trough, 284–322 m, seep.
17. *Serradonta vestimentifericola* Okutani et al. 1992. Okutani et al. 1992: 139, figs. 2–6; Warén and Bouchet 2001: appendix 2; Sasaki et al. 2005: 109, fig. 12B–C. WP: Off Hatsushima, Sagami Bay, 1,110–1,200 m, seep.

Family Lepetidae

Bathylepeta sp. [mistake for *Iothia*] Sellanes et al. 2008: table 2. Concepción Methane Seep Area, Chile, 740–870 m, seep and regular deep-sea bottom?

Clade Vetigastropoda
Family uncertain

18. *Sahlingia xandaros* Warén and Bouchet 2001. Warén and Bouchet 2001: 129, figs. 7d–f, 8a–b, 9a–d, 15d, 16h, appendix 2; Kiel 2004: 171: fig. 10. EP: Aleutian Trench, 4,800–4,900 m, seep.

Superfamily Fissurelloidea
Family Fissurellidae

[19.] *Clathrosepta becki* McLean and Geiger 1998. McLean and Geiger 1998: 14, fig. 7; Warén and Bouchet 2001: appendix 2. WP: Manus Basin, Vienna Woods, 2,494 m, vent.
[20.] *Clathrosepta depressa* McLean and Geiger 1998. McLean and Geiger 1998: 13, fig. 6; Warén and Bouchet 2001: appendix 2. EPR: 13°N, 1,160 m, vent.
[21.] *Cornisepta levinae* McLean and Geiger 1998. McLean and Geiger 1998: 22, fig. 13; Warén and Bouchet 2001: appendix 2. EPR: 13°N, 1,775 m, vent.
[22.] *Cornisepta verenae* McLean and Geiger 1998. McLean and Geiger 1998: 23, fig. 14; Warén and Bouchet 2001: appendix 2. EP: JdF, Axial Seamount, 1,530 m, vent.
23. *Puncturella parvinobilis* Okutani et al. 1993. Okutani et al. 1993: 128, figs. 8–12; Okutani and Fujiwara 2000: 123, fig. 4; Warén and Bouchet 2001: appendix 2; Sasaki et al. 2003: 197, fig. 10; Sasaki et al. 2005: 111, fig. 13A. WP: Okinawa Trough: Minami Ensei Knoll (type locality), North Knoll of Iheya Ridge, Izena Hole, 690–1,340 m, vent.
24. *Puncturella rimaizenaensis* Okutani et al. 1993. Okutani et al. 1993: 126, figs. 2–7; Warén and Bouchet 2001: appendix 2; Sasaki et al. 2005: 112, fig. 13B. WP: Okinawa Trough: Izena Hole, 1,340 m, vent.
25. *Puncturella similis* Warén and Bouchet 2009. Warén and Bouchet 2009: 2335, figs. 2H–I, 6H–J. Guiness Site, West Africa, 750 m, seep.
26. *Puncturella solis* Beck 1996. Beck 1996: pl. 3, figs. 1–9; Warén and Bouchet 2001: appendix 2. WP: Edison Seamount, 1,492 m, vent.

Puncturella sp. 1. Sellanes et al. 2008: table 2. Concepción Methane Seep Area, Chile, 740–870 m, seep and regular deep-sea bottom?

***Puncturella* sp. 2. Sellanes et al. 2008: table 2. Concepción Methane Seep Area, Chile, 740–870 m, seep and regular deep-sea bottom.

Superfamily Lepetelloidea
Family Lepetellidae

**Lepetella* sp. Olu-Le Roy et al. 2004: table 2. Anaximander area, Eastern Mediterranean Sea, ca. 1,700–2,000 m, seep, on vestimentiferan tube.

Family Pseudococculinidae

27. *Amphiplica gordensis* McLean 1991. McLean 1991: 44, fig. 25–32; Warén and Bouchet 2001: appendix 2. Gorda Ridge, 3,305 m, vent.
28. *Tentaoculus granulatus* Warén and Bouchet 2009. Warén and Bouchet 2009: 2333, figs. 5A–H. Off Angola, West Africa, 1,307 m, seep.

**Tentaoculus?* sp. Warén and Bouchet 2001: appendix 2. New Zealand, seep.

Family Pyropeltidae

29. *Pyropelta bohlei* Beck 1996. Beck 1996: text-fig. 4, pl. 4, figs. 1–7; Warén and Bouchet 2001: appendix 2; Warén et al. in Desbruyères et al. 2006: 93 WP: Edison Seamount, 1,483 m, vent.
30. *Pyropelta corymba* McLean and Haszprunar 1987. McLean and Haszprunar 1987: 200, figs. 9–11; McLean 1992b: 409, figs. 17–24; Warén and Bouchet 2001: 125, figs. 4c–d, 6c–f, 7a–b, appendix 2; Warén et al. in Desbruyères et al. 2006: 93. Guaymas Basin, seep (type locality) and Oregon Margin (Warén et al. in Desbruyères et al. 2006); off California, whale-fall (McLean 1992b), 2,022–524 m.
31. *Pyropelta musaica* McLean and Haszprunar 1987. McLean and Haszprunar 1987: 198, figs. 1–8, 9A; McLean 1992b: 406, figs. 9–16; Warén and Bouchet 1993: 2; Warén and Bouchet 2001: appendix 2; Kiel 2004: 171, figs. 8–9; Warén et al. in Desbruyères et al. 2006: 93. JdF: Axial Seamount, 1,546–1,575 m, vent (type locality) and Jalisco Block, seep (Warén and Bouchet 2001) and off California, 1,240 m, whal-fall (McLean 1992b).

**Pyropelta* cf. *musica* McLean and Haszprunar 1987. Warén and Bouchet 2001: 125, figs. 6a–b, 7c. JdF, Axial Seamount, off California 33–36°N, possibly also Jalisco Block Seeps, 20°N, 940–1,575 m, possibly also 3,000–3,775 m, vent, whale skeletons.

32. *Pyropelta oluae* Warén and Bouchet 2009. Warén and Bouchet 2009: 2332, figs. 4G–J. Regab site, West Africa, 3,150 m, seep.
33. *Pyropelta ryukyuensis* Sasaki et al. 2008. Sasaki et al. 2008: 309, figs. 2–9. WP: Okinawa Trough: Hatoma Knoll (type locality) and Daiyon Yonaguni Knoll, 1,451–1,523 m, vent.
34. *Pyropelta sibuetae* Warén and Bouchet 2009. Warén and Bouchet 2009: 2332, figs. 4A–F, 13B–D. Guiness Site, West Africa, 750 m, seep.
35. *Pyropelta yamato* Sasaki et al. 2003. Sasaki et al. 2003: 197, fig. 9; Sasaki et al. 2005: 112, fig. 12G; Warén et al. in Desbruyères et al. 2006: 93. WP: Sumisu Caldera, 676 m, vent.

Superfamily Lepetodriloidea
Family Lepetodrilidae

36. *Clypeosectus curvus* McLean 1989. McLean 1989b: 21, figs. 10A–H, 11F; Haszprunar 1989a: 5, figs. 3, 4B–C, E, 5–6; McLean 1989b: 21, figs. 10A–H, 11F; Warén and Bouchet 2001: 155, appendix 2; Warén et al. in Desbruyères et al. 2006: 86. JdF: Axial Seamount (type locality), Endeavour Segment; Explorer Ridge; 1,575–2,212 m, vent.
37. *Clypeosectus delectus* McLean 1989. McLean 1989b: 18, figs. 7A–H, 8A–H, 9A–F, 11A–E; Haszprunar 1989a: 5, fig. 4A, D; Mullineaux et al. 1996: fig. 2J–L; Warén et al. in Desbruyères et al. 2006: 86. EPR: Galapagos Spreading Center (type locality), 21°N–17°S, ca. 2,400–2,600 m, vent.
38. *Gorgoleptis emarginatus* McLean 1988. McLean 1988a: 22, figs. 21–24, 84–92; Fretter 1988: 58, figs. 16–17; Mullineaux et al. 1996: fig. 2H–I; Warén and Bouchet 2001: 154, fig. 17b, appendix 2; Warén et al. in Desbruyères et al. 2006: 87. EPR: 21–9°N (type locality: 21°N), ca. 2,600 m, vent.
39. *Gorgoleptis patulus* McLean 1988. McLean 1988a: 24, figs. 98–102; Fretter 1988: 65, figs. 18b–c; Warén and Bouchet 2001: 154, fig. 17a, appendix 2; Warén et al. in Desbruyères et al. 2006: 87. EPR: Galapagos Spreading Center (type locality), 13°N, ca. 2,480–2,600 m, vent.
40. *Gorgoleptis spiralis* McLean 1988. McLean 1988a: 23, figs. 93–97; Fretter 1988: 64, fig. 18b–c; Warén and Bouchet 2001: 155, fig. 17a, appendix 2; Warén et al. in Desbruyères et al. 2006: 87. EPR: 13°N, ca. 2,630 m, vent.
41. *Lepetodrilus atlanticus* Warén and Bouchet 2001. Warén and Bouchet 2001: 143, figs. 14d, f, 15i, 19a–f, 20a–d, 21a, b, appendix 2; Warén et al. in Desbruyères et al. 2006: 88. MAR: 38–23°N (type locality: Menez Gwen), 850–3,500 m, vent.
42. *Lepetodrilus corrugatus* McLean 1993. McLean 1993: 29, fig. 11–16; Warén and Bouchet 2001: appendix 2; Warén et al. in Desbruyères et al. 2006: 88. JdF, 2,420 m, vent. Only known from the holotype.
43. *Lepetodrilus cristatus* McLean 1988. McLean 1988a: 15, figs. 13–14, 57–61; Fretter 1988: 53, figs. 11–12; Warén and Bouchet 2001: 151, fig. 22a, appendix 2; Warén et al. in Desbruyères et al. 2006: 88. EPR: 21–13°N (type locality: 21°N), Galapagos Spreading Center, ca. 2,500–2,600 m, vent.
44. *Lepetodrilus elevatus* McLean 1988. McLean 1988a: 18, figs. 17–20, 75–83; McLean 1993: 32, figs. 17–25; Fretter 1988: 49, figs. 7–8; Beck 1993: 175, figs. 5–6, 36–44; Mullineaux et al. 1996: fig. 1E, H; Warén and Bouchet 2001: 151, fig. 22c, appendix 2; Warén et al. in Desbruyères et al. 2006: 88. EPR: 21°N–17°S (type locality: 21°N), Galapagos Spreading Center, WP: North Fiji Basin, Lau Basin, Mariana Basin, ca. 2,400–2,700 m, vent. *Remarks:* Johnson et al. (2008) showed that records from localities other than EPR are misidentifications; at the EPR and Galapagos spreading Centre there are probably 4–6 sympatric and cryptic species.
45. *Lepetodrilus fucensis* McLean 1988. McLean 1988a: 18, figs. 17–20, 75–83; Fretter 1988: 56, fig. 15; McLean 1993: 32; Warén and Bouchet 2001: 154, appendix 2; Warén et al. in Desbruyères et al. 2006: 88; Johnson et al. 2006:

figs. 3E–H, 4B–C, 5A–B, 5D–F. JdF (type locality: Endeavor Segment), ca. 1,500–2,200 m, vent. *Remarks:* Records from the Gorda Ridge are based on *L. gordensis* (Johnson et al. 2006).

46. *Lepetodrilus galriftensis* McLean 1988. McLean 1988a: 13, figs. 7–8, 45–50 [as *L. elevatus galriftensis*]; Mullineaux et al. 1996: fig. 1D, G [as *L. e. galriftensis*]; Warén et al. in Desbruyères et al. 2006: 88. EPR 9°N, Galapagos Spreading Center, 2,451 m, vent.
47. *Lepetodrilus gordensis* Johnson et al. 2006. Johnson et al. 2006: 151, figs. 3A–D, I–J, A, D–I, 5C, G–L. Gorda Ridge, 2,696–2,716 m, vent.
48. *Lepetodrilus guaymasensis* McLean 1988. McLean 1988a: 16, figs. 15–16, 66–74; Fretter 1988: 54, figs. 13–14; McLean 1993: 32; Warén and Bouchet 2001: 153, appendix 2; Warén et al. in Desbruyères et al. 2006: 88. Guaymas Basin, 2,000–2,019 m, seep.
49. *Lepetodrilus japonicus* Okutani et al. 1993. Okutani et al. 1993: 134, figs. 30–33; Warén and Bouchet 2001: appendix 2; Sasaki et al. 2005: 113, fig. 14A; Warén et al. in Desbruyères et al. 2006: 88. WP: Okinawa Trough: Minami Ensei Knoll, 700–710 m, vent.
50. *Lepetodrilus nux* (Okutani et al. 1993). Okutani et al. 1993: 132 [as *Rynchopelta*?]; Sasaki 1998: 94, figs. 62–65; Warén and Bouchet 2001: appendix 2; Sasaki et al. 2003: 199, fig. 11; Sasaki et al. 2005: 114, fig. 14B; Warén et al. in Desbruyères et al. 2006: 88. WP: Okinawa Trough: Izena Hole (type locality) and Iheya Ridge, 990–1,390 m, vent.
51. *Lepetodrilus ovalis* McLean 1988. McLean 1988a: 14, figs. 9–12, 51–56; Fretter 1988: 50, figs. 8–10; Mullineaux et al. 1996: fig. 1F, I; Warén and Bouchet 2001: 153, fig. 22b, appendix 2; Warén et al. in Desbruyères et al. 2006: 88. EPR: 21°N–17°S (type locality: 21°N), Galapagos Spreading Center, ca. 2,500–2,600 m, vent.
52. *Lepetodrilus pustulosus* McLean 1988. McLean 1988a: 8, figs. 1–4, 25–35; Fretter 1988: 35, figs. 1–6; Warén and Bouchet 2001: 153, fig. 22d–e, appendix 2; Kiel 2004: 173, figs. 15–16; Warén et al. in Desbruyères et al. 2006: 88. EPR: Galapagos Spreading Center (type locality), 21°N–17°S, ca. 2,600 m, vent.
53. *Lepetodrilus schrolli* Beck 1993. Beck 1993: 171, figs. 2–3, pls. 1–5; Warén and Bouchet 2001: appendix 2; Warén et al. in Desbruyères et al. 2006: 88. WP: Manus Back-Arc Basin, 1,450–2,505 m, vent.
54. *Lepetodrilus shannonae* Warén and Bouchet 2009. Warén and Bouchet 2009: 2335, figs. 2F–G, 6A–F, 7A–F. Regab site (type locality) and Diapir site, West Africa, 2,300–3,150 m, seep.
55. *Lepetodrilus tevnianus* McLean 1993. McLean 1993: 28, figs. 1–10; Warén and Bouchet 2001: appendix 2; Warén et al. in Desbruyères et al. 2006: 88. EPR: 11°N, 2,536 m, vent.

**Lepetodrilus* sp. Warén and Bouchet 2001: appendix 2. WP: Manus, vent.

**Lepetodrilus* spp. 20 species were barcoded by Johnson et al. (2008) for an ongoing revision of the group. At least 6 species were not available for sequencing. This indicates the group to be even more species rich than generally assumed.

56. *Pseudorimula marianae* McLean 1989. McLean 1989b: 24, figs. 12A–H, 13A–D; Haszprunar 1989a: 5, figs. 7–9; Warén and Bouchet 2001: appendix 2; Warén et al. in Desbruyères et al. 2006: 91. WP: Mariana Back-Arc Basin, 3,640–3,660 m, vent.
57. *Pseudorimula midatlantica* McLean 1992. McLean 1992a: 116, figs. 1–7; Warén and Bouchet 2001: 155, figs. 15h, 18c, 23a–h, appendix 2; Kiel 2004: 173; Warén et al. in Desbruyères et al. 2006: 91 MAR: 38–15°N, ca. 1,500–3,500 m, vent.

Family Sutilizonidae (Ongoing work indicates these species to belong to the Lepetodrilidae.)

58. *Sutilizona pterodon* Warén and Bouchet 2001. Warén and Bouchet 2001: 141, figs. 15g, 17c, e, g–h, 18a, f, appendix 2; Warén et al. in Desbruyères et al. 2006: 98. MAR: Snake Pit, 3,470–3,520 m, vent.
59. *Sutilizona theca* McLean 1989. McLean 1989b: 15, figs. 3A–C, 4D–F, 5E–F; Haszprunar 1989a: 3, fig. 2; Kiel 2004: 173, fig. 14; Warén and Bouchet 2001: 141, fig. 17f, I, appendix 2; Warén et al. in Desbruyères et al. 2006: 98. EPR: 12–13°N, ca. 2,500–2,700 m, vent.
60. *Sutilizona tunnicliffae* Warén and Bouchet 2001. Warén and Bouchet 2001: 143, figs. 15f, 17d, 18b, d, g, appendix 2; Warén et al. in Desbruyères et al. 2006: 98. JdF: Endeavour Segment, 2,202 m, vent.
61. *Temnocinclis euripes* McLean 1989. McLean 1989b: 7, figs. 1A–H, 4A, 5A–B; Haszprunar 1989a: 3, fig. 1; Warén and Bouchet 2001: 143, appendix 2; Warén et al. in Desbruyères et al. 2006: 99. JdF, ca. 44–50°N (type locality: Axial Seamount), ca. 1,500–2,300 m, vent.
62. *Temnozaga parilis* McLean 1989. McLean 1989b: 9, figs. 2A–H, 4B–C, 5C–D; Haszprunar 1989a: 3; Warén and Bouchet 2001: appendix 2; Warén et al. in Desbruyères et al. 2006: 100 (figured). EPR: 21–13°N (type locality: 21°N), ca. 2,600 m, vent.

Superfamily Scissurelloidea
Family Scissurellidae

**Sinezona* sp. Warén and Bouchet 1989: 68, figs. 3–4. EP: off Mexico, 25.50N, 109,06W, 2,633 m, vent.

**Anatoma* sp.: Warén and Bouchet 1993: 7 [as Scissurellidae] WP: Lau Basin, Hine Hina, vent.

Superfamily Seguenzioidea
Family uncertain

63. *Adeuomphalus collinsi* Kano et al. 2009. Kano et al. 2009: 408, figs. 4C–H, 6. Manus Basin, off Papua New Guinea, ca. 1,440 m, vent.
64. *Adeuomphalus elegans* Kano et al. 2009. Kano et al. 2009: 404, fig. 3. Lau Basin, South Pacific, 1,737 m, vent.

65. *Adeuomphalus trochanter* Warén and Bouchet 2001. Warén and Bouchet 2001: 132, figs. 8f, 15l, 16d, appendix 2. Kano et al. 2009: 404, fig. 2D–F. EP: Jdf: CoAxial Segment, 2,060 m, vent.

**Moelleriopsis* sp. Warén and Bouchet 1989: 86, fig. 91 [family uncertain, Seguenzioidea(?)]; Warén and Bouchet 2001: appendix 2. EPR: 13°N, 2,635 m, vent.

66. *Akritogyra conspicua* (Monterosato 1880). Olu-Le Roy et al. 2004: table 2. Olimpi and Anaximander areas, Eastern Mediterranean Sea, ca. 1,700–2,000 m, seep and regular deep-sea bottom?

67. *Ventsia tricarinata* Warén and Bouchet 1993. Warén and Bouchet 1993: 31, figs. 18I, 23C–F, 24B–F, 25A–C; Warén and Bouchet 2001: appendix 2. WP: Lau Basin, 1,750–1,900 m, vent.

**Ventsia* sp. aff. *tricarinata* Warén and Bouchet 1993. Warén and Bouchet 2001: appendix 2. WP: Marianas, vent.

Family Cataegidae

68. *Cataegis meroglypta McLean and Quinn 1987. McLean and Quinn 1987: 115,* figs. 3–4, 7–8; Warén and Bouchet 1993: 19, figs. 14A–G, 15A–B; Warén and Bouchet 2001: 133, appendix 2. Louisiana Slope to Barbados Prism, 421–1,135 m (type locality: off Louisiana, 845–858 m), seep.

Family Chilodontidae

[69.] *Bathybembix macdonaldi* (Dall 1891). Sellanes et al. 2008: table 2. Concepción Methane Seep Area, Chile, 740–870 m, seep and regular deep-sea bottom.

**Calliotropis* sp. Sellanes et al. 2008: table 2. Concepción Methane Seep Area, Chile, 740–870 m, seep and regular deep-sea bottom.

70. *Putzeysia wiseri* (Calcara 1841). Olu-Le Roy et al. 2004: table 2. Olimpi and Anaximander areas, Eastern Mediterranean Sea, ca. 1,700–2,000 m, seep and regular deep-sea bottom.

Family Seguenziidae

[71.] *Bathymargarites symplector* Warén and Bouchet 1989. Warén and Bouchet 1989: 91, figs. 92–95, 100–101, 104–107, 110; Warén and Bouchet 1993: 11, figs. 10A–E, 11A–B; Warén and Bouchet 2001: 132, appendix 2; Kiel 2004: 173, figs. 17–18; Warén et al. in Desbruyères et al. 2006: 101, figured [as Chilodontidae]. MAR: 13–21°N (type locality: 21°N), ca. 2,500–2,600 m, vent.

Superfamily Trochoidea
Family Calliostomatidae

72. *Calliostoma chilena* Rehder 1971. Sellanes et al. 2008: table 2. Concepción Methane Seep Area, Chile, 740–870 m, seep and regular deep-sea bottom [not seep dependent?].

[73.] *Falsimargarita nauduri* Warén and Bouchet 2001. Warén and Bouchet 2001: 133, figs. 8e, 10a–d, 11a–b, 14e, 16c, appendix 2. EPR: 17°S, 2,578 m, vent deep-sea bottom [not seep dependent?].

[74.] *Otsukaia crustulum* Vilvens and Sellanes 2006. Vilvens and Sellanes 2006: 16, figs. 1–5; Sellanes et al. 2008: table 2 [as *Calliostoma*]. Off Central Chile, Concepción Methane Seep Area, 728–870 m, seep and regular deep-sea bottom [not seep dependent].

Family Solariellidae

75. *Zetela alphonsi* Vilvens 2002. Sellanes et al. 2008: table 2. Concepción Methane Seep Area, Chile, 740–870 m, seep and regular deep-sea bottom [presumably not seep dependent].

Family Trochidae

76. *Clelandella myriamae* Gofas 2005. Olu-Le Roy et al. 2004: table 2 [as *Clelandella* sp.]. Gofas 2005: 141, figs. 11, 12E–F, 13H. Anaximander (type locality) and Olimpi seamounts, Eastern Mediterranean Sea, 1,700–2,030 m, seep and regular deep-sea bottom?

Family Turbinidae

77. *Gaza fisheri* Dall 1889. Warén and Bouchet 1993: 13; Warén and Bouchet 2001: appendix 2. Louisiana Slope and Caribbean Sea, 600–1,061 m, seep.
78. *Margarites huloti* Vilvens and Sellanes 2006. Vilvens and Sellanes 2006: 16, figs. 1–5; Sellanes et al. 2008: table 2. Off Central Chile, 843–728 m, seep.
79. *Margarites ryukyuensis* Okutani et al. 2000. Okutani et al. 2000: 273, fig. 6; Sasaki et al. 2005: 114, fig. 15A. WP: Okinawa Trough: North Knoll of Iheya Ridge, 968–1,053 m, vent.
80. *Margarites shinkai* Okutani, Tsuchida and Fujikura 1992. Okutani et al. 1992: 142, figs. 12–16; Okutani et al. 1993: 132, figs. 20–21; Warén and Bouchet 2001: appendix 2; Sasaki et al. 2005: 115, fig. 15B. WP: Sagami Bay (type locality) and Okinawa Trough, 1,110–1,340 m, seep/vent.

**Margarites* sp. Warén and Bouchet 2001: appendix 2. WP: Manus, vent.

Subfamily Skeneinae, Family Turbinidae

81. *Bruceiella athlia* Warén and Bouchet 2001. Warén and Bouchet 2001: 139, figs. 8d, 11d–f, 13a–d, 15e, 16j, appendix 2; Kiel 2004: 173, fig. 13. Aleutian Trench, ca. 4,800 m, seep.
82. *Bruceiella globulus* Warén and Bouchet 1993. Warén and Bouchet 1993: 27, figs. 17C–F, 18E–F, 21A–C, 22A–D; Warén and Bouchet 2001: appendix 2; Warén et al. in Desbruyères et al. 2006: 94. WP: North Fiji (type locality) and Lau Back-Arc Basins, 1,750–2,443 m, vent.
83. *Bruciella wareni* Okutani et al. 2004. Okutani et al. 2004: 2, fig. 2A–F. IO: Kairei Field, Rodriguez Triple Junction, 2,422–2,443 m, vent.
84. *Fucaria mystax* Warén and Bouchet 2001. Warén and Bouchet 2001: 136, figs. 8c, 11c, 12a–d, 16e, appendix 2 [as Trochidae]; Warén et al. in Desbruyères et al. 2006: 95. WP: Edison Seamount, 1,483 m, vent.

***Fucaria* sp. Warén and Bouchet 2001: appendix 2 [as Trochidae]. Florida Escarpment, seep.

85. *Fucaria striata* Warén and Bouchet 1993. Warén and Bouchet 1993: 16, figs. 9E–H, 12A–C, 13A–C, 38C [as Trochidae]; Warén and Bouchet 2001: 136, appendix 2 [as Trochidae]; Warén et al. in Desbruyères et al. 2006: 95. JdF, 2,425 m, vent.
86. *Iheyaspira lequios* Okutani et al. 2000. Okutani et al. 2000: 268, figs. 2–5; Sasaki et al. 2005: 115, fig. 15C. WP: Okinawa Trough: North Knoll of Iheya Ridge, 968–1,053 m, vent.
87. *Protolira thorvaldssoni* Warén 1996. Warén 1996: 201, figs. 3E–F, 4A–D [as *thorvaldsoni*]; Warén and Bouchet 2001: 138, appendix 2; Warén et al. in Desbruyères et al. 2006: 97. MAR: Snake Pit to Iceland (type locality: off southwestern Iceland), ca. 850–3,700 m, vents and whale bone.
88. *Protolira valvatoides* Warén and Bouchet 1993. Warén and Bouchet 1993: 22, figs. 17A–B, 18D, G, 19A–C, 20A–F; Warén and Bouchet 2001: 138, appendix 2; Kiel 2004: 173, figs. 11–12; Warén et al. in Desbruyères et al. 2006: 97. MAR: 23°N, 3,478 m (Warén and Bouchet 1993: type locality) and Menez Gwen to Lucky Strike, 850–1,800 m (Warén and Bouchet 2001), vent.

Family Collonidae

89. *Cantrainea jamsteci* (Okutani and Fujikura 1990). Okutani and Fujikura 1990: 85, figs. 2–6 [as *Thermocollonia*]; Warén and Bouchet 2001: appendix 2; Sasaki et al. 2005: 115, fig. 15D. WP: Okinawa Trough: Minami-Ensei Knoll, 700–720 m, vent.
90. *Cantrainea macleani* Warén and Bouchet 1993. Warén and Bouchet 1993: 8, figs. 4A–C, 5B, F–G, 6A; Warén and Bouchet 2001: 132, appendix 2. Louisiana Slope, ca. 500–1,000 m, seep.
91. *Cantrainea nuda* Okutani 2001. Okutani 2001: 122, fig. 4. WP: Okinawa Trough, 730 m, vent.
92. *Cantrainea panamense* (Dall 1908). Sellanes et al. 2008: table 2 [as *Homalopoma*]. Concepción Methane Seep Area, Chile, 740–870 m, seep and regular deep-sea bottom.

Clade Neompahlina
Family uncertain

93. *Helicrenion reticulatum* Warén and Bouchet 1993. Warén and Bouchet 1993: 45, figs. 32C–D, 33A, C, 36A–C [as Peltospiridae]; Warén and Bouchet 2001: appendix 2; Warén et al. in Desbruyères et al. 2006: 102 (figured). WP: Lau Back-Arc Basin, 1,900 m, vent.
94. *Retiskenea diploura* Warén and Bouchet 2001. Kiel 2004: 174, fig. 19; Warén and Bouchet 2001: 158, figs. 15j, 16g, 18e, 24c, appendix 2; Okutani and Fujikura 2002: 212, fig. 2A–I; Sasaki et al. 2005: 114. Aleutian Trench (type locality) and Japan Trench, ca. 4,800–6,290 m, seep.

***Retiskenea* sp. Warén and Bouchet 2001: 158, fig. 25a–e, appendix 2. Oregon Margin, seep.

95. *Vetulonia phalcata* Warén and Bouchet 1993. Warén and Bouchet 1993: 11, figs. 7A–B, 8E–F, 11C, 18C; Warén and Bouchet 2001: appendix 2; Warén

et al. in Desbruyères et al. 2006: 103 (figured). WP: North Fiji Back-Arc Basin, 2,000 m, vent.

Family Melanodrymiidae

96. *Leptogyra costellata* Warén and Bouchet 2009. Warén and Bouchet 2009: 2337, figs. 6H, 8A–D. Regab site, West Africa, 3,150 m, seep.
97. *Leptogyra inflata* Warén and Bouchet 1993. Warén and Bouchet 1993: 47, figs. 29E–F, 32A–B, 37A–C, 38A–B, D [as Peltospiridae]; Warén and Bouchet 2001: appendix 2; Warén et al. in Desbruyères et al. 2006: 96. WP: Lau Basin, 1,750 m, vent.
98. *Melanodrymia aurantiaca* Hickman 1984. Hickman 1984: 20, figs. 1–2; Israelsson 1998: 105; Haszprunar 1989b: 175, figs. 1–11; Warén and Bouchet 1989: 75, figs. 26–27, 32–33, 35–36, 75, 82 [as Peltospiridae]; Warén and Bouchet 1993: 41 [as Peltospiridae]; Mullineaux et al. 1996: fig. 4A, C–D; Warén and Bouchet 2001: 159, appendix 2 [Neomphalidae]; Kiel 2004: 174, fig. 20; Warén et al. in Desbruyères et al. 2006: 106 (figured). EPR: 21°N–17°S (type locality: 29°50′N), ca. 2,500–2,600 m, vent.
99. *Melanodrymia brightae* Warén and Bouchet 1993. Warén and Bouchet 1993: 43, figs. 34A–C, 35A–B; Warén and Bouchet 2001: 159, appendix 2; Warén et al. in Desbruyères et al. 2006: 106. JdF: Endeavour Segment, 2,200 m, vent.
100. *Melanodrymia galeronae* Warén and Bouchet 2001. Warén and Bouchet 2001: 161, 15k, 16f, 21c, 14a–b, d, appendix 2; Warén et al. in Desbruyères et al. 2006: 106 (figured). EPR: 13°N, ca. 2,600 m, vent.

**Melanodrymia* sp. Warén and Bouchet 2001: appendix 2. Galapagos Rift, vent.

**Melanodrymia* sp. "rust covered". Warén and Bouchet 2001: 159, figs. 21d, 24e–f, appendix 2; Warén et al. in Desbruyères et al. 2006: 106. EPR: 13°N, vent.

Family Neomphalidae

101. *Cyathermia naticoides* Warén and Bouchet 1989. Warén and Bouchet 1989: 70, figs. 6–10, 15, 16, 18, 21–23, 71, 80; Warén and Bouchet 1993: 33 [as Cyathermidae]; Kiel 2004: 175, fig. 21; Mullineaux et al. 1996: fig. 2A–C; Warén and Bouchet 2001: 158, appendix 2; Warén et al. in Desbruyères et al. 2006: 104 (figured). EPR: 21–9°N (type locality: 13°N), ca. 2,600 m, vent.
102. *Lacunoides exquisitus* Warén and Bouchet 1989. Warén and Bouchet 1989: 72, figs. 13–14, 17, 20, 72, 81; Warén and Bouchet 2001: appendix 2; Warén et al. in Desbruyères et al. 2006: 105 (figured). EPR: Galapagos Spreading Center, 2,447–2,518 m, vent.
103. *Lacunoides vitreus* Warén and Bouchet 2001. Warén and Bouchet 2001: 162, fig. 29a–e, appendix 2; Warén et al. in Desbruyères et al. 2006: 105 (figured). JdF: Axial Seamount, 1,543 m, vent.

104. *Neomphalus fretterae* McLean 1981. McLean 1981: 294, figs. 1–10; Warén and Bouchet 1989: 69, figs. 5, 11, 12, 19, 79; Warén and Bouchet 1993: 33; Mullineaux et al. 1996: fig. 2D–F; Warén and Bouchet 2001: 162, appendix 2; Warén et al. in Desbruyères et al. 2006: 107 (figured). EPR: 21–9°N, Galapagos Spreading Center, ca. 2,400–2,600 m, vent.
105. *Planorbidella depressa* Warén and Bouchet 1993. Warén and Bouchet 1993: 38, figs. 30B–D, 31C–D, F; Warén and Bouchet 2001: appendix 2; Warén et al. in Desbruyères et al. 2006: 109. WP: Lau Basin, Hine Hina, vent.
106. *Planorbidella planispira* (Warén and Bouchet 1989). Warén and Bouchet 1989: 81, figs. 48–50, 53–54, 74, 84 [as *Depressigyra*]; Warén and Bouchet 1993: 37, fig. 29D [as Peltospiridae]; Warén and Bouchet 2001: 162, appendix 2; Kiel 2004: 175; Warén et al. in Desbruyères et al. 2006: 109 (figured). EPR: 21°N–17°S (type locality: 21°N), ca. 2,500–2,600 m, vent.
107. *Solutigyra reticulata* Warén and Bouchet 1989. Warén and Bouchet 1989: 82, figs. 44, 57–58, 59–62, 77, 89–90 [as Peltospiridae]; Warén and Bouchet 1993: 32 [as Skeneidae]; Warén and Bouchet 2001: appendix 2. EPR: 21–13°N (type locality: 21°N), ca. 2,000–2,600 m, vent.
108. *Symmetromphalus hageni* Beck 1992. Beck 1992a: 246, figs. 2–6, pls. 1–6; Warén and Bouchet 2001: appendix 2; Warén et al. in Desbruyères et al. 2006: 110. WP: Manus Basin, 2,488–2,500 m, vent.
109. *Symmetromphalus regularis* McLean 1990. McLean 1990a: 79, figs. 1–17; Warén and Bouchet 2001: appendix 2; Warén et al. in Desbruyères et al. 2006: 110 (figured). WP: Mariana Back-Arc Basin, 3,640 m, vent.

**Symmetromphalus* sp. Warén and Bouchet 2001: appendix 2. WP: North Fiji and Lau Basins, vent.

Family Peltospiridae

110. *Ctenopelta porifera* Warén and Bouchet 1993. Warén and Bouchet 1993: 34, figs. 26D–G, 27A–F, 28A–D; Warén and Bouchet 2001: 170, appendix 2; Kiel 2004: 175, figs. 24–25; Warén et al. in Desbruyères et al. 2006: 111 (figured). EPR: 13°N, ca 2,600 m vent
111. *Depressigyra globulus* Warén and Bouchet 1989. Warén and Bouchet 1989: 80, figs. 30–31, 45–47, 51–52, 73, 83; Warén and Bouchet 1993: 35; Warén and Bouchet 2001: 173, fig. 15o, appendix 2; Warén et al. in Desbruyères et al. 2006: 112 (figured). JdF, 1,500–2,400 m, vent.
112. *Echinopelta fistulosa* McLean 1989. McLean 1989a: 60, figs. 41–48; Gustafson and Lutz 1994: figs. 4.7–4.9; Warén and Bouchet 2001: 170, fig. 28b, appendix 2; Warén et al. in Desbruyères et al. 2006: 113 (figured). EPR: 21–13°N (type locality: 21°N), ca. 2,600 m vent
113. *Hirtopelta hirta* McLean 1989. McLean 1989a: 62, figs. 49–55; Warén and Bouchet 1993: 35, fig. 26A–C; Warén and Bouchet 2001: 169, appendix 2; Warén et al. in Desbruyères et al. 2006: 114 (figured). EPR: 21–13°N (type locality: 13°N), ca. 2,500–2,600 m vent

114. *Hirtopelta tufari* Beck 2002. Beck 2002: 252, figs. 1–15; Desbruyères et al. 2006: 114. EPR: 21°S, 2,802 m, vent.
115. *Lirapex costellata* Warén and Bouchet 2001. Warén and Bouchet 2001: 170, figs. 15m, 16i, 30a–c, 31a–c, appendix 2; Kiel 2004: 176; Warén et al. in Desbruyères et al. 2006: 115 (figured). MAR: Lucky Strike (type locality), ca. 1,600–1,700 m, and Snake Pit, vent.
116. *Lirapex granularis* Warén and Bouchet 1989. Warén and Bouchet 1989: 85, figs. 68–70, 88; Mullineaux et al. 1996: fig. 3G–I; Warén and Bouchet 2001: appendix 2; Warén et al. in Desbruyères et al. 2006: 115. EPR: 21–9°N (type locality: 21°N), ca. 2,600 m, vent.
117. *Lirapex humata* Warén and Bouchet 1989. Warén and Bouchet 1989: 84, figs. 43, 55–56, 63–65, 78, 87, appendix 2; Warén and Bouchet 2001: appendix 2; Warén et al. in Desbruyères et al. 2006: 115. EPR: 21°N, 2,616 m, vent.

   **Lirapex* sp. Warén and Bouchet 2001: 171, appendix 2. MAR: 23°N, vent.

118. *Nodopelta heminoda* McLean 1989. McLean 1989a: 53, figs. 17–23; Warén and Bouchet 2001: 169, figs. 27g, j. 28e, appendix 2; Warén et al. in Desbruyères et al. 2006: 116 (figured). EPR: 21–13°N (type locality: 21°N), ca. 2,600 m, vent.
119. *Nodopelta rigneae* Warén and Bouchet 2001. Warén and Bouchet 2001: 169, figs. 21e–f, 27a–b, f, i, 28a, d, appendix 2; Warén et al. in Desbruyères et al. 2006: 116 (figured). EPR: 13°N, 2,630 m, vent.
120. *Nodopelta subnoda* McLean 1989. McLean 1989a: 56, figs. 24–31; Warén and Bouchet 2001: 169, figs. 27e, h, 28c, appendix 2; Warén et al. in Desbruyères et al. 2006: 116 (figured). EPR: 13°N–17°S (type locality: 13°N), ca. 2,500–2,600 m, vent
121. *Pachydermia laevis* Warén and Bouchet 1989. Warén and Bouchet 1989: 76, figs. 28–29, 34, 37–40, 41, 42, 76, 85–86 [as Peltospiridae]; Warén and Bouchet 1993: 40 [as Peltospiridae]; Israelsson 1998: 95, figs. 1–7; Kiel 2004: 175, fig. 22; Warén and Bouchet 2001: 161, appendix 2; Warén et al. in Desbruyères et al. 2006: 108 (figured). EPR: 21°N–17°S (type locality: 13°N), ca. 2,500–2,600 m, vent.
122. *Pachydermia sculpta* Warén and Bouchet 1993. Warén and Bouchet 1993: 41, figs. 29A–C, 30E–F, 31A–B, E, 32E, 35C; Warén and Bouchet 2001: appendix 2; Warén et al. in Desbruyères et al. 2006: 108. WP: North Fiji Basin (type locality) and Lau Basin, 1,750–2,000 m, vent.

   **Pachydermia* sp. aff. *sculpta* Warén and Bouchet 1993. Warén and Bouchet 2001: appendix 2. WP: Mariana Back-Arc Basin, vent.

123. *Peltospira delicata* McLean 1989. McLean 1989a: 53, figs. 9–16; Warén and Bouchet 2001: 168, fig. 29f, appendix 2; Warén et al. in Desbruyères et al. 2006: 118 (figured). EPR: 13–9°N, ca. 2,500–2,600 m, vent.
124. *Peltospira lamellifera* Warén and Bouchet 1989. Warén and Bouchet 1989: 74, fig. 24; Warén and Bouchet 2001: 168, figs. 16k, 46b, 48c–d, appendix 2; Warén et al. in Desbruyères et al. 2006: 118. EPR: 13°N, ca. 2,600 m, vent.

125. *Peltospira operculata* McLean 1989. McLean 1989a: 51, figs. 1–8; Mullineaux et al. 1996: fig. 3E–F; Warén and Bouchet 2001: 168, appendix 2; Warén et al. in Desbruyères et al. 2006: 118 (figured). EPR: 21–9°N, 17°S, ca. 2,500–2,600 m, vent.
126. *Peltospira smaragdina* Warén and Bouchet 2001. Warén and Bouchet 2001: 165, figs. 15n, 26a–d, 27c–d, 31e–f, appendix 2; Kiel 2004: 175, fig. 23; Warén et al. in Desbruyères et al. 2006: 118 (figured) MAR: 38–15°N (type locality: Lucky Strike at 37°N), 850–3,500 m, vent
127. *Rhynchopelta concentrica* McLean 1989. McLean 1989a: 58, figs. 32–40; Mullineaux et al. 1996: fig. 3A–C; Warén et al. in Desbruyères et al. 2006: 119 (figured). EPR: 21°N–17°S, ca. 2,500–2,600 m, vent.

*"Scaly foot gastropod". Warén et al. 2003: fig. 1; Goffredi et al. 2004: fig. 1; Warén et al. in Desbruyères et al. 2006: 120 (figured); Suzuki et al. 2006c: fig. 1. IO: Kairei Field, Rodriguez Triple Junction, ca. 2,420 m, vent.

*Peltospiridae sp. A. Warén and Bouchet 1993: 41, figs. 30A, 32F, 33B. WP: Lau Basin, Hine Hina, vent.

Clade Neritimorpha
Family Neritidae

128. *Bathynerita naticoidea* Clarke 1989. Clarke 1989: 125, figs. 3–4, text figs.; Warén and Bouchet 1993: 3, figs. 1–2, 3A–C; Warén and Bouchet 2001: 177, fig. 30d–f, appendix 2; Kiel 2004: 176, figs. 26–27. Louisiana Slope (type locality) and Barbados Prism, 550–1,135 m, seep.

Family Phenacolepadidae

**Olgasolaris* sp. Warén et al. in Desbruyères et al. 2006: 121. WP: North Fiji and Lau Back-Arc Basins, vent.

129. *Olgasolaris tollmanni* Beck 1992. Beck 1992b: 266, figs. 2–5, pls. 4–6; Warén and Bouchet 2001: appendix 2; Warén et al. in Desbruyères et al. 2006: 121 (figured). WP: Manus Back-Arc Basin, 2,489–2,505 m, vent.
130. *Shinkailepas briandi* Warén and Bouchet 2001. Warén and Bouchet 2001: 174, figs. 16a–b, 31d, 32a–h, 33a–e, appendix 2; Kiel 2004: 177, fig. 28–29; Warén et al. in Desbruyères et al. 2006: 122 (figured). MAR: Menez Gwen (type locality), Lucky Strike, and Logatchev (15°N), 850–3,500 m, vent.
131. *Shinkailepas kaikatensis* Okutani et al. 1989. Okutani et al. 1989: 225, figs. 2–8, 10–15; Warén and Bouchet 2001: appendix 2; Sasaki et al. 2005: 116, fig. 16A; Warén et al. in Desbruyères et al. 2006: 122. WP: Kaikata Seamout, off Ogasawara Islands, Japan, 470 m, vent.
132. *Shinkailepas myojinensis* Sasaki et al. 2003. Sasaki et al. 2003: 201, figs. 12–13; Sasaki et al. 2005: 117, fig. 16B; 2006: figs. 1–16; Warén et al. in Desbruyères et al. 2006: 122. WP: Myojin Knoll, Japan, 1,260–1,340 m, vent.
133. *Shinkailepas tufari* Beck 1992. Beck 1992b: 262, fig. 5C, pls. 1–3; Warén and Bouchet 2001: appendix 2; Warén et al. in Desbruyères et al. 2006: 122. WP: Manus Back-Arc Basin, 2,450–2,505 m, vent.

**Shinkailepas* sp. Warén et al. in Desbruyères et al. 2006: 122. WP: North Fiji Basin, vent.
**Shinkailepas* sp. Sasaki et al. 2003: 203, fig. 14; Sasaki et al. 2005: 117. WP: Okinawa Trough: North Knoll of Iheya Ridge, 976 m, vent.
**Shinkailepas* sp. Warén and Bouchet 2001: appendix 2. WP: Manus, vent.

Clade Caenogastropoda
Family Capulidae

[134.] *Hyalorisia galea* (Dall 1889). Warén and Bouchet 1993: 76; Warén and Bouchet 2001: appendix 2. Louisiana Slope to Barbados, 329–768 m (Warén and Bouchet 1993), seep. Probably not a really seep favoured species.

Family Cerithiopsidae

[135.] *Speculator cariosus* Warén and Bouchet 2001. Warén and Bouchet 2001: 180, figs. 34c–d, 36c, 49g, appendix 2; Warén et al. in Desbruyères et al. 2006: 126 (figured). JdF: Explorer Ridge, 1,762 m, vent. Probably not a really seep favoured species.

Family Elachisinidae

136. *Laeviphitus desbruyeresi* Warén and Bouchet 2001. Warén and Bouchet 2001: 182, figs. 34e–f, 36a, 37b, 49e, appendix 2; Warén et al. in Desbruyères et al. 2006: 127 (figured). MAR: Menz Gwen to Rainbow (type locality: Lucky Strike), 850–2,300 m, vent.
137. *Laeviphitus japonicus* Okutani et al. 1993. Okutani et al. 1993: 135, figs. 37–43; Sasaki et al. 2005: 119; Warén and Bouchet 2001: appendix 2. WP: Kaikata Seamount, Japan, 440 m, vent

**Laeviphitus* sp. Warén and Bouchet 2001: appendix 2. New Zealand, seep.
**Laeviphitus* sp. (veligers only). Warén and Bouchet 2001: appendix 2. EPR: 13–10°N, vent.

Family Provannidae

138. *Alviniconcha hessleri* Okutani and Ohta 1988. Okutani and Ohta 1988: 2, figs. 1–9; Kiel 2004: 178, fig. 33; Warén and Bouchet 1993: 56, figs. 44A–C, 45A–C, 46, 47, 48A–E, 49, 50C–D; Warén and Bouchet 2001: appendix 2; Warén in Desbruyères et al. 2006: 128 (figured). WP: Mariana (type locality: 3,600–3,680 m), North Fiji and Lau Back-Arc Basins, 1,750–3,680 m, vent. *Remarks*: These records possibly include several different biological species (see review herein).

**Alviniconcha aff. hessleri* Okutani and Ohta 1988. Okutani et al. 2004: 6, figs. 5A–B, 6; Suzuki et al. 2005a: 5442, fig. 1. Kairei Field, Rodriguez Triple Junction, 2,420–2,454 m, vent.

139. *Cordesia provannoides* Warén and Bouchet 2009. Warén and Bouchet 2009: 2341, figs. 9E, 11A–F. Regab site, West Africa, 3,150 m (type locality) and Florida Escarpment, seep.

140. *Desbruyeresia cancellata* Warén and Bouchet 1993. Warén and Bouchet 1993: 71, figs. 44E, 54D, F–H, 55B; Warén and Bouchet 2001: appendix 2; Warén et al. in Desbruyères et al. 2006: 129 (figured). WP: Lau Basin (type locality) and North Fiji Basin, 1,750–2,000 m, vent.
141. *Desbruyeresia marianaensis* (Okutani 1990). Okutani 1990: 22, figs. 3–4, 7–8, 10 [as *Provanna*]; Warén and Bouchet 2001: appendix 2; Warén et al. in Desbruyères et al. 2006: 129 [not (Okutani and Fujikura 1990)]. WP: Mariana Back-Arc Basin, 3,670–3,680 m, vent.
142. *Desbruyeresia marisindica* Okutani et al. 2004. Okutani et al. 2004: 4, figs. 3A–C, 4A–B; Warén et al. in Desbruyères et al. 2006: 129. Kairei Field, Rodriguez Triple Junction, 2,422–2,454 m, vent.
143. *Desbruyeresia melanoides* Warén and Bouchet 1993. Warén and Bouchet 1993: 71, figs. 54C, 55A, 56D, 57C; Warén and Bouchet 2001: appendix 2; Warén et al. in Desbruyères et al. 2006: 129 (figured). WP: Lau Basin (type locality) and Manus Basin, 1,900 m, vent.
144. *Desbruyeresia spinosa* Warén and Bouchet 1993. Warén and Bouchet 1993: 72, figs. 44D, 54A, 56C–D; Warén and Bouchet 2001: appendix 2; Warén et al. in Desbruyères et al. 2006: 129 (figured). WP: North Fiji Back-Arc Basin, 1,900–2,765 m, vent.

Desbruyeresia aff. *spinosa* Warén and Bouchet 1993. Warén and Bouchet 2001: appendix 2. WP: Mariana Back-Arc Basin, vent.

Desbruyeresia sp. aff. *spinosa* Warén and Bouchet 1993. Warén et al. in Desbruyères et al. 2006: 129. WP: Mariana Back-Arc Basin, vent.

145. *Ifremeria nautilei* Bouchet and Warén 1991. Bouchet and Warén 1991: 498, figs. 1A–C, 2A–E; Beck 1991: 282, fig. 2, pls. 1–4 (as *Olgaconcha tufari*); Warén and Bouchet 1993: 64, figs. 25D, 45D–F, 50A–B, 51A–B, 52, 53A–B, 55C–D; Warén and Bouchet 2001: appendix 2; Warén et al. in Desbruyères et al. 2006: 130 (figured). Lau Basin, 1,750 m (type locality), North Fiji Basin, 2,000 m and Manus Back-Arc Basin, 2,500 m, vent.
146. *Provanna abyssalis* Okutani and Fujikura 2002. Okutani and Fujikura 2002: 214, fig. 3A–C; Sasaki et al. 2005: 118. WP: Japan Trench, 5,379 m, seep.
147. *Provanna admetoides* Warén and Ponder 1991. Warén and Ponder 1991: 53, figs. 4B, figs. 20A–B, 22D, 23A; Warén and Bouchet 2001: appendix 2. Florida Escarpment, 624–631 m, seep.
148. *Provanna buccinoides* Warén and Bouchet 1993. Warén and Bouchet 1993: 74, figs. 54B, E, 56A–B; Warén and Bouchet 2001: appendix 2; Warén et al. in Desbruyères et al. 2006: 132 (figured). WP: Lau Basin (type locality) and North Fiji Basin, 1,900–2,765 m, vent.
149. *Provanna chevalieri* Warén and Bouchet 2009. Warén and Bouchet 2009: 2340, figs. 9A, B, 10E–H. Regab site, West Africa, 3,150 m, seep.
150. *Provanna glabra* Okutani et al. 1992. Okutani et al. 1992: 143, figs. 17–21; Warén and Bouchet 2001: appendix 2; Sasaki et al. 2005: 118; Warén et al. in Desbruyères et al. 2006: 132. WP: Sagami Bay: off Hatsushima, 1,110–1,200 m, seep.

151. *Provanna goniata* Warén and Bouchet 1986. Warén and Bouchet 1986: 163, figs. 5, 6, 19, 25; Warén and Ponder 1991: 51; Warén and Bouchet 2001: 178, appendix 2. Guaymans Basin, 2,000–2,020 m, seep.
152. *Provanna ios* Warén and Bouchet 1986. Warén and Bouchet 1986: 162, figs. 8–9, 20, 22; Warén and Ponder 1991: 51; Warén and Bouchet 2001: 178, appendix 2; Warén et al. in Desbruyères et al. 2006: 132 (figured). EPR: 21°N–17°S (type locality: 12°49′N), Galapagos Spreading Center, 2,450–2,620 m, vent.
153. *Provanna laevis* Warén and Ponder 1991. Warén and Ponder 1991: 53, figs. 3C–D, 4A, 20H–I, 21C–D, 23D; Warén and Bouchet 2001: 178, fig. 6c, appendix 2; Warén et al. in Desbruyères et al. 2006: 132 (figured). Gulf of California, Guaymas Basin (type locality), ca. 2,000 m, to Oregon Margin, 500–600 m, JdF, ca. 1,500 m (Warén and Bouchet 2001), vent/seep.
154. *Provanna lomana* (Dall 1918). Warén and Bouchet 1986: 161, figs. 1–2, 18, 23; Warén and Ponder 1991: 53, fig. 12; Warén and Bouchet 2001: 178, appendix 2. Oregon Margin, ca. 450–1,200 m, seep.
155. *Provanna macleani* Warén and Bouchet 1989. Warén and Bouchet 1989: 94, figs. 111–114; Warén and Bouchet 2001: 179, appendix 2. Oregon Margin, 2,713–2,750 m, seep and sunken drift wood.
156. *Provanna muricata* Warén and Bouchet 1986. Warén and Bouchet 1986: 163, figs. 3, 12, 21; Warén and Ponder 1991: 51; Warén and Bouchet 2001: appendix 2; Warén et al. in Desbruyères et al. 2006: 132. EPR: 21°N, 2,615 m and Galapagos Spreading Center, 2,451–2,457 m (Warén and Ponder 1991), and also North Fiji and Lau Back-Arc Basins (Warén et al. in Desbruyères et al. 2006), vent. Type locality: Galapagos Rift, 2,450 m. Remarks: Records from Fiji and Lau are wrong.
157. *Provanna nassariaeformis* Okutani 1990. Okutani 1990: 20, figs. 1–2, 5–6, 9; Warén and Bouchet 2001: appendix 2; Warén et al. in Desbruyères et al. 2006: 132. Mariana Back-Arc Basin, 3,670–3,680 m, vent (type locality: Okutani 1990) and also Manus Back-Arc Basins, vent (Warén et al. in Desbruyères et al. 2006).
158. *Provanna reticulata* Warén and Bouchet 2009. Warén and Bouchet 2009: 2339, figs. 9C–D, 10I–L. Regab, Guiness and MPS 1-Congo sites, West Africa, 750–3,150 m, seep.
159. *Provanna sculpta* Warén and Ponder 1991. Warén and Ponder 1991: 53, figs. 2, 3A–B, 8, 14, 18, 20C, E, 21A–B, 23B; Warén and Bouchet 2001: 179, appendix 2. Louisiana Slope, ca. 550 m, seep.
160. *Provanna segonzaci* Warén and Ponder 1991. Warén and Ponder 1991: 51, figs. 20D, G, 22A–C, 23C; Warén and Bouchet 1993: 75, fig. 57A–B; Warén et al. in Desbruyères et al. 2006: 132; Warén and Bouchet 2001: appendix 2. WP: Lau Back-Arc Basin, 1,750–1,900 m, vent.
161. *Provanna shinkaiae* Okutani and Fujikura 2002. Okutani and Fujikura 2002: 216, fig. 4A–C; Sasaki et al. 2005: 118. WP: Japan Trench, 5,343 m, seep.
162. *Provanna variabilis* Warén and Bouchet 1986. Warén and Bouchet 1986: 163, figs. 13–15, 26–28; Warén and Ponder 1991: 53, figs. 5–7, 9–11, 13, 15–17; Warén and Bouchet 1993: 76; Gustafson and Lutz 1994: figs. 4.13–4.15;

Warén and Bouchet 2001: 179, appendix 2; Kiel 2004: 178, figs. 30–32; Warén et al. in Desbruyères et al. 2006: 132 (figured). JdF: Endeavour Segment (type locality), Gorda Ridge, Oregon Margin, 675–2,200 m, vent, seep.

Provanna sp. Okutani et al. 1993: 135, fig. 5 [as *P. glabra*]; Okutani and Fujiwara 2000: 124, figs. 34–36 [as *P. glabra*]; Sasaki et al. 2005: 118. WP: Okinawa Trough, 710–1,049 m, vent.

Provanna sp. Warén and Ponder 1991: 51. Galapagos Spreading Center, 2,494 m and EPR: 21°N, 2,617 m, vent.

Provanna sp. 1. Warén and Bouchet 1986, 2001: 179, appendix 2. WP: Edison Seamount, Manus, vent.

Provanna sp. 2. Warén and Bouchet 1986, 2001: 179, appendix 2. Aleutian Trench, seep.

Provanna sp. 3. Warén and Bouchet 1986, 2001: 179, appendix 2. Jalisco Block, vent.

Provanna spp. Warén and Bouchet 2001: appendix 2. New Zealand, seep.

Family Naticidae

[*] *Natica* sp. Sellanes et al. 2008: table 2. Concepción Methane Seep Area, Chile, 740–870 m, seep and regular deep-sea bottom.

Family Ranellidae

[163.] *Fusitriton magellanicus* (Röding 1798). Sellanes et al. 2008: table 2. Concepción Methane Seep Area, Chile, 740–870 m, seep and regular deep-sea bottom.

Family Rissoidae

164. *Alvania stenolopha* Bouchet and Warén 1993. Bouchet and Warén 1993: 653, figs. 1147, 1482–4; Warén and Bouchet 2001: appendix 2. Following records as *Alvania* cf. *stenolopha*. Warén and Bouchet 2001: 186, figs. 34a–b, 36b, 37a; Warén and Bouchet in Desbruyères et al. 2006: 133 (figured). MAR: Lucky Strike, Menez Gwen and surroundings (type locality: off Azores), 650–1,850 m, vent.

165. *Pseudosetia azorica* Bouchet and Warén 1993. Bouchet and Warén 1993: 690, figs. 1603–4, 1615–7; Warén and Bouchet 2001: 186. Off Azores (type locality), ca. 350–850 m; MAR: Menez Gwen, vent.

[166.] *Benthonella tenella* (Jeffreys 1869). Olu-Le Roy et al. 2004: table 2. Anaximander area, Eastern Mediterranean Sea, ca. 1,700–2,000 m, seep and regular deep-sea bottom.

Family Vitrinellidae

[167.] *Neusas marshalli* (Sykes 1925). Warén and Bouchet 2001: 186, figs. 35a–c, 36d–f, 49f, appendix 2; Warén et al. in Desbruyères et al. 2006: 134 (figured). MAR: Menez Gwen (38°N), 850–2,000 m.

Clade Neogastropoda

Remarks: It seems likely that many or most of the larger caenogastropods recorded from seeps and vents are generalist scavengers or predators that are attracted by the concentration of organic material.

Family Buccinidae

168. *Bayerius arnoldi* (Lus 1981). Lus 1981: 140, figs. 1–4 [as *Tactia*]; Warén and Bouchet 2001: 189, figs. 38b–e, 39b, 49c, appendix 2; Okutani and Fujikura 2002: 217, fig. 5A–B; Okutani and Fujikura 2005: 121, fig. 1; Sasaki et al. 2005: 120. Aleutian Trench, Kuril Trench, Japan Trench (type locality: northeastern Pacific, 37°56′N, 146°24′E), 4,800–7,434 m, seep and regular deep-sea bottom.
169. *Bayerius peruvianus* Warén and Bouchet 2001. Warén and Bouchet 2001: 187, figs. 38h, 39c, 49d, appendix 2. Off Peru, 5,996–5,385 m, seep.

*-*Bayerius* sp. Warén and Bouchet 2001: 189, appendix 2. Jalisco Block Seeps, seep

170. *Buccinum soyomaruae* Okutani 1977. Okutani et al. 1992: 145, figs. 22–23; Warén and Bouchet 2001: appendix 2; Sasaki et al. 2005: 119. WP: Sagami Bay: off Hatsushima, Japan, 1,110–1,500 m, seep and normal muddy sea bottom.
171. *Buccinum thermophilum* Harasewych and Kantor 2002. Warén and Bouchet 1993: 77 as *Buccinum* sp.; Warén and Bouchet 2001: 190 as *Buccinum* sp.; Harasewych and Kantor 2002: 40, figs. 1–3, 5–17; Martell et al. 2002: fig. 3. JdF: Endeavour Segment, 2,192–2,225 m, vent. *Remarks: Buccinum* sp. (*viridum* Dall 1890?) in Warén and Bouchet (2001: appendix 2) also indicates this species. *Buccinum viridum* (Dall 1890) is common on methane bottoms and whale carcasses off California and Oregon (Warén, unpubl.).
172. *Calliloncha nankaiensis* Okutani and Iwasaki 2003. Okutani and Iwasaki 2003: 3, figs. 3A–C, 4A–B; Sasaki et al. 2005: 120, fig. 17A. WP: Nanaki Trough: off Cape Muroto, Japan, 3,540–3,571 m, seep.

Costaria? sp. Warén and Bouchet 1993: 77, fig. 58A–B; Sasaki et al. 2005: 120. WP: Nankai Trough: Tenryu Canyon, Japan, 3,800–4,020 m, seep.

173. *Eosipho auzendei* Warén and Bouchet 2001. Warén and Bouchet 2001: 191, figs. 38j–k, 39a, 49a, appendix 2; Bouchet in Desbruyères et al. 2006: 124. EPR: 17–23°S (type locality: 17°S), Pacific-Antarctic Ridge: 31–38°S, vent.
174. *Eosipho canetae* (Clench and Aguayo 1944). Warén and Bouchet 1993: 76, fig. 59D (as "*Buccinum*"); 2001: 191, figs. 38i, 39d, 49b, appendix 2. Louisiana Slope, 420–850 m, seep.
175. *Eosipho desbruyeresi* Okutani and Ohta 1993. Okutani and Ohta 1993: 217, figs. 1–5; Warén and Bouchet 2001: 191, appendix 2; Bouchet in Desbruyères et al. 2006: 124 (figured). WP: Lau Basins (type locality) and North Fiji, 1,750–2,000 m (Okutani and Ohta 1993), also Mariana Basin (Bouchet in Desbruyères et al. 2006), vent.

176. *Eosipho desbruyeresi nipponensis* Okutani and Fujiwara 2000. Okutani and Fujiwara 2000: 125, figs. 6–10; Sasaki et al. 2005: 121, fig. 17B; Bouchet in Desbruyères et al. 2006: 124. WP: Okinawa Trough (Norht Knoll of Iheya Ridge: type locality), Myojin Knoll and Sumisu Caldera, Japan, 676–1,362 m, vent.
177. *Neptunea acutispiralis* Okutani 1968. Okutani et al. 1993: 137, figs. 51–54; Warén and Bouchet 2001: appendix 2; Sasaki et al. 2005: 119. WP: Sagami Bay, 1,100–1,380 m, seep and normal muddy sea bottom.
178. *Neptunea insularis* (Dall 1895). Warén and Bouchet 2001: appendix 2. WP: Okinawa Trough, vent.
179. *Neptunea pribiloffensis* (Dall 1919). Warén and Bouchet 1993: 78. Off northern California, seep.
180. *Neptunea amianta* (Dall 1890). Warén and Bouchet 2001: 224, Oregon Margin, seep.

Neptunea sp. cf. *insularis* (Dall 1895). Okutani et al. 1993: 138, fig. 55 [as *N. incularis* var.]; Sasaki et al. 2005: 119. WP: Okinawa Trough, 690–710 m, vent.

[181.] *Kryptos explorator* Fraussen and Sellanes 2007. Fraussen and Sellanes 2007: 102, figs. 5–6, 16–25; Sellanes et al. 2008: table 2. Concepción Methane Seep Area, Chile, 708–850 m, seep and regular deep-sea bottom.

Family Cancellariidae

182. *Admete tenuissima* Okutani and Fujikura 2002. Okutani and Fujikura 2002: 218, fig. 6A–B; Sasaki et al. 2005: 123. Japan Trench, 5,343–5,379 m, seep.
183. *Cancellaria rosewateri* Petit 1983. Warén and Bouchet 1993: 78; Warén and Bouchet 2001: appendix 2. Louisiana Slope, off Mississippi and off Alabama, 360–580 m, seep.
184. *Iphinopsis boucheti* Okutani et al. 2004. Okutani et al. 2004: 9, fig. 5C–D. Kairei Field, Rodriguez Triple Junction, 2,422–2,442 m, vent.

Family Muricidae

Coronium cf. *wilhelmense*. Sellanes et al. 2008: table 2. Concepción Methane Seep Area, Chile, 740–870 m, seep and regular deep-sea bottom.

[185.] *Trophon condei* Houart 2003: Houart 2003: 102, figs. 1–2, 23, 31; Houart and Sellanes 2006: 55, figs. 1–4, 33; Sellanes et al. 2008: table 2. Concepción Methane Seep Area, Chile, 728–1,350 m, seep and regular deep-sea bottom.
[186.] *Trophon ceciliae* Houart 2003: Houart 2003: 104, text fig. A, figs. 3–5, 24, 31; Houart and Sellanes 2006: 56, figs. 5–10; Sellanes et al. 2008: table 2. Concepción Methane Seep Area, Chile, 728–1,350 m, seep and regular deep-sea bottom.
[187.] *Pagodula concepcionensis* Houart and Sellanes 2006: Houart and Sellanes 2006: 59, figs. 15–19, 34–35; Sellanes et al. 2008: 2008: table 2. Concepción Methane Seep Area, Chile, 726–930 m, seep and regular deep-sea bottom.

Pagodula cf. concepcionensis Houart and Sellanes 2006: Houart and Sellanes 2006: 61, figs. 20–21. Off Chile, Concepción seep area, 930 m, possibly seep-associated.

Trophon sp. 1. Warén and Bouchet 2001: appendix 2. Barbados Prism, seep.

Trophon sp. 2. Warén and Bouchet 2001: appendix 2. Barbados Prism, seep.

Family Conidae ("Turridae")

[*] *Aforia* cf. *goniodes* (Watson 1881). Sellanes et al. 2008: table 2. Concepción Methane Seep Area, Chile, 740–870 m, seep and regular deep-sea bottom. Certainly not seep related.

188. *Bathybela papyracea* Warén and Bouchet 2001. Warén and Bouchet 2001: 200, fig. 42f, appendix 2. Jalisco Block, 3,662 m, seep.

Bathybela spp. Warén and Bouchet 1993: 79. Off Barbados, 4,935 m; WP: Nankai Trough, Tenryu Canyon, 3,835 m, seep.

[189.] *Benthomangelia macra* (Watson 1881). Olu-Le Roy et al. 2004: table 2. Anaximander area, Eastern Mediterranean Sea, ca. 1,700–2,000 m, seep and regular deep-sea bottom.

Benthomangelia sp. Okutani and Iwasaki 2003: 7, fig. 2C; Sasaki et al. 2005: 12. WP: Nankai Trough: off Cape Muroto, Japan, 3,581 m, seep.

[190.] *Drilliola loprestiana* (Calcara 1841). Olu-Le Roy et al. 2004: table 2. Anaximander area, Eastern Mediterranean Sea, ca. 1,700–2,000 m, seep and regular deep-sea bottom (very comman on normal bottom).

191. *Gymnobela extensa* (Dall 1881). Warén and Bouchet 1993: 78, fig. 59A–C; Warén and Bouchet 2001: 199, appendix 2. Yucatan Channel to off Louisiana, 512–1,463 m, seep.

Gymnobela sp. A. Warén and Bouchet 2001: 199, fig. 42d, appendix 2. EPR: 13°N, vent.

Gymnobela? sp. B. Warén and Bouchet 2001: 199, fig. 42c, appendix 2. MAR: 23°N, vent.

192. *Oenopota ogasawarana* Okutani et al. 1993. Okutani et al. 1993: 138, figs. 47–50; Warén and Bouchet 2001: appendix 2; Sasaki et al. 2005: 121, fig. 17C. WP: Kaikata Seamount, off Ogasawara Islands, Japan, 440–450 m, vent.

193. *Oenopota sagamiana* Okutani and Fujikura 1992. Okutani and Fujikura 1992: 2, figs. 2–3, pl. 1; Warén and Bouchet 2001: appendix 2 [as *Gymnobela*]; Sasaki et al. 2005: 121, fig. 17D. WP: Sagami Bay: off Hatsushima, 1,170 m, seep.

Oenopota sp. Okutani and Fujiwara 2000: 126; Sasaki et al. 2005: 121. WP: Okinawa Trough: North Knoll of Iheya Ridge, 1,049 m, vent.

[*] *Phymorhynchus* aff. *alberti* (Dautzenberg and Fischer 1906). Warén and Bouchet 2001: 199, fig. 42e, appendix 2. Northeastern Atlantic, 16–47°N, 1,850–4,800 m, seep.

194. *Phymorhynchus buccinoides* Okutani et al. 1992. Okutani et al. 1993: 140, figs. 44–46; Warén and Bouchet 2001: 199, fig. 42e, appendix 2. WP: Sagami Bay: off Hatsushima, Japan, 1,160 m, seep.
195. *Phymorhynchus carinatus* Warén and Bouchet 2001. Warén and Bouchet 2001: 195, figs. 42g–h, 43g–I, appendix 2; Desbruyères et al. 2006: 138 (figured). MAR: 23–15°N (type locality: Logatchev site at 14°45′N, 3,040 m), vent.
196. *Phymorhunchus cingulatus* Warén and Bouchet 2009. Warén and Bouchet 2009: 2344, figs. 9F, 10A. Regab site, West Africa, 3,150 m, seep.
197. *Phymorhunchus coseli* Warén and Bouchet 2009. Warén and Bouchet 2009: 2344, figs. 9G–K, 10B–D, 12L. Regab site, West Africa, 3,150 m, seep.
198. *Phymorhynchus hyfifluxi* Beck 1996. Beck 1996: text-fig. 1d, pl. 1, figs. 1–8; Warén and Bouchet 2001: appendix 2; Okutani in Desbruyères et al. 2006: 138. WP: North Fiji Basin, 2,003 m, vent.
199. *Phymorhynchus major* Warén and Bouchet 2001. Warén and Bouchet 2001: 197, figs. 41h, 43a, appendix 2; Okutani in Desbruyères et al. 2006: 138. EPR: 13–9°N (type locality: 9°50′N), ca. 2,500–2,600 m, vent.
200. *Phymorhynchus moskalevi* Sysoev and Kantor 1995. Sysoev and Kantor 1995: 22, figs. 1H–I, 4, 5D–F; Warén and Bouchet 2001: 195, figs. 41a–d, 43e–f, appendix 2; Okutani in Desbruyères et al. 2006: 138 (figured). MAR: 26–23°N, 3,400–3,700 m, vent.
201. *Phymorhynchus ovatus* Warén and Bouchet 2001. Warén and Bouchet 2001: 194, figs. 41e–g, 42a–b, 43b–d, appendix 2; Okutani in Desbruyères et al. 2006: 138 (figured). MAR: 37–15°N (type locality: Logatchev site at 14°45′N), 1,600–3,500 m, vent.
202. *Phymorhynchus starmeri* Okutani and Ohta 1993. Okutani and Ohta 1993: 220, figs. 6–7; Warén and Bouchet 2001: appendix 2; Okutani in Desbruyères et al. 2006: 138. WP: North Fiji Basin, 2,750 m (type locality), and Manus Back-Arc Basin, vent.
203. *Phymorhynchus turris* Okutani and Iwasaki 2003. Okutani and Iwasaki 2003: 6, fig. 3D; Sasaki et al. 2005: 123, fig. 17F. WP: Nankai Trough: off Cape Muroto, Japan, 3,540–3,581 m, seep.
204. *Phymorhynchus wareni* Sysoev and Kantor 1995. Sysoev and Kantor 1995: 19, figs. 1A–D, G, 2, 3, 5A–C; Warén and Bouchet 2001: appendix 2; Okutani in Desbruyères et al. 2006: 138. WP: Edison Seamount, 1,483 m, vent.

**Phymorhynchus* sp. Warén and Bouchet 1989: 95, figs. 115–116. [= *P. major* Warén and Bouchet 2001]

**Phymorhynchus* sp. 1 Warén and Bouchet 1993: 79, fig. 59E. MAR: Snake Pit, 3,478 m, vent.

**Phymorhynchus* sp. 2 Warén and Bouchet 1993: 79. [= *P. carinatus* Warén and Bouchet 2001]

**Phymorhynchus* sp. 3 Warén and Bouchet 1993: 79. JdF: Endeavour Ridge, vent.

[205.] *Taranis moerchi* (Malm 1861). Olu-Le Roy et al. 2004: table 2. Olimpi and Anaximander areas, Eastern Mediterranean Sea, ca. 1,700–2,000 m, seep and regular deep-sea bottom (mainly non-seep areas)

*Turrid sp. Warén and Bouchet 1993: 79. EPR: 13°N, vent.
[*] Turridae, gen. sp. 1. Sellanes et al. 2008: table 2. Concepción Methane Seep Area, Chile, 740–870 m, seep and regular deep-sea bottom.
[*] Turridae, gen. sp. 2. Sellanes et al. 2008: table 2. Concepción Methane Seep Area, Chile, 740–870 m, seep and regular deep-sea bottom.

Family Volutidae

[206.] *Tractolira sparta* Dall 1896. Poppe and Goto 1992: 121, pl. 39, figs. 7–8 (holotype); Warén and Bouchet 2001: 192, fig. 38a, appendix 2. Off Peru, 3,100–5,300 m, off Acapulco, Mexico to Gulf of Panama, 3,043–4,062 m, seep.
[207.] *Miomelon philippiana* (Dall 1890). Sellanes et al. 2008: table 2. Concepción Methane Seep Area, Chile, 740–870 m, seep and regular deep-sea bottom.

Clade Heterobranchia
Family Hyalogyrinidae

208. *Hyalogyra vitrinelloides* Warén and Bouchet 1993. Warén and Bouchet 1993: 52, figs. 39A–C, 40A, D, F, 41A, 42C; Warén and Bouchet 2001: appendix 2. WP: Lau Basin, 1,900 m, vent
209. *Hyalogyrina globularis* Warén and Bouchet 2001. Warén and Bouchet 2001: 202, figs. 37f, 44a, 46a, 47e–f, appendix 2; Warén et al. in Desbruyères et al. 2006: 135 (figured). JdF, Endeavour Segment, 2,200–2,250 m, vent.
210. *Hyalogyrina grasslei* Warén and Bouchet 1993. Warén and Bouchet 1993: 49, figs. 39D–G, 40B–C, E, 41B–D, 42E; Warén and Bouchet 2001: appendix 2; Warén et al. in Desbruyères et al. 2006: 135. Guaymas Basin, ca. 2,000 m, seep
211. *Hyalogyrina rissoella* Warén and Bouchet 2009. Warén and Bouchet 2009: 2345, figs. 8E, 12A–K. Regab site (type locality) and Guiness site, West Africa, 600–3,150 m, seep.
212. *Hyalogyrina umbellifera* Warén and Bouchet 2001. Warén and Bouchet 2001: 200, figs. 37e, g, 44h–j, 45a–d, 47d, 48a–b, 49i, appendix 2; Kiel 2004: 179, figs. 35–36. Aleutian Trench, 4,800 m, seep.

**Hyalogyrina?* sp.: Warén and Bouchet 2001: appendix 2. New Zealand, vent.

Family Orbitestellidae

213. *Lurifax japonica* Sasaki and Okutani 2005: Sasaki and Okutani 2005: 121, fig. 1A–D. WP: Sumisu Caldera, Japan, 676 m, vent.
214. *Lurifax vitreus* Warén and Bouchet 2001: Warén and Bouchet 2001: 207, figs. 37c–d, 44e–g, 46c–d, 47a–b, appendix 2; Warén et al. in Desbruyères et al. 2006: 136 (figured). Mid-Atlantic Ridge: Menez Gwen (type locality)

and Lucky Strike, 38–36°N, ca. 850–1,800 m, vent. Olu-Le Roy et al. 2004: table 2. Anaximander area, Eastern Mediterranean Sea, ca. 1,700–2,000 m, seep. *Remarks:* Smriglio and Mariottini (2002) reported a single empty shell from Western Mediterranean Sea.

**Lurifax* sp.: Warén and Bouchet 2001: appendix 2. New Zealand, seep.

Family Pyramidellidae

215. *Eulimella lomana* (Dall 1908). Warén and Bouchet 1993: 55, fig. 59F–G. Southern California and Gulf of California, 1,168–2,008 m, seep.

**Eulimella* sp. Okutani and Fujiwara 2000: 126, fig. 11; Sasaki et al. 2005: 124. WP: Okinawa Trough, North Knoll of Iheya Ridge, 1,049 m, vent.

[*] *Odostomia* sp. Olu-Le Roy et al. 2004: table 2. Anaximander area, Eastern Mediterranean Sea, ca. 1,700–2,000 m, seep. No evidence for seep relations.

Family Xylodisculidae

216. *Xylodiscula analoga* Warén and Bouchet 2001. Warén and Bouchet 2001: 208, figs. 44b–d, 47c, 49h, appendix 2; Kiel 2004: 178, fig. 34; Warén et al. in Desbruyères et al. 2006: 137 (figured). MAR: Lucky Strike (type locality) and Menez Gwen, 37–36°N, 850–1,728 m, vent.
217. *Xylodiscula major* Warén and Bouchet 1993. Warén and Bouchet 1993: 53, figs. 42A–B, 43A–E; Warén and Bouchet 2001: appendix 2. WP: North Fiji Basin, 2,000 m, vent.

**Xylodiscula* sp. Warén and Bouchet 2001: appendix 2. New Zealand, seep.

**Xylodiscula* sp. Olu-Le Roy et al. 2004: table 2. Anaximander area, Eastern Mediterranean Sea, ca. 1,700–2,000 m, seep and regular deep-sea bottom?

Family Dendronotidae

218. *Dendronotus comteti* Valdés and Bouchet 1998. Valdés and Bouchet 1998: 320, figs. 1–3; Warén and Bouchet 2001: appendix 2; Bouchet and Segonzac in Desbruyères et al. 2006: 140 (figured). MAR: Lucky Strike, 37°N, 1,685 m, vent.

References

Aktipis SW, Giribet G, Lindberg DR et al (2008) Gastropoda: An overview and analysis. In: Ponder WF, Lindberg DR (eds) Phylogeny and evolution of the mollusca. University of California Press, Berkeley, CA

Bates AE (2007a) Feeding strategy, morphological specialisation and presence of bacterial episymbionts in lepetodrilid gastropods from hydrothermal vents. Mar Ecol Prog Ser 347:87–99

Bates AE (2007b) Persistence, morphology, and nutritional state of a gastropod hosted bacterial symbiosis in different levels of hydrothermal vent flux. Mar Biol 152:557–568

Batten RL (1984) Shell structure of the Galapagos rift limpet *Neomphalus fretterae* McLean, 1981, with notes on muscle scars and insertions. Am Mus Novitates 2276:1–13

Beck LA (1991) *Olgaconcha tufari* n. gen. et n. sp. – an new mesogastropod (Gastropoda: Prosobranchia) from hydrothermal vents in the Manus Back-Arc Basin (Bismarck Sea, Papua New Guinea). Ann Naturhist Mus Wien B92:277–286

Beck LA (1992a) *Symmetromphalus hageni* sp. n., a new neomphalid gastropod (Prosobranchia: Neomphalidae) from hydrothermal vents at the Manus Back-Arc Basin (Bismarck Sea, Papua New Guinea). Ann Naturhist Mus Wien B93:243–257

Beck LA (1992b) Two new neritacean limpets (Gastropoda: Prosobranchia: Neritacea: Phenacolepadidae) from active hydrothermal vents at Hydrothermal Field 1 "Wienerwald" in the Manus Back Arc Basin (Bismarck Sea, Papua New Guinea). Ann Naturhist Mus Wien B93:259–275

Beck LA (1993) Morphological and anatomical studies on a new lepetodrilacean limpet (Gastropoda, Prosobranchia) from hydrothermal vents at the Manus Back-Arc Basin (Bismarck Sea, Papua New Ginea). Ann Naturhist Mus Wien B94(95):167–179

Beck LA (1996) Morphology and anatomy of new species of neolepetopsid, acmaeid, fissurellid and pyropeltid limpets from Edison Seamount off Lihir Islands (West Pacific). Arch Mollusk 125:87–103

Beck LA (2002) *Hirtopelta tufari* sp. n., a new gastropod species from hot vents at the East Pacific Rise (21degree S) harbouring endocytosymbiotic bacteria in its gill: (Gastropoda: Rhipidoglossa: Peltospiridae). Arch Mollusk 130:249–257

Bergquist DC, Eckner JT, Urcuyo IA et al (2007) Using stable isotopes and quantitative community characteristics to determine a local hydrothermal vent food web. Mar Ecol Prog Ser 330:49–65

Bouchet P, Warén A (1991) *Ifremeria nautilei*, a new gastropod from hydrothermal vents, probably associated with symbiotic bacteria. C R Acad Sci III 312:495–501

Bouchet P, Warén A (1993) Revision of the Northeast Atlantic bathyal and abyssal Mesogastropoda. Boll Malac Suppl 3:580–840

Bouchet P, Fryda J, Hausdorf B et al (2005) Working classification of the Gastropoda. In: Bouchet P, Rocroi JP (eds) Classification and nomenclator of gastropod families. Malacologia 47:240–283

Braby CE, Rouse GW, Johnson SB et al (2007) Temporal variation among *Osedax* boneworms and related megafauna on whale-falls in Monterey Bay, California. Deep Sea Res Pt I 54:1773–1791

Campbell KA, Peterson DE, Alfaro AC (2008) Two new species of *Retiskenea*? (Gastropoda: Neomphalidae) from lower Cretaceous hydrocarbon-seep carbonates of northern California. J Paleontol 82:140–153

Clarke AH (1989) New mollusks from under-sea oil seep sites off Louisiana. Malac Data Net 2:122–134

Colgan DJ, Ponder WF, Beacham E, Macaranas JM (2007) Molecular phylogenetics of Caenogastropoda (Gastropoda: Mollusca). Mol. Phyloget Evol 42:717–737

Craddock C, Lutz RA, Vrijenhoek RC (1997) Patterns of dispersal and larval development of archaeogastropod limpets at hydrothermal vents in the eastern Pacific. J Exp Mar Biol Ecol 210:37–51

Cruz R, Farina M (2005) Mineralization of major lateral teeth in the radula of a deep-sea hydrothermal vent limpet (Gastropoda: Neolepetopsidae). Mar Biol 147:163–168

Dattagupta S, Martin J, Liao SM et al (2007) Deep-sea hydrocarbon seep gastropod *Bathynerita naticoidea* responds to cues from the habitat-providing mussel *Bathymodiolus childressi*. Mar Ecol 28:193–198

Declerck C (1995) The evolution of suspension feeding in gastropods. Biol Rev 70:549–569

Denis F, Jollivet D, Moraga D (1993) Genetic separation of two allopatric populations of hydrothermal snails *Alviniconcha* spp. (Gastropoda) from two South Western Pacific Back-arc Basins. Biochem Syst Ecol 21:431–440

Desbruyères D, Segonzac M, Bright M (eds) (2006) Handbook of deep-sea hydrothermal vent fauna. Denisia 18:1–544

Distel DL, Baco AR, Chuang E et al (2000) Do mussels take wooden steps to deep-sea vents? Nature 403:725–726

Fraussen K, Sellanes J (2007) Three new buccinid species (Gastropoda: Neogastropoda) from Chilean deep-water, including one from a methane seep. Veliger 50(2):97–106

Fretter V (1988) New archaeogastropod limpets from hydrothermal vents, superfamily Lepetodrilacea. II. Anatomy. Philos T Roy Soc B319:33–82

Fretter V (1989) The anatomy of some new archaeogastropod limpets (Superfamily Peltospiracea) from hydrothermal vents. J Zool 218:123–169

Fretter V (1990) The anatomy of some new archaeogastropod limpets (Order Patellogastropoda, Suborder Lepetopsina) from hydrothermal vents. J Zool 222:529–555

Fretter V, Graham A, McLean JH (1981) The anatomy of the Galapagos Rift limpet, *Neomphalus fretterae*. Malacologia 21:337–361

Fuchigami T, Sasaki T (2005) The shell structure of the Recent Patellogastropoda (Mollusca: Gastropoda). Paleontol Res 9:143–168

Fujikura K, Sasaki T, Yamanaka T et al (2009) Turrid whelk, *Phymorhynchus buccinoides* feeds on *Bathymodiolus* mussels at a seep site in Sagami Bay, Japan. Plankton Benthos Res 4:23–30

Geiger DL, Thacker CE (2006) Molecular phylogeny of basal gastropods (Vetigastropoda) show stochastic colonization of chemosynthetic habitats at least from the mid Triassic. Cah Biol Mar 47:343–346

Geiger DL, Nützel A, Sasaki T (2008) Vetigastropoda. In: Lindberg DR, Ponder WF (eds) Phylogeny and evolution of the Mollusca. University of California Press, Berkeley, CA

Gofas S (2005) Geographical differentiation in *Clelandella* (Gastropoda: Trochidae) in the Northeastern Atlantic. J Moll Stud 71:133–144

Goffredi SK, Warén A, Orphan VJ et al (2004) Novel forms of structural integration between microbes and a hydrothermal vent gastropod from the Indian Ocean. Appl Environ Microb 70:3082–3090

Gustafson RG, Lutz RA (1994) Molluscan life history traits at deep-sea hydrothermal vents and cold methane/sulfide seeps. In: Young CM, Eckelbarger KJ (eds) Reproduction, larval biology, and recruitment of the deep-sea benthos. Columbia University Press, New York

Harasewych MG, Kantor YI (2002) *Buccinum thermophilum* (Gastropoda: Neogastropoda: Buccinidae), a new species from the Endeavour vent field of the Juan de Fuca Ridge. J Moll Stud 68:39–44

Harasewych MG, McArthur AG (2000) A molecular phylogeny of the Patellogastropoda (Mollusca: Gastropoda). Mar Biol 137:183–194

Hasegawa K (1997) Sunken wood-associated gastropods collected from Suruga Bay, Pacific side of the central Honshu, Japan, with descriptions of 12 new species. Nat Sci Mus Monogr 12:59–123

Haszprunar G (1989a) New slit-limpets (Scissurellacea and Fissurellacea) from hydrothermal vents. Part 2. Anatomy and relationships. Contrib Sci Nat Hist Mus Los Angeles Co 408:1–17

Haszprunar G (1989b) The anatomy of *Melanodrymia aurantiaca* Hickman, a coiled archaeogastropod from the East Pacific hydrothermal vents (Mollusca, Gastropoda). Acta Zool 70:175–186

Healy JM (1990) Taxonomic affinities of the deep-sea genus *Provanna* (Caenogastropoda): new evidence from sperm ultrastructure. J Moll Stud 56:119–122

Heß M, Beck F, Gensler H et al (2008) Microanatomy, shell structure and molecular phylogeny of *Leptogyra*, *Xyleptogyra* and *Leptogyropsis* (Gastropoda: Neomphalida: Melanodrymiidae) from sunken wood. J Moll Stud 74:381–401

Hickman CS (1984) A new archaeogastropod (Rhipidoglossa, Trochacea) from hydrothermal vents on the East Pacific Rise. Zool Scr 13:19–25

Hickman CS, McLean JH (1990) Systematic revision and suprageneric classification of trochacean gastropods. Nat Hist Mus Los Angeles County Sci Ser 35:1–169

Hodgson AN, Healy JM, Tunnicliffe V (1997) Spermatogenesis and sperm structure of the hydrothermal vent prosobranch gastropod *Lepetodrilus fucensis* (Lepetodrilidae, Mollusca). Invertebr Reprod Dev 31:87–97

Hodgson AN, Eckelbarger KJ, Young CM (1998) Sperm morphology and spermiogenesis in the methane-seep mollusc *Bathynerita naticoidea* (Gastropoda: Neritacea) from the Louisiana slope. Invertebr Biol 117:199–207

Hodgson AN, Eckelbarger KJ, Young CM (2009) Sperm ultrasptructure and spermatogenesis in the hydrothermal vent gastropod *Rhynchopelta concentrica* (Peltospiridae). J Moll Stud 75:159–165

Houart R (2003) Description of three new species of *Trophon* s. 1. Montfort, 1810 (Gastropoda: Muricidae) from Chile. Novapex 4:101–110

Houart R, Sellanes J (2006) New data on recently described Chilean trophonines (Gastropoda: Muricidae), description of a new species and notes of their occurrence at cold seep site. Zootaxa 1222:53–68

Houbrick RS (1979) Classification and systematic relationships of the Abyssochrysidae, a relict family of bathyal snails (Prosobranchia: Gastropoda. Smithsonian Contrib Zool 290:1–21

Hunt S (1992) Structure and composition of the shell of the archaeogastropod limpet *Lepetodrillus elevatus elevatus* (McLean, 1988). Malacologia 34:129–141

Israelsson O (1998) The anatomy of *Pachydermia laevis* (Archaeogastropoda: 'Peltospiridae'). J Moll Stud 64:93–109

Jablonski D, Lutz RA (1983) Larval ecology of marine benthic invertebrates: paleobiological implications. Biol Rev 58:21–89

Jenkins RG, Kaim A, Hikida Y (2007a) Antiquity of the substrate choice among acmaeid limpets from Late Cretaceous chemosynthesis-based communities. Acta Palaeontol Pol 52:369–373

Jenkins RG, Kaim A, Hikida Y et al (2007b) Methane-flux-dependent lateral faunal changes in a Late Cretaceous chemosymbiotic assemblage from the Nakagawa area of Hokkaido, Japan. Geobiology 5:127–139

Johnson SB, Young CR, Jones WJ et al (2006) Migration, isolation, and speciation of hydrothermal vent limpets (Gastropoda; Lepetodrilidae) across the Blanco Transform Fault. Biol Bull 210:140–157

Johnson SB, Warén A, Lee RW, Kaim A, Davis A, Kano Y, Strong EE, Vrijenhoek RC. in press. Living fossils: *Rubyspira*, a new genus of bone-eating snails from the deep-sea. *Biol. Bull.*

Johnson SB, Warén A, Vrijenhoek RC (2008) DNA barcoding of *Lepetodrilus* limpets reveals cryptic species. J Shellfish Res 27:43–51

Kaim A (2004) The evolution of conch ontogeny in Mesozoic open sea gastropods. Palaeontol Pol 62:6–183

Kaim A, Jenkins RG, Warén A (2008a) Provannid and provannid-like gastropods from the Late Cretaceous cold seeps of Hokkaido (Japan) and the fossil record of the Provannidae (Gastropoda: Abyssochrysoidea). Zool J Linn Soc 154:421–436

Kaim A, Kobayashi Y, Echizenya H et al (2008b) Chemosynthesis-based associations on Cretaceous plesiosaurid carcasses. Acta Palaeontol Pol 53:97–104

Kano Y (2006) Usefulness of the opercular nucleus for inferring early development in neritimorph gastropods. J Morphol 267:1120–1136

Kano Y (2008) Vetigastropod phylogeny and a new concept of Seguenzioidea: independent evolution of copulatory organs in the deep-sea habitats. Zool Scr 37:1–21

Kano Y, Haga T (2010) Marine ecosystems: Sulfide rich environments. In: Bouchet P, Le Guyader H, Pascal O (eds) The natural history of Santo. Patrimoines Naturels. 69 Muséum National d'Histoire Naturelle, Paris

Kano Y, Chiba S, Kase T (2002) Major adaptive radiation in neritopsine gastropods estimated from 28S rRNA sequences and fossil records. P Roy Soc Lond B269:2457–2465

Kano Y, Chikyu E, Warén A (2009) Morphological, ecological and molecular characterization of the enigmatic planispiral snail genus *Adeuomphalus* (Vetigastropoda: Seguenzioidea). J Moll Stud 75:397–418

Kelly NE, Metaxas A (2007) Influence of habitat on the reproductive biology of the deep-sea hydrothermal vent limpet *Lepetodrilus fucensis* (Vetigastropoda: Mollusca) from the Northeast Pacific. Mar Biol 151:649–662

Kelly NE, Metaxas A (2008) Population structure of two deep-sea hydrothermal vent gastropods from the Juan de Fuca Ridge, NE Pacific. Mar Biol 153:457–471

Kiel S (2004) Shell structures of selected gastropods from hydrothermal vents and seeps. Malacologia 46:169–183

Kiel S (2006) New records and species of molluscs from Tertiary cold-seep carbonates in Washington State, USA. J Paleontol 80:121–137

Kiel S, Goedert JL (2006) A wood-fall association from late Eocene deep-water sediments of Washington State, USA. Palaios 21:548–556

Kiel S, Goedert JL (2007) New mollusks associated with biogenic substrates in Cenozoic deep-water sediments of Washington State. Acta Palaeontol Pol 52:41–52

Kiel S, Little CTS (2006) Cold-seep mollusks are older than the general marine mollusk fauna. Science 313:1429–1431

Kiel S, Campbell KA, Elder WP et al (2008) Jurassic and Cretaceous gastropods from hydrocarbon seeps in forearc basin and accretionary prism settings, California. Acta Palaeontol Pol 53:679–703

Kojima S, Segawa R, Fujiwara Y et al (2000) Genetic differentiation of populations of a hydrothermal vent-endemic gastropod, *Ifremeria nautilei*, between the North Fiji Basin and the Manus Basin revealed by nucleotide sequences of mitochondrial DNA. Zool Sci 17:1167–1174

Kojima S, Segawa R, Fijiwara Y et al (2001) Phylogeny of hydrothermal-vent-endemic gastropods *Alviniconcha* spp. from the western Pacific revealed by mitochondrial DNA sequences. Biol Bull 200:298–304

Kojima S, Fujikura K, Okutani T et al (2004) Phylogenetic relationship of *Alviniconcha* gastropods from the Indian Ocean to those from the Pacific Ocean (Mollusca: Provannidae) revealed by nucleotide sequences of mitochondrial DNA. Venus 63:65–68

Kunze T, Heß M, Brückner M et al (2008) Skeneimorph gastropods in Neomphalina and Vetigastropoda – A preliminary report. Zoosymposia 1:119–131

Levesque C, Juniper SK, Limen H (2006) Spatial organization of food webs along habitat gradients at deep-sea hydrothermal vents on Axial Volcano, Northeast Pacific. Deep Sea Res Pt I 53:726–739

Levin LA, Bridges TS (1995) Pattern and diversity in reproduction and development. In: McEdward L (ed) Ecology of marine invertebrate larvae. CRC Press, Boca Raton, FL

Limén H, Levesque C, Juniper SK (2007) POM in macro-/meiofaunal food webs associated with three flow regimes at deep-sea hydrothermal vents on Axial Volcano, Juan de Fuca Ridge. Mar Biol 153:129–139

Lindberg DR (2008) Patellogastropoda, Neritimorpha, and Cocculinoidea. The low-diversity gastropod clades. In: Ponder WF, Lindberg DR (eds) Phylogeny and Evolution of the Mollusca. University of California Press, Berkeley, CA

Lindberg DR, Hedegaard C (1996) A deep water patellogastropod from Oligocene water-logged wood of Washington State, USA (Acmaeoidea: *Pectinodonta*). J Moll Stud 62:299–314

Little CRS, Cann JR (1999) Late Cretaceous hydrothermal vent communities from the Troodos ophiolite, Cyprus. Geology 27:1027–1030

Lus VJ (1981) New species of *Tacita* (Prosobranchia, Buccinidae) with wide distribution in the north-western part of the Pacific Ocean. Trudy Instit Okeanol P P Shirshov 115:140–154 (in Russian)

McArthur AG, Tunnicliffe V (1998) Relics and antiquity revisited in the modern vent fauna. In: Mills RA, Harrison K (eds), Modern Ocean floor processes and the geological record. Geol Soc Lond Spec Pub 148:271–291

Marshall BA (1986) Recent and Tertiary Cocculinidae and Pseudococculinidae (Mollusca: Gastropoda) from New Zealand and New South Wales. New Zeal J Zool 12:505–546

Marshall BA (1988) Skeneidae, Vitrinellidae, and Orbitestellidae (Mollusca: Gastropoda) associated with biogenic substrata from bathyal depth off New Zealand and New South Wales. J Nat Hist 22:949–1004

Marshall BA (1995) Callistomatidae (Gastropoda: Trochoidea) from New Caledonia, the Loyalty Islands, and the northern Lord Howe Rise. Mem Mus Nat Hist Natur 167:381–458

Martell KA, Tunnicliffe V, MacDonald IR (2002) Biological features of a buccinid whelk (Gastropoda, Neogastropoda) at the Endeavour ventfields of Juan de Fuca Ridge, Northeast Pacific. J Moll Stud 68:45–53

Martin JW, Haney TA (2005) Decapod crustaceans from hydrothermal vents and cold seeps: a review through 2005. Zool J Linn Soc 145:445–552

Matabos M, Thiébaut E, Le Guen D, Sadosky F, Jollivet D, Bonhomme F (2007) Geographic clines and stepping-stone patterns detected along the East Pacific Rise in the vetigastropod *Lepetodrilus elevatus* reflect species crypticism. *Mar Biol* 153:545–563

McArthur AG, Harasewych MG (2003) Molecular systematics of the major lineages of the Gastropoda. In Lydeard EC, Lindberg DR, (eds) Molecular Systematics and Phylogeography of Mollusks. Smithsonian Institution Press, Washington, DC

McArthur AG, Koop BF (1999) Partial 28S rDNA sequences and the antiquity of the hydrothermal vent endemic gastropods. Mol Phyloget Evol 13:255–274

McEdward L (ed) (1995) Ecology of Marine invertebrate larvae. CRC Press, Boca Raton, FL

McLean JH (1981) The Galapagos Rift limpet *Neomphalus*: relevance to understanding the evolution of a major Paleozoic-Mesozoic radiation. Malacologia 21:291–336

McLean JH (1988a) New archaeogastropod limpets from hydrothermal vents, superfamily Lepetodrilacea. Part I: Systematic descriptions. Philos T Roy Soc Lond B319:1–32

McLean JH (1988b) Three new limpets of the family Pseudococculinidae from abyssal depths (Mollusca, Archaeogastropoda). Zool Scr 17:155–160

McLean JH (1989a) New archaeogastropod limpets from hydrothermal vents: new family Peltospiridae, new superfamily Peltospiracea. Zool Scr 18:49–66

McLean JH (1989b) New slit-limpets (Scissurellacea and Fissurellacea) from hydrothermal vents. Part 1. Systematic descriptions and comparisons based on shell and radular characters. Contrib Sci Nat Hist Mus Los Angeles County 407:1–29

McLean JH (1990a) A new genus and species of neomphalid limpet from the Mariana vents with a review of current understanding of relationships among Neomphalacea and Peltospiracea. Nautilus 104:77–86

McLean JH (1990b) Neolepetopsidae, a new docoglossate limpet family from hydrothermal vents and its relevance to patellogastropod evolution. J Zool 222:485–528

McLean JH (1991) Four new pseudococculinid limpets collected by the deep-submersible Alvin in the Eastern Pacific. Veliger 34:38–47

McLean JH (1992a) A new species of *Pseudorimula* (Fissurellacea: Clypeosectidae) from hydrothermal vents of the Mid-Atlantic Ridge. Nautilus 106:115–118

McLean JH (1992b) Cocculiniform limpets (Cocculinidae and Pyropeltidae) living on whale bone in the deep sea off California. J Moll Stud 58:401–414

McLean JH (1993) New species and records of *Lepetodrilus* (Vetigastropoda: Lepetodrilidae) from hydrothermal vents. Veliger 36:27–35

McLean JH (2008) Three new species of the family Neolepetopsidae (Patellogastropoda) from hydrothermal vents and whale falls in the northeastern Pacific. J Shellfish Res 27:15–20

McLean JH, Geiger DL (1998) New genera and species having the *Fissurisepta* shell form, with a generic level phylogenetic analysis (Gastropoda: Fissurellidae). Contrib Sci Nat Hist Mus Los Angeles County 475:1–32

McLean JH, Harasewych MG (1995) Review of western Atlantic species of cocculinid and pseudococculinid limpets, with descriptions of new species (Gastropoda: Cocculiniformia). Contrib Sci Nat Hist Mus Los Angeles County 453:1–33

McLean JH, Haszprunar G (1987) Pyropeltidae, a new family of cocculiniform limpets from hydrothermal vents. Veliger 30:196–205

McLean JH, Quinn JF Jr (1987) *Cataegis*, a new genus of three new species from the continental slope (Trochidae: Cataeginae new subfamily). Nautilus 101:111–116

Mills SW, Beaulieu SE and Mullineaux LS (2007). Photographic identification guide to larvae at hydrothermal vents in the eastern Pacific. Published at: http://www.whoi.edu/science/B/vent-larvel-id

Mullineaux LS, Kim SL, Pooley A, Lutz RA (1996) Identification of archaeogastropod larvae from a hydrothermal vent community. Mar Biol 124:551–560

Nakano T, Ozawa T (2007) Worldwide phylogeography of limpets of the older Patellogastropoda: Molecular, morphological and palaeontological evidence. J Moll Stud 73:79–99

O'Sullivan JB, McConnaughey RR, Huber ME (1987) A blood sucking snail – the cooper nutmeg, *Cancellaria cooperi* Gabb, parasitizes the California electric ray, *Torpedo californica* Ayres. Biol Bull 172:362–366
Okutani T (1990) Two new species of *Provanna* (Gastropoda: Cerithiacea) from "snail pit" in the hydrothermal vent site at the Mariana Back-arc Basin. Venus 49:19–24
Okutani T (2001) Six new bathyal and shelf trochoidean species in Japan. Venus 60(3):121–127
Okutani T, Fujikura K (1990) A new turbinid gastropod collected from the warm seep site in the Minami-Ensei Knoll west of the Amami-Oshima Island, Japan. Venus 49:83–92
Okutani T, Fujikura K (1992) A new turrid gastropod associated with metachromatic seabed near the *Calyptogena* community in Sagami Bay. Venus 51:1–7
Okutani T, Fujikura K (2002) Abyssal gastropods and bivalves collected by Shinkai 6500 on slope of the Japan Trench. Venus 60:211–224
Okutani T, Fujikura K (2005) Distribution of a deep-sea whelk *Bayerius arnoldi* (Gastropoda: Buccinidae). Chiribotan 35:121–124 (in Japanse)
Okutani T, Fujiwara Y (2000) Gastropod fauna of a thermal vent site on the North Knoll of Iheya Ridge, Okinawa Trough. Venus 59:123–128
Okutani T, Iwasaki N (2003) Noteworthy abyssal mollusks (excluding vesicomyid bivalves) collected from the Nankai Trough off Shikoku by the ROV Kaiko of the Japan Marine Science and Technology Center. Venus 62:1–10
Okutani T, Ohta S (1988) A new gastropod mollusk associated with hydrothermal vents in the Mariana Back-arc Basin, Western Pacific. Venus 47:1–10
Okutani T, Ohta S (1993) New buccinid and turrid gastropods from North Fiji and Lau Basins. Venus 52:217–221
Okutani T, Saito H, Hashimoto J (1989) A new neritacean limpet from a hydrothermal vent site near Ogasawara Islands, Japan. Venus 48:223–230
Okutani T, Tsuchida E, Fujikura K (1992) Five bathyal gastropods living within or near the *Calyptogena* community of the Hatsushima Islet, Sagami Bay. Venus 51:137–148
Okutani T, Fujikura K, Sasaki T (1993) New taxa and new distribution records of deepsea gastropods collected from or near the chemosynthetic communities in the Japanese waters. Bull Nat Sci Mus Ser A (Zool) 19:123–143
Okutani T, Sasaki T, Tsuchida S (2000) Two additional new species to gastropod fauna of chemosynthetic site on North Knoll of Iheya Ridge, Okinawa Trough. Venus 59:267–275
Okutani T, Hashimoto J, Sasaki T (2004) New gastropod taxa from hydrothermal vent (Kairei Field) in the central Indian Ocean. Venus 63:1–11
Olu-Le Roy K, Sibuet M, Fiala-Médioni A et al (2004) Cold seep communities in the deep eastern Mediterranean Sea: composition, symbiosis and spatial distribution on mud volcanoes. Deep Sea Res Pt I 51:1915–1936
Plouviez S, Shank TM, Faure B et al (2009) Comparative phylogeography among hydrothermal vent species along the East Pacific Rise reveals vicariant processes and population expansion in the South. Mol Ecol 18:3903–3917
Ponder WF, Lindberg DR (1997) Towards a phylogeny of gastropod mollusks: a preliminary analysis using morphological characters. Zool J Linn Soc 19:83–265
Pradillon F, Shillito B, Young CM et al (2001) Developmental arrest in vent worm embryos. Nature 413:698–699
Puillandre N, Samadi S, Boisselier MC et al (2008) Starting to unravel the toxoglossan knot: Molecular phylogeny of the "turrids" (Neogastropoda: Conoidea). Mol Phylogenet Evol 47:1122–1134
Robertson R (1989) Spermatophores of aquatic non-stylommatophoran gastropods: a review with new data on *Heliacus* (Architectonicidae). Malacologia 30:341–364
Robertson R (2007) Taxonomic occurrences of gastropod spermatozeugmata and non-stylommatophoran spermatophores updated. Am Malacol Bull 23:11–16
Salvini-Plawen Lv, Steiner G (1996) Synapomorphies and plesiomorphies in higher classification of Mollusca. In: Taylor JD (ed) Origin and evolutionary radiation of the Mollusca. Oxford University Press, Oxford

Sasaki T (1998) Comparative anatomy and phylogeny of the Recent Archaeogastropoda (Mollusca: Gastropoda). Univ Mus Univ Tokyo Bull 38:1–223

Sasaki T (2008) Micromolluscs in Japan: taxonomic composition, habitats, and future topics, In: Geiger DL, Ruthensteiner B (eds) Micromolluscs: methodological challenges – exciting results. Zoosymposia 1:147–232

Sasaki T, Ishikawa H (2002) The first occurrence of a neritopsine gastropod from phreatic community. J Moll Stud 68:286–288

Sasaki T, Okutani T (2005) A new species of *Lurifax* (Gastropoda: Heterobranchia: Orbitestellidae) from Sumisu Caldera, southern Japan. Venus 63:121–124

Sasaki T, Okutani T, Fujikura K (2003) New taxa and new records of patelliform gastropods associated with chemoautosynthesis-based communities in Japanese waters (Mollusca: Gastropoda). Veliger 46:189–210

Sasaki T, Okutani T, Fujikura K (2005) Molluscs from hydrothermal vents and cold seeps in Japan: A review of taxa recorded in twenty recent years (1984–2004). Venus 64:87–133

Sasaki T, Okutani T, Fujikura K (2006a) Anatomy of *Bathyacmaea secunda* Okutani, Fujikura and Sasaki, 1993 (Patellogastropoda: Acmaeidae). J Moll Stud 72:295–309

Sasaki T, Okutani T, Fujikura K (2006b) Anatomy of *Shinkailepas myojinensis* Sasaki, Okutani and Fujikura, 2003 (Gastropoda: Neritopsina). Malacologia 48:1–26

Sasaki T, Fujikura K, Okutani T (2007) Molluscs collected in the cruise NT06-04 of R/V *Natsushima* from methane-seeps off Hatsushima Island. Sagami Bay. Chiribotan 37:197–207 (in Japanese)

Sasaki T, Okutani T, Fujikura K (2008) A new species of *Pyropelta* (Lepetelloidea: Pyropeltidae) from hydrothermal vents off Yaeyama Islands, Okinawa, Japan. J Moll Stud 74:309–316

Sellanes J, Quiroga E, Neira C (2008) Megafauna community structure and trophic relationships at the recently discovered Concepción Methane Seep Area, Chile, 36°S. ICES J Mar Sci 65:1102–1111

Shuto T (1974) Larval ecology of prosobranch gastropods and its bearing on biogeography and paleontology. Lethaia 7:239–256

Smith CR, Baco AR (2003) Ecology of whale falls at the deep-sea floor. Oceanogr Mar Biol Ann Rev 41:311–354

Smith CR, Kukert H, Wheatcroft RA et al (1989) Vent fauna on whale remains. Nature 34:127–128

Smriglio C, Mariottini P (2002) *Lurifax vitreus* Warén and Bouchet, 2001 (Gastropoda, Orbitestellidae), first report from Western Mediterranean Sea. Boll Malacol 38:45–47

Stein JL, Cary SC, Hessler RR et al (1988) Chemoautotrophic symbiosis in a hydrothermal vent gastropod. Biol Bull 174:373–378

Suzuki Y, Sasaki T, Suzuki M et al (2005a) Novel chemoautotrophic endosymbiosis between a member of the e-Proteobacteria and the hydrothermal-vent gastropod *Alviniconcha* aff. *hessleri* (Gastropoda: Provannidae) from the Indian Ocean. Appl Environ Microb 71:5440–5450

Suzuki Y, Sasaki T, Suzuki M et al (2005b) Molecular phylogenetic and isotopic evidence of two lineages of chemoautotrophic endosymbionts distinct at the subdivision level harbored in one host-animal type: the genus *Alviniconcha* (Gastropoda: Provannidae). FEMS Microb Lett 249:105–112

Suzuki Y, Kojima S, Sasaki T et al (2006a) Host-symbiont relationships in hydrothermal vent gastropods of the genus *Alviniconcha* from the Southwest Pacific. Appl Environ Microb 72:1388–1393

Suzuki Y, Kojima S, Watanabe H et al (2006b) Single host and symbiont lineages of hydrothermalvent gastropods *Ifremeria nautilei* (Provannidae): biogeography and evolution. Mar Ecol Prog Ser 325:167–175

Suzuki Y, Kopp RE, Kogure T et al (2006c) Sclerite formation in the hydrothermal-vent "scaly-foot" gastropod – possible control of iron sulfide biomineralization by the animal. Earth Planet Sci Lett 242:39–50

Sysoev AV, Kantor YI (1995) Two new species of *Phymorhynchus* (Gastropoda, Conoidea, Conidae) from the hydrothermal vents. Ruthenica 5:17–26

Tamburri MN, Barry JP (1999) Adaptations for scavenging by three diverse bathyal species, *Eptatretus stouti*, *Neptuea amianta* and *Orchomene obtusus*. Deep Sea Res Pt I 46:2079–2093

Thorson G (1950) Reproductive and larval ecology of marine bottom invertebrates. Biol Rev 25:1–45

Tunnicliffe V (1991) The biology of hydrothermal vents ecology and evolution. Oceanogr Mar Biol Ann Rev 29:319–407

Tunnicliffe V (1992) The nature and origin of the modern hydrothermal vent fauna. Palaios 7:338–350

Tyler PA, Young CM (2003) Dispersal at hydrothermal vents: a summary of recent progress. Hydrobiologia 503:9–19

Tyler PA, Pendlebury S, Mills SW et al (2008) Reproduction of gastropods from vents on the East Pacific Rise and the Mid-Atlantic Ridge. J Shellfish Res 27:107–118

Valdés À, Bouchet P (1998) Naked in toxic fluids: A nudibranch mollusc from hydrothermal vents. Deep sea Res Pt II 45:319–327

Van Dover CL (2000) The ecology of deep-sea hydrothermal vents. Princeton University Press, Princeton, NJ

Van Gaest AL, Young CM, Young JJ et al (2007) Physiological and behavioral responses of *Bathynerita naticoidea* (Gastropoda: Neritidae) and *Methanoaricia dendrobranchiata* (Polychaeta: Orbiniidae) to hypersaline conditions at a brine pool cold seep. Mar Ecol 28:199–207

Vermeij GJ (1993) A natural history of shells. Princeton University Press, Princeton, NJ

Vilvens C, Sellanes J (2006) Descriptions of *Otukaia crustulum* new species (Gastropoda: Trochoidea: Calliostomatidae) and *Margarites huloti* new species (Gastropoda: Trochoidea: Trochidae) from a methane seep area off Chile. Nautilus 120:15–20

Voight JR (2000a) A deep-sea octopus (*Graneledone* cf. *boreopacifica*) as a shell-crushing hydrothermal vent predator. J Zool 252:335–341

Voight JR (2000b) A review of predators and predation at deep-sea hydrothermal vents. Cah Biol Mar 41:155–166

Voight JR, Sigwart JD (2007) Scarred limpets at hydrothermal vents: evidence of predation by deep-sea whelks. Mar Biol 152:129–133

Warén A (1992) New and litte known "skeneimorph" gastropods from the Mediterranean Sea and the adjacent Atlantic Ocean. Boll Malacol 27:149–248

Warén A (1996) New and little known Mollusca from Iceland and Scandinavia, Part 3. Sarsia 81:197–245

Warén A, Bouchet P (1986) Four new species of *Provanna* Dall (Prosobranchia, Cerithiacea) from East Pacific hydrothermal sites. Zool Scr 15:157–164

Warén A, Bouchet P (1989) New gastropods from east Pacific hydrothermal vents. Zool Scr 18:67–102

Warén A, Bouchet P (1993) New records, species, genera, and a new family of gastropods from hydrothermal vents and hydrocarbon seeps. Zool Scr 22:1–90

Warén A, Bouchet P (2001) Gastropoda and Monoplacophora from hydrothermal vents and seeps; new taxa and records. Veliger 44:116–231

Warén A, Bouchet P (2009) New gastropods from deep-sea hydrocarbon seeps off West Africa. Deep Sea Res Pt II 57:2326–2349

Warén A, Ponder WF (1991) New species, anatomy and systematic position of the hydrothermal vent and hydrocarbon seep gastropod family Provannidae fam. n. (Caenogastropoda). Zool Scr 20:27–56

Warén A, Bengtson S, Goffredi SK et al (2003) A hot-vent gastropod with iron sulfide dermal sclerites. Science 302:1007

Watanabe H, Fujikura K, Kinoshita G et al (2009) Egg capsule of *Phymorhynchus buccinoides* (Gastropoda: Turridae) in a deep-sea methane seep site in Sagami Bay, Japan. Venus 67:181–188

Williams ST, Ozawa T (2006) Molecular phylogeny suggests polyphyly of both the turban shells (family Turbinidae) and the superfamily Trochoidea (Mollusca: Vetigastropoda). Mol Phylogenet Evol 39:33–51

Williams ST, Karube S, Ozawa T (2008) Molecular systematics of Vetigastropoda: Trochidae, turbinidae and trochoidea redefined. Zool Scr 37:483–506

Windoffer R, Giere O (1997) Symbiosis of the hydrothermal vent gastropod *Ifremeria nautilei* (Provannidae) with endobacteria – structural analyses and ecological considerations. Biol Bull 193:381–392

Chapter 8
The Fossil Record of Vent and Seep Mollusks

Steffen Kiel

Contents

8.1 Introduction

Mollusks have by far the most extensive and best-investigated fossil record of all organism groups inhabiting hydrothermal vents and hydrocarbon seeps. More than 250 bivalve, gastropod, and polyplacophoran species have been reported from ancient vents and seeps, nearly 200 of them from the Cenozoic alone. Members of at least five bivalve families live in symbiosis with sulfur- or methane-oxidizing bacteria, and among the gastropods at least three groups took this path of adaptation. Mollusks are common at vent communities of Mesozoic and Paleozoic age, but appear to be less common in seep communities of this age. It is generally believed that brachiopods were the dominant taxon at Mesozoic and Paleozoic seeps; however, an increasing number of Paleozoic and Mesozoic seep sites with mollusks have been discovered in recent years.

Mollusks have often been used to formulate hypotheses on the evolutionary history of the vent and seep fauna, because they dominate many modern sites and

S. Kiel (✉)
Institut für Geowissenschaften, Christian-Albrechts-Universität, Ludewig-Meyn-Str. 10, 24118, Kiel, Germany
e-mail: steffen.kiel@gmx.de

S. Kiel (ed.), *The Vent and Seep Biota*, Topics in Geobiology 33,
DOI 10.1007/978-90-481-9572-5_8, © Springer Science+Business Media B.V. 2010

because of their high degree of adaptation. Their rich fossil record has been used to test such hypotheses and to outline a paleobiologic view of their evolutionary history. A wealth of taxonomic work on mollusks especially from seeps during the past 5 years significantly enhanced the quality of the paleontologic data. However, many faunas remain to be described and revised. This chapter provides an outline and a critical evaluation of the current knowledge of the fossil history of the major mollusk groups at ancient vents and seeps. It also presents certain aspects of the evolution of the vent and seep fauna from a paleontologic point of view, and indicates gaps and not-well-understood aspect of the fossil record of these faunas.

8.2 Bivalvia

Bathymodiolins. The two bivalve groups that dominate vents and seeps today, the Vesicomyidae (Fig. 8.1) and Bathymodiolinae (Figs. 8.2a, b), are geologically speaking young groups. They first appear in the fossil record in the middle Eocene. Compared to the vesicomyids which dominated fossil seeps since the late Eocene, the bathymodiolins apprently made a slow start. They are common at a few late Eocene and Oligocene seep deposits in Washington State, USA, where they are also found on fossil whale- and wood-falls (Goedert and Squires 1990; Squires and Goedert 1991; Kiel 2006; Kiel and Goedert 2006a). But otherwise only few bathymodiolins have been found at some Oligocene to Pliocene seep sites in the Caribbean, Italy, and Japan (Taviani 1994; Gill et al. 2005; Majima et al. 2005). This sparse Neogene record is surprising considering the ubiquity of bathymodiolins at modern vents and seeps. It is

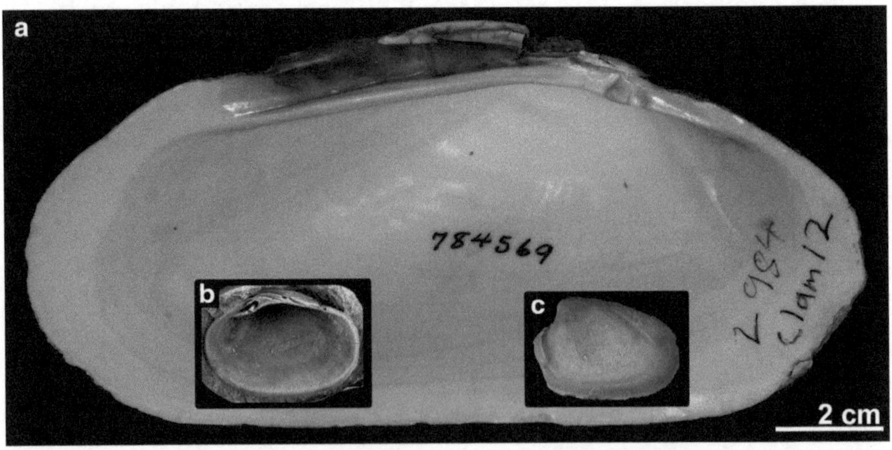

Fig. 8.1 Vesicomyid bivalves, (**a**) Recent *Calyptogena magnifica* from a hydrothermal vent on the Galapagos Rift; (**b**) The oldest *Calyptogena*, a new species from Early Oligocene dark shales the Katalla district in southern Alaska; (**c**) The oldest vesicomyid: *Archivesica* cf. *tschudi* from a middle Eocene seep deposit in western Washington State, USA (Images from Kiel and Amano [2010; b] and Amano and Kiel [2007; c])

Fig. 8.2 Bathymodiolins, solemyids and *Nucinella*. (**a**), (**b**) The oldest fossil bathymodiolin, *Bathymodiolus willapaensis*, from Eocene seep deposits in Washington State, USA; adult shell (**a**) and juvenile shell (**b**). (**c**), (**d**) Solemyids from an Early Miocene seep deposit in Washington State, showing the outer organic layer (periostracum) that overlaps the calcareous shell. (**e**) Internal mold of the Early Cretaceous *Acharax mikasaensis* from Hokkaido, Japan, showing characteristic features of the anterior muscle attachment scar. (**f**) Late Cretaceous *Nucinella gigantea* from Hokkaido, Japan, with internal radial striations (lower *left*) (Images from Kiel [2006; b], Kiel [2010; d] and Kiel et al. [2008a; e, f])

also at odds with the suggestion of Miyazaki et al. (2008) that bathymodiolins spread worldwide during the early/middle Miocene climate optimum. There is molecular and morphologic evidence that several lineages of large bathymodiolins exist and that these lineages adapted independently to seeps and vents, possibly at different geologic times (Jones et al. 2006; Cosel and Janssen 2008; Chapter 6, this volume). The geologically oldest bathymodiolins, *Bathymodiolus willapaensis* Squires and Goedert 1991 and *Idas*? *olympicus* Kiel and Goedert 2007 have been identified based on the shape of their early juvenile shells and their shell microstructure (Kiel 2006; Kiel and Goedert 2007; Génio and Kiel 2010), but to clarify to which of the modern lineages they belong, if at all, needs further investigation. Muscle and mantle attachment scars in fossil bathymodiolins might provide further insight to their identity and to the evolutionary history of bathymodiolins. The only pre-Cenozoic record of a putative seep-related mytilid, *Modiola major* from late Jurassic to early Cretaceous seep deposits in California, is in fact a member of the extinct bivalve clade Modiomorphida (Stewart 1930; Kiel and Peckmann 2008).

Vesicomyids. The recorded history of the vesicomyids is very different from that of the bathymodiolins, and can probably be called the most successful adaptation to vents and seeps by any mollusk group. After their first appearance in seep deposits of the middle Eocene Humptulips Formation in Washington State, vesicomyids are found, often in large numbers, in virtually every seep deposit world-wide from the Late Eocene onward. The family is not restricted to vents and seeps, there is a series of records from turbidite deposits, organic-rich shales, and whale-falls (Goedert and Squires 1993; Amano et al. 2001; Amano and Kanno 2005; Amano and Little 2005; Kiel and Amano 2010). Generic distinctions within the family have long been neglected, but systematic improvements have recently begun (Cosel and Salas 2001; Krylova and Sahling 2006; Amano and Kiel 2007; Kiel 2007) and are likely to reveal more about the evolutionary and biogeographic history of this family. Previous reports of vesicomyids from the Cretaceous period could not be confirmed. These had been described from Japanese seep deposits. Recent re-investigations of these sites and their fossils showed that an Albian species does in fact belong to the solemyids (Kiel et al. 2008a), a Cenomanian record belongs to a new large and veneriform lucinid genus (Amano et al. 2008), and a large Campanian specimen previously coined '*Calyptogena*' most likely belongs to the extinct bivalve clade Modiomorphida (Amano and Kiel 2007; Kiel and Peckmann 2008).

Further bivalve families that are frequently encountered at modern vents and seeps are the solemyids, tyasirids, and lucinids. The species of these families live buried within the sediment and are thus less obvious than the surface-dwelling vesicomyids and bathymodiolins. In the fossil record, however, where seep carbonates can easily be broken up to recover fossils, these families are quite common.

Solemyids. The Solemyidae are among the most basal families among the Bivalvia with a fossil record dating back to the Ordovician. They are well-represented – although never abundant – at seeps from the later Mesozoic onwards, and there are also a couple of Paleozoic records. Modern and Mesozoic-Cenozoic solemyids are usually differentiated into two genera depending on the position of the ligament: in *Solemya* the ligament is internal, in *Acharax* it is external. Of the two genera, *Acharax* appears to be more common at seeps, perhaps because it is adapted to greater water depth (Chapter 5, this volume). Modern solemyids with their elongate shells and characteristic radial ribs (Figs. 8.2c–e) first appear at seeps in the late Jurassic of California (Campbell and Bottjer 1993; Campbell 2006).

At modern seep sites, solemyids are usually found in the periphery of the seep where sulfide flux and concentrations are low (Sahling et al. 2002). The same pattern of distribution was reported from a Late Cretaceous (Campanian) seep site on Hokkaido, Japan (Jenkins et al. 2007b) and from an Early Cretaceous seep site in eastern Greenland (Kelly et al. 2000) where solemyids occur in patches at the periphery of the seep carbonates. Also at some of the Paleogene seep sites in Washington State, USA, peripheral clusters of solemyids were found, whereas at other sites, solemyids were rare and scattered through the deposit (own observation). In contrast, at an early Cretaceous seep site on Hokkaido (the Albian Ponbetsu site) solemyids, along with thyasirids and lucinids are common throughout the deposit (Kiel et al. 2008a). This site, however, consists of two silty carbonate blocks with few authigenic cements which most likely resulted from low seepage rates and diffuse seepage (Peckmann et al. 2009) which could explain the relative abundance of solemyids.

An Early Carboniferous record from a seep in the Harz Mountains in Germany has been questioned because the specimens found so far are poorly preserved internal molds lacking the characteristic radial ribs and muscle attachment scars (Peckmann et al. 2001; Goedert et al. 2003). From a middle Devonian seep deposit in Morocco known as Hollard Mound, a potential solemyid was reported as *Dystactella*? (Peckmann et al. 1999; Aitken et al. 2002). The description and figure of this species does indeed resemble that of other well-known Paleozoic members of this genus (Pojeta 1988). Solemyids might thus have been exploiting sulfides at seeps for nearly 400 million years.

Nucinella. Phylogenetically related to solemyids is *Nucinella* (family Manzanellidae), a genus of generally very small deep-sea species that rarely exceed a few millimeters in length. Recently, comparatively large specimens of *Nucinella*, reaching nearly 2 cm in length, have been reported from Late Cretaceous seep deposits on Hokkaido, Japan (Fig. 8.2f; Amano et al. 2007a; Kiel et al. 2008a) and a similar shell has been found in a late Triassic seep site in Oregon (Peckmann et al. 2010). Their large size, internal radiating striae similar to those in the chemosymbiotic lucinids and thyasirids, and the lack of a gut in the large extant *Nucinalla maxima* lead Amano et al. (2007a) to suggest the possibility of chemosymbiosis in such large *Nucinella* species (see also Chapter 5, this volume).

Thyasirids. Living members of the Thyasiridae are apparently able to host both sulfur- and methane-oxidizing symbionts, although small species may have no symbionts at all (Kamenev et al. 2001; Dufour 2005; Chapter 5, this volume). In fossil seeps thyasirids usually occur in small clusters or scattered throughout the deposit. Despite their basal position among the heterodont bivalves, thyasirids have a relatively short fossil record (Taylor et al. 2007). The oldest species with typical thyasirid characters is from the Early Cretaceous (Hauterivian) of Europe, the oldest record from a seep deposit is of Albian age (Kiel et al. 2008a) and it has been reported from a late Cretaceous wood-fall (Kiel et al. 2009). These early records can be attributed to the genus *Thyasira*. From the Maastrichtian onward the large genus *Conchocele* is more commonly found (cf., Kiel et al. 2008a). There is a tendency among paleontologists to lump large Cenozoic thyasirids from the North Pacific rim into a single species: *Conchocele bisecta* (Fig. 8.3b). However, Kiel and Peckmann (2007) and Kiel and Goedert (2007) indicated that thyasirid specimens from Eocene to Oligocene seeps and organic substrates in this region show distinctive shell features and do in fact belong to more than one species.

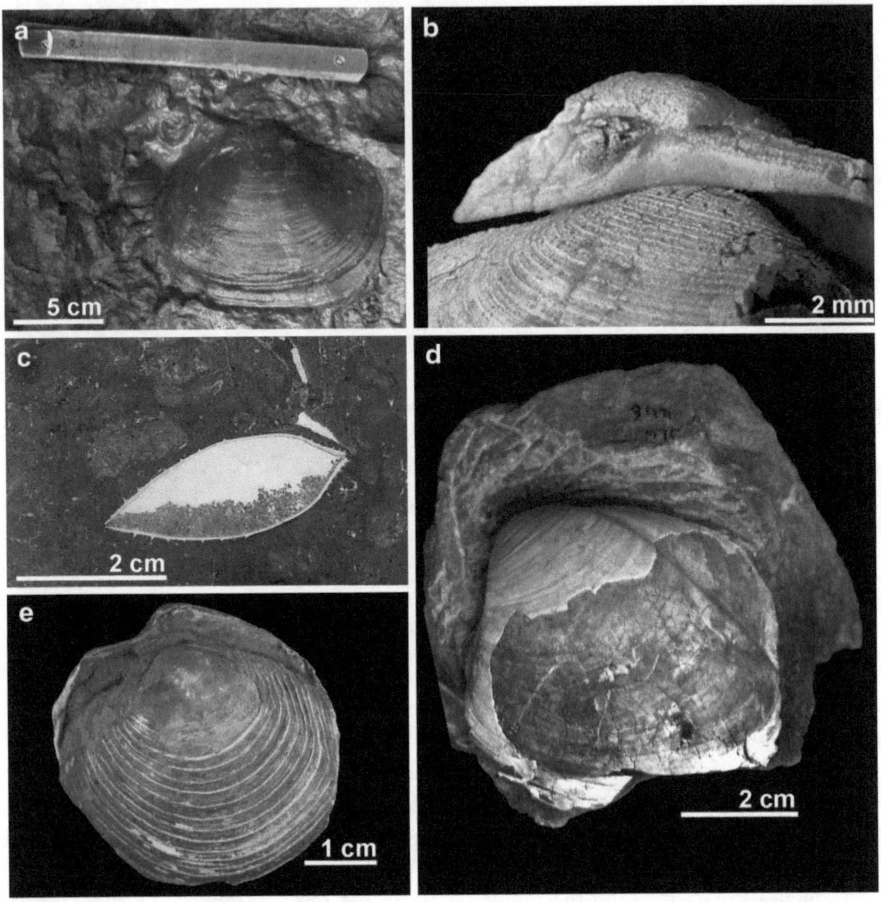

Fig. 8.3 Lucinids and thyasirids. (**a**) The lucinid *Nipponothracia ponbetsensis* from an Albian (Early Cretaceous) seep site on Hokkaido, Japan. (**b**) Close-up on the edentulous hinge of the lucinid *Nipponothracia yezoensis* from a Cenomanian (Late Cretaceous) seep site on Hokkaido, Japan. (**c**) Cross-section of an articulated specimen of *Lucinoma* spp. in an Early Miocene seep limestone in Washington State, USA. (**d**) *Lucinoma acutilineata*, a common species in Oligocene seep deposits in Washington. (**e**) The thyasirid *Conchocele bisecta* from the Early Miocene Astoria Formation in Washington (Images from Kiel et al. [2008a; b], and with courtesy of Jörn Peckmann [c])

Lucinids. Lucinids live in symbiosis with sulfur-oxidizing bacteria. They do not, however, rely entirely on their endosymbionts for nutrition but are also capable of suspension-feeding (see also Chapter 5, this volume). In contrast to the thyasirids the fossil history of lucinids can be traced back into the Silurian and it is assumed that the chemosymbiotic lifestyle is as old as the family (Taylor and Glover 2006). It is thus surprising that lucinids are found at ancient seeps only from the Late Jurassic onward (Gaillard et al. 1985, 1992). They are usually found as articulated shells, often in large clusters or shell beds, but also as scattered individuals. At modern vents and seeps they appear to be rare, but this might be a sampling artifact because most attention has been given to the larger epifaunal animals with less sampling effort devoted to the deeper-burrowing infauna (Hashimoto et al. 1995; Salas and Woodside 2002; Glover et al. 2004).

In Late Jurassic to Eocene seeps lucinids are taxonomically diverse and many species appear to be restricted to individual seep sites or regions. *Nymphalucina occidentalis* is one such example: it occurs exclusively in the Western Interior Seaway of the United States, a shallow seaway that connected the Gulf of Mexico and the Arctic Ocean during the Late Cretaceous. Here *Nymphalucina* is especially abundant in methane seep deposits known as Tepee Buttes (Speden 1970; Kauffman et al. 1996). In the Eocene of the Pacific coast of North America, *Cryptolucina* occurs in methane seep deposits in Washington State and *Epilucina* in oil seeps in southern California (Saul et al. 1996; Squires and Gring 1996; Kiel and Peckmann 2007). Older seep deposits in this region also include some lucinids, but these are generally rare and poorly preserved and have so far escaped a modern systematic evaluation (cf., Campbell 2006). Several as-yet poorly identified lucinids are known from European Jurassic and Cretaceous seep deposits (Lemoine et al. 1982; Gaillard et al. 1992; Kiel and Peckmann 2008) as well as from Cretaceous seeps in Japan (Hikida et al. 2003; Kiel et al. 2008a).

In contrast, members of the edentulous, large and elongate-oval lucinid genus *Nipponothracia* (Figs. 8.3a, b) are geographically widespread, with occurrences in Japan, the western United States, and Greenland. In the western Pacific (Japan, Indonesia and the Philippines) *Nipponothracia* apparently survived until Pliocene time (Kase et al. 2007). Despite being taxonomically diverse, lucinids at Late Jurassic to Eocene seeps consistently lack ornamentation. This is in contrast to the sharply ribbed *Lucinoma* (Figs. 8.3c, d), which is the most common lucinid at seeps from the Oligocene onwards. *Lucinoma* is known from numerous Neogene seep deposits in Japan (Majima et al. 2005), from Oligocene seeps in the Caribbean and Peru (Kiel and Peckmann 2007), from Miocene seep deposits in New Zealand (Campbell 2006), from several Oligocene and Neogene sites in Washington (Campbell 1992; Goedert and Campbell 1995; Peckmann et al. 2002), and from the Miocene 'Calcare a *Lucina*' in Italy (Moroni 1966). Lucinids other than *Lucinoma* in post-Eocene seep deposits have hesitantly been assigned to *Myrtea*, like specimens from the Oligocene of Peru and Cuba (Cooke 1919; Olsson 1931; Kiel and Peckmann 2007), although their systematic affiliations are still unclear. Likewise, the affinities of *Lucina hoernea* which dominates many seep carbonates in the Italian Miocene 'Calcare a *Lucina*' have yet to be evaluated.

Modiomorphids. As the name suggests, members of this clade resemble modiolins (family Mytilidae) in general shell shape and certain seep-related modiomorphids have indeed been mistaken for members of the mytilids in the past. The Modiomorphida are an extinct clade of bivalves that is commonly associated with the Anomalodesmata. Modiomorphids had sporadically been reported from ancient vents and seeps over the past 10 years. Records include *Sibaya ivanovi* from a middle Devonian hydrothermal vent deposit in the Ural Mountains (Little et al. 1999b). The identification was based on external shell morphology only and remains to be confirmed. Recent research interest focuses on the Late Jurassic to Cretaceous genus *Caspiconcha*, first described from seep deposits in eastern Greenland (Kelly et al. 2000). Modiomorphids are extinct and thus the reconstruction of the mode of life of vent- and seep-inhabiting species – especially the question whether they harbored chemotrophic symbionts – is not as straight-forward as in the groups discussed so far. The current knowledge on the live habits of vent and seep modiomorphids is here outlined using the relatively well-studied *Caspiconcha*.

Species of *Caspiconcha* are usually quite large, reaching 20–30 cm in length. They are found exclusively at ancient seeps where they are associated with a variety of other seep-related taxa, including tube worms, the brachiopod *Peregrinella*, various gastropods, and solemyid and lucinid bivalves (Kelly et al. 2000; Hikida et al. 2003; Kiel et al. 2008b; Kiel and Peckmann 2008; Jenkins et al. 2009). At the seep sites on Greenland and on the Crimean peninsula, *Caspiconcha* occurs in clusters and was found positioned at an angle of about 30° relative to the bedding plane, with the posterior side pointing upwards, which suggests a semi-infaunal mode of life (Kelly et al. 2000; Kiel and Peckmann 2008). *Caspiconcha* species from different parts of the world apparently share another feature: the interior of their shells show traces of burrowing organisms, especially in their posterior part.

Non-chemosymbiotic groups. A few other, non-chemosymbiotic bivalves frequently occur at vents and seeps. Most commonly found are nuculids and nuculanids (Protobranchia). *Nuculana grasslei* occurs in large numbers in a Recent, sediment-hosted vent in the Guaymas Basin. The detailed description of its anatomy, however, showed no evidence for any particular adaptation to this environment. Rather than being chemosymbiotic, *Nuculana grasslei* is a suspension- and deposit-feeder like nuculanids in general (Allen 1993). Nearly identical specimens have been recorded from seeps as old as Eocene along with other nuculanid bivalves (Kiel 2006). The ribbed nuculid *Acila* and its subgenus *Truncacila* are frequently collected at Cretaceous to Neogene seeps in Japan and a few other sites of the Pacific rim (Majima et al. 2005; Kiel et al. 2008a). Surprisingly, *Acila* has not been reported from any of the many well-studied modern seeps in this region.

Also commonly encountered are certain pectinoids. Two species of small, filter-feeding propeamussiids, *Bathypecten vulcani* and *Sinepecten segonzaci*, have been found at Recent hydrothermal vents (Schein-Fatton 1985; Schein 2006). The supposedly basal anatomy of *Bathypecten vulcani* has recently been questioned (Dijkstra and Gofas 2004) and these authors also indicated that *Bathypecten* is a synonym of the deep-water genus *Catillopecten*. A member of *Catillopecten* is known from an Oligocene seep deposits in Washington State, USA (Kiel 2006). Deep- and cold-water pectinids frequently found in Late Jurassic-Early Cretaceous seep deposits are *Buchia* and *Aucellina*. These two closely related genera occur in mudstones of this age throughout the northern Hemisphere and have been reported from geographically distant seeps in California (Campbell and Bottjer 1993; Campbell 2006) and the southern Ukraine (Kiel and Peckmann 2008).

Inoceramids. The inoceramids are a group of pteriomorph bivalves that originated in the Permian and became extinct at the end of the Mesozoic. Inoceramids lived in a wide range of habitats, including oxygen-poor basins, where they are often quite common and large. Due to this unusual mode of occurrence, Kauffman (1988) considered them as chemosymbiotic, a view that was supported by a comparison of stable oxygen and carbon isotopes of inoceramid shells and those of planktic and benthic foraminifera (MacLeod and Hoppe 1992). Grossman (1993) challenged MacLeod and Hoppe's (1992) interpretation of their stable isotope data and considered inoceramids as non-chemosymbiotic. The question has not yet been resolved. However, I find the idea of chemosymbiosis in inoceramids based on their occurrence in organic-rich, oxygen-poor sediments not convincing. Suspended food would have been abundant for filter-feeding inoceramids in this environment, and due to the absence of any other taxa they were save from predation and competitors. This could easily explain their large and often extremely thin shells, as well as their abundance.

8.3 Gastropoda

The high degree of endemism encountered at vents and seeps has long fascinated researchers, and indeed, many classes of animals include genera and families living exclusively in these environments. But only among the gastropods is an entire order apparently endemic to chemosynthetic ecosystems: the Neomphalina (McLean 1981; Warén and Bouchet 2001; Heß et al. 2008; Chapter 7, this volume). Most of the living neomphalids are found at hydrothermal vents, a single genus has so far been reported from seeps. In addition, a fair number of taxa living on sunken driftwood (wood falls) have recently been identified as neomphalids (Kiel and Goedert 2006b; Heß et al. 2008). Whereas neomphalids are not known from ancient vent deposits, the seep-inhabiting genus *Retiskenea* has a fossil record going back to the Early Cretaceous (Figs. 8.4a–c; Kiel 2006; Campbell et al. 2008b). A potential neomphalid was described as *Lithomphalus enderlini* from Early Cretaceous seeps in California (Kiel and Campbell 2005), although its affiliation with the Neomphalina needs confirmation.

Provannids. The Provannidae is a caenogastropod family that lives exclusively in chemosynthetic habitats, including vents, seeps, whale- and wood-falls. Most provannids are relatively small and graze on biofilms, but the two large genera *Alviniconcha* and *Ifremeria* harbor sulfur-oxidizing symbionts (Warén and Bouchet 2001; see also Chapter 7, this volume). Fossil provannids (Fig. 8.5a) are known from a similar range of habitats, including records from Late Cretaceous and Eocene wood-falls, Miocene whale-falls, Late Cretaceous plesiosaur-falls, and Late Cretaceous and Paleogene cold seeps (Squires 1995; Amano and Little 2005; Gill et al. 2005; Kiel and Goedert 2006b; Amano et al. 2007b; Kaim et al. 2008a, b; Kiel et al. 2009). Provannids may even occur at a fossil vent deposit: Little et al. (1999a) described several high-spired gastropods from volcanogenic massive sulfide deposits of the Late Cretaceous Troodos Ophiolite on Cyprus, which were subsequently interpreted as potential provannids (Little 2002; Kaim et al. 2008a).

Relatives of provannids. There is an assortment of globular to high-spired gastropods known from Late Jurassic to Oligocene seep deposits that might be related to

Fig. 8.4 Neomphalid and limpet gastropods. (**a–c**) The neomphalid *Retiskenea statura* from an Oligocene seep deposit in Washington, note fine tuberculate sculpture on the protoconch on Fig. c. (**d, e**) Patellogastropod *Serradonta* cf. *vestimentifericola* from a Late Cretaceous seep deposit on Hokkaido. (**f, g**) Patellogastropod *Bathyacmaea* cf. *subnipponica* from a Late Cretaceous seep deposit on Hokkaido. (**h**) Vetigastropod *Fissurella bipunctata* from an Early Cretaceous seep deposit in California. (**i**) Two specimens of the vetigastropod *Pyropelta* sp. from a Mid-Eocene seep deposit in Washington (Images from Kiel [2006; a, c. i] and Jenkins et al. [2007a; d–g])

the modern provannids. The more slender examples had previously been identified as abyssochrysids, zygopleurids, or cerithioids. Based on Late Cretaceous material from Japan, in which the taxonomically important larval shells were preserved, Kaim et al. (2008a) introduced the family Hokkaidoconchidae for these high-spired taxa (Figs. 8.5d, e). This family can now be traced back into late Jurassic time (Kaim and Kelly 2009; Kiel et al. 2010). A large and high-spired and due to its two apertural notches somewhat unusual genus that may either belong to the hokkaido-conchids or the abyssochrysids is *Humptulipsia*. Originally described from Eocene

Fig. 8.5 Provannids and related gastropods. (**a**) Silicified early ontogenetic shell of the Late Cretaceous *Provanna tappuensis* from a seep deposit on Hokkaido, Japan, note the fine cancellate pattern on the larval shell. (**b, c**) *Paskentana paskentaensis* from a Late Jurassic seep deposit in California. (**d, e**) *Hokkaidoconcha tahemaensis* from the same seep deposit. (**f**) Early Cretaceous *Atresius liratus* from California. (**g, h**) *Humptulipsia raui*, a possible abyssochrysid from middle Eocene seep limestones in Washington State (Images from Kiel [2008b] and Kiel et al. [2008b])

seep sites in the Humptulips River area in western Washington (Kiel 2008b), another species was recently identified at an early Cretaceous seep site in southern France (Kiel et al. 2010). At these sites, *Humptulipsia* is abundant and is associated with large lucinid bivalves. The more globular provannid-like species belong to the genera *Atresius* (Fig. 8.5f) and *Paskentana* (Figs. 8.5b, c), and larval shell data indicate a relationship to provannids or zygopleuroids also for these two genera. Just as provannids today, hokkaidoconchids as well as *Atresius* and *Paskentana* are often found in large numbers in ancient seep deposits (Kaim et al. 2008a; Kiel et al. 2008b; Kaim and Kelly 2009).

Limpets. Gastropod limpets are well-represented at Recent and fossil vents and seeps. Limpets are not a monophyletic group but have evolved repeatedly in many different gastropod clades. They have few diagnostic shell characters and those they have often require exceptional preservation to be recognized in the fossil record. However, there are promising exceptions.

The extant Japanese patellogastropod *Serradonta vestimentifericola* lives at seeps attached to vestimentiferan tubes (Okutani et al. 1992). Jenkins et al. (2007a) showed that a similar, if not identical, species had the same habitat preference already in the Late Cretaceous, about 80 million years ago (Figs. 8.4d, e). Based on this observation, Jenkins et al. (2007a) argued that the geographic limitation of extant *Serradonta* to Japanese waters is a result of this habitat preference: it could disperse only through vent and seep sites that had already been inhabited by vestimentiferan tube worms. In contrast, members of the patellogastropod genus *Bathyacmaea* (Figs. 8.4f, g) have a wider geographic range, and their ability to inhabit a wide range of substrates appears to be as old as the adaptation of *Serradonta* to worm tubes (Jenkins et al. 2007a). Pyropeltids are cocculiniform limpets known to inhabit a broad range of chemosynthetic habitats (McLean and Haszprunar 1987; Warén and Bouchet 2001; Sasaki et al. 2003) and they have been reported from ancient seeps as old as Eocene (Fig. 9.4i; Kiel 2006).

Limpets with a rather poor fossil record at vents and seeps are slit- and keyhole-limpets. Several groups of slit- and keyhole-limpets are endemic to modern vents and seeps (McLean 1988, 1989, 1992). But the few examples that have been found at fossil seeps are poorly identified and their relations to the endemic modern groups are unclear. Stanton (1895) described *Fissurella bipunctata* from Early Cretaceous limestone nodules in California (Fig. 8.4h), which have subsequently been identified as ancient seep deposits (Campbell and Bottjer 1993; Birgel et al. 2006). The same rocks yield two new fissurellids and sutilizonids (Kiel et al. 2010). Further fissurellids were reported by Goedert and Squires (1990) from Eocene seeps in Washington and by Gill et al. (2005) from poorly dated (Eocene to Miocene) seep deposits on Barbados. But these specimens are likely to be limpets with the apical tip corroded away rather than true keyhole limpets (J.L. Goedert, pers. comm. and own observations).

Neritids. Neritids are a distinctive gastropod group typically found on hard substrate in the intertidal zone, but a few taxa adapted to cold seeps. These are *Bathynerita* in the Gulf of Mexico and the Caribbean (Warén and Bouchet 2001) and the limpets of the family Shinkailepadidae (Okutani 1989; Sasaki et al. 2003). The latter lack a fossil record, but shells resembling *Bathynerita* have been found at Miocene and Oligocene seeps and are usually referred to as *Thalassonerita*; an Eocene record has been questioned (Squires and Goedert 1996; Warén and Bouchet 2001). Neritids found at Late Jurassic and Early Cretaceous seeps (Kiel and Peckmann 2008; Kiel et al. 2010) appear to be unrelated to modern *Bathynerita*. This scenario is consistent with molecular age estimates for shinkailepadids which indicate an early Cenozoic origin of this group (Kano et al. 2002).

8.4 Mono- and Polyplacophorans

A single monoplacophoran species, *Rokopella segonzaci*, is known from a vent site on the Mid-Atlantic ridge. This species is less than a millimeter long, its radula does not indicate a particular adaptation to the vent environment, and it is not clear whether it is a regular member of the vent fauna (Warén and Bouchet 2001). Large conical shells (height >5 cm) from a Silurian vent site in the Ural Mountains attracted attention when they were described as ancient monoplacophorans

(Little et al. 1999b). However, the specimens lack the serial muscle attachment scars that are diagnostic for monoplacophorans. They could thus just as well be limpet-shaped gastropods or bellerophontids, as suggested earlier (Kuznetsov et al. 1993; Zaikov et al. 1995).

Polyplacophorans (commonly known as chitons) have so far only been reported from four modern vent and seeps sites in Japan, California, the Caribbean, and the Gulf of Mexico (Saito and Okutani 1990; Barry et al. 1996; Olu et al. 1996; Chapter 10, this volume). This wide geographic distribution together with several fossil specimens of *Leptochiton* from Eocene to Oligocene seep deposits in Washington State (Goedert and Campbell 1995; Peckmann et al. 2002) make it likely that chitons occur at more vent and seep sites than currently appreciated. However, the known species show no particular adaptation to the vent and seep habitat and may be deep-sea generalists (Saito and Okutani 1990).

8.5 Ancient Mollusks in Evolutionary Biology

The highly specialized and endemic fauna at vents and seeps has often been regarded as a relict fauna. Apart from the high degree of endemism, the idea was based on comparison to related, or supposedly related, fossil taxa. One such example is the iconic gastropod limpet *Neomphalus fretterae*, which McLean (1981) considered as a living relative of the Paleozoic euomphalids (the name *Neomphalus* was coined to indicate this affinity). Although the link to these Paleozoic gastropods is unlikely considering their different early ontogenetic shells (cf., Bandel and Frýda 1998), Jim McLean certainly gets credit for recognizing the neomphalids as a clade that must have split from all other gastropod groups very early in the evolutionary history of gastropods (cf., Heß et al. 2008). Whether this implies that already the earliest neomphalids inhabited vents and seeps, as assumed by Geiger and Thacker (2005) remains to be tested; their recorded history in these environments goes only back to the Early Cretaceous (Campbell et al. 2008b).

Warén and Bouchet (2001) used the fossil record of the respective sister or higher taxa of vent and seep gastropods to test the relict fauna hypothesis. Their results indicate that new taxa have continuously adapted to vents and seeps through earth history. Kiel and Little (2006) reached a similar conclusion when they compared the geologic ranges of the extant mollusk genera at seeps with those of the marine mollusk fauna in general. This study yielded another result that is promising from a paleontologists' perspective. Despite the intuitive assumption that deep-sea fossils are rare, the major taxa at modern vents and seeps in fact do have a fossil record, and those taxa without a fossil record are very rare also in the modern oceans. Furthermore, there appears to be very little bias in the fossil record of seep mollusks towards large-sized taxa (Kiel and Little 2006).

Evidence for a geologically young vent and seep fauna comes from molecular biologists who use molecular clocks to infer a Cenozoic origin of most major clades inhabiting vents and seeps today (Baco et al. 1999; Shank et al. 1999; Distel et al. 2000;

Kano et al. 2002). In case of the two major bivalve groups, vesicomyids and bathymodiolins, Little et al. (1999b) and Little and Vrijenhoek (2003) noted that these young ages appear to be contradicted by the longer geologic record of these families at seeps. However, none of these older records could be confirmed by a critical evaluation. In fact, both molecular age estimates and fossil data indicate an Eocene origin of the two groups in question (Kiel 2006; Amano and Kiel 2007; Amano et al. 2008; Kiel et al. 2008a; Kiel and Peckmann 2008).

Since there is a growing body of evidence that the vent and seep fauna is not the relict fauna it was regarded as initially, another hypothesis on its origin attracted a lot of attention. When vent-type taxa were found on decaying whale carcasses in the deep sea (Smith et al. 1989) it was suggested that whale carcasses could act as dispersal stepping stones for the vent fauna that enabled it to radiate much further than before the origin of whales. This hypothesis was supported by molecular age estimates of the bivalve family Vesicomyidae that dated the origin of this family at about 40 million years ago, coinciding roughly with the origin of whales (Baco et al. 1999; Smith 2007). One implication of this hypothesis is that vesicomyids inhabited whale falls from the Eocene onward, but this appears not to be the case. A survey of fossil whale-fall communities from the North Pacific rim showed that vesicomyids colonized whale falls only from the middle Miocene onward. The Eocene and Oligocene whale falls lacked vesicomyids, despite the fact that vesicomyids were abundant in cold-seep deposits of the same sediments (Kiel and Goedert 2006a). The reason for this appears to lie in the bone physiology of the whales. Whale bones today consist of up to 65% of lipids and the anaerobic decay of these lipids sets free the sulfides that fuel the chemosynthetic fauna at whale falls (Smith and Baco 2003). Thus Kiel and Goedert (2006a) suggested that either an increase in the size of whales or an increase in the relative lipid content of their bones should have caused the observed changes in whale fall communities from the Oligocene to the Miocene. Since these suggestions were published, two further Miocene whale falls with abundant vesicomyids were found, one on Hokkaido (Amano et al. 2007b) and one in California (Pyenson and Haasl 2007). In both cases the whales had an estimated length of about 4 m, which is considerably smaller than the up to 6 m long Oligocene whales that lacked vesicomyids. It was thus concluded that a relative increase in the lipid content of the bones was the more likely cause for the change from sulfide-poor Oligocene whale falls to sulfide-rich Miocene ones (Pyenson and Haasl 2007; Kiel 2008a).

Yet another hypothesis that triggered a lot of research is the ‘wooden steps to deep-sea vents’-hypothesis (Distel et al. 2000). It claims that bathymodiolin mussels first adapted to habitats with low sulfide levels like wood- and whale-falls in the deep-sea and colonized sulfide-rich environments like seeps and vents from there. It is based on molecular evidence showing that the wood- and bone-inhabiting bathymodiolins *Idas* and *Adipicola* are basal to the larger taxa at vents and seeps (Distel et al. 2000; Jones et al. 2006; Samadi et al. 2007; see also Chapter 6, this volume). These habitats and their respective mussel inhabitants are preserved in the fossil record. Thus comparing the first occurrences of mussels at the respective habitats could be a way to test this hypothesis. Presently the first occurrences are in the opposite sequence as predicted by the hypothesis: in the mid-Eocene at seeps, in the late Eocene at wood. However, these occurrences are stratigraphically very close to each other. It seems possible that the time between the adaptation to wood and bone on the one hand, and from there to seeps and vents on the other hand, was so short that it cannot be resolved in the fossil record.

8.6 Fossil Mollusks and Seep Biogeography

The first modern groups of seep mollusks started to appear in the mid-Cretaceous, and from the Late Eocene onwards most of the principle modern endemic seep taxa had reached a world-wide distribution (Majima et al. 2005; Campbell 2006; Kiel 2006; Kiel and Little 2006; Amano and Kiel 2007; Jenkins et al. 2007a; Kaim et al. 2008a). It has recently become apparent that a globally distributed, seep-restricted fauna existed also in older geologic time. Late Jurassic to Early Cretaceous seeps were inhabited by a distinct set of taxa, including the modiomorphid *Caspiconcha*, the gastropods *Paskentana* and *Hokkaidoconcha*, and, though only shortly, the brachiopod *Peregrinella*. Perhaps the most interesting aspect of this fauna is that even on the family level it shows no close phylogenetic relation to the modern vent and seep fauna (Campbell and Bottjer 1995; Kaim et al. 2008a; Kiel et al. 2008a, b, 2010; Kiel and Peckmann 2008; Kaim and Kelly 2009).

The modern Mediterranean seep fauna, especially its mollusks, is surprisingly distinct from that of the western Atlantic. Whereas the latter includes *Bathymodiolus*, large vesicomyids, provannids, lepetodrilids, and other gastropods typical for seep faunas around the world (Olu-Le Roy et al. 2007; Warén and Bouchet 2009), the Mediterranean seep fauna is dominated by the small bathymodiolin mussel *Idas*, a vesicomyid genus (*Isorropodon*) that is unusual for seeps, and by local lucinid bivalves and trochid gastropods (Olu et al. 2004). This is an interesting contrast to the late Miocene seep fauna of the Mediterranean Sea, which includes large vesicomyids, bathymodiolins, buccinids, and neritids, and in general resembles that of the Gulf of Mexico today (Taviani 1994). This difference probably lies in the geologic history of the Mediterranean Sea. The entire deep-water fauna of the Mediterranean, including its seep fauna, had been wiped out by the Messinian salinity crisis about 6 million years ago. In the subsequent re-colonization local and shallow-water species had an advantage because the relatively shallow Strait of Gibraltar seems to be an effective barrier for deep Atlantic seep taxa: *Bathymodiolus* has apparently not managed to re-colonize the Mediterranean Sea until the present day, despite its occurrence on mud volcanoes in the Gulf of Cadiz, just outside the Strait of Gibraltar (Génio et al. 2008). Instead, its place has been taken by the small bathymodiolin *Idas*. *Idas* species can live also in comparably shallow water of a few hundred meter depth and thus the Strait of Gibraltar poses no barrier for them (Dell 1987; Warén and Carrozza 1990; Jones et al. 2006).

An area with perhaps a similar history of local extinction and unsuccessful repopulation is the Japan Sea. The vesicomyid genus *Calyptogena* lived here since the middle Miocene (Kanno et al. 1989; Amano et al. 2001; Amano and Kanno 2005) and disappeared during the Pleistocene glaciations, when this isolated basin became anoxic (Tada 1994; Amano 2001). Apparently, *Calyptogena* has not yet repopulated the Japan Sea, possibly because the seaways linking the Japan Sea with the Pacific Ocean and the Okhotsk Sea are too shallow (see also Chapter 12, this volume).

8.7 Conclusions and Outlook

Mollusks have a long and relatively well-researched history at ancient vents and seeps. In contrast to the early 'relict fauna' hypothesis, the fossil record of mollusks shows that new taxa have more-or-less continuously adapted to vents and seeps through Earth history. The first modern seep mollusks started to appear in the mid-Cretaceous (Fig. 8.6), and seep faunas from the Late Eocene onwards have a largely modern appearance. This Late Eocene age seems to support the idea that the rise of whales played a role in the radiation of the seep faunas, but the fossil record contradicts this hypothesis because it shows that the major groups of seep mollusks did not colonize whale-falls before the Miocene. The existence of a globally distributed fauna of seep

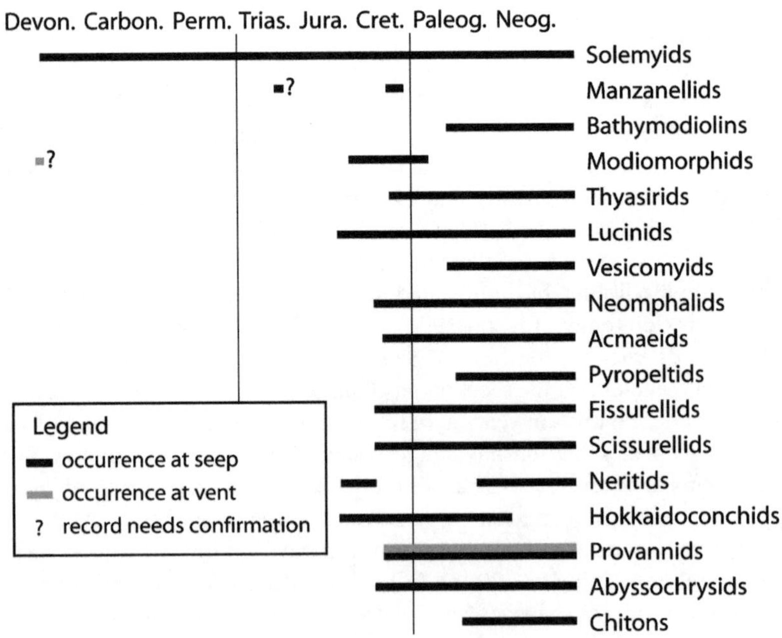

Fig. 8.6 Geologic ranges of selected mollusks at vents and seeps

obligates is not unique to the modern ocean. Also Late Jurassic to Early Cretaceous seeps world-wide were inhabited by a distinct set of taxa. Interestingly, these taxa show no close phylogenetic relation to the modern vent and seep fauna.

All extant chemosymbiotic bivalve families adapted to vents and seeps at or shortly after their first appearance in the geologic record. The only exception seems to be the Lucinidae. This family took up a chemosymbiotic lifestyle as early as Silurian (cf., Taylor and Glover 2006) but is known from ancient seep deposits only from the Late Jurassic onward. This is most likely a sampling artifact and it might be worthwhile to re-investigate lucinid occurrence of pre-Jurassic age. There are indications that some species of the manzanellids (incl. *Nucinella* spp.) might be chemosymbiotic (cf., Amano et al. 2007a) and indeed, manzanellids start to appear in Mesozoic seep deposits. Members of this group might turn up more frequently at ancient seeps when more Mesozoic seep site become known. A group that is also likely to have played a more important role at Paleozoic and Mesozoic seep than currently appreciated is the modiomorphids. The modiomorphid genus *Caspiconcha* is now known from late Mesozoic seeps world-wide and there are potential modiomorphids at seeps and vents as old as the Devonian. Surely, more is to be found.

There is a great interest in the biogeographic patterns seen in the modern vent and seep fauna, and many of these patterns are likely to have an historic basis (e.g. Tunnicliffe and Fowler 1996; Bachraty et al. 2009). Recent observations and hypotheses on the biogeography of vent and seep mollusks include multiple trans-Pacific migrations of vesicomyid clams (Kojima et al. 2004), amphi-Atlantic species complexes among bathymodiolins (Olu et al. 2007), dispersal of the vesicomyid *Calyptogena* along continental margins (Krylova and Sahling 2006), and an accumulation of biodiversity on southwest Pacific seamounts (Krylova and Janssen 2006; Cosel and Janssen 2008). The fossil record can provide further evidence for the origin and history of such pattern, although Cenozoic seep fossils are certainly not equally distributed around the globe. The tectonically active Pacific Rim is already the region with the best-investigated Cenozoic record of seep faunas. Within this region, the most promising areas with known deep-water deposits that may yield further seep faunas are Alaska (cf. Kiel and Amano 2009), the West Coast of South America from which Oligocene seep fossil have been described (cf. Olsson 1931; Kiel and Peckmann 2007), and the Philippines that recently revealed early Pliocene seep deposits (Majima et al. 2007). Work on Miocene seep faunas from New Zealand that include bathymodiolins, vesicomyids, lucinids, and other typical seep mollusks is in progress (Campbell 2006; Campbell et al. 2008a).

Much more problematic in this respect is the Atlantic Ocean. It is bordered by passive continental margins only and there is little in respect to tectonic processes that would uplift deep-water sediments here. Consequently, evidence for the Cenozoic evolution of the Atlantic seep fauna needs to be inferred from seep deposits in the adjacent Mediterranean basin to the East and from the Caribbean and Gulf of Mexico to the West. Likewise, the Indian Ocean is largely bordered by passive continental margins. Only the Indo-Malayan archipelago on its eastern side is tectonically active, and has been so throughout the Cenozoic. It has thus the potential to yield fossil seep faunas. The late Miocene asphalt deposits of Buton which produced small vesicomyids and lucinids (Beets 1942, 1953) have sometimes been considered as an ancient seep deposit but this suggestion needs confirmation.

Acknowledgments I appreciate the constructive criticism of Kazutaka Amano and Sven N. Nielsen on earlier drafts of this chapter.

References

Aitken SA, Collom CJ, Henderson CM, Johnston PA (2002) Stratigraphy, paleoecology, and origin of Lower Devonian (Emsian) carbonate mud buildups, Hamar Laghdad, eastern Anti-Atlas, Morocco, Africa. Bull Can Petrol Geol 50:217–243

Allen JA (1993) A new deep-water hydrothermal species of *Nuculana* (Bivalvia: Protobranchia) from the Guaymas Basin. Malacologia 35:141–151

Amano K (2001) Pliocene molluscan fauna of Japan Sea borderland and the paleoceanographic conditions. Biol Sci (Tokyo) 53:178–184 (in Japanese)

Amano K, Kanno S (2005) *Calyptogena* (Bivalvia: Vesicomyidae) from Neogene strata in the Joetsu district, Niigata Prefecture, central Japan. Venus 47:202–212

Amano K, Kiel S (2007) Fossil vesicomyid bivalves from the North Pacific region. Veliger 49:270–293

Amano K, Little CTS (2005) Miocene whale-fall community from Hokkaido, northern Japan. Palaeogeogr Palaeocl 215:345–356

Amano K, Hamuro T, Hamuro M, Fujii S (2001) The oldest vesicomyid bivalves from the Japan Sea borderland. Venus 60:189–198

Amano K, Jenkins RG, Hikida Y (2007a) A new gigantic *Nucinella* (Bivalvia: Solemyoida) from the Cretaceous cold-seep deposit in Hokkaido, northern Japan. Veliger 49:84–90

Amano K, Little CTS, Inoue K (2007b) A new Miocene whale-fall community from Japan. Palaeogeogr Palaeocl 247:236–242

Amano K, Jenkins RG, Kurihara Y, Kiel S (2008) A new genus for *Vesicomya inflata* Kanie and Nishida, a lucinid shell convergent with that of vesicomyids, from Cretaceous strata of Hokkaido, Japan. Veliger 50:255–262

Bachraty C, Legendre P, Desbruyères D (2009) Biogeographic relationships among deep-sea hydrothermal vent faunas at global scale. Deep Sea Res I 56:1371–1378

Baco AR, Smith CR, Peek AS, Roderick GK et al (1999) The phylogenetic relationships of whale-fall vesicomyid clams based on mitochondrial COI DNA sequences. Mar Ecol Prog Ser 182:137–147

Bandel K, Frýda J (1998) The systematic position of the Euomphalidae (Gastropoda). Senckenb Lethaea 78:103–131

Barry JP, Greene G, Orange DL, Baxter CH et al (1996) Biologic and geologic characteristics of cold seeps in Monterey Bay, California. Deep Sea Res I 43:1739–1762

Beets C (1942) Beiträge zur Kenntnis der angeblich oberoligocänen Mollusken-Fauna der Insel Buton, Niederländisch-Ostindien. Leidse Geol Meded 13:255–328

Beets C (1953) Reconsideration of the so-called Oligocene fauna in the asphaltic deposits of Buton (Malay Archipelago). Leidse Geol Meded 17:237–258

Birgel D, Thiel V, Hinrichs K-U, Elvert M et al (2006) Lipid biomarker patterns of methane-seep microbialites from the Mesozoic convergent margin of California. Org Geochem 37:1289–1302

Campbell KA (1992) Recognition of a Mio-Pliocene cold seep setting from the Northeast Pacific Convergent Margin, Washington, U.S.A. Palaios 7:422–433

Campbell KA (2006) Hydrocarbon seep and hydrothermal vent paleoenvironments and paleontology: past developments and future research directions. Palaeogeogr Palaeocl 232:362–407

Campbell KA, Bottjer DJ (1993) Fossil cold seeps. Natl Geogr Res Explor 9:326–343

Campbell KA, Bottjer DJ (1995) *Peregrinella*: an Early Cretaceous cold-seep-restricted brachiopod. Paleobiology 24:461–478

Campbell KA, Francis DA, Collins M, Gregory MR et al (2008a) Hydrocarbon seep-carbonates of a Miocene forearc (East Coast Basin), North Island, New Zealand. Sediment Geol 204:83–105

Campbell KA, Peterson D, Alfaro AC (2008b) Two new species of *Retiskenea*? (Gastropoda: Neomphalidae) from Lower Cretaceous hydrocarbon seep-carbonates of northern California. J Paleontol 82:140–153

Cooke CW (1919) Contributions to the geology and paleontology of the West Indies IV. Tertiary mollusks from the leeward islands and Cuba. Carnegie Inst Wash Publ 291:103–156

Cosel Rv, Janssen R (2008) Bathymodioline mussels of the *Bathymodiolus* (s. l.) *childressi* clade from methane seeps near Edison Seamount, New Ireland, Papua New Guinea (Bivalvia: Mytilidae). Archiv Mollus 137:195–224

Cosel Rv, Salas C (2001) Vesicomyidae (Mollusca: Bivalvia) of the genera *Vesicomya*, *Waisiuconcha*, *Isorropodon* and *Callogonia* in the eastern Atlantic and the Mediterranean. Sarsia 86:333–366

Dell RK (1987) Mollusca of the family Mytilidae (Bivalvia) associated with organic remains from deep water off New Zealand, with revisions of the genera *Adipicola* Dautzenberg, 1927 and *Idasola* Iredale, 1915. Natl Mus N Z Rec 3:17–36

Dijkstra HH, Gofas S (2004) Pectinoidea (Bivalvia: Propeamussiidae and Pectinidae) from some northeastern Atlantic seamounts. Sarsia 89:33–78

Distel DL, Baco AR, Chuang E, Morrill W et al (2000) Do mussels take wooden steps to deep-sea vents? Nature 403:725–726

Dufour SC (2005) Gill anatomy and the evolution of symbiosis in the bivalve family Thyasiridae. Biol Bull 208:200–212

Gaillard C, Bourseau J-P, Boudeulle M, Pailleret P et al (1985) Les pleudo-biohermes de Beauvoisin (Drôme): un site hydrothermal sur lar marge téthysienne a l'Oxfordien? Bull Soc Géol Fr 8:69–78

Gaillard C, Rio M, Rolin Y (1992) Fossil chemosynthetic communities related to vents or seeps in sedimentary basins: the pseudobioherms of southeastern France compared to other world examples. Palaios 7:451–465

Geiger DL, Thacker CE (2005) Molecular phylogeny of Vetigastropoda reveals non-monophyletic Scissurellidae, Trochoidea, and Fissurelloidea. Mollus Res 25:47–55

Génio L, Johnson SB, Vrijenhoek RC, Cunha MR et al (2008) New record of *Bathymodiolus mauritanicus* Cosel from the Gulf of Cadiz (NE Atlantic) mud volcanoes. J Shell Res 27:53–61

Gill FL, Harding IC, Little CTS, Todd JA (2005) Palaeogene and Neogene cold seep communities in Barbados, Trinidad and Venezuela: an overview. Palaeogeogr Palaeocl 227:191–209

Glover EA, Taylor JD, Rowden AA (2004) *Bathyaustriella thionipta*, a new lucinid bivalve from a hydrothermal vent on the Kermadec Ridge, New Zealand and its relationship to shallow-water taxa (Bivalvia: Lucinidae). J Mollus Stud 70:283–295

Goedert JL, Campbell KA (1995) An Early Oligocene chemosynthetic community from the Makah Formation, northwestern Olympic Peninsula, Washington. Veliger 38:22–29

Goedert JL, Squires RL (1990) Eocene deep-sea communities in localized limestones formed by subduction-related methane seeps, southwestern Washington. Geology 18:1182–1185

Goedert JL, Squires RL (1993) First Oligocene record of *Calyptogena* (Bivalvia: Vesicomyidae). Veliger 36:72–77

Goedert JL, Thiel V, Schmale O, Rau WW et al (2003) The late Eocene 'Whiskey Creek' methane-seep deposit (western Washington State) Part I: geology, palaeontology, and molecular geobiology. Facies 48:223–240

Grossman EL (1993) Evidence that inoceramid bivalves were benthic and harbored chemosynthetic symbionts: comment and reply. Geology 21:94–96

Hashimoto J, Fujikura K, Fujiwara Y, Tanishima M et al (1995) Observations of a deep-sea biological community co-dominated by lucinid bivalve, *Lucinoma spectabilis* (Yokoyama, 1920) and vestimentiferans at the Kanesu-no-Se Bank, Enshu-Nada Central Japan. JAMSTEC J Deep Sea Res 11:211–217

Heß M, Beck F, Gensler H, Kano Y et al (2008) Microanatomy, shell structure, and molecular phylogeny of *Leptogyra*, *Xyleptogyra* and *Leptogyropsis* (Gastropoda, Neomphalida, Melanodrymiidae) from sunken wood. J Mollus Stud 74:383–401

Hikida Y, Suzuki S, Togo Y, Ijiri A (2003) An exceptionally well-preserved seep community from the Cretaceous Yezo forearc basin in Hokkaido, northern Japan. Paleontol Res 7:329–342

Jenkins RG, Kaim A, Hikida Y (2007a) Antiquity of the substrate choice among acmaeid limpets from the Late Cretaceous chemosynthesis-based communities. Acta Palaeontol Pol 52:369–373

Jenkins RG, Kaim A, Hikida Y, Tanabe K (2007b) Methane-flux-dependent lateral faunal changes in a Late Cretaceous chemosymbiotic assemblage from the Nakagawa area of Hokkaido, Japan. Geobiology 5:127–139

Jenkins RG, Kaim A, Iba Y, Tanabe K (2009) On *Caspiconcha* bivalves from the early Cretaceous methane-seep deposits in Hatonosu District, Hokkaido and California, U.S.A. 158th Regular meeting, Palaentological Society of Japan (in Japanese, title translated)

Jones WJ, Won Y-J, Maas PAY, Smith PJ et al (2006) Evolution of habitat use by deep-sea mussels. Mar Biol 148:841–851

Kaim A, Kelly SRA (2009) Mass occurrence of hokkaidoconchid gastropods in the Upper Jurassic methane seep carbonate from Alexander Island, Antarctica. Antarct Sci 21:279–284

Kaim A, Jenkins RG, Warén A (2008a) Provannid and provannid-like gastropods from Late Cretaceous cold seeps of Hokkaido (Japan) and the fossil record of the Provannidae (Gastropoda: Abyssochrysoidea). Zool J Linn Soc 154:421–436

Kaim A, Kobayashi Y, Echizenya H, Jenkins RG et al (2008b) Chemosynthesis-based associations on Cretaceous plesiosaurid carcasses. Acta Palaeontol Pol 53:97–104

Kamenev GM, Nadtochy VA, Kuznetsov AP (2001) *Conchocele bisecta* (Conrad, 1849) (Bivalvia: Thyasiridae) from cold-water methane-rich areas of the Sea of Okhotsk. Veliger 44:84–94

Kanno S, Amano K, Ban H (1989) *Calyptogena* (*Calyptogena*) *pacifica* Dall (Bivalvia) from the Neogene system in the Joetsu district, Niigata prefecture. Trans Proc Palaeontol Soc Jpn 153:25–35

Kano Y, Chiba S, Kase T (2002) Major adaptive radiation in neritopsine gastropods estimated from 28S rRNA sequences and fossil records. Proc R Soc B 269:2457–2465

Kase T, Kurihara Y, Hagino K (2007) Middle Miocene chemosynthetic thraciid *Nipponothracia gigantea* (Shikama, 1968) from central Japan is a large lucinid bivalve (Lucinoidea; Mollusca). Veliger 49:294–302

Kauffman EG (1988) The case of the missing community: low-oxygen adapted Paleozoic and Mesozoic bivalves ("flat-clams") and bacterial symbioses in typical Phanerozoic oceans. Geol Soc Am Abstr Prog 20:A48

Kauffman EG, Arthur MA, Howe B, Scholle PA (1996) Widespread venting of methane-rich fluids in Late Cretaceous (Campanian) submarine springs (Tepee Buttes), Western Interior seaway, U.S.A. Geology 24:799–802

Kelly SRA, Blanc E, Price SP, Withham AG (2000) Early Cretaceous giant bivalves from seep-related limestone mounds, Wollaston Forland, Northeast Greenland. In: Harper EM, Taylor JD, Crame JA (eds) The evolutionary biology of the Bivalvia. Geological Society of London (Special Publication), London, pp 227–246

Kiel S (2006) New records and species of mollusks from Tertiary cold-seep carbonates in Washington State, USA. J Paleontol 80:121–137

Kiel S (2007) Status of the enigmatic fossil vesicomyid bivalve *Pleurophopsis*. Acta Palaeontol Pol 52:639–642

Kiel S (2008a) Fossil evidence for micro- and macrofaunal utilization of large nekton-falls: examples from early Cenozoic deep-water sediments in Washington State, USA. Palaeogeogr Palaeocl 267:161–174

Kiel S (2008b) An unusual new gastropod genus from an Eocene hydrocarbon seep in Washington State, USA. J Paleontol 82:188–191

Kiel S (2010) On the potential generality of depth-related ecologic structure in cold-seep communities: Evidence from Cenozoic and Mesozoic examples. Palaeogeogr Palaeoclima Palaeoeco 295:245–257

Kiel S, Amano K (2010) Oligocene and Miocene vesicomyid bivalves from the Katalla district in southern Alaska, USA. Veliger 51:76–84

Kiel S, Campbell KA (2005) *Lithomphalus enderlini* gen. et sp. nov. from cold-seep carbonates in California – a Cretaceous neomphalid gastropod? Palaeogeogr Palaeocl 227:232–241

Kiel S, Goedert JL (2006a) Deep-sea food bonanzas: early Cenozoic whale-fall communities resemble wood-fall rather than seep communities. Proc R Soc B 273:2625–2631

Kiel S, Goedert JL (2006b) A wood-fall association from Late Eocene deep-water sediments of Washington State, USA. Palaios 21:548–556

Kiel S, Goedert JL (2007) Six new mollusk species associated with biogenic substrates in Cenozoic deep-water sediments in Washington State, USA. Acta Palaeontol Pol 52:41–52

Kiel S, Little CTS (2006) Cold seep mollusks are older than the general marine mollusk fauna. Science 313:1429–1431

Kiel S, Peckmann J (2007) Chemosymbiotic bivalves and stable carbon isotopes indicate hydrocarbon seepage at four unusual Cenozoic fossil localities. Lethaia 40:345–357

Kiel S, Peckmann J (2008) Paleoecology and evolutionary significance of an Early Cretaceous *Peregrinella*-dominated hydrocarbon-seep deposit on the Crimean Peninsula. Palaios 23:751–759

Kiel S, Amano K, Jenkins RG (2008a) Bivalves from Cretaceous cold-seep deposits on Hokkaido, Japan. Acta Palaeontol Pol 53:525–537

Kiel S, Campbell KA, Elder WP, Little CTS (2008b) Jurassic and Cretaceous gastropods from hydrocarbon-seeps in forearc basin and accretionary prism settings, California. Acta Palaeontol Pol 53:679–703

Kiel S, Amano K, Hikida Y, Jenkins RG (2009) Wood-fall associations from Late Cretaceous deep-water sediments of Hokkaido, Japan. Lethaia 42:74–82

Kiel S, Campbell KA, Gaillard C (2010) New and little known mollusks from ancient chemosynthetic environments. Zootaxa 2390:26–48

Kojima S, Fujikura K, Okutani T (2004) Multiple trans-Pacific migrations of deep-sea vent/seep-endemic bivalves in the family Vesicomyidae. Mol Phylogenet Evol 32:396–406

Krylova EM, Janssen R (2006) Vesicomyidae from Edison Seamount (South West Pacific: Papua New Guinea: New Ireland fore-arc basin). Archiv Mollus 135:231–261

Krylova EM, Sahling H (2006) Recent bivalve molluscs of the genus *Calyptogena* (Vesicomyidae). J Mollus Stud 72:359–395

Kuznetsov AP, Maslennikov VV, Zaikov VV (1993) The near-hydrothermal fauna of the Silurian paleoocean in the south Ural. Izv Akad Nauk Ser Biol 4:525–534 (in Russian)

Lemoine M, Arnaud-Vanneau A, Arnaud H, Létolle R et al (1982) Indices possibles de paléo-hydrothermalisme marin dans le Jurassique et le Crétacé des Alpes occidentales (océan téthysien et sa marge continentale européenne): essai d'inventaire. Bull Soc Géol Fr 24:641–647

Little CTS (2002) The fossil record of hydrothermal vent communities. Cah Biol Mar 43:313–316

Little CTS, Vrijenhoek RC (2003) Are hydrothermal vent animals living fossils? Trends Ecol Evol 18:582–588

Little CTS, Cann JR, Herrington RJ, Morisseau M (1999a) Late Cretaceous hydrothermal vent communities from the Troodos ophiolite, Cyprus. Geology 27:1027–1030

Little CTS, Maslennikov VV, Morris NJ, Gubanov AP (1999b) Two Palaeozoic hydrothermal vent communities from the southern Ural Mountains, Russia. Palaeontology 42:1043–1078

MacLeod KG, Hoppe KA (1992) Evidence that inoceramid bivalves were benthic and harbored chemosynthetic symbionts. Geology 20:117–120

Majima R, Nobuhara T, Kitazaki T (2005) Review of fossil chemosynthetic assemblages in Japan. Palaeogeogr Palaeocl 227:86–123

Majima R, Kase T, Kawagata S, Aguilar YM et al (2007) Fossil cold-seep assemblages from Leyte Island, Philippines. J Geogr 116:643–652

McLean JH (1981) The Galapagos Rift limpet *Neomphalus*: relevance to understanding the evolution of a major Paleozoic-Mesozoic radiation. Malacologia 21:291–336

McLean JH (1988) New archaeogastropod limpets from hydrothermal vents; superfamily Lepetodrilacea. I. Systematic descriptions. Philos Trans R Soc Lond B 319:1–32

McLean JH (1989) New slit-limpets (Scissurellacea and Fissurellacea) from hydrothermal vents. Part 1. Systematic descriptions and comparisons based on shell and radular characters. Contrib Sci 407:1–29

McLean JH (1992) A new species of *Pseudorimula* (Fissurellacea: Clypeosectidae) from hydrothermal vents of the Mid-Atlantic Ridge. Nautilus 106:115–118

McLean JH, Haszprunar G (1987) Pyropeltidae, a new family of cocculiniform limpets from hydrothermal vents. Veliger 30:196–205

Miyazaki J-I, Matsumoto H, Fujita Y (2008) Evolution and phylogeny of bathymodiolins. In: Fujikura K, Okutani T, Maruyama T (eds) Deep-sea life - Biological observations using research submersibles. Tokai University Press, Hatano, Japan, pp 126–128 (in Japanese, title translated)

Moroni MA (1966) Malacofauna del "Calcare a Lucine" di S. Sofia – Forlì. Palaeontogr Italica 60:69–87

Okutani T (1989) A new neritacean limpet from a hydrothermal vent site, near Ogasawara Islands, Japan. Venus 48:223–230

Okutani T, Tsuchida S, Fujikura K (1992) Five bathyal gastropods living within or near the *Calyptogena*-community of the Hatsushima Islet, Sagami Bay. Venus 51:137–148

Olsson AA (1931) Contributions to the Tertiary paleontology of northern Peru: Part 4, The Peruvian Oligocene. Bull Am Paleontol 17:97–264

Olu K, Sibuet M, Harmegnies F, Foucher J-P et al (1996) Spatial distribution of diverse cold seep communities living on various diapiric structures of the southern Barbados prism. Prog Oceanogr 38:347–376

Olu K, Sibuet M, Fiala-Médoni A, Gofas S et al (2004) Cold seep communities in the deep eastern Mediterranean Sea: composition, symbiosis and spatial distribution on mud volcanoes. Deep Sea Res I 51:1915–1936

Olu K, Rv C, Hourdez S, Carney SL et al (2007) Amphi-Atlantic cold-seep *Bathymodiolus* species complexes across the equatorial belt. Deep Sea Res I 54:1890–1911

Olu-Le Roy K, Caprais J-C, Fifis A, Fabri M-C et al (2007) Cold-seep assemblages on a giant pockmark off West Africa: spatial patterns and environmental control. Mar Ecol 28:115–130

Peckmann J, Walliser OH, Riegel W, Reitner J (1999) Signatures of hydrocarbon venting in a middle Devonian carbonate mound (Hollard Mound) at the Hamar Laghdad (Antiatlas, Morocco). Facies 40:291–296

Peckmann J, Gischler E, Oschmann W, Reitner J (2001) An Early Carboniferous seep community and hydrocarbon-derived carbonates from the Harz Mountains, Germany. Geology 29:271–274

Peckmann J, Goedert JL, Thiel V, Michaelis W et al (2002) A comprehensive approach to the study of methane-seep deposits from the Lincoln Creek Formation, western Washington State, USA. Sedimentology 49:855–873

Peckmann J, Birgel D, Kiel S (2009) Molecular fossils reveal fluid composition and flow intensity at a Cretaceous seep. Geology 37:847–850

Pojeta J (1988) The origin and Paleozoic diversification of solemyoid pelecypods. New Mex Bur Mines Miner Resour Mem 44:201–271

Pyenson ND, Haasl DM (2007) Miocene whale-fall from California demonstrates that cetacean size did not determine the evolution of modern whale-fall communities. Biol Lett 3:709–711

Sahling H, Rickert D, Lee RW, Linke P et al (2002) Macrofaunal community structure and sulfide flux at gas hydrate deposits from the Cascadia convergent margin, NE Pacific. Mar Ecol Prog Ser 231:121–138

Saito H, Okutani T (1990) Two new chitons (Mollusca: Polyplacophora) from a hydrothermal vent site of the Iheya Small Ridge, Okinawa Trough, east China Sea. Venus 49:165–179

Salas C, Woodside J (2002) *Lucinoma kazani* n. sp (Mollusca: Bivalvia): evidence of a living benthic community associated with a cold seep in the Eastern Mediterranean Sea. Deep Sea Res I 49:991–1005

Samadi S, Quéméré E, Lorion J, Tillier A et al (2007) Molecular phylogeny in mytilids supports the wooden steps to deep-sea vents hypothesis. CR Biol 330:446–456

Sasaki T, Okutani T, Fujikura K (2003) New taxa and new records of patelliform gastropods associated with chemoautosynthesis-based communities in Japanese waters. Veliger 46:189–210

Saul LR, Squires RL, Goedert JL (1996) A new genus of cryptic lucinid? bivalve from Eocene cold seeps and turbidite-influenced mudstone, western Washington. J Paleontol 70:788–794

Schein E (2006) A new deep-sea pectinid bivalve from thermal vents of Manus back-arc Basin (south-western Pacific), *Sinepecten segonzaci* n. gen., n. sp. (Pectinoidea: Pectinidae), and its relationships with the genera *Bathypecten* and *Catillopecten*. Zootaxa 1135:1–27

Schein-Fatton E (1985) Découverte sur la ride du Pacifique oriental à 13° N d'un Pectinidae (Bivalvia, Pteriomorpha) d'affinités paléozoiques. CR Acad Sci Paris 301:491–496

Shank TM, Black MB, Halanych KM, Lutz RA et al (1999) Miocene radiation of deep-sea hydrothermal vent shrimp (Caridea: Bresiliidae): evidence from mitochondrial cytochrome oxidase subunit I. Mol Phylogenet Evol 13:244–254

Smith CR (2007) Bigger is better: the role of whales as detritus in marine ecosystems. In: Estes JA, DeMaster DP, Doak DF, Williams TM, Brownel RL (eds) Whales, whaling and marine ecosystems. University of California Press, Berkeley, CA, pp 286–300

Smith CR, Baco AR (2003) Ecology of whale falls at the deep-sea floor. Oceanogr Mar Biol 41:311–354

Smith CR, Kukert H, Wheatcroft RA, Jumars PA et al (1989) Vent fauna on whale remains. Nature 341:27–28

Speden IG (1970) The type Fox Hills Formation, Cretaceous (Maestrichtian), South Dakota. Part 2. Systematics of the Bivalvia. Peabody Mus Nat Hist Bull 33:1–222

Squires RL (1995) First fossil species of the chemosynthetic-community gastropod *Provanna*: localized cold-seep limestones in Upper Eocene and Oligocene rocks, Washington. Veliger 38:30–36

Squires RL, Goedert JL (1991) New Late Eocene Mollusks from localized limestone deposits formed by subduction-related methane seeps, southwestern Washington. J Paleontol 65:412–416

Squires RL, Goedert JL (1996) A new species of *Thalassonerita*? (Gastropoda: Neritidae?) from a Middle Eocene cold-seep carbonate in the Humptulips Formation, Western Washington. Veliger 39:270–272

Squires RL, Gring MP (1996) Late Eocene chemosynthetic? bivalves from suspect cold seeps, Wagonwheel Mountain, central California. J Paleontol 70:63–73

Stanton TW (1895) Contributions to the Cretaceous paleontology of the Pacific coast: the fauna of the Knoxville beds. US Geol Surv Bull 133:1–132

Stewart RB (1930) Gabb's California Cretaceous and tertiary type lamellibranchs. Acad Nat Sci Phila (Special Publication) 3:1–314

Tada R (1994) Paleoceanographic evolution of the Japan Sea. Palaeogeogr Palaeocl 108:487–508

Taviani M (1994) The "calcari a *Lucina*" macrofauna reconsidered: deep-sea faunal oases from Miocene-age cold vents in the Romagna Apennine, Italy. Geo Mar Lett 14:185–191

Taylor JD, Glover EA (2006) Lucinidae (Bivalvia) – the most diverse group of chemosymbiotic molluscs. Zool J Linn Soc 148:421–438

Taylor JD, Williams ST, Glover EA (2007) Evolutionary relationships of the bivalve family Thyasiridae (Mollusca: Bivalvia), monophyly and superfamily status. J Mar Biol Assoc UK 87:565–574

Tunnicliffe V, Fowler MR (1996) Influence of sea-floor spreading on the global hydrothermal vent fauna. Nature 379:531–533

Warén A, Bouchet P (2001) Gastropoda and Monoplacophora from hydrothermal vents and seeps; new taxa and records. Veliger 44:116–231

Warén A, Bouchet P (2009) New gastropods from deep-sea hydrocarbon seeps off West Africa. Deep Sea Res I 56:2326–2349

Warén A, Carrozza F (1990) *Idas ghisottii* sp.n., a new mytilid bivalve associated with sunken wood in the Mediterranean. B Malacol 26:19–24

Zaikov VV, Shadlun TN, Maslennikov VV, Bortnikov NC (1995) Yaman-Kasy sulfide deposit – ancient black smoker in the floor of the Uralian Paleoocean. Geol Rudnih Mestorogdeniy 37:511–529 (in Russian)

Chapter 9
Brachiopods from Ancient Hydrocarbon Seeps and Hydrothermal Vents

Michael R. Sandy

Contents

9.1 Introduction

Brachiopods are an enigmatic group of organisms. With a fossil record extending back for over 500 million years these marine invertebrates are well-known for their Palaeozoic peak in diversity and abundance (Fig. 9.1). Their lesser abundance in the post-Palaeozoic world after the Permo-Triassic extinctions has meant that they have frequently received scant mention or been neglected altogether in introductory marine invertebrate textbooks (Rudwick 1970). From the rhynchonelliform brachiopods several brachiopod orders survived into the post-Palaeozoic world (Figs. 9.1 and 9.2) and have undergone phases of radiation themselves, notably the rhynchonellids and the terebratulids (both the "short-looped" terebratulids and the "long-looped" terebratellids). These are the best-known living orders but the

M.R. Sandy (✉)
Department of Geology, University of Dayton, 300 College Park, Dayton, Ohio 45469-2364, USA
e-mail: michael.sandy@notes.udayton.edu

S. Kiel (ed.), *The Vent and Seep Biota*, Topics in Geobiology 33,
DOI 10.1007/978-90-481-9572-5_9, © Springer Science+Business Media B.V. 2010

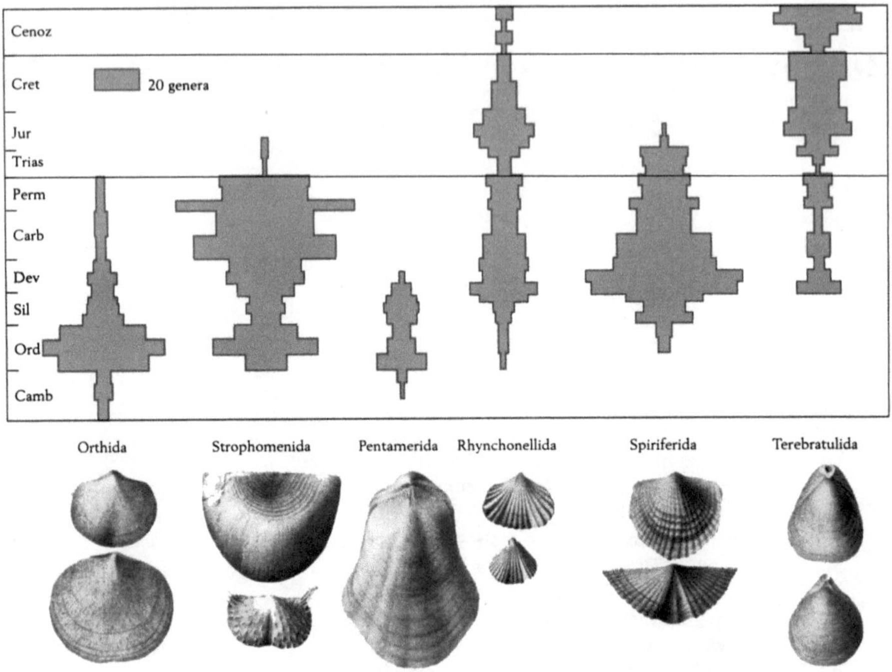

Fig. 9.1 Plot of genus-abundance of some important articulate brachiopod orders through time. Strophomenida plot includes Strophomenida and Productida; Spiriferida includes the spire-bearing Spiriferinida, Atrypida and Athyridida (see Fig. 9.2 for example) (From Stanley 2005)

diminutive thecideids are also extant. They are generally considered to be derived from the spiriferids (e.g. Carlson and Leighton 2001) and are distinctive not only for their small size and cemented lifestyle, but also as the only order considered to have a post-Palaeozoic origin (Fig. 9.2). Subsequent to restricted diversity during the Permo-Triassic extinctions the spire-bearing brachiopods (Athyridida and Spiriferinida) had periods of evolutionary radiation and diversification during the Triassic although neither survived past the Early Jurassic (Fig. 9.2). Amongst brachiopods associated with ancient chemosynthesis-dominated environments it is the rhynchonelliforms that are the most numerous in terms of both numbers of taxa and known localities. Of the "inarticulate" brachiopod subphyla both the Craniiformea and Linguliformea range from the Cambrian to the present (Fig. 9.2); it is only the latter that is currently known to have a representative at an ancient hydrothermal vent deposit (*Pyrodiscus* in Little et al. 1999). This low representation could be a reflection of the difficulty of studying and identifying fossils found at hydrothermal vents (Little et al. 1999, 2004).

The focus herein will be the fossil record of brachiopods from hydrocarbon seep environments where there is a growing, significant body of knowledge contributing to our understanding of brachiopods. Since hydrothermal vents and hydrocarbon seeps are generally considered approximate "end members" along something of a variable environmental continuum (e.g. Gaillard et al. 1992, see discussion in

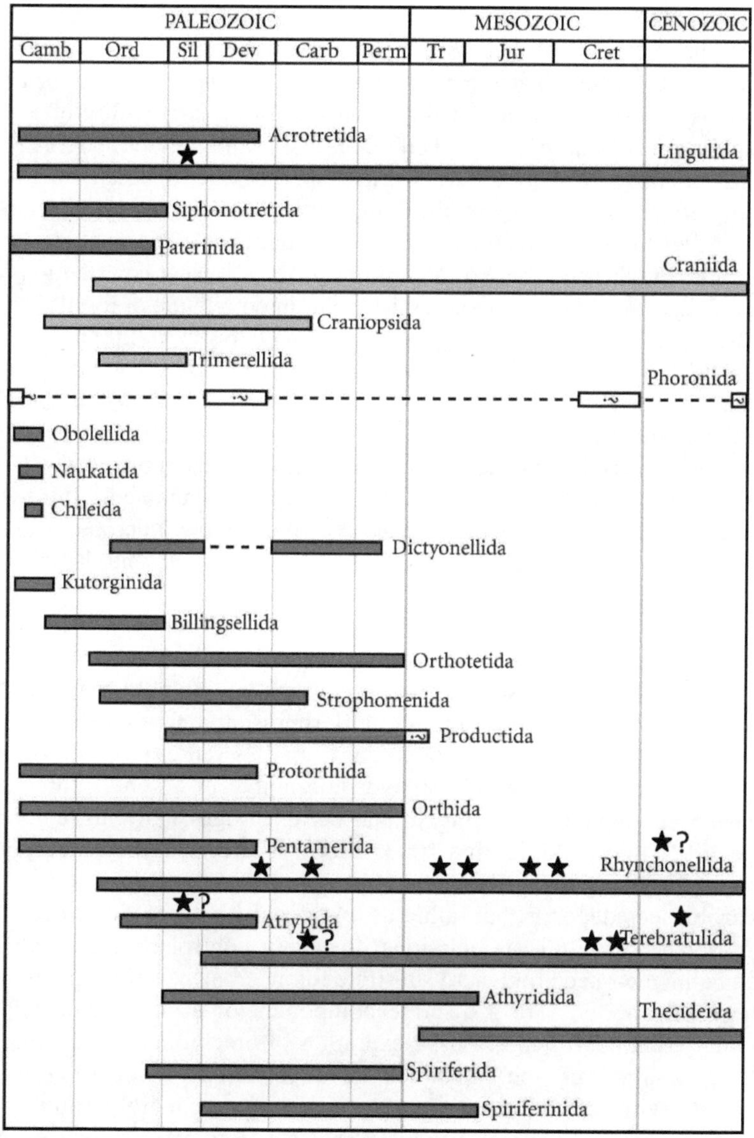

Fig. 9.2 Classification and stratigraphic distribution of brachiopod orders in the Phylum Brachiopoda showing occurrence of brachiopods with ancient chemosynthesis-dominated environments. Key: ★ = brachiopod occurrence or occurrences associated with ancient chemosynthesis-based deposits referred to in Table 9.1 and discussed in text; ? = doubtful record (discussed in text). The plotted points may represent just one locality with one taxon represented or a few contemporaneous taxa from single or multiple localities (e.g. compare with greater stratigraphic detail of Fig. 9.5) (Adapted from Carlson in Benton and Harper 2009)

Campbell 2006, pp. 363, 383–384) I will consider the association of brachiopods with both ancient seeps and vents in this account. There are references to brachiopods from modern chemosynthesis-dominated environments (e.g., Barry et al. 1996;

Zezina 2003, 2008) but these occurrences are marginal to chemosynthesis-dominated environments and it has not been demonstrated that they are specifically members of a chemosynthesis-dominated environment (obligate) as opposed to located near to a chemosynthesis-dominated environment due to the availability of a suitable substrate for attachment (e.g. Campbell 2006) and availability of nutrients.

As we continue to explore the world's shelf-seas and oceans we are still continuing to learn more about the biogeographic distributions, diversity and ecology of living brachiopods (e.g. reviews of Richardson 1997a, b; Emig 1997a, b; Zezina 2003, 2008; paper by Kowalewski et al. 2002). New contributions to our understanding of living brachiopods assist in our interpretation of fossil representatives. There are five extant brachiopod orders (Williams et al. 2000; Carlson and Leighton 2001) of which at present three (Rhynchonellida, Terebratulida, Lingulida) are considered to have fossil representatives associated with ancient chemosynthesis-based communities (Table 9.1, Fig. 9.2). The Atrypida is the only extinct order that has been reported from ancient chemosynthesis-based communities (Barbieri et al. 2004). However, this latter record has not been substantiated and the isotopic data indicate methanogenesis, the formation of methane by microbes, rather than methane oxidation at a hydrocarbon seep (e.g. Buggisch and Krumm 2005; Himmler et al. 2008).

Representatives of the Phylum Brachiopoda were discovered and known to science from the fossil record before they were described as living organisms (Muir-Wood 1955). The association of brachiopods with cold-seep and hydrothermal vent communities could potentially represent a parallel story. As yet it seems that brachiopods have not been confirmed as representatives of modern hydrocarbon seep ("cold seep") or hydrothermal vent ("vent") communities. They have been found in the vicinity of mid-oceanic ridges and vents (e.g. Gregory and Lee 1995; Lee 2000; Zezina 2003, 2008). However, these brachiopods do not appear to be occurring in mass-abundances as recorded for some ancient hydrocarbon seep deposits, but rather as low-density dispersed occurrences that may in fact be fortuitous associations; the main control on the brachiopods' location being their need for a hard substrate for attachment rather than indicating that these brachiopods are a discrete component of a chemosynthesis-based community (Campbell 2006). Fossil examples of brachiopods in the interstices of pillow basalts may be examples of a fortuitous association of fossil brachiopods with volcanic activity, the brachiopods merely looking for a suitable substrate (e.g. the athyridide *Pexidella* from the Triassic of Alaska, Sandy et al. 2001).

If we consider the number of brachiopod taxa at the generic level that are considered to be associated with ancient chemosynthesis-based communities, or considered likely candidates for further investigation due to a number of indicators ("seep-" or "vent-suspect") the number is not large, approximately a dozen, although growing (Campbell and Bottjer 1995a; Sandy 1995a; Kaim et al. 2010, Table 9.1 herein). There are surely other brachiopod taxa that are identifiable using the schema for identifying chemosynthesis-based communities in the fossil record that await identification, investigation and discussion. Callender and Powell (1992)

Table 9.1 Taxonomic rank, environment, community placement and geologic range of brachiopod genera that are associated with or suspected to be associated with chemosynthetic environments

Taxonomic rank	Env	Loc	Geological range of taxon	Locations of HCS or HTV
Phylum Brachiopoda			Lower Cambrian – Holocene	
Subphylum Rhynchonelliformea			Lower Cambrian (Atdabanian) – Holocene	
Order Rhynchonellida			Lower Ordovician (Llanvirn) – Holocene	
Superfamily Dimerelloidea			Upper Devonian (Famennian) – Holocene)	
Family Dimerellidae			U Triassic (Norian) – Upper Jurassic (Tithonian)	
Subfamily Dimerellinae			Upper Triassic (Norian)	
Dimerella	UNK		Upper Triassic (Norian)	
Subfamily Rhynchonellininae			Upper Triassic (Norian) – Upper Jurassic (Tithonian)	
**Rhynchonellina*	HCS	CC	Upper Triassic (Norian) – Lower Jurassic (Pliensbachian)	Alpine localities?
**Sulcirostra*	HCS	CC/PP	Upper Triassic (Norian) – Lower Jurassic (Pliensbachian)	Oregon, USA; Alpine localities?
Cooperrhynchia	HCS	CC	Upper Jurassic (Tithonian)	California, USA
Subfamily Peregrinelloideinae			Lower Jurassic (Pliensbachian-Sinemurian)	
Peregrinelloidea	HCS?	CC?	Lower Jurassic (Pliensbachian)	Russia?
Anarhynchia	HCS HTV	CC/PP	Lower Jurassic (Pliensbachian-Sinemurian)	California, Oregon, USA
Family Peregrinellidae			U Devonian (Famennian) – L Cretaceous (Hauterivian)	
Subfamily Peregrinellinae			U Triassic (?Norian, Rhaetian) – L Cret (Hauterivian)	
Peregrinella	HCS	CC	Lower Cretaceous (Berriasian – Hauterivian)	Europe, Tibet, USA
Peregrinella (*Peregrinella*)	HCS	CC	Lower Cretaceous (Berriasian – Hauterivian)	Europe, Tibet, USA
Peregrinella (*Peregrinellina*)	HCS	CC	Lower Cretaceous (Valanginian – Hauterivian)	Tibet
?*Carapezzia*	HCS?	CC?	Upper Triassic (?Norian, Rhaetian) – Lower Jurassic	Alpine localities?
?Family Halorellidae			Upper Triassic (?upper Carnian, Norian)	
***Halorella*	HCS	CC?	Upper Triassic (?upper Carnian, Norian)	Oregon; Alpine localities?
***Halorelloidea*	HCS?	CC?	Upper Triassic (Norian)	Alpine localities?

(continued)

Table 9.1 (continued)

Taxonomic rank	Env	Loc	Geological range of taxon	Locations of HCS or HTV
Subfamily Dzieduszyckiinae			U Devonian (Famennian– L Carboniferous (upper Visean)	
Dzieduszyckia	HCS HTV?	CC	Upper Devonian (Famennian)	Morocco, Nevada, USA; other localities?
Ibergirhynchia	HCS	CC	Lower Carboniferous (upper Visean)	Germany
??Family Cryptoporidae			Upper Cretaceous (upper Campanian) – Holocene	
Cryptopora	–		Paleogene (lower Danian) – Neogene (Pliocene)	
Aulites	–		Paleogene (upper Oligocene) – Holocene	
?*Cryptoporella*	–		U Cretaceous (upper Campanian – lower Maastrichtian)	
Family Basiliolidadae			Upper Triassic (Carnian) – Holocene	
Subfamily Basiliolinae			Lower Jurassic (?Sinemurian, Pliensbachian) – Holocene	
Probolarina	?HCS?	CC	Upper Cretaceous – Paleogene (Eocene)	Mexico (Paleocene)
Subfamily Lacunosellinae			M Jurassic (Aalenian) – Lower Cretaceous (Barremian)	
Lacunosella			M Jurassic (Bathonian) – Lower Cretaceous (Barremian)	Hungary (Valanginian); others?
Lacunosella (*Lacunosella*)	HTV?		U Jurassic (Oxfordian) – Lower Cretaceous (Barremian)	
Lacunosella (*Dichotomosella*)			M Jurassic (Bathonian) – Upper Jurassic (Kimmeridgian)	
Order Atrypida			Ordovician (Llandeilo) – Upper Devonian (Frasnian)	
Suborder Lissatrypoidea			Ordovician (Llandeilo) – Upper Devonian (Frasnian)	
Superfamily Lissatrypoidea			Ordovician (Caradoc) – Middle Devonian (Eifelian)	
Family Septatrypidae			Ordovician (Caradoc)–L Dev (Emsian), M Dev (?Eifelian)	
Subfamily Septatrypinae			Silurian (Llandovery) – L Devonian (Frasnian), M Dev (?Eifelian)	
Septatrypa		CC	Silurian (Aeronian) – Lower Devonian (Pragian)	Morocco
****Septatrypa* (*Herciniscа*)	?HCS?		Silurian (middle Wenlock)–Lower Devonian (Lochkovian)	
****Septatrypa* (*Septatrypa*)			Silurian (Aeronian) – Lower Devonian (Pragian)	

Order Terebraulida			Lower Devonian – Holocene	
Suborder Terebratellidina			Upper Triassic – Holocene	
Family Zeillerioidea			Lower Triassic (Induan) – Holocene	
Subfamily Vectellinae			Middle Triassic – Lower Cretaceous (Albian)	
Modestella	HCS	PP	Lower Cretaceous (Albian)	NWT, Canadian Arctic (probably Albian)
Suborder Terebratulidina			Lower Devonian – Holocene	
Superfamily Cancellothyridoidea			Lower Jurassic (Pliensbachian) – Holocene	
Family Childonophoridae			Lower Jurassic (Pliensbachian) – Holocene	
Subfamily Eucalathinae			Upper Cretaceous – Holocene	
Eucalathis	HCS	PP	Upper Cretaceous – Holocene	Japan (Campanian)
Superfamily Dielasmatidae			Carboniferous (Mississippian) – U Triassic, ? L Jurassic	
Family Beecheriidae			Carboniferous (L Mississippian) – Permian (Lopingian)	
Beecheria	HCS?	CC	Carboniferous (L Mississippian) – Permian (Lopingian)	Newfoundland, Canada (Mississippian)
Superfamily Terebratuloidea			?U Jurassic, Lower Cretaceous (Berriasian) – Holocene	
Family Terebratulidae			Paleogene (Paleocene) – Holocene	
Terebratulinae			Paleogene (Paleocene) – Holocene	
Liothyrella?	HCS	PP	Paleogene (Eocene) – Holocene	New Zealand (Miocene)
"*Terebratula*" spp.	HCS	PP	Miocene	New Zealand (Miocene)
Superfamily Dyscolioidea			Lower Jurassic – Holocene	
Family Nucleatidae			Lower Jurassic – Upper Cretaceous (Cenomanian)	
Nucleata	HTV?		?M Jur (Bathonian), U Jur (Oxfordian) – U Cret (Ceno)	Hungary (Valanginian); others?
Family Pygopidae			L Jurassic (Pliensbachian) – L Cretaceous (Barremian)	
Subfamily Pygopinae			U Jurassic (Kimmeridgian) – L Cretaceous (Barremian)	
Pygope	?HCS?	?	U Jurassic (Kimmeridgian) – L Cretaceous (Barremian)	Alpine localities?

(continued)

Table 9.1 (continued)

Taxonomic rank	Env	Loc	Geological range of taxon	Locations of HCS or HTV
Subphylum Linguiformea			Lower Cambrian (Tommotian) – Holocene	
Class Lingulata			Lower Cambrian (upper Atdabanian) – Holocene	
Order Lingulida			Lower Cambrian (upper Atdabanian) – Holocene	
Family Uncertain				
Pyrodiscus	HTV		Silurian	Russia

Taxonomic rank: taxononic ranks given for the brachiopod genera discussed herein. Orders are shown in bold; ? = uncertain placement of family or genus; use of asterisk *, **, ***, indicates morphological couplets identified among these genera. Classification mostly based on Williams et al. 2000, 2002, 2006, 2007, with a few minor modifications

Env: palaeoenvironment: HCS = hydrocarbon seep; HTV = hydrothermal vent; ? = seep or vent suspect, requires further investigation; ?HCS? = speculative assignment to hydrocarbon seep; UNK = unknown if associated with chemosynthesis-based environments

Loc: location and role that the brachiopod is thought to occupy within the chemosynthetic community. This is inferred specifically for hydrocarbon seeps and not hydrothermal vents and some taxa may have more than one inferred location (from different seeps e.g. *Anarhynchia*). CC = central core or core species, in a central location in the community, very abundant to dominant numerically in the community, considered comparable to an obligate species (e.g., Barry et al. 1996) although not considered chemosymbiotic; PP = peripheral position or peripheral species, typically most common at the margins or flanks of the community, may be high-density and abundant to low-density accumulations; considered comparable to non-obligate species (e.g., Barry et al. 1996)

Geological Range of Taxon: the range given for the taxonomic ranks in the Treatise of Invertebrate Paleontology (Williams et al. various dates). For genus it is the range of the taxon as given in the Treatise, not the specific age of the chemosynthetic community from which the brachiopods have been recorded. However in some cases where the genus range is restricted to a single stage this gives the age for the chemosynthetic environment by default. Ages of chemosynthetic communities are generally given in text and/or in the Locations column where helpful

Locations of HCS or HTV = location of hydrocarbon seep or hydrothermal vent community considered that three criteria needed to be met to consider a deposit as a possible ancient chemosynthesis-based deposit: low diversity, almost monospecific in nature and with abundant large individuals within faunally impoverished formations (although large-size is however not a prerequisite). The reality is that this has transformed into an interdisciplinary study when assessing any potential deposit that ideally includes stable isotope geochemistry, including the recognition of distinctive stable isotopic signatures of carbon, oxygen or sulphur in authigenic precipitates and the recognition of lipid biomarkers (e.g. Peckmann and Thiel 2004; Campbell 2006). At the genus-level one taxon may be represented by numerous sites that are confirmed or suspected to be of chemosynthetic origin. *Peregrinella* (Figs. 9.3a–g and 9.5a, Plates 18d, f, 19a, b, e, f) would be the example *par excellence* with over one dozen localities identified as hydrocarbon seeps or seep-suspect (Campbell and Bottjer 1995b) based on characteristics including the occurrence of this brachiopod typically represented by numerous (tens- to hundreds-of-thousands to millions) individual brachiopods of a single species in a limestone lens in a clastic-dominated sequence commonly in association with chemosynthetic-related molluscan taxa.

To date brachiopods have most commonly been recorded associated with ancient hydrocarbon seep deposits compared to ancient hydrothermal vents. The pervasive mineralization at hydrothermal vents has resulted in significant problems in identifying fossil taxa at such settings as original shell material is often absent and/or replaced (e.g., Little et al. 1999, 2004). However, pervasive mineralization, mineral overgrowths and replacement are also challenges when dealing with fossil material from ancient hydrocarbon seeps. Numerically the published record of fossil hydrocarbon seeps with brachiopods dominates those of published records from fossil hydrothermal vent faunas by approximately an order of magnitude. This numerical dominance reflects the general situation when comparing the number of fossil-bearing hydrocarbon seep deposits compared to fossil-bearing hydrothermal vent deposits (Campbell 2006). It is certainly likely that the number of faunas from ancient chemosynthesis-based communities with brachiopods associated will increase with some rapidity over the next decade or so as new faunas and taxa are discovered, confirmed and described.

9.2 A Non-uniformitarian Approach

The association between extinct taxa of brachiopods with hydrothermal vent and hydrocarbon seep deposits has been recognized during the past three decades (e.g. Macsotay 1980; Lemoine et al. 1982; Campbell et al. 1993; Little et al. 1998). Chemosynthesis-dominated environments have been in existence for hundreds of millions and even billions of years (at least three billion is one estimate based on the oldest finds to date, references in Campbell 2006) and the growing literature on their fossil deposits indicates that certain eukaryotic life-forms flourished at times at some of these sites during the Phanerozoic. A variety of invertebrate

Fig. 9.3 Brachiopods from Mesozoic and Palaeozoic hydrocarbon seeps. (**a**) *Peregrinella multicarinata* (Lamarck) Neocomian, Châtillon-en-Dois, France, USNM 75595, width of specimen 60.8 mm. (**b**) *Peregrinella multicarinata* (Lamarck) Acc. 319165, Châtillon-en-Dois (Drôme) Collection Thomel, width = 59.3 mm. (**c–f**) *Peregrinella whitneyi* (Gabb), Cretaceous, Wilbur Springs, California. (**c**) Width = 56.0 mm, (**d**) width = 54.7 mm, (**e**) USNM 23264, width = 20.7 mm, (**f**) Width = 18.4 mm. (**g**) *Peregrinella chisana* Sandy and Blodgett, Cretaceous, Alaska, length = 29.2. (**h**) *Halorella*, Triassic, Morgan Mountain, Grant Co., Oregon, width = 56.8 mm. (**k**) *Halorella*, Triassic, Oregon, LSJU 3586, width = 53.3 mm. (**i, j**) *Anarhynchia*, Jurassic, Seneca, Oregon, showing crura encrsted with minerals; encrsted crura arch just over 8 mm towards the ventral valve, width of specimen approximately 3 cm. Collected Jim and Gail Goedert. (**l**) *Dzieduszyckia* from Morocco, width of specimen estimated at 8 cm (Specimen photographed during field work, photograph courtesy of Jörn Peckmann)

taxa have clearly adapted to and been successful in chemosynthesis-dominated environments through the Phanerozoic (e.g. Little et al. 1998, 1999; McArthur and Tunnicliffe 1998; Little and Vrijenhoek 2003; Campbell 2006) and the brachiopods are one such group. The fact that they are more numerous in the fossil record, at times forming shell-bed accumulations, with no currently-confirmed modern analogue in such environments is of course intriguing. The fossil hydrocarbon seeps dominated by brachiopods (Table 9.1) (with other associated taxa) became the focus of a paradigm shift with no modern analogue (Bottjer et al. 1995). For example, in the Great Valley Group of California, the *Peregrinella* and *Cooperrhynchia*-bearing carbonates were initially considered by Anderson (1945) to be shallow-water "reefs" in deltaic settings, but are now considered seep carbonates in deeper-water settings of the Great Valley Group (Campbell et al. 1993, 2002; Campbell and Bottjer 1993, 1995b; Sandy and Campbell 1994; Bottjer et al. 1995). The *Peregrinella*-bearing limestone blocks famous in southern and Alpine Europe were considered to be transported olistolith blocks (Ager 1965). Some of these may certainly genuinely be olistoliths, but a hydrocarbon seep carbonate origin may still prove to be applicable (Campbell and Bottjer 1995b); however a hydrocarbon seep limestone in a slope setting is going to look very much like an olistolith (and vice versa). Obviously careful study of field relations may (or may not) be able to clarify origins in complex tectonic regimes. Isotope geochemistry is a valuable tool in attempting to answer these questions. However, certain instances of brachiopod fossil occurrences that are suspected to be from ancient chemosynthesis-based deposits may prove difficult to confirm from a geochemical point of view as thermal maturity and diagenetic alteration may alter original oxygen isotope signatures for the deposit. For example, with the type locality for the Jurassic rhynchonellid brachiopod species *Anarhynchia gabbi* from Bedford Canyon (Fig. 9.4a–c, Plates 18a, c, 19g), Santa Ana Mountains, California (KA Campbell, personal communication, 2000); similarly with a Jurassic *Sulcirostra* locality (Fig. 9.4e–h, Plates 18c, 19g), Seneca, Oregon (J Peckmann, personal communication, 2006). A few occurrences of the Triassic rhynchonellids *Halorella* and its sister taxon *Halorelloidea* have been specifically recorded as from neptunean dykes in Alpine Europe (e.g. Ager 1965). These certainly merit further investigation as possible fault-bounded settings from which hydrocarbons could emanate and/or brines be retained and lead to the development of isolated chemosynthesis-dominated environments. Mass occurrences of *Halorella* from the Triassic of Oregon (Figs. 9.3h, k and 9.5b, Plate 19d) are currently under investigation (Peckmann, Goedert, Kiel, Sandy) and may confirm the chemosynthesis-based origin of this taxon as speculated by Sandy (2001).

It seems almost a prerequisite that a locality with brachiopods that is considered a possible candidate as a chemosynthesis-based community is difficult to give a geologic age to (e.g. Biernat's discussion of occurrences of *Peregrinella*, 1957, pp. 25–27). This reflects the fact that ancient hydrocarbon seep and hydrothermal vent deposits have meter-scale (up to decimetre lateral extent) dimensions and are readily recognized as they are "out of place" (as a fossil-rich carbonate, for example) with respect to their embedding stratigraphy (e.g. Callender and Powell 1992; Campbell et al. 2002). The brachiopods themselves within the deposit may not have

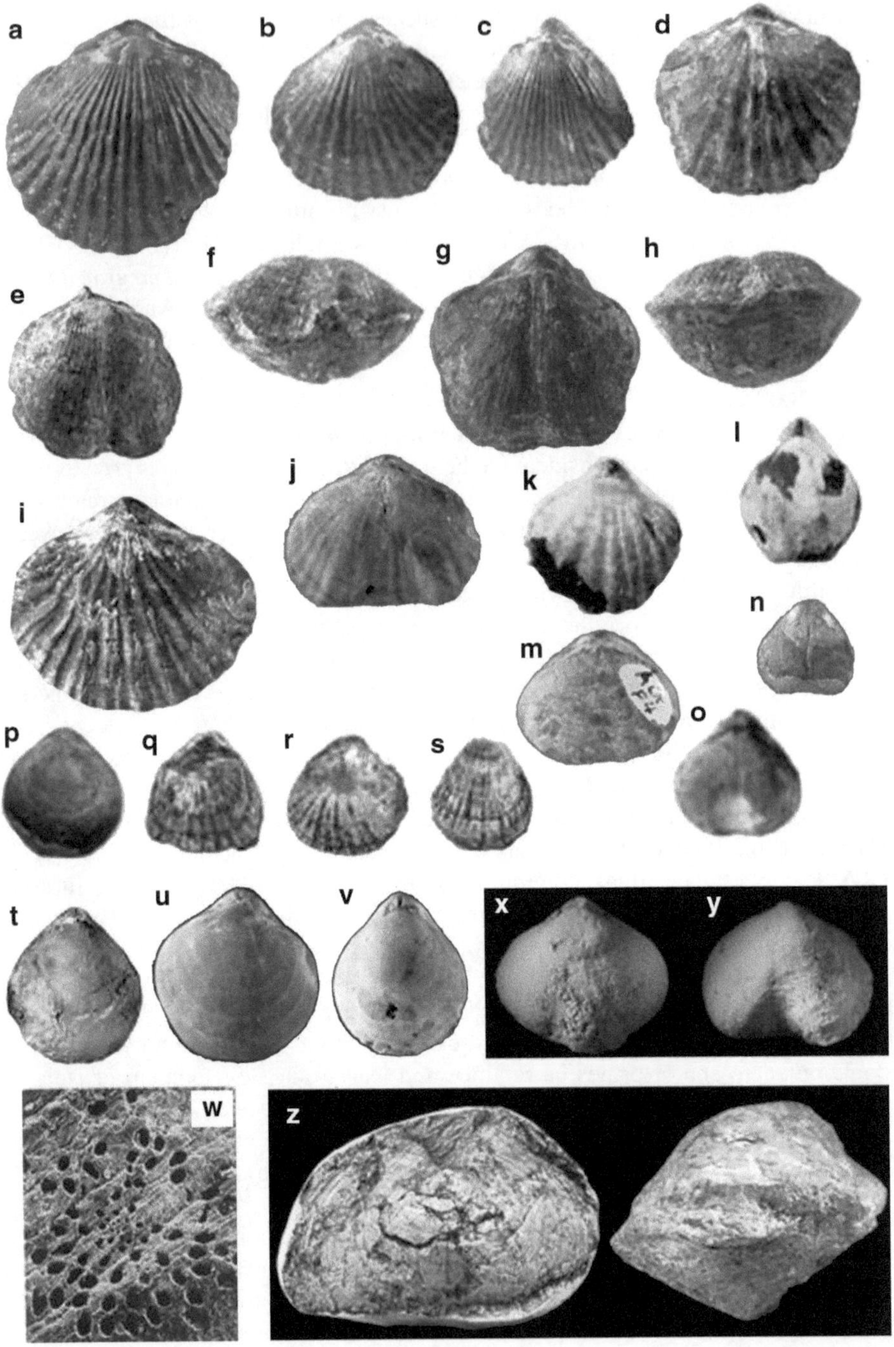

Fig. 9.4 Brachiopods primarily from Mesozoic and Palaeozoic hydrocarbon seeps and a Palaeozoic hydrothermal vent and a brachiopod-produced trace fossil. (**a–c**) *Anarhynchia gabbi* Ager, Jurassic, Bedford Canyon, California. (**a**) Length = 35.8 mm, (**b**) length 26.0 mm, (**c**) length = 30.0 mm. (**d**) *Anarhynchia* sp., Jurassic, Seneca, Oregon, LSJU 3586. (**e, f**) *Sulcirostra* sp.,

great stratigraphic utility and therefore it is other taxa likely from surrounding matrix that are typically relied upon to provide a geological age (e.g. Little et al. 2004 for a Jurassic hydrothermal vent deposit). Of course, if the carbonate is interpreted as a transported block (olistolite) rather than an *in situ* development, then age relationships become more difficult to interpret (e.g. Sulser and Furrer 2008). The age relationships in zones of Alpine-type tectonics can be difficult to determine at the best of times.

9.3 History of Recognition of Brachiopods Associated with Ancient Chemosynthesis-Based Communities

The discovery of hydrothermal vents in 1977 (Lonsdale 1977; Corliss et al. 1979) and hydrothermal seep communities a few years later (Paull et al. 1984) were major geological and biological discoveries of the latter part of the twentieth century. Subsequent to their discovery it was not long before interpretations in the geological record were influenced by these discoveries.

Macsotay (1980) considered that the occurrence of *Peregrinella* with associated molluscs in the "*Peregrinella* Beds" (Fig. 9.3a, b, Plate 18d, 19a) from the Early Cretaceous of southern France may be the product of hydrothermal activity; this appears to be the first direct link between fossil brachiopods and chemosynthesis-dominated environments although such unusual occurrences had certainly attracted much attention and speculation beforehand (e.g. Ager 1965; Ager et al. 1976). The association of the Devonian rhynchonellid brachiopod *Dzieduszyckia* with extensive barite mineralization (Dubé 1988; Poole 1988) was interpreted as a hydrothermal vent deposit. However, these "hydrothermal deposits" have more recently been reinterpreted as hydrocarbon seeps (Torres et al. 2003; Campbell 2006). Other discoveries of chemosynthesis-based communities (with brachiopods associated) followed such as hydrocarbon seep deposits in the Cretaceous of the Canadian Arctic (Beauchamp et al. 1989) with the terebratulid *Modestella* (Sandy 1990; Fig. 9.4p)

Fig. 9.4 (continued) Jurassic, Seneca, Oregon, width = 26.2 mm. (**g, h**) *Sulcirostra* sp., Jurassic, Seneca, Oregon, width = 32.3 mm. (**i**) *Anarhynchia* sp., Jurassic, Seneca, Oregon width = 36.2 mm. (**j**) *Ibergirhynchia contraria*, Early Carboniferous (latest Visean), Iberg Reef, Germany, width = 19.7 mm. Specimen collected Eberhard Gischler. (**k, l**) *Probolarina neoleonensis*, Potrerillos Formation, Palaeocene, Mexico; doubtfully of hydrocarbon seep affinities. (**m**) *Cooperrhynchia*, Knoxville Formation, Jurassic, St. Helena Quadrangle, Napa County, California, UCMP A-4312, width = 21.4 mm (**n**) *Cooperrhynchia schucherti* (Stanton), Tithonian, Jurassic, Paskenta, California, collected Kathy Campbell, W = 15.1 mm. (**o**) *Cooperrhynchia schucherti* (Stanton), Tithonian, Jurassic, Paskenta, California. (**p**) *Modestella*, Cretaceous, Canada. (**q–s**) *Eucalathis methanophila* Bitner, Late Cretaceous Omagari site, Japan. (**t**) Terebratulid brachiopod from the Miocene, New Zealand. Photograph courtesy of Kris Saether. (**u**) *Beecheria*, Carboniferous, Newfoundland, length = 20.7 mm. (**v**) *Beecheria*, Carboniferous, Newfoundland, length =20.3 mm. (**w**) *Podichnus* Jurassic, France. (**x, y**) *Septatrypa*, Silurian, Morocco; doubtfully of hydrocarbon seep affinities. From Barbieri et al. (2004). (**z**) *Pyrodiscus*, Silurian, Russia; hydrothermal vent (Little et al. 1999) (Photograph courtesy of Cris Little)

Fig. 9.5 Hand specimens of brachiopods from hydrocarbon seep deposits. (**a**) *Peregrinella*-rich block of limestone, Romania, Photograph courtesy I Lazar. (**b**) *Halorella*-rich block of limestone, Triassic, Oregon, large specimen 39 mm wide. Collected Jim and Gail Goedert. (**c, d**) *Anarhynchia*-rich limestone blocks, Jurassic, Bedford Canyon, California. (**e**) *Anarhynchia*-rich sediment sample, Jurassic, Seneca, Oregon, large specimen 44 mm wide. Collected J and G Goedert.

and hydrothermal vent communities in the Carboniferous of Newfoundland, Canada (Von Bitter et al. 1990, 1992) with another terebratulid *Beecheria* associated (Fig. 9.4u, v, Plate 19n, o). Whether this occurrence of *Beecheria* was part of a chemosynthesis-dominated environment has been questioned (Dix and Edwards 1996; Campbell 2006).

In terms of understanding brachiopods and highlighting their association with "problematic" environments in the fossil record the work of Campbell and co-workers is important (Campbell et al. 1993; Campbell and Bottjer 1993) on the Jurassic and Cretaceous limestones of the Great Valley Group, California. Jurassic and Cretaceous carbonates yielded a variety of molluscan taxa but brachiopods were also present and abundant at certain localities (Campbell et al. 1993). The brachiopods were first described by Gabb (1864) and Stanton (1895) and two rhynchonellid brachiopod taxa were identified, *"Rhynchonella" whitneyi* (Fig. 9.3c–f, Plate 19b, e, f) and "*Rhynchonella*" *schucherti* (Figs. 9.4m–o, Plates 19i, j), the former a species referred to *Peregrinella*, the latter placed in the genus *Cooperrhynchia* (Sandy and Campbell 1994). The papers of Campbell et al. (1993) and Campbell and Bottjer (1993, 1995b) were significant because they tied together the diverse strands of the palaeoecology and palaeobiogeography of *Peregrinella*. *Peregrinella* was (and still is) an enigmatic brachiopod (Fig. 9.3a–g, Plates 18d, f; 19a, b, e, f), being the largest of the Mesozoic brachiopods and also highly disjunct in its distribution. The distribution of *Peregrinella* and its palaeoecology has been much discussed through the years (e.g. Ager 1967, 1968, 1973, 1986; Ager and Sun 1989; Biernat 1957; Bujtor 2007; Campbell et al. 1993; Campbell and Bottjer 1995b; Kiel and Peckmann 2008; Middlemiss 1973; Owen 1973; Posenato and Morsilli 1999; Ruggiero 1997; Sandy 1991; Sandy and Campbell 1994; Sun 1986; Thieuloy 1972). To explain the occurrence of forms such as *Peregrinella* and its occurrence in limestones set among "deeper water" sequences Thieuloy (1972) and Ager et al. (1976) considered shoals in deeper water basins could explain this distribution. Ager (1965) had previously put forward the suggestion that *Peregrinella* and some other brachiopods with similar occurrences (such as *Dzieuduszykia*, *Halorella*, *Carapezzia*) occupied rocky, shallow-water coastlines and were only preserved by transport downslope. However, the new interpretation of *Peregrinella* and *Cooperrhynchia* as members of chemosynthesis-based communities (Campbell et al. 1993; Campbell and Bottjer 1993, 1995b, Sandy and Campbell 1994), supported by geological evidence from palaeontology, sedimentology, stratigraphy and stable isotope geochemistry seemed to make perfect sense as it explained the palaeobiogeographical and palaeoecological conundrums posed by *Peregrinella* and other "problematic" brachiopods. Of course each case must be assessed on the basis of its own merits

Fig. 9.5 (continued) (**f**) *Probolarina*-rich limestone, Palaeocene, Mexico, large specimen in top left corner 15 mm wide. Collected Francisco Vega. (**g**) Cut and polished slab containing *Dzieduszyckia* showing geopetal cement infills. Sidi Amar locality, Morocco of Peckmann et al. 2007a (Photograph courtesy of Jörn Peckmann) and supporting evidence. Some suspected occurrences may just end up having to be "suspect" until additional evidence, if available, can confirm or deny a chemosynthesis-based community association.

The first (and to date only) occurrence of a lingulid brachiopod from a hydrothermal vent deposit was described from a Silurian-aged locality in the Ural Mountains, Russia (Little et al. 1999; Fig. 9.4z herein). This brachiopod belongs to the Subphylum Linguliformea and is therefore distinct from any other records to date as the others all belong to the articulate brachiopods (Subphylum Rhynchonelliformea; Fig. 9.2 herein). Little et al. (1999, 2004) also reported on the occurrence of the Jurassic rhynchonellid brachiopod *Anarhynchia* from a hydrothermal vent deposit in California. *Anarhynchia* was first described (Ager 1968) from an isolated carbonate in the Great Valley Group of Bedford Canyon, California (Silberling et al. 1961), which is considered a likely hydrocarbon seep limestone (Stefanoff and Sandy 1998; Sandy 2001; Sandy and Campbell 2003). This form was previously been named "*Halorella*" (Cooper in Silberling et al. 1961) and was therefore thought to indicate a Late Triassic age for the unit here. However an Early Jurassic age seems most reasonable based on ammonites (Silberling et al. 1961) and age-associations for *Anarhynchia* recorded from the Figueroa hydrothermal vent community of the Franciscan Complex, California (Pliensbachian based on radiolarian biostratigraphy, Little et al. 2004). It therefore appears that *Anarhynchia* could be the first brachiopod to be recorded from both hydrocarbon seep and hydrothermal vent deposits. *Dzieduszyckia* from the Palaeozoic may have the same distinction but as previously noted the barite deposits from Nevada and Mexico with which it is associated have been reinterpreted as hydrocarbon seep deposits (Torres et al. 2003; Campbell 2006).

There are few records of brachiopods from chemosynthesis-dominated environments from the Cenozoic (Table 9.1). However, the number of records is sure to increase. One interesting occurrence from the Palaeocene of Mexico is represented by a carbonate lens packed with thousands of representatives of the rhynchonellid brachiopod *Probolarina* (Klosterman et al. 2007; Figs. 9.4k, 1 and 9.5f herein). These lenses developed along the flank of a rising salt diapir and there is a strong possibility that hydrocarbon seeps along the flank of this diapir could have given rise to chemosynthesis-based communities dominated, in this case, by brachiopods. The abundance of *Probolarina* is similar to the abundances reached by other rhynchonellids in the Palaeozoic and Mesozoic (e.g. *Dzieduszyckia*, *Sulcirostra*, *Peregrinella*). Geochemical analysis of these carbonates has not been able to confirm a chemosynthesis-based origin for this deposit and their genesis as associated with a chemosynthesis-dominated environment has to remain in doubt (S Kiel, J Peckmann, F Vega, personal communications, 2009).

Several terebratulid brachiopods have been identified from ancient chemosynthesis-dominated environments (Table 9.1) and these range through the Late Mesozoic (Cretaceous) – Cenozoic (Miocene) (to Recent?) and are therefore very important in understanding the role of brachiopods in chemosynthesis-based settings especially as rhynchonellids currently dominate the records of brachiopods from hydrocarbon seeps from the Late Devonian – Early Cretaceous.

These terebratulids include:

- *Modestella* identified from hydrocarbon seep carbonates from the Early Cretaceous (probably Albian) of the Canadian Arctic associated with a hydrocarbon province (Beauchamp et al. 1989; Beauchamp and Savard 1992; Sandy 1990). This is the first known occurrence of a brachiopod associated with a chemosynthesis-dominated environment post-*Peregrinella*; it seems that it could be significant that it should be a terebratulid suggesting perhaps a "new order" (sic) had been established among the brachiopods now found in chemosynthesis-based settings
- *Nucleata* from sediments considered to be possibly from a hydrothermal vent from the Early Cretaceous of Hungary (Butjor 2007) in which the average size of specimens and of the associated rhynchonellid *Lacunosella* was larger than in comparable "normal" marine shelf-environments. This would make this form contemporaneous with some occurrences of *Peregrinella.*
- A mass-occurrence of *Lacunosella* in a ca. 1.6 m thick bed from the Late Jurassic of Romania that is currently under investigation may prove to be a hydrocarbon seep-carbonate (Lazar et al. 2009a, b). If so, it would signify very interesting Jurassic-Cretaceous palaeoecological associations for *Lacunosella.*
- Small terebratulids that can be considered micromorphs (typically <6 mm in length) from hydrocarbon seep carbonates from the Late Cretaceous of Japan (Hikida et al. 2003; Kaim et al. 2010) referred to *Eucalathis methanophila* (Fig. 9.4q–s herein)
- And *Liothyrella*? from the Miocene of New Zealand (Campbell et al. 2008)

Perhaps the easiest way for brachiopods to colonize chemosynthesis-dominated environments is as "settler-peripherals" just as modern forms that are found on the sea floor near vents and oceanic ridges, and were subsequently able to colonize closer to, and then as part of the "core" of the community (as with for example, *Dzieduszyckia, Anarhynchia, Sulcirostra, Cooperrhynchia, Peregrinella*).

9.4 The Dimerelloid Rhynchonellid Brachiopods – A Chemosynthetic Dynasty?

I was not planning to review in detail all of the occurrences of fossil brachiopods that have been reported associated with ancient hydrocarbon seeps or hydrothermal deposits here. However, it seems appropriate to discuss the dimerelloid rhynchonellids because it has been suggested that they could represent a lineage of brachiopods ranging from the Devonian to the Cretaceous (Fig. 9.6) that may be exclusively(?) associated with chemosynthesis-dominated environments (Sandy and Campbell 1994; Campbell and Bottjer 1995a, b; Sandy 1995a; Gischler et al. 2003; Little and Vrijenhoek 2003; Peckmann et al. 2007a; Kiel and Peckmann 2008).

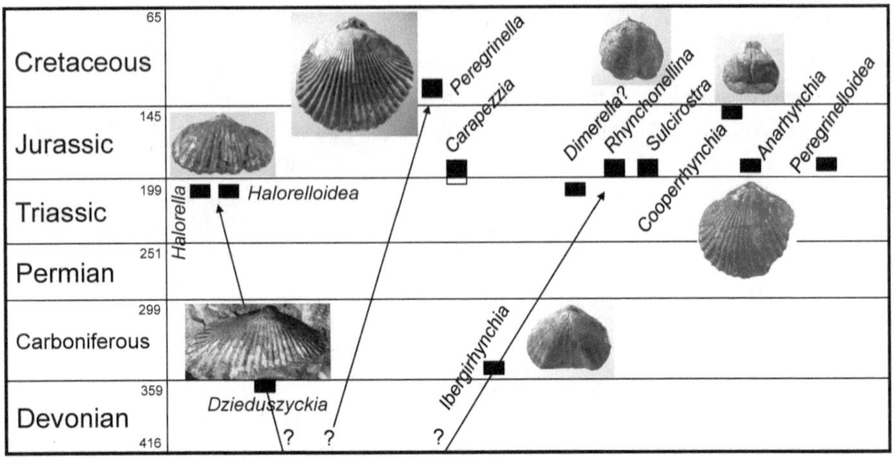

Fig. 9.6 Stratigraphic distribution and conjectural representation of evolutionary relationships of Palaeozoic-Mesozoic representatives of the Superfamily Dimerelloidea (primarily after Mancenido et al. 2002) that are considered to be associated with chemosynthesis-dominated environments, mostly cold-seeps but also hydrothermal vents (see Table 9.1 and text for further discussion). This represents a revision and expansion of Fig. 9.10 by Sandy in Gischler et al. 2003. Illustrated taxa, clockwise from top right corner: *Cooperrhynchia*; *Anarhynchia*; *Ibergirhynchia*; *Dzieduszyckia*; *Halorella*; *Peregrinella*; *Sulcirostra*; these specimens are figured elsewhere herein (Figs. 9.3–9.6, Plates 18, 19) for further details on their localities

While this is not the place to go into the intricacies of brachiopod classification and family and subfamily placement of genera it is relevant to comment on some of the problems in identifying the Superfamily Dimerelloidea and the limits of the dimerelloid stock as this is critical to identifying if a long-ranging evolutionary association with chemosynthesis-dominated environments is apparent or real; is it a single stock of related brachiopods or several unrelated lineages? It is interesting that Mancenido and Owen (2001) and Mancenido et al. (2002) had some reservations about placing the Triassic Family Halorellidae (includes *Halorella* and *Halorelloidea* (Table 9.1)) within the Superfamily Dimerelloidea. Also, they revised the placement of the Family Cryptoporidae and placed it in the Dimerelloidea with some reservations. The significance of this is that if correct the purported Palaeozoic-Mesozoic lineage discussed here would have some continuity beyond *Peregrinella* (the last dimerelloid associated with chemosynthesis-based environments). To date no members of the Cryptoporidae have been recorded from seeps or vents although representatives are recorded from deep waters (e.g. Zezina 2008). In addition, they would represent dwarfism from the "typical" dimerelloid forms associated with chemosynthesis-dominated environments (Fig. 9.5). Subsequent to *Peregrinella*, large shell size does not appear to be a characteristic of brachiopods associated with chemosynthesis-dominated environments (Table 9.1) although this might prove simply to be a collecting failure to date. However, for the time being to consider the Cryptoporidae as derived from earlier dimerelloid stock being considered herein (Fig. 9.6) is, I think, highly speculative.

It is tempting to consider that iterative evolution may be a factor for these brachiopods that frequent Palaeozoic-Mesozoic chemosynthesis-based settings. However, the fact that brachiopods considered to be related at least at the Superfamily, Family, or Subfamily taxonomic rank such as *Dzieduszyckia*, *Ibergirhynchia*, *Halorella*, *Peregrinella*, *Sulcirostra* and *Anarhynchia* appear to be associates of chemosynthesis-dominated environments lends strong support to some evolutionary-ecological linkage between these genera especially when based on a consideration of their internal characters too (Fig. 9.6; various aspects discussed by Sandy and Campbell 1994; Campbell and Bottjer 1995a, b; Sandy 1995a, 2001, Gischler et al. 2003; Little and Vrijenhoek 2003; Peckmann et al. 2007a).

The Brachiopoda have a long geological history as evidenced by their fossil record. However, given the fact that brachiopods have two valves encasing the animal, over the long time span the phylum has been in existence there are numerous cases of homoeomorphy within and between various brachiopod orders. One way to try to unravel homoeomorphy within the Brachiopoda is by a study of shell structure and the internal structural supports for the lophophore (brachidium) in the terebratulids and in the rhynchonellids a simpler single-pair of fork-like processes (crura). It is common for serial sections (or serial grindings) to be taken through specimens to determine their internal structures; or if preservation is suitable it may be possible to excavate infilling matrix (Figs. 9.3i, j shows a matrix-free interior with long ventrally-curving crura (ensiform), a feature that is typical of many dimerelloid genera. The general curvature of the paired crura can be determined but the crura have been encrusted and overgrown and details obscured). The incorporation of the study of internal structures in investigations of the dimerelloid lineage is one important tool that will continue to contribute to clarifying relationships within this lineage.

An important question in piecing together the evolutionary history of the Dimerelloidea is whether they are a monophyletic lineage or a polyphyletic lineage within the Order Rhynchonellida. The evolutionary convergence of both internal and external characters would indeed be a difficult problem to solve; after all some of these lineages potentially span hundreds of millions of years. However, these brachiopods do not occur in isolation and the potential to recognize evolutionary and associated palaoeobiogeographic patterns exists.

Peregrinella and *Peregrinelloidea* are strikingly similar externally. When considering the internal structures of the latter it does not possess a median septum or dental plates in the pedicle valve when compared with *Peregrinella*. This suggests that these two genera are not closely related. The degree of variability of internal structures within a brachiopod species or genus is not well understood though. The presence or absence of a median septum might allow distinction between genera and place them in different families (such as *Dzieduszyckia* and *Peregrinella* versus *Anarhynchia* and *Sulcirostra*). The presence or absence of a median septum appears to be a variable character allowable in the definition of the genus *Halorella* (Ager 1968); this assumes a genus diagnosis "broad enough" to encompass this variability. *Peregrinelloidea* is known from an isolated carbonate lens in a clastic sequence (AS Dagys, personal communication, 1993) so it may be a candidate for occurring in a chemosynthesis-based setting. Dagys considered *Anarhynchia* to be a synonym of *Peregrinelloidea* and this could, in fact, make sense at first glance. I have observed in *Sulcirostra* the horizontal hinge plates that are seen in *Peregrinelloidea* and this suggests that it is very close to this part of the dimerelloid lineage. A problem in accepting Dagys' suggestion is that I have not seen horizontal hinge plates in specimens of *Anarhynchia* I have investigated (personal observation and Sandy 2001; Gischler et al. 2003), whereby crural plates rise from the floor of the dorsal valve (serial sections were given through specimens by Ager 1968 and Sandy 2001, photomicrograph in Gischler et al. 2003, Fig. 9.11). In addition, internal characters appear to be variable in other representatives of the Dimerellidae such as the development of septalial plate-like structures in *Sulcirostra* and "buttresses" in *Cooperrhynchia* (discussed in Sandy and Campbell 1994, p. 1250) in the floor of the dorsal valve. This variability may be due to variability within taxa or the fact that we are in fact dealing with different taxa at the rank of genus. It seems that in chemosynthesis-based settings brachiopods may have secreted additional calcite thickening in the interior of their shells adding to variability in observed internal structures which may not necessarily have had any functional significance and therefore forming overgrowths that mask early juvenile growth phases or lead to precipitation of additional plate-like structures. Alternatively these structures could be aragonitic early diagenetic deposits (J Peckmann, personal communication, 2010) which would render them likely to be of no taxonomic value.

It is interesting that there is a cluster of dimerelloid genera in the Early Jurassic (Fig. 9.6) considered as likely associates of chemosynthesis-dominated environments. However, the differentiation of *Sulcirostra* from *Rhynchonellina* appears to be one of semantics, differentiating smooth-shelled *Rhynchonellina* from capillate *Sulcirostra*; in addition if *Anarhynchia* proves to be a synonym of *Peregrinelloidea* as considered by Dagys, then this Early Jurassic "explosion" or "burst" will have been rather quashed. What this does suggest however is that there are many additional taxa that could be part of this dimerelloid lineage. By contrast the lineage containing the lingulid *Pyrodiscus* (Fig. 9.4z) is restricted to one known genus at present.

One characteristic of many of the dimerelloid brachiopods (Fig. 9.6) is their very patchy distribution. Some of the Early Jurassic occurrences of the Dimerellidae are good candidates deserving further investigation as possible associates with hydrocarbon seeps. Ruggiero (1997) described some species of *Rhynchonellina* and *Sulcirostra* from carbonate lenses within a more regularly stratified carbonate sequence from the Early Jurassic of Italy. Sulser and Furrer (2008) described *Sulcirostra* and *Carapezzia* from the presumed Early Jurassic of Switzerland. Sulser and Furrer (2008) do not favour the chemosynthesis-based origins of the brachiopod occurrences they describe but they note that they are associated with hardgrounds and are restricted to a very small geographic area. They favour transport of these brachiopods by turbidity currents downslope.

Peregrinella is considered to have two subgenera (e.g. Mancenido et al. 2002). This is based on the nature of ornamentation with *Peregrinella* (*Peregrinellina*) having a denser ornament of costae that tend to bifurcate and would appear to be an arbitrary distinction within *Peregrinella* especially as the designation of a subgenus does not appear to be supported by any internal structural details such as differences in the median septum, dental plates, or development and shape of the crura. The forms considered as *Peregrinella* (*Peregrinellina*) are believed to have similar ecological associations with hydrocarbon seeps as noted for *Peregrinella* (*Peregrinella*). It would appear quite reasonable that differences in costation would develop in populations of *Peregrinella*. Perhaps such differences in ornamentation will be useful in understanding palaeobiogeographic and evolutionary trends within the genus. If so, then perhaps the subgenus could be worth retaining.

The highly disjunct distribution of a number of brachiopod taxa was one contributing factor that led to questions about their paleoecology and their recognition as being associated with hydrocarbon seep environments (*Peregrinella*, Campbell and Bottjer 1995b). Campbell and Bottjer (1995b) compiled a list of localities from which *Peregrinella* was known. More recently Posenato and Morsilli (1999) described a new species of *Peregrinella* from southern Italy and Sandy and Blodgett (1996) described a new species from Alaska. Posenato and Morsilli (1999) comment that the *Peregrinella* assemblage from southern Italy has all the characteristics of cold-seep communities proposed by Callender and Powell (1992): low diversity; almost monospecific; abundant large individuals within faunally impoverished formations. However, they did not include any isotope geochemistry which is one important line of evidence in support of carbonates that accumulate in hydrocarbon seep environments. This geochemical evidence is not available for all investigated occurrences of *Peregrinella*, including the Alaskan material which came from an isolated block and was first reported by Capps (1916).

9.5 Ancient Brachiopods as Living Animals – Aspects of Palaeoecology

Brachiopods appear to occupy at least a couple of palaeoecological niches within hydrocarbon seep environments based on their abundance and placement within these fossilized communities. (1) As "core components" of the community, based on hyperabundance of brachiopods, in rock-forming quantities, comprising the main component of carbonates that range from less than a meter to a few meters in vertical thickness, and up to several meters or more in lateral extent. This abundance and location has only been identified for fossil hydrocarbon seep localities to date, hydrothermal vent localities with fossil brachiopods having a lesser lateral and vertical extent (e.g. Little et al. 2004). This is the type of occurrence that large ribbed rhynchonellids appear to most commonly frequent (dimerelloids) in the

Palaeozoic and Mesozoic (morphotype 1 of Sandy and Campbell 1995). This would mean that the brachiopods would have functioned as, or comparable to "obligate species" (which in modern settings implies contact with the seep, e.g. Barry et al. 1996). Whether these taxa are in fact restricted to chemosynthesis-dominated environments has still to be determined. (2) In a peripheral position whereby they do not form thick shell-beds. This may be at the margins of the main site of hydrocarbon seepage or associated with early stages of formation (Beauchamp et al. 1989) or brecciation (Hikida et al. 2003; Campbell et al. 2008, *Sulcirostra* in the Jurassic of Oregon occurs in breccias and a shell-bed). This appears to be the main type of niche that brachiopods occupy in a "post-*Peregrinella*" world, in the Late Cretaceous (Hikida et al. 2003) and Miocene (Campbell et al. 2008). It would seem most likely that this would be the niche to expect living brachiopods to occupy in modern hydrocarbon seeps. This would be perhaps most comparable to "non-obligate species" as identified in modern hydrocarbon seeps (cf. Barry et al. 1996) but this term is used for species that are not restricted to the seep locality. The fossil brachiopods recorded from hydrothermal vent settings (*Pyrodiscus* and *Anarhychia*) presumably occupied a peripheral location to the central core of the vent, rather than the high-temperature core (Little et al. 1999, 2004).

There are a few things that can be anticipated as work continues on brachiopods from fossil chemosynthesis-dominated environments such as: evidence of epifauna; evidence of predation; interactions with other fauna; a fuller picture of the evolution and palaeobiogeography of brachiopods from chemosynthesis-dominated environments. Obviously fossil brachiopods were once living organisms – perhaps not the most dynamic of invertebrates. The taxa under consideration herein were living a basic sedentary existence, leading a life attached to substrate (including neighbouring brachiopods) by a pedicle although in some of these taxa the pedicle atrophied during life and the pedicle opening (foramen) closed in adult life (as noted for populations of *Peregrinella* from Poland by Biernat 1957 and from the Ukraine by Kiel and Peckmann 2008). Presumably the large-sized adult *Peregrinella* would remain in place in what would be a relatively quiescent environment (at least in terms of a presumed lack of strong currents) and these accumulations of brachiopods could provide a cluster of crowded, self-supporting brachiopods. The presence of serpulid worm tubes within the interiors of specimens of *Peregrinella* from the Crimea is considered to represent parasitic infestations as the worm tubes are overgrown and encased within the brachiopod shell (Kiel 2008). It would be interesting to see if this association is identified elsewhere.

Did some fossil brachiopods, perhaps most notably or obviously the large forms, harbour chemosymbionts? This is an interesting question but difficult to answer with certainty at present and may be difficult to prove ultimately. The large size of *Peregrinella* has led to speculation that it might have had an atypical biology compared to other brachiopods and perhaps a chemosythetic component to its biology (e.g. Campbell and Bottjer 1995b, Sandy 1995a). However, at present this remains speculative.

9.5.1 Epifauna – "Brachiopods on Brachiopods"

Presumably brachiopods are attaching to hard substrates in seep environments after their motile larval stage, including other brachiopods. It would therefore seem likely that the trace fossil *Podichnus* (Fig. 9.4w), the brachiopod pedicle attachment scar, should be found at some stage during the investigation of brachiopods and other suitable substrates. An example of *Podichnus* is shown from a brachiopod from a "normal" shallow shelf Jurassic environment; this trace fossil is presumed to be the result of terebratulid attachment to this terebratulid brachiopod (as was recorded by Nekvasilova 1975 among Cretaceous terebratulids). It seems reasonable that traces similar to this should be expected to be found in fossil seep environments due to the numerous individuals present and providing shell material is suitably preserved. In addition, asymmetry and deformation of shell outline is considered a reflection of the high settlement densities and crowding that are presumably commonplace in these brachiopod-dominated communities (e.g. as illustrated by Balinski and Biernat 2003); asymmetry may also result from predation attacks on the brachiopod.

9.5.2 Predation on Brachiopods – Durophagy

Predation on brachiopods is a well-known phenomenon (e.g. Alexander 1981; Ruggiero 1991; Brett 2003; Harper 2005). Durophagous attacks (as opposed to drilling by gastropods or other organisms) have been recorded from some brachiopods considered to be from chemosynthesis-dominated environments: *Dzieduszyckia* (Balinski and Biernat 2003), *Peregrinella* (Biernat 1957; Posenato and Morsilli 1999; Kiel and Peckmann 2008) and *Sulcirostra* from the Jurassic of Oregon (Sandy et al. 2005). When comparing predation rates from brachiopod faunas from chemosynthesis-dominated environments to brachiopod faunas from "normal" shelf environments predation attacks (for the most part sublethal predation whereby continued shell growth and/or repair indicates brachiopod "survival" after the attack) are much more common in the chemosynthetic setting (Fig. 9.4e–h, Plate 191, m); from collections of *Sulcirostra* from the Jurassic of Central Oregon 10% of specimens are considered to show evidence of predation; a high percentage of specimens of *Peregrinella* from the Ukraine show evidence of predation (S Kiel, personal communication, 2009). This could be a preservational effect with a higher percentage of brachiopods preserved (with predation scars) in a chemosynthesis-dominated environment compared to a fauna from a "typical" shallow shelf that may be subjected to current activity, abrasion, sorting and attrition of individuals. However, if there was a high level of predation or predation attacks in shelf environments it would still proportionately be well represented in the fossil fauna. It is likely that the "oasis" aspect of chemosynthesis-dominated environments, just as they have been opportunities for brachiopods, also represent opportunities for predators. The originator of the attack is an obvious question. The predation scars on *Sulcirostra* from Oregon are symmetrical with damage on both valves of the brachiopod implying either a cephalopod or crustacean attack (Fig. 9.4e–h).

Decapod crustaceans, specifically galatheid crabs, are most probably the most prevalent predators and scavengers at modern hydrothermal seeps although carnivores are less obvious at cold seeps (Senowbari-Daryan et al. 2007 and references therein). In addition, the abundance of faecal pellets at some seep sites indicates the common occurrence of decapods at these sites (Palik 1965; Bujtor 2007; Peckmann et al. 2007b). An early recognition of the association of decapod microcoprolites with mineralized deposits now considered hydrothermal in origin was from the Early Cretaceous Mecsek Mountains of southern Hungary (Palik 1965). Palik identified fragmentary crab remains from the deposit. Subsequently large-sized representatives of the brachiopods *Lacunosella* and *Nucleata* have been recorded from these deposits (Bujtor 2007). It would seem a reasonable inference that decapods and brachiopods were part of this community and that the brachiopods would have been targets for predation. It is, I think, very likely that the association of microcoprolites with chemosynthesis-dominated environments will be more widely recognized with further studies (e.g., Peckmann et al. 2007b). Senowbari-Daryan et al. (2007) comment that microcoprolites provide the best chance to understanding the role of crustaceans as predators and scavengers at ancient vent and seep environments; it seems very likely that decapods will prove to have been important predators on brachiopods in ancient chemosynthesis-dominated environments.

9.5.3 Refuges, Extinctions and Geological Distribution

Have hydrocarbon seeps and hydrothermal vents acted as refuges for brachiopods? The lack of brachiopods to date from modern chemosynthesis-based communities would indicate this is not the case. Little and Vrijenhoek (2003) commented that modern vents do not contain Palaeozoic representatives and so does not support modern vent faunas as relic faunas or refuges. However, it would appear that the dimerelloid rhynchonellids contain lineages that are on the order of hundreds of millions of years in duration (Fig. 9.6). This does not qualify them as refuges during that time interval necessarily, but a long-ranging lineage yes. Little and Vrijenhoek (2003) indicated that a number of modern taxa from vents are first known in the fossil record from seeps. Accepting that the fossil record of vents is much more difficult to work with, the only brachiopod currently considered likely to be from both vents and seeps is *Anarhynchia* from the Jurassic. There is not enough known about this genus at present to be able to draw any conclusions about the origins of *Anarhynchia* in chemosynthesis-based communities (i.e., did it migrate from one environment to the other?).

However, the deep-water setting of many ancient seeps and vents would presumably offer respite from perturbations and extinctions in shallower-water faunas

(MacArthur and Tunnicliffe 1998). If the as yet unknown ancestral stock to *Dzieduszyckia* inhabited chemosynthesis-dominated environments (although Ager 1968 considered forms in the Silurian as possibly ancestral) it would presumably have been well-placed to survive the effects of the Frasnian-Famennian extinctions (comparable to vents acting as refuges as discussed by MacArthur and Tunnicliffe 1998). Alternatively does such an extinction event pressure forms to extend their range and inhabit "new" environments such as chemosynthesis-dominated environments if they are able to settle and inhabit them? How the Permo-Triassic extinctions affected brachiopod faunas from seeps and vents is also speculative due to a lack of information about these environments during the Late Palaeozoic and Early Mesozoic. However, with the formation of Pangaea by the Late Palaeozoic there would presumably be hydrocarbon seeps and hydrothermal vent communities living in and around the Panthalassic Ocean. Not all seeps and vents are located in deep water and the abilities of organisms to disperse and colonize them, particularly sedentary organisms such as brachiopods, bivalves and crinoids (that mostly depend on their larval stage for dispersal), would presumably be influenced by factors such as water depth and ocean current-circulation patterns.

The Late Triassic – Early Jurassic is an important time interval in the post-Palaeozoic history of the brachiopods. The Late Triassic is a time of diversification among a number of brachiopod orders (e.g. Sandy 1995b). There are enough well-known brachiopod faunas by the latter part of the Triassic that there is certainly a "monographic burst" in terms of their diversity too. The remnants of the orders Spiriferinida and Athyridida are lost during the Late Triassic – Early Jurassic interval (Figs. 9.1 and 9.2). Among the brachiopods considered to be associated with seeps and vents during this time interval (Table 9.1, Figs. 9.2 and 9.5) this is in fact a significant time interval. This may well simply be a reflection of the fact that brachiopods that are known to meet the criteria for being identified as potential candidates from seep or vent deposits are well-known during this time interval importantly through the work of Ager and his various discussions of brachiopod ecology and occurrence primarily from Alpine Europe (e.g. Ager 1965) and subsequently from the USA (Ager 1968) and Africa (Ager et al. 1976 for Palaeozoic occurrences in Morocco). However this does appear to be an interesting time interval for brachiopods associated with chemosynthesis-based communities with closely-related forms (some synonymous?) that merit further field and laboratory investigation. What happens beyond the Early Jurassic? *Cooperrhynchia* appears to pick up the trail in the Late Jurassic after a Middle Jurassic "gap" that awaits filling. Then in the Early Cretaceous *Peregrinella* "bursts" on the scene as the most widely represented brachiopod genus in terms of numbers of recorded localities associated with chemosynthesis-based settings (Campbell and Bottjer 1995b, Table 1).

The effect of the Cretaceous – Tertiary extinction event on brachiopods has been debated. Some authors note a major extinction event among the brachiopods at the K-T boundary (Surlyk and Johansen 1984; Johansen 1987) but this appears to be more apparent rather than real as facies changes at the Cretaceous-Tertiary boundary signal (or magnify) apparent extinction (Owen in MacLeod et al. 1997).

Brachiopod diversity is affected at the generic level but unlike the Permo-Triassic extinctions no brachiopod orders are lost (Figs. 9.1 and 9.2). Overall the K-T extinction is considered to have had no major effect on the diversity of Late Mesozoic – Cenozoic brachiopods (Owen in MacLeod et al. 1997). If the K-T event was a set-back to the Brachiopoda it does not appear to have lasted long and the terebratulids regained generic diversities that are highly comparable to those they enjoyed during the Cretaceous (Curry 2007). Based on an analysis using data from the recently published revision of the brachiopod Treatise, Curry (2007) comments that there is a 53% reduction of brachiopod genera in the Early Palaeogene compared to the Late Cretaceous. However he considers this to be very likely an overestimate in part due to taxonomic differentiation and partitioning among brachiopodologists as one crosses the K-T boundary. By this time the dimerelloid association with seep environments appears lost, based on the current small number of post-Early Cretaceous brachiopod taxa recorded from seep environments (Table 9.1). However, based on just these few records both terebratulids (Beauchamp et al. 1989; Beauchamp and Savard 1992; Sandy 1990; Hikida et al. 2003; Kaim et al. 2010) in the Cretaceous, and rhynchonellids (possibly) (Klosterman et al. 2007) and terebratulids (Campbell et al. 2008) in the Cenozoic play a role in Late Cretaceous – Cenozoic hydrocarbon seep environments. The Cretaceous forms have not been traced into Cenozoic hydrocarbon seep environments.

9.6 Brachiopod Morphology at Seeps and Vents

Large brachiopod size is one characteristic of brachiopods associated with chemosynthesis-based communities; examples being *Pyrodiscus* (up to 8 cm in diameter) from Silurian hydrothermal vents and the large ribbed *Dzieduszyckia*, *Halorella* and *Peregrinella* (up to 10 cm in width and/or length) from hydrocarbon seeps (Figs. 9.3–9.5, Plates 18, 19). Sandy and Campbell (1995) referred these large ribbed rhynchonellids to "morphotype 1". However, small, smooth, sulcate forms generally less than 1 cm in diameter are also associates of chemosynthesis-based environments, such as the rhynchonellid *Cooperrhynchia* from the Late Jurassic and the terebratulid *Modestella* from the Early Cretaceous (Figs. 9.4m–p, Plate 19i, j; morphotype 2 Sandy and Campbell 1995). However the reality may simply be that a variety of brachiopod taxa with differing gross morphologies from a variety of different brachiopod orders have at some point colonized chemosynthesis-based communities. Various other morphologies are now known from chemosynthesis-based communities such as small terebratulids (*Eucalathis methanophila* from the Late Cretaceous of Japan, Fig. 9.4q–s) and small rhynchonellids (possibly related to hydrocarbon seeps, *Probolarina* from the Palaeocene of Mexico, Figs. 9.4k, l and 9.5f); a variety of morphologies are illustrated herein (Figs. 9.3–9.6, Plates 18, 19).

9.6.1 Morphological Couplets

A recurring theme among the brachiopods considered herein are morphological couplets (identified in Table 9.1 by paired asterisks: e.g. *). These are forms that have often originally been described as a single genus but are subsequently divided by later workers. The taxa are very closely related, contemporaneous or at least overlapping in range, and part of the same evolutionary lineage based on a consideration of their overall morphology (their internal and external characters). However, in terms of their external morphology they can be divided into a smooth-shelled form and a finely- or coarsely-ribbed costate form.

Such couplets are evident in the brachiopods being considered here:

- *Rhynchonellina* and *Sulcirostra*, the latter being erected by Muir-Wood and Cooper (1951) and split off from *Rhynchonellina*
- *Halorella* and *Halorelloidea* represent another couplet with Ager (1960) describing the latter to distinguish it from the more coarsely-costate *Halorella*
- The subgenera of *Septatrypa* can be considered a third such couplet although *Septatrypa* (*Herciniscа*) is questionably associated with hydrocarbon seep carbonates
- *Peregrinella* has also been split into subgenera but here the differentiation is made on the style of costation rather than there being a costate and smooth-shelled subgenus

Whether there is in fact any evolutionary or ecological significance to these couplets has yet to been determined and this phenomenon is restricted to the taxa being considered herein (and both *Septatrypa* and *Probolarina* are questionably hydrocarbon seep associates). Other examples of such morphological couplets come from, for example, Cretaceous marine shelf environments and include: the biplicate terebratulid *Sellithyris* and its costate counterpart *Glosseudesia*; and the smooth-shelled terebratellid *Psilothyris* and its costate (at the commissure) counterpart *Timacella*.

9.6.2 Micromorphs

As noted earlier there has been some discussion regarding the placement of the Cretaceous-Recent brachiopod family Cryptoporidae (Mancenido and Owen 2001; Mancenido et al. 2002). However, if these brachiopods are truly dimerelloids they take on significance as being related to the enigmatic Palaeozoic and Mesozoic members of this family. In terms of Recent brachiopods that are found on oceanic rises and seamounts, Zezina (2008) commented that *Cryptopora* is among these but there are no records of this being part of modern chemosynthesis-based communities and *Cryptopora* is recorded (dredged) from "normal" marine environments. In addition the Cryptoporidae and indeed *Cryptopora* can be considered micromorphic where adult species attain lengths of 5 mm or less (1–2 mm). Zezina (2003) suggested that stress is a factor that leads to micromorphic species (dwarfism and/or paedaeomorphosis) in the modern ocean brachiopod fauna. Did the large dimerelloids give rise to micromorphic species? Would this happen if these brachiopods were to seek refuge? Compared to the potentially abundant food supply at an active hydrocarbon seep or hydrothermal vent (which would suggest they are not refuges) the oceanic abyssal plain would in fact appear to be a refuge, an environment of lesser nutrient supply where a brachiopod could survive. As a hydrocarbon seep or hydrothermal vent evolves and seepage or venting lessens the brachiopods would potentially be under stress. Presumably through their reproductive cycle larvae could disperse to new sites if a hydrocarbon seep or vent becomes inactive. A reduction in seepage or venting at a critical time in their life-cycle such as spawning might have a disastrous effect on a brachiopod population. Micromorphs that are associated with hydrocarbon seeps include terebratulid forms from the Late Cretaceous of Japan (Hikida et al. 2003; Kaim et al. 2010, Fig. 9.4q–s herein).

9.7 Palaeobiogeography

Given the fact that currently there are few records of ancient seep and vent deposits with brachiopods it might be some time before a coherent view of palaeobiogeography among vent and seep brachiopods emerges. One of the pointers in identifying brachiopods from ancient seeps and vents was that they had a very disjointed, stochastic distribution and therefore did not conform to traditional palaeobiogeographic models of faunal provinces (Sandy 1995a; Bujtor 2007). This new interpretation meant that, for example, *Peregrinella* and other taxa known only from low-palaeolatitudes could no longer be interpreted as classic Tethyan Realm indicators (Sandy 1995a) as contemporaneous hydrocarbon seep distributions would not have been restricted to the Tethyan Realm. However, it does not mean that these seep and vent brachiopods are without their own palaeobiogeography, a scenario made more complex if they behaved as suggested for other seep and vent taxa, having originated outside seep and vent settings and migrated in and out of such settings through time (as suggested by Little et al. 1998 and Little and Vrijenhoek 2003 for other components of such communities).

9.8 Evolutionary Trends

Various taxa appear to have migrated in and out of chemosynthesis-based communities through time (Little et al. 1998; Little and Vrijenhoek 2003) and it would appear that brachiopods are not using hydrocarbon seeps or hydrothermal vents as refuges. Indeed brachiopods from up to four Orders are known from ancient chemosynthesis-dominated environments and are spectacularly abundant at certain times. Based on an assessment of the Palaeozoic and Mesozoic representatives and record, patchy though it is at the moment, it would appear that the brachiopods are evolving and diversifying in such environments and taking an opportunistic role. In the Palaeozoic and early to mid-Mesozoic they are often a more numerous (Campbell and Bottjer 1995a; Little and Vrijenhoek 2003) although less diverse component of ancient hydrocarbon seeps than molluscs. Campbell and Bottjer (1995a) comment that replacement of brachiopods in chemosynthesis-based communities by bivalve and gastropod families with modern affinities was beginning by the Late Jurassic. The prolific records of *Peregrinella*, relative to any other Mesozoic brachiopod considered to be associated with chemosynthesis-based communities would seem to mark the last significant "stand" by brachiopods associated with chemosynthesis-based communities. However, *Peregrinella* is a large brachiopod (it may reach 10 cm in length and width); it is very easy to spot in the field if it is present, and if it happens to be present in a carbonate block that is sitting in the middle of a fossil-barren deeper-water clastic interval, then perhaps even more obvious to track down. *Halorella* in the Triassic and *Dzieudszyckia* in the Devonian are perhaps the other brachiopod taxa that could come close to *Peregrinella* in terms of being known from a relatively large number of localities (including some that are isolated blocks indicating highly disjunct distributions in places). Other records for taxa such as *Lacunosella* and *Modestella* that are considered to be associated with hydrocarbon seeps or hydrothermal vent settings occur in other "normal marine" environments too. This shows that specific brachiopod taxa, at least at the generic level, may be found in both chemosynthesis-based, and non-chemosynthesis-dominated environments. *Anarhynchia* is considered to be represented in both hydrocarbon seeps and hydrothermal vents.

Have brachiopods once so common at times at ancient hydrocarbon seeps (Table 9.1, Fig. 9.7) been replaced by bivalves or are they just excluded from modern hydrocarbon seeps? It appears that bivalves made significant inroads into seep environments in the Jurassic (Campbell and Bottjer 1995a; Little and Vrijenhoek 2003; Campbell 2006). However Palaeozoic carbonate mounds are known to contain bivalves (Peckmann et al. 1999). If this were a simple replacement of brachiopods by bivalves it is discordant with the Permo-Triassic brachiopod extinction and molluscan radiation associated with the Early Mesozoic. The brachiopods found in Mesozoic and younger chemosynthesis-dominated environments were apparently undergoing their own phases of radiation and diversification. This would mirror the post-Palaeozoic history of the Brachiopoda which records periods of modest diversification among the rhynchonellids and terebratulids (e.g. Sandy 1995b, Figs. 9.1 and 9.2 herein). In a post-*Peregrinella* world (post Hauterivian) it would appear that brachiopods have not been eradicated from chemosynthesis-based communities but typically play a more peripheral/marginal role in them compared to their shell-bed forming potential in the Palaeozoic and Early-Middle Mesozoic although there could be exceptions even to this generalization (e.g. if *Probolarina* from the Palaeocene of Mexico represents a hydrocarbon seep associate).

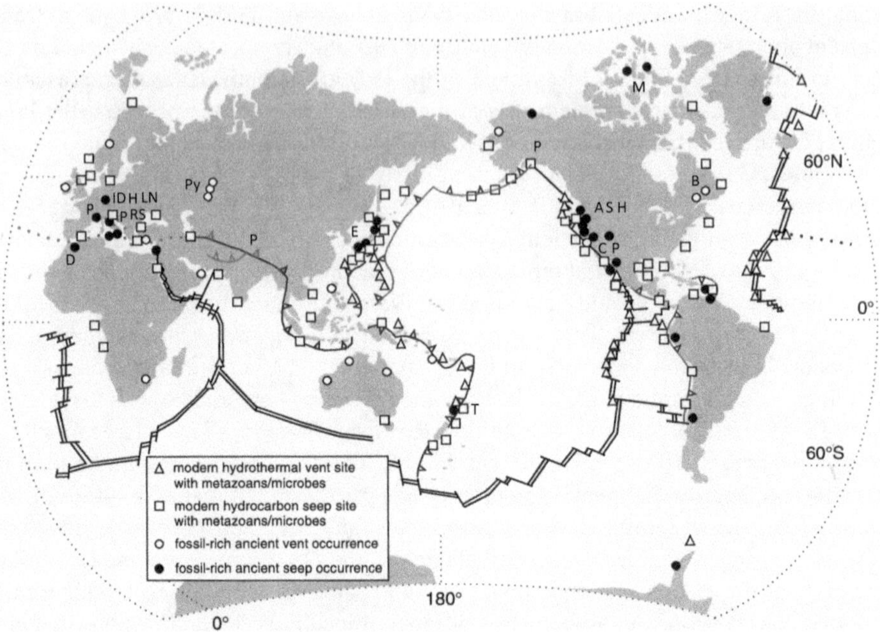

Fig. 9.7 Map showing approximate location of selected brachiopod occurrences with ancient hydrocarbon seeps and hydrothermal vents. These are plotted on a distribution map of Archean-Recent chemosynthesis-dominated environments that have associated metazoan and/or microbial signatures from Campbell 2006 (wherein references may be found). Key to brachiopod taxa: A = *Anahynchia*; B = *Beecheria*; C = *Cooperrhynchia*; D = *Dzieduszyckia*; E = *Eucalathis*; H = *Halorella*; I = *Ibergirhynchia*; L = *Lacunosella*; M = *Modestella*; N = *Nucleata*; P = *Peregrinella*; R = *Rhynchonellina*; Py = *Pyrodiscus*; S = *Sulcirostra*; T = terebratulids from Miocene of New Zealand. Note that because of the scale the locations of brachiopod localities is very approximate and not all occurrences are plotted (Modified from Campbell (2006). With permission from Elsevier)

9.9 Some Concluding Remarks

It is clear that the number of brachiopod taxa recorded from ancient seeps and vents will continue to grow (Table 9.1, Figs. 9.2 and 9.7). Whether they are totally devoid from modern seeps is probably unlikely given the resilience and ability of the Phylum Brachiopoda to inhabit a wide range of environments, even in the post-Palaeozoic world. Brachiopods are often well-camouflaged by epizoans and cryptic forms such as the thecidicians are likely to be difficult to find by visual surveys from submersibles alone.

The fossil record of brachiopods associated with chemosynthesis-dominated environments is characterized by variety with several brachiopod orders recorded at various times throughout the Phanerozoic and represented by various-sized brachiopods from large (10 cm) to small (5–10 mm) to micromorphic forms, and various morphologies from coarsely-costate to finely-costellate to smooth-shelled and sulcate forms. However, it is the large-sized rhynchonellids that have been most readily identified to date as members of ancient hydrocarbon seep communities. To date among living brachiopods the craniids and thecidians have not been reported from fossil communities but they would find hard substrates for direct cementation in chemosynthesis-based settings. If brachiopods are found as components of living chemosynthesis-based communities it might be representatives of these cemented, cryptic brachiopods that will be found as well as representatives of the Terebratulida and Rhynchonellida that have been recorded from Cenozoic chemosynthesis-based communities. While the Palaeozoic-Early Cretaceous story of brachiopods associated with chemosynthesis-based communities has been one dominated by records of rhynchonellids (Table 9.1) forming core species it appears that the post-Early Cretaceous torch is taken up by the terebratulids (Table 9.1), the most diverse group of articulate brachiopods during this time interval, and other rhynchonellids; these records are interpreted mostly of brachiopods as peripheral species. Terebratulids are the most diverse group of living brachiopods and do have a broad latitudinal and bathymetric distribution in the world's oceans (e.g. Richardson 1997a, b; Zezina 2003, 2008) and would be anticipated to be represented at least in some measure from modern hydrocarbon seeps. We still do have very few data points with regard to fossil brachiopods from chemosynthesis-based communities and this fascinating story will continue to unfold with continued research.

Acknowledgements For assistance with this paper in many ways including discussion, donation of samples or photographs thanks to Kathy Campbell, Eberhard Gischler, Jim and Gail Goedert, Steffen Kiel, Iuliana Lazar, Cris Little, Jörn Peckmann, Kris Saether, Francisco Vega. Also many thanks to Steffen Kiel, Jörn Peckmann and an anonymous reviewer for reviews that helped improve the manuscript.

References

Ager DV (1960) Nomenclatural problems in the Mesozoic Rhynchonelloidea. Geol Mag 97: 157–162

Ager DV (1965) The adaptation of Mesozoic brachiopods to different environments. Palaeogeogr Palaeoclimatol Palaeoecol 1:143–172

Ager DV (1967) Some Mesozoic brachiopods in the Tethys Region. In: Adams CC and Ager DV (eds) Aspects of tethyan biogeography. The Systematics Association Publication 7:135–151

Ager DV (1968) The supposedly ubiquitous Tethyan brachiopod *Halorella* and its relations. J Paleont Soc India 5–9(1960–1964):54–70

Ager DV (1973) Mesozoic Brachiopoda. In: Hallam A (ed) Atlas of palaeobiogeography. Elsevier, Amsterdam, pp 431–436

Ager DV (1986) Migrating fossils, moving plates and an expanding earth. Mod Geol 10:377–390

Ager DV, Sun D (1989) Distribution of Mesozoic brachiopods on the northern and southern shores of Tethys. Palaeontol Cathayana 4:23–51

Ager DV, Cossey SPJ, Mullin PR, Walley CD (1976) Brachiopod ecology in mid-Palaeozoic sediments near Khenifra, Morocco. Palaeogeogr Palaeoclimatol Palaeoecol 20:171–185

Alexander RR (1981) Predation scars preserved in Chesterian brachiopods: Probable culprits and evolutionary consequences for the articulates. J Paleont 58:192–202

Anderson FM (1945) Knoxville series in the California Mesozoic. GSA Bull 56:909–1014
Baliński A, Biernat G (2003) New observations on rhynchonelloid brachiopod *Dzieduszyckia* from the Famennian of Morocco. Acta Palaeont Polonica 48:463–474
Barbieri R, Ori GG, Cavalazzi B (2004) A Silurian cold-seep ecosystem from the Middle Atlas, Morocco. Palaios 19:527–542
Barry JP, Greene HG, Orange DL, Baxter CH, Robison BH, Kochevar RE, Nybakken JW, Reed DL, McHugh CM (1996) Biologic and geologic characteristics of cold seeps in Monterey Bay, California. Deep Sea Res I 43(11–12):1739–1762
Beauchamp B, Savard M (1992) Cretaceous chemosynthetic carbonate mounds in the Canadian Arctic. Palaios 7:434–450
Beauchamp B, Krouse HR, Harrison JC, Nassichuk WW, Eliuk LS (1989) Cretaceous cold-seep communities and methane-derived carbonates in the Canadian Arctic. Science 244:53–56
Benton MJ, Harper DAT (2009) Introduction to paleobiology and the fossil record, 1st edn. Wiley-Blackwell, Oxford
Biernat G (1957) On *Peregrinella multicarinata* (Lamarck) (Brachiopoda). Acta Palaeontol Pol 2:19–50
Bottjer DJ, Campbell KA, Schubert JK, Droser ML (1995) Palaeoecological models, non-uniformitarianism, and tracking the changing ecology of the past. In: Bosence WJ, Allison PA (eds) Marine palaeoenvironmental analysis from fossils, vol 83. Geological Society (special publication), London, pp 7–26
Brett CE (2003) Durophagous predation in Paleozoic marine benthic assemblages. In: Kelley PH, Kowalewski M, Hansen TA (eds) Predator-prey interactions in the fossil record. Topics in geobiology, vol 20. Kluwer/Plenum, New York, pp 401–432
Buggisch B, Krumm K (2005) Palaeozoic cold seep carbonates from Europe and North Africa – an integrated isotopic and geochemical approach. Facies 50:566–583
Bujtor L (2007) A unique Valanginian paleoenvironment at an iron ore deposit near Zengõvárkony (Mecsek Mts, South Hungary), and a possible genetic model. Cent Eur Geol 50(3):183–198
Callender WR, Powell EN (1992) Taphonomic signature of petroleum seep assemblages on the Louisiana upper continental slope: recognition of autochthonous shell beds in the fossil record. Palaios 7:388–408
Campbell KA (2006) Hydrocarbon seep and hydrothermal vent paleoenvironments and paleontology: past developments and future research directions. Palaeogeogr Palaeoclimatol Palaeoecol 232:362–407
Campbell KA, Bottjer DJ (1993) Fossil cold seeps (Jurassic–Pliocene) along the convergent margin of western North America. Natl Geogr Res Exploration 9:326–343
Campbell KA, Bottjer DJ (1995a) Brachiopods and chemosymbiotic bivalves in Phanerozoic hydrothermal vent and cold-seep paleoenvironments. Geology 23:321–324
Campbell KA, Bottjer DJ (1995b) *Peregrinella*: an Early Cretaceous cold-seep-restricted brachiopod. Paleobiology 21:461–478
Campbell KA, Carlson C, Bottjer DJ (1993) Fossil cold seep limestones and associated chemosymbiotic macroinvertebrate faunas, Jurassic–Cretaceous Great Valley Group, California. In: Graham S, Lowe D (eds) Advances in the sedimentary geology of the Great Valley Group, vol 73. Pacific Section, Society of Economic Paleontologists and Mineralogists, Los Angeles, CA, pp 37–50
Campbell KA, Farmer JD, Des Marais D (2002) Ancient hydrocarbon seeps from the Mesozoic convergent margin of California: carbonate geochemistry, fluids and palaeoenvironments. Geofluids 2:63–94
Campbell KA, Francis DA, Collins M, Gregory MR, Nelson CS, Greinert J, Aharon P (2008) Hydrocarbon seep-carbonates of a Miocene forearc (East Coast Basin), North Island, New Zealand. Sed Geol 204:83–105
Capps SR (1916) The Chisana-White River District. Alaska, USGS Bulletin 630, 130 p
Carlson SJ, Leighton LR (2001) The phylogeny and classification of the Rhynchonelliformea. In: Carlson SJ, Sandy MR (eds) Brachiopods ancient and modern. Paleontological Society Papers 7, Yale University, New Haven, CT, pp 27–51
Cooper GA, Muir-Wood HM (1951) Brachiopod homonyms. J Wash Acad Sci 41:195–196

Corliss JB, Dymond J, Gordon LI, Edmond JM, von Herzen RP, Ballard RD, Green K, Williams D, Bainbridge A, Crane K, van Andel TH (1979) Submarine thermal Springs on the Galápagos Rift. Science 203(4385):1073–1083

Curry GB (2007) Geological history of Brachiopods (poster). Paleontological Institute, Lawrence, KS. 2 sides

Dix GR, Edwards C (1996) Carbonate-hosted, shallow-submarine and burial hydrothermal mineralization in the Upper Mississippian Big Cove Formation, Port au Port Peninsula, western Newfoundland. Econ Geol 91:180–203

Dubé TE (1988) Tectonic significance of Upper Devonian igneous rocks and bedded barite, Roberts Mountain allochthon, Nevada, U.S.A. In: McMillan SM, Embry AF, Glass DJ (eds) Devonian of the world (sedimentation), vol II. Canad Soc Petrol Geol, Calgary, pp 235–249

Emig CC (1997a) Biogeography of inarticulated brachiopods. In: Kaesler RL (ed) Treatise on invertebrate paleontology. Part H, Brachiopoda revised, vol. 1: Introduction. Geological Society of America, Boulder, CO, pp 497–502

Emig CC (1997b) Ecology of inarticulated brachiopods. In: Kaesler RL (ed) Treatise on Invertebrate Paleontology. Part H, Brachiopoda revised, vol. 1: Introduction. Geological Society of America, Boulder, CO, pp 473–495

Gabb WM (1864) Palaeontology volume 1. Geological Survey of California, 243 p. Geological Survey of California II, pp 1–299

Gaillard C, Rio M, Rolin Y, Roux M (1992) Fossil chemosynthetic communities related to vents or seeps in sedimentary basins: the pseudobioherms of southeastern France compared to other world examples. Palaios 7:451–465

Gischler E, Sandy MR, Peckmann J (2003) *Ibergirhynchia contraria* (F. A. Roemer, 1850) an Early Carboniferous seep-related rhynchonellide brachiopod from the Harz Mountains, Germany – a possible successor to *Dzieduszyckia*? J Paleont 77:293–303

Gregory MR, Lee DE (1995) A new cancellothyrid brachiopod from the abyssal seafloor, Southeast Indian Ridge, a cryptic survivor. Geol Soc New Zealand Misc Publ 81A:113

Harper EM (2005) Evidence of predation damage in Pliocene *Apletosia maxima* (Brachiopoda). Palaeontology 48:197–208

Hikida Y, Suzuki S, Togo Y, Ijiri A (2003) An exceptionally well-preserved fossil seep community from the Cretaceous Yezo Group in the Nakagawa area, Hokkaido, northern Japan. Paleont Res 7(4):329–342

Himmler T, Freiwald D, Stollhofen H, Peckmann J (2008) Late Carboniferous hydrocarbon-seep carbonates from the glaciomarine Dwyka Group, southern Namibia. Palaeogeogr Palaeoclimatol Palaeoecol 257:185–197

Johansen MB (1987) Brachiopods from the Maastrichtian-Danian boundary sequence at Nye Klov, Jylland, Denmark. Fossils Strata 20:1–99

Kaim A, Bitner MA, Jenkins RG, Hikida Y (2010) A monospecific assemblage of terebratulide brachiopods in the Upper Cretaceous seep deposits of Omagari, Hokkaido, Japan. Acta Palaeont Pol 55:73–84

Kiel S (2008) Parasitic polychaetes in the Early Cretaceous hydrocarbon seep-restricted brachiopod *Peregrinella multicarinata*. J Paleont 82:1215–1217

Kiel S, Peckmann J (2008) Paleoecology and evolutionary significance of an Early Cretaceous *Peregrinella*-dominated hydrocarbon-seep deposit on the Crimean Peninsula. Palaios 23:751–759

Klosterman S, Sandy MR, Vega FJ, Giles KA, Graf K, Shelley D, Sole J (2007) New Paleocene rhynchonellide brachiopods from the Potrerillos Formation, northeastern Mexico. J Paleont 81:483–489

Kowalewski MM, Simoes MG, Carroll M, Rodland DL (2002) Abundant brachiopods on a tropical, upwelling-influenced shelf (southeast Brazilian Bight, South Atlantic). Palaios 17:277–286

Lazar I, Grigore D, Sandy MR (2009a) Upper Jurassic brachiopod assemblages from the Haghimas Mountains (Eastern Carpathians, Romania) – taxonomy, paleoecology and paleobiogeographical significance. 9th North American Paleontological Convention Abstracts, Cincinnati Museum Center Scientific Contributions, No. 3, p 206

Lazar I, Sandy M, Grigore D, Panaiotu C (2009b) Taxonomy, paleoecology and paleobiogeographical significance of the *Lacunosella* brachiopod assemblage from Upper Jurassic of Hasmas Mountains (Eastern Carpathians, Romania). 8th Symposium of I.G.C.P. 506, Marine and non-marine Jurassic: correlation and major geological events, Abstracts and Field Guide, pp 20–21

Lee DE (2000) A new deep-sea cancellothyrid brachiopod from volcanic substrates of the mid-oceanic ridge system. The Millenium Brachiopod Congress, Abstracts. Natural History Museum, London, 1 p

Lemoine M, Arnaud-Vanneau A, Arnaud H, Letolle R, Mevel C, Thieuloy JP (1982) Indices possible de paléo-hydrothermalisme marin dans le Jurassique et le Crétacé des Alpes occidentals (océan téthysien et sa marge continentale européenne): essai d'inventaire. Bull Soc géol France 24:641–647

Little CTS, Vrijenhoek RC (2003) Are hydrothermal vent animals living fossils? Trends Ecol Evol 18:582–588

Little CTS, Herrington RJ, Maslennov ZVV (1998) The fossil record of hydrothermal vent communities. In: Mills RA, Harrison K (eds) Modern ocean floor processes and the geological record, vol 149. Geological Society, London, pp 259–270

Little CTS, Maslennikov VV, Morris NJ, Gubanov AP (1999) Two Palaeozoic hydrothermal vent communities from the southern Ural Mountains, Russia. Palaeontology 42:1043–1078

Little CTS, Danelian T, Herrington RJ, Haymon RM (2004) Early Jurassic hydrothermal vent community from the Franciscan Complex. J Paleont 78:542–559

Lonsdale P (1977) Clustering of suspension-feeding macrobenthos near abyssal hydrothermal vents at oceanic spreading centers. Deep-Sea Research 24:857–863

MacLeod N, Rawson PF, Forey PL, Banner FT, Boudagher-Fadel MK, Bown PR, Burnett JA, Chambers P, Culver S, Evans SE, Jeffery C, Kaminski MA, Lord AR, Milner AC, Milner AR, Morris N, Owen EF, Rosen BR, Smith AB, Taylor PD, Urquhart E, Young JR (1997) The Cretaceous-Tertiary biotic transition. J Geol Soc (London) 154:265–292

Macsotay O (1980) Mollusques benthiques du cretace inferieur: une method de correlation entre la Tethys Mesogeenne et le domaine Paleo-Caraibe (Venezuela). Ph.D. thesis, University of Lyon

Mancenido MO, Owen EF (2001) Post-Palaeozoic Rhynchonellida (Brachiopoda): classification and evolutionary background. In: Brunton CHC, Cocks LRM, Long SL (eds) Brachiopods past and present: The Systematics Association special vol, 63. Taylor & Francis, London, pp 189–200

Mancenido MO, Owen EF, Savage NM, Dagys AS (2002) Dimerelloidea. In: Kaesler RL (ed) Treatise on invertebrate paleontology, Part H, Brachiopoda revised, Vol. 4: Rhynchonelliformea (part). Geological Society of America, Boulder, CO, pp 1236–1245

McArthur AG, Tunnicliffe V (1998) Relics and antiquity revisited in the modern vent fauna. In: Mills R, Harrison K (eds) Modern ocean floor processes and the geological record, vol 14. Special Publication of the Geological Society, London, pp 271–291

Middlemiss FA (1973) The geographical distribution of Lower Cretaceous Terebratulacea in Western Europe. In: Casey R, Rawson PF (eds) The Boreal Lower Cretaceous. Seel House Press, Liverpool, pp 111–120

Muir-Wood HM (1955) A history of the classification of the Phylum Brachiopoda. Publication No. 78, British Museum (Natural History), London, 124 p

Nekvasilova O (1975) The etching traces produced by pedicles of Upper Cretaceous brachiopods from Bohemia (Czechoslovakia). Casopis pro mineralogii a geologii 20:69–74

Owen EF (1973) The distribution of Lower Cretaceous (Berriasian-Barremian) rhynchonellid and terebratelloid brachiopods in the Northern Hemisphere. In: Casey R, Rawson PF (eds) The Boreal lower Cretaceous. Seel House Press, Liverpool, pp 121–130

Palik P (1965) Remains of crustacean excrement from the Lower Cretaceous of Hungary. Micropaleont 11:98–104

Paull CK, Hecker B, Commeau R, Freeman-Lynde RP, Neumann AC, Corso WP, Golubic S, Hook J, Sikes E, Curray J (1984) Biological communities at Florida Escarpment resemble hydrothermal vent communities. Science 226:965–967

Peckmann J, Thiel V (2004) Carbon cycling at ancient methane-seeps. Chem Geol 205:443–467

Peckmann J, Walliser OH, Riegel W, Reitner J (1999) Signatures of hydrocarbon venting in a Middle Devonian carbonate mound (Hollard Mound) at the Hamar Laghdad (Antiatlas Morocco). Facies 40:281–296

Peckmann J, Campbell KA, Walliser OH, Reitner J (2007a) A Late Devonian hydrocarbon-seep deposit dominated by dimerelloid brachiopods, Morocco. Palaios 22:114–122

Peckmann J, Senowbari-Daryan B, Birgel D, Goedert JL (2007b) The crustacean ichnofossil *Palaxius* associated with callianassid body fossils in an Eocene methane-seep limestone, Humptulips Formation, Olympic Peninsula, Washington. Lethaia 40:273–280

Poole FG (1988) Stratiform barite in Paleozoic rocks of the western United States. In: Zachrisson E (ed) Proceedings of the Seventh Quadrennial IAGOD Symposium. E Schweizerbart'sche Verlagsbuchhandlung, Stuttgart, pp 309–319

Posenato R, Morsilli M (1999) New species of *Peregrinella* (Brachiopoda) from the Lower Cretaceous of the Gargano Promontory (southern Italy). Cretaceous Research 20:641–654

Richardson JR (1997a) Biogeography of articulated brachiopods. In: Kaesler RL (ed) Treatise on invertebrate paleontology. Part H, Brachiopoda revised, Vol. 1: Introduction. Geological Society of America, Boulder, CO, pp 463–472

Richardson JR (1997b) Ecology of articulated brachiopods. In: Kaesler RL (ed) Treatise on invertebrate paleontology. Part H, Brachiopoda revised, vol. 1: Introduction. Geological Society of America, Boulder, CO, pp 441–462

Rudwick MJS (1970) Living and fossil brachiopods. Hutchinson University Library, London

Ruggiero ET (1991) A study of damage evidence in brachiopod shells. In: Mackinnon DL, Lee DE, Campbell JD (eds) Brachiopods through time. AA Balkema, Rotterdam, pp 203–210

Ruggiero ET (1997) Rhynchonellininae (Brachiopoda) from the Lower Jurassic (Domerian) of Gran Sasso. Palaeopelagos 7:113–120

Sandy MR (1990) A new Early Cretaceous articulate brachiopod from the Northwest Territories, Canada, and its paleobiogeographic significance. J Paleont 64:367–372

Sandy MR (1991) Aspects of Middle to Late Jurassic-Cretaceous Tethyan brachiopod biogeography in relation to tectonic and paleoceanographic developments. Palaeogeogr Palaeoclimatol Palaeoecol 87:137–154

Sandy MR (1995a) A review of some Palaeozoic and Mesozoic brachiopods as members of cold seep chemosynthetic communities: "unusual" palaeoecology and anomalous palaeobiogeographic patterns explained. Földtani Közlöny Bull Hung Geol Soc 125:241–258

Sandy MR (1995b) Early Mesozoic (Late Triassic-Early Jurassic) Tethyan brachiopod biofacies: possible evolutionary intra-phylum niche replacement within the Brachiopoda. Paleobiology 21:479–495

Sandy MR (2001) Mesozoic articulated brachiopods from the Western Cordillera of North America: their significance for palaeogeographic and tectonic reconstruction, palaeobiogeography and palaeoecology. In: Brunton CHC, Cocks LRM, Long SL (eds) Brachiopods past and present. The Systematics Association special vol, 63. Taylor & Francis, London, pp 394–410

Sandy MR, Blodgett RB (1996) *Peregrinella* (Brachiopoda; Rhynchonellida) from the Early Cretaceous, Wrangellia Terrane, Alaska. In: Copper P, Jin J (eds) Brachiopods. AA Balkema, Rotterdam

Sandy MR, Campbell KA (1994) New rhynchonellid brachiopod genus from Tithonian (Upper Jurassic) cold seep deposits of California and its paleoenvironmental setting. J Paleont 68(6):1243–1252

Sandy MR, Campbell KA (1995) Brachiopod morphotypes associated with cold seep and hydrothermal vent environments. GSA Abstr Programs 27(6):445

Sandy MR, Campbell KA (2003) *Anarhynchia* (Jurassic Brachiopoda) in a possible seep deposit from Bedford Canyon, California, USA. GSA Abstracts with Programs 35(6)

Sandy MR, Blodgett RB, Fryda J (2001) Paleobiogeographic signatures for Upper Triassic brachiopods and gastropods from Kuiu Island and adjacent Keku Strait, SE Alaska (Alexander Terrane). Cordilleran Section, GSA Abstracts with Programs, 33:A-53

Sandy MR, Goedert JL, Peckmann J, Taylor D (2005) Brachiopods under attack – evidence of predation on a mass-occurrence of the rhynchonellid brachiopod *Sulcirostra* from an autochtonous cold-seep limestone – Early Jurassic of Oregon, U.S.A. Abstracts, The 5th National Symposium of Paleontology, University of Bucharest, Romania

Senowbari-Daryan B, Gaillard C, Peckmann J (2007) Crustacean microcoprolites from Jurassic (Oxfordian) hydrocarbon-seep deposits of Beauvoisin, southeastern France. Facies 53:229–238

Silberling NJ, Schoellhamer JE, Gray CH Jr, Imlay RW (1961) Upper Jurassic fossils from Bedford Canyon Formation, southern California. Bull Am Assoc Petroleum Geol 45: 1746–1748

Stanley SM (2005) Earth system history, 2nd edn. WH Freeman, New York

Stanton TW (1895) Contributions to the Cretaceous paleontology of the Pacific coast: the fauna of the Knoxville beds. USGS Bull 133:1–132

Stefanoff M, Sandy MR (1998) Evolutionary relationships of *Anarhynchia*, a possible chemosynthetic Jurassic brachiopod from North America. GSA Abstr Programs 30(2):A-72–73

Sulser H, Furrer H (2008) Dimerelloid rhynchonellide brachiopods in the Lower Jurassic of the Engadine (Canton Graubünden, National Park, Switzerland). Swiss J Geosci 101:203–222

Sun DL (1986) Discovery of Early Cretaceous *Peregrinella* (Brachiopoda) in Xizang (Tibet) and its significance. Palaeontol Cathayana 2:211–227

Surlyk F, Johansen MB (1984) End-Cretaceous brachiopod extinctions in the Chalk of Denmark. Science 223:1174–1177

Thieuloy JP (1972) Biostratigraphie des lentilles a peregrinelles (brachiopodes) de l'Hauterivien de Rottier (Drome, France). Geobios 5:5–53

Torres ME, Bohrmann G, Dubé TE, Poole FG (2003) Formation of modern and Paleozoic stratiform barite at cold methane seeps on continental margins. Geology 31:897–900

Von Bitter PH, Scott SD, Schenk PE (1990) Early Carboniferous low-temperature hydrothermal vent communities from Newfoundland. Nature 344:145–148

Von Bitter PH, Scott SD, Schenk PE (1992) Chemosynthesis: An alternate hypothesis for Carboniferous biotas in bryozoan/microbial mounds, Newfoundland, Canada. Palaios 7:451–165

Williams A et al (2000) Treatise on Invertebrate Paleontology. Part H, Brachiopoda Revised, Vol. 2: Linguliformea, Craniiformea, and Rhynchonelliformea (part). In: Kaesler RL (ed) Treatise on invertebrate paleontology. Geological Society of America, Boulder, CO

Williams A et al (2002) Treatise on Invertebrate Paleontology. Part H, Brachiopoda Revised, Vol. 4: Rhynchonelliformea (part). In: Kaesler RL (ed) Treatise on invertebrate paleontology. Geological Society of America, Boulder, CO

Williams A et al (2006) Part H, Brachiopoda Revised, Vol. 5: Rhynchonelliformea (part). In: Kaesler RL (ed) Treatise on invertebrate paleontology. Geological Society of America, Boulder, CO

Williams A et al (2007) Part H, Brachiopoda Revised, Vol. 6: Supplement. In: Selden PA (ed) Treatise on invertebrate paleontology. Geological Society of America, Boulder, CO

Zezina ON (2003) On the ecological, morphological, and evolutionary features of brachiopods living in marginal and extreme environments. Paleont J 37(3):263–269

Zezina ON (2008) Biogeography of the recent brachiopods. Paleont J 42(8):830–858

Chapter 10
Unusual Habitats and Organisms Associated with the Cold Seeps of the Gulf of Mexico

Erik E. Cordes, Stéphane Hourdez, and Harry H. Roberts

Contents

10.1 Unusual Habitats

10.1.1 Geological Setting

The northern Gulf of Mexico continental slope is part of perhaps the most dynamic and structurally complex passive margin in today's oceans. The geologic framework promotes expulsion of fluids and gases from the deep subsurface to the modern seafloor. This framework results from the deposition of enormous volumes

E.E. Cordes (✉)
Biology Department, Temple University, 1900 N 12th St, Philadelphia, PA 19122, USA
e-mail: ecordes@temple.edu

S. Hourdez
Station Biologique de Roscoff, CNRS-UPMC, Roscoff, France

H.H. Roberts
Coastal Studies Institute, Louisiana State University, Baton Rouge, LA, USA

S. Kiel (ed.), *The Vent and Seep Biota*, Topics in Geobiology 33,
DOI 10.1007/978-90-481-9572-5_10, © Springer Science+Business Media B.V. 2010

of river-borne sediment over thick and malleable salt deposits. As part of the Triassic rifting that separated North America, Africa, and South America, the Gulf of Mexico was created. During the mid-to-late Jurassic (~175–145 million years ago) a thick salt unit, the Louann Salt, was deposited in the Gulf (Brooks et al. 1987). This period was characterized by an arid climate. At that time the embryonic Gulf was composed of a complex of individual basins and structural arches, much like the present Red Sea and its gulfs. A high evaporation rate and restricted marine basins led to the creation of thick salt deposits. After salt formation, the Gulf became a shallow basin with extensive algal-rich carbonate tidal flats, similar to today's Persian Gulf and its flanking environments. These late Jurassic and early Cretaceous deposits (~150–100 million years ago), now deeply buried, are known source rocks for a significant part of Gulf's rich oil reserves (Salvador 1987). From the late Cretaceous period which followed to modern times, rivers draining the North American continental interior have delivered large volumes of sediment to the northwestern and northern Gulf. The initiation of this siliciclastic sedimentation during late Cretaceous times corresponded to the rise of the Rocky Mountains in western North America. As a consequence of this massive influx of sediment, underlying salt deposits were deformed into diapirs, tabular salt masses, salt tongues, thrust sheets, and many other geometries (Humphris 1979; Peel et al. 1995). Salt deformation is still an on-going process. Thickening of the southward building siliciclastic sediment wedge and on-going salt deformation create both deep-seated and shallow faults (Brooks et al. 1987; Aharon et al. 1992). These faults function as migration pathways for hydrocarbons generated in the deep subsurface to shallower subsurface reservoirs (Kennicutt et al. 1988). In addition, when these faults provide a clear migration route to the seafloor, they are manifest as hydrocarbon seeps or more dynamic fluid vents.

Across the northern Gulf's continental slope, the major geomorphic elements are smooth, intraslope sedimentary basins separated and bordered by higher relief areas of complex seafloor consisting of ridges and domes usually supported by salt in the shallow subsurface (Fig. 10.1, Plate 20). Features resulting from fluid and gas expulsion are of a much smaller scale than the intraslope basins and bordering ridges and domes. These seep- and vent-related features and the faults that function as the hydrocarbon conduits are concentrated along the flanks of the intraslope sedimentary basins. Features such as mounds of various types, mud volcanoes, mud flows, pockmarks, brine pools, and asphalt seeps result from the expulsion process (Roberts 2001). Using the vast 3D-seismic database that has been developed in support of oil and gas exploration and production in the northern Gulf of Mexico, many sites of fluid-gas expulsion have been localized on the seafloor. As is illustrated by the surface reflectivity map of Fig. 10.2, which was generated from 3D-seismic data, the areas of seepage are "bright spots" of high reflectivity. They are distinct from the surrounding seafloor and result from hard bottom conditions. These sites may take on many forms, including some of the more spectacular occurrences of brine pools, asphalt seeps, and barite chimneys. The presence of such a diversity of geological settings helps to establish the Gulf of Mexico as an excellent natural laboratory for the study of seep systems that may serve as a reference for future work in other known and as-yet-to-be discovered seep systems around the world.

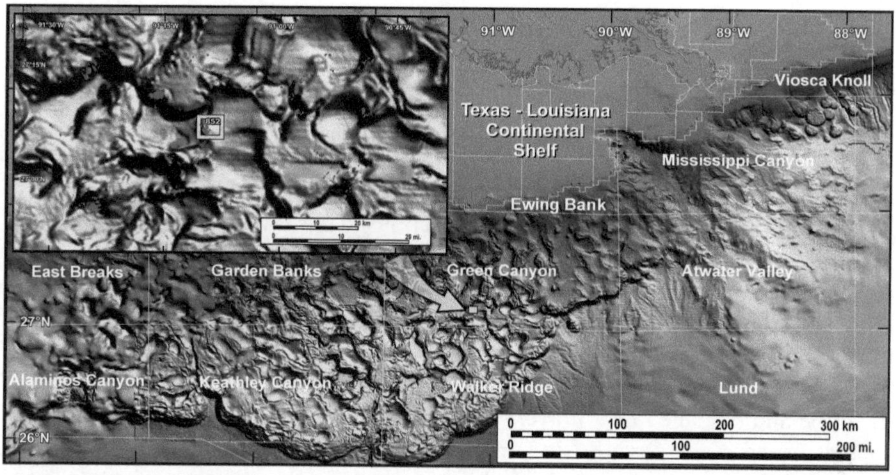

Fig. 10.1 Multibeam bathymetry of the northern Gulf's continental slope, which has been enhanced for a 3D-effect, illustrates the complex of intraslope sedimentary basins (smooth areas) and flanking rough topography comprised of domes and ridges. A color plate of this figure can be found in Appendix (Plate 20)

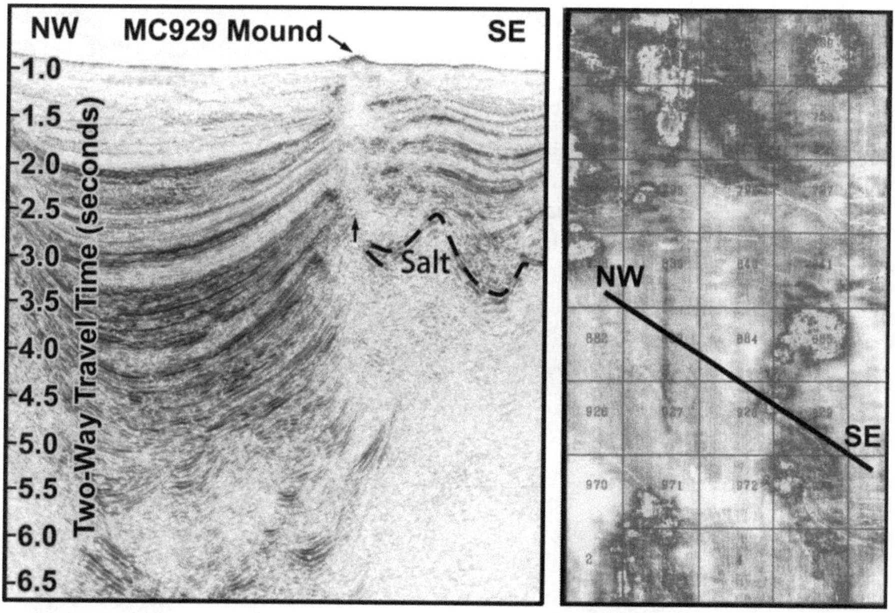

Fig. 10.2 The 3D-seismic surface reflectively map of this figure illustrates a series of seafloor "bright spots" that represent hydrocarbon seep sites. These surface reflectively anomalies are arranged at the flanks of sedimentary basins. The NW-SE oriented seismic cross section illustrates a thick sedimentary basin and the MC 929 mound on the flank of the basin. A distinct vertical fluid-gas migration pathway is shown as an acoustically amorphous zone from the edge of subsurface salt to the seafloor. The MC 929 mound has resulted from the expulsion process. A color plate of this figure can be found in Appendix (Plate 21)

10.1.2 Brine Pools

Hypersaline brine is present on the seafloor of the Gulf of Mexico in numerous locations. This brine is formed from the interaction of pore waters with relatively shallow salt bodies (Reilly et al. 1996). The mobile salt compresses the overlying sediment layers resulting in de-watering of the sediments and a subsequent increase in both salinity and hydrocarbon concentration as the fluids migrate towards the surface, or laterally along density gradients in areas of high bathymetric relief (Roberts and Carney 1997). When these fluids reach the seafloor they may form streams of brine usually in the form of darkly colored anoxic sediments, but may also pool in local depressions formed by salt withdrawal or pockmark craters. The degree of mixing with the overlying seawater is a function of the salinity of the brine and seepage rate with very slow seepage resulting in gradients of salinity such as found at the Orca Basin (Brooks et al. 1990), and high salinity leading to distinct density interfaces such as the brine pool NR1 in Green Canyon lease block 233 (MacDonald et al. 1990) and the brine lake in Alaminos Canyon 601 (Roberts et al. 2007).

The Orca Basin site is a diffuse brine basin descending from 2,000 to 2,400 m depth (Brooks et al. 1990). Below approximately 2,200 m, the salinity gradually increases until it reaches over 200 ppt at the bottom of the basin with very gradual mixing by diffusion with the overlying water column. The megafauna surrounding the basin decreases as the salinity increases and oxygen concentration declines until the community is dominated almost exclusively by hexactinellid sponges with a few holothurians (*Benthodytes typica*) also occasionally present. At salinities above 65 ppt, even these organisms are excluded and the epibenthos appeared entirely depauperate.

The most well-known brine pool of the Gulf is the brine pool NR1. This feature is a crater filled with a slowly seeping dense brine of approximately 120 ppt. The brine maintains a temperature anomaly to the seafloor and is supersaturated in methane (Joye et al. 2005), with visible bubble streams emanating from the center of the pool at times (MacDonald et al. 1990). The pool is surrounded by a ring of *Bathymodiolus childressi* mussels. The mussels of the inner edge of the bed appear to be in better physiological condition, while the mussels of the outer edges of the pool are generally larger but have lower glycogen content and slower growth rates (Nix et al. 1995; Smith et al. 2000). The communities associated with the mussel beds are highly homogenous and consist primarily of the commensal gastropod *Bathynerita naticoidea*, the "snot worm" *Methanoaricia dendrobranchiata*, an undescribed galatheoid *Munidopsis* sp. 1, and the shrimp *Alvinocaris stactophila* (Bergquist et al. 2005).

The Brine Lake in Alaminos Canyon 601 is a larger feature that also contains a distinct interface at the surface of the pool (Plate 23a). It is a highly distinct feature on the seafloor (Fig. 10.3, Plate 22) that was first described from a dive of the deep submergence vehicle *Alvin* in June 2006 (Roberts et al. 2007). The brine interface and the shoreline were quite distinct, but there also appeared to be two former

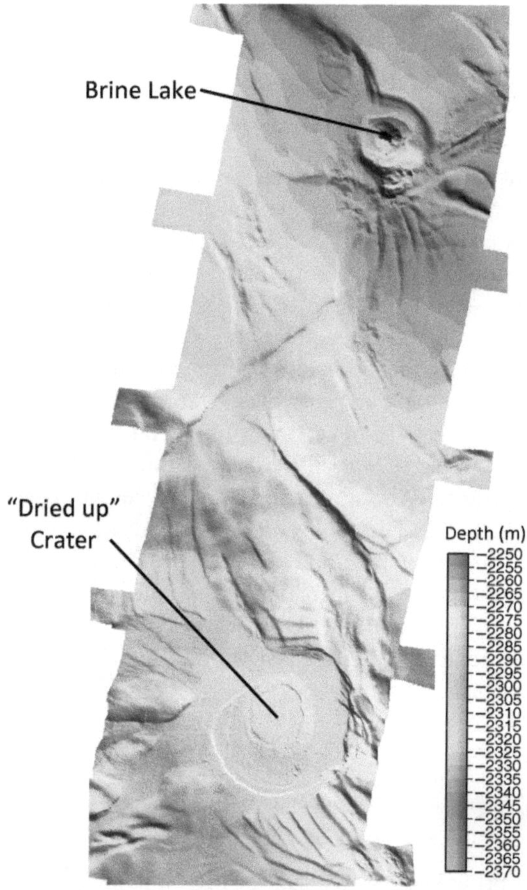

Fig. 10.3 Multibeam bathymetry of the Alaminos Canyon 601 lease block. Bathymetry data acquired by the Hugin AUV operated by C&C Technologies as part of the Expedition to the Deep Slope study funded by the U.S. Minerals Management Service and NOAA Office of Ocean Exploration. In the northern section of the image, the well-defined depression of the brine lake is apparent. In the southern section of the image, the outline of the dried up brine lake can be seen along with possible flows from large expulsion events or drainage of the brine from the crater at the top of the feature. A color plate of this figure can be found in Appendix (Plate 22)

shorelines where the lake contained greater volumes of fluids in the recent geologic past. There were occasional mussel beds composed of *Bathymodiolus brooksi* surrounding the lake interspersed with beds of the heart urchin (spatangid echinoid) *Sarsiaster greigi* (Plate 25). The sediments beneath these communities exhibited low sulfate reduction rates, but high concentrations of hydrogen sulfide. There was also a new species of octopus collected from the shoreline of the Lake that appeared in a well-publicized photograph from Roberts et al. (2007). The fluids of the lake have a salinity of up to 90‰ and exhibited low rates of anaerobic methane oxidation and sulfate reduction. There were barite "rafts" suspended within the brine throughout the lake, suggesting that the majority of the abundant barium contained in seeping pore fluids undergoes rapid precipitation when coming in contact with the sulfate contained in the brine lake.

During another Gulf of Mexico research cruise with the ROV Jason II in July 2007, the large feature to the south of the brine lake was investigated (Fig. 10.3, Plate 22). This 200 m diameter crater was apparent on the AUV multibeam survey conducted, but was not visited during the 2006 cruise. The ROV conducted a survey around the perimeter of the lake, which was a distinct shoreline with small brine pools and brine-stained sediments. On the northern edge of this second feature was an extensive mussel bed estimated to be nearly 3,000 m^2, among the largest continuous biogenic habitats ever observed in the Gulf of Mexico. Although by all appearances this was another brine lake, when a brine sample was attempted near the northern shore of the lake, the sinking bottle encountered the bottom almost immediately. As the ROV transited across the center of the "lake" the brine disappeared as the sediments became more oxidized, including areas of an apparent iron-oxide crust (Plate 23). These observations suggest that this feature is a large crater that fills with brine during active periods that are punctuated by dormant phases (like the one observed) when the lake would "dry up." This could occur either through slow diffusive mixing of the brine with the overlying seawater, or perhaps by a catastrophic failure of the shoreline following the accumulation of a thick layer of dense brine.

10.1.3 Asphalt Flows

In addition to the brine pools and lakes of the Gulf of Mexico, another recently discovered anomalous geological feature are the asphalt seeps of the Gulf. These features were first described from a series of sites in the Chapapote Knolls region of the southern Gulf (MacDonald et al. 2004). The asphalt flows at this site covered large areas of the seafloor and resembled volcanic lava flows. In the vicinity of the flows and within the cracks of the asphalt were siboglinid tubeworms similar in appearance to *Escarpia laminata* from the northern Gulf as well as symbiotic vesicomyid and bathymodioline bivalves.

The source of the heavy asphaltine hydrocarbons is a matter of some debate. It was suggested that the abundance of asphalt at the seafloor was a result of rapid transport from a deep-seated source by supercritical fluid (Hovland et al. 2005). This hypothesis relies on a narrow fault running directly to the crust-sediment interface 13 km below the seafloor where fluids may become heated over 400°C. The buoyant fluid could transport relatively unaltered hydrocarbons to the surface where they would rapid devolatize and form a hardened crust and lava-like appearance while maintaining a heated core of fluid and liquid hydrocarbons. This process relies on the fluids and hydrocarbons remaining over 100°C at the seafloor under pressures over 280 ATM (2,800 m water depth).

The alternative hypothesis for the formation of the Chapopote asphalt flows is the hydrocarbon trap-leakage model (Ding et al. 2008). This involves the presence of a shallow coarse sand reservoir that accumulates hydrocarbons along with intrusions of these hydrocarbons into the overlying thin sediment layers and subsequent leakage to the seafloor. In this large shallow reservoir, only 100–200 m below the seafloor, the hydrocarbons mature through chemical, physical, or biological removal of lighter hydrocarbons. The hydrocarbons are able to appear at the seafloor as asphalt because of the very short, direct connections provided by salt-related faulting of the overlying trap.

During the surveys of the northern Gulf of Mexico in 2007, an asphalt site was discovered at 950 m in Garden Banks 647. The asphalt flows were not as extensive as the Chapopote site, but isolated flows were abundant and spread throughout an area of approximately 500 by 700 m. These were associated with authigenic carbonates, gorgonian corals (*Placogorgia* sp. and *Villogorgia* sp.), and siboglinid tubeworms (*Seepiophila jonesi* and *Escarpia laminata*). Upon collection of one of the tubeworms, a thin string of hydrocarbons came out of the sediments attached to the posterior, buried portion of the tube suggesting that heavy hydrocarbons were abundant in the shallow subsurface even if they were not visible on the seafloor. Some of the asphalt nodules were collected and left on the lab bench following sampling. As they warmed to 20–25°C overnight, one of the nodules cracked and a thick, viscous tar leaked out of the outer crust (see Plate 23c). The shallow depth of this site, the presence of abundant hydrocarbons in surface sediments, and the large area over which evidence of asphalt seepage was apparent support the hydrocarbon trap-leakage model rather than the supercritical fluid hypothesis for asphalt seep formation.

10.1.4 Barite Chimneys

At most hydrocarbon seep sites the bottom is largely composed of Ca-Mg carbonate that forms as a by-product of the chemistry created from microbial oxidation of hydrocarbons. However, in the case of the Mississippi Canyon (MC) 929 mound, imaged on the seismic profile and surface reflectivity map of Fig. 10.2, the hard bottom is composed of primarily barium sulfate or barite (Roberts and Aharon 1994). At selected sites, along with hydrocarbons, barium-rich waters are fluxed from an anaerobic subsurface environment to the aerobic sediment-seawater interface where barite is precipitated. Barite cones (see Plate 23d) have developed over the top of the MC 929 mound along with broad areas of sheet-like barite deposition on the surrounding seafloor. Bathymodiolin mussels are found on the cones as well as scattered on the surrounding seafloor. Associated with these mussels is a very low diversity community consisting of gastropods (primarily *Cataegis meroglypta* and *Bathynerita naticoidea*), a few species of polychaetes (*Nereis* sp. being the most abundant) and unusually high abundances of chitons (*Ischnochiton mexicanus* and *Lepidopleura* spp.) (EEC unpublished data).

10.2 Seep Organisms

10.2.1 Tubeworms

The diversity of vestimentiferan siboglinid tubeworms in the Gulf of Mexico is the highest in the world, with currently five to six species reported (see Fig. 10.4 and Plate 24 for images of *Escarpia laminata* and *Lamellibrachia luymesi*). Three species are found at the shallow sites between approximately 500 and 1,000 m (*Seepiophila jonesi*, *L. luymesi*, and an undescribed species of escarpiid, the latter being very uncommon), and two or three others at depths greater than approximately 1,200 m (*E. laminata*, and one or two species of undescribed *Lamellibrachia*, one common, the other one rare). The common *Lamellibrachia* species on the lower slope appears morphologically distinct, but does not exhibit significant genetic differentiation from *L. luymesi* in the 16S or COI sequences obtained to date (Miglietta et al. in press). Microsatellite data indicate that there is sufficient gene flow among the upper slope (<1,000 m) seep sites in the Gulf of Mexico to prevent genetic drift among populations of *L. luymesi* and *S. jonesi* (McMullin 2003), but additional data is required to examine patterns in gene flow among the populations of the upper and lower continental slope species.

Although there appears to be high gene flow among populations of tubeworms in the Gulf of Mexico, there is an apparent structure in the populations of the symbionts. In the Gulf of Mexico, there are three main clades of sulfide-oxidizing symbionts at the sites surveyed to date, according to depth and/or location: one was found in the Upper Louisiana Slope (550–650 m), another was found at Alaminos Canyon (2,200 m), and the last one at the Florida Escarpment (3,300 m; McMullin et al. 2003). Tubeworm symbionts are transmitted environmentally rather than vertically from one generation to the next (Nelson and Fisher 2000; Nussbaumer et al. 2006). The relationship among symbiont types is not congruent with host phylogeny since at some sites, different siboglinid species may harbor the same symbiont phylotype (e.g. *Lamellibrachia* sp. and *Escarpia laminata* at Alaminos Canyon). This suggests that there is no specific choice of symbiont by the host but rather the acquisition of the local strain of the symbionts from the environment.

Lamellibrachia luymesi was originally described from off the coast of Guyana, and based on morphological characters the common shallow-water species in the Gulf of Mexico has been identified as *L. luymesi* (Gardiner and Hourdez 2003). Two specimens of another species, *L. victori*, were collected off Uruguay (Mañe-Garzon and Montero 1985). Based on morphological characters alone, this species could actually correspond to *L. luymesi* since the morphological variability observed in *L. luymesi* encompasses the morphological characters and proportions described for *L. victori*. Another species of *Lamellibrachia* has been collected off the coast of Chile, although the taxonomic identification of this species remains unresolved at this time (Sellanes et al. 2008). A species of

Fig. 10.4 Fauna of the Gulf of Mexico seeps. (**a**) Habitat formed by the vestimentiferan *Escarpia laminata* and the mussel *Bathymodiolus brooksi* from Alaminos Canyon 818 at approximately 2,750 m depth. Also shown are the shrimp *Alvinocaris muricola*, the galatheid crab *Munidopsis* sp., and the holoturian *Chirodota heheva*. (**b**) A small portion of the 3,000 m^2 mussel bed at the southern crater in Alaminos Canyon 601. The mussels are primarily *Bathymodiolus brooksi* and a worm-like individual of *Chirodota heheva* is present in the upper right corner. This part of the mussel bed lies at the shoreline of the crater where there are still pools of brine. (**c**) The iceworm *Hesiocaeca methanicola* on an outcropping piece of methane hydrate at 525 m depth in Green Canyon 234. (**d**) The tubeworm-associated clam *Acesta oophagia* enclosing the tips of the tubeworm *Lamellibrachia luymesi* at approximately 460 m in Mississippi Canyon 751 (Images A and B courtesy of the Expedition to the Deep Slope investigators, image C courtesy of Chuck Fisher and Erik Cordes, image D courtesy of the Lophelia II: Reefs, Rigs, and Wrecks investigators). A color plate of this figure can be found in Appendix (Plate 24)

Lamellibrachia closely related to the Gulf of Mexico *L. luymesi* has also been found in the Mediterranean (Southward et al. submitted). Based on all of this evidence, *L. luymesi* could be very widespread and inhabit a much larger bathymetric range than previously thought.

A study of communities of the Barbados Accretionary prism also revealed the presence of tubeworms, *Escarpia* cf. *laminata*, and *Lamellibrachia* sp. (Olu et al. 1996). In the REGAB area off Western Africa, another species of *Escarpia* (*E. southwardae*) forms dense bushes in areas with abundant hard substrata (Andersen et al. 2004). Interestingly, this species is genetically indistinguishable from *E. laminata* in the Gulf of Mexico based on COI data (Andersen et al. 2004). In addition, *E. spicata* from the Pacific coast of North America is also nearly identical in COI sequence to these two species (McMullin et al. 2003). It is possible that there is ongoing genetic dispersal throughout this wide biogeographic province. Although connectivity among all of these widely dispersed sites is unlikely, additional exploration for stepping-stone seep sites along the South American margin may help resolve this issue.

The close relatedness of all of these species groups suggests that larvae may be able (either currently or in the past) to travel wide distances over their lifespan. Siboglinids appear to have internal fertilization with the presence of spermathecae observed in females along with the release of fertilized embryos (Hilario et al. 2005). The eggs develop into trochophore larvae that can spend up to several weeks in the water column, potentially permitting a wide dispersal and colonization of active seep sites (Young et al. 1996). This is corroborated by evidence that the larvae of *L. luymesi* are capable of surviving in temperatures as high as 17°C (CM Young 2009, personal communication), and the presence of siboglinid sequences collected in the plankton at depths as shallow as 5 m in the Mediterranean (Marie et al. 2006). This would allow rapid transport of larvae in the higher velocity currents of the surface layers of the ocean and could potentially lead to gene flow among the *Lamellibrachia* and *Escarpia* species throughout the Atlantic and potentially around South America to the Pacific.

Alternatively, there is simply a lack of resolution in the majority of the genetic markers commonly used for phylogeography in this group. It is possible that the evolutionary rate has decreased in at least some lineages of vesitmentiferan tubeworms. Interestingly, the groups for which the genetic markers commonly used fail to discriminate closely related species all inhabit cold seep areas, a type of environment usually stable over thousands of years. These species are characterized by very long life-spans (several hundreds of years), which results in an overlap of many generations of worms, all potentially able to interbreed. This may in fact reduce the apparent mutation rate.

10.2.2 Mussels

The bathymodiolin mussels are also very diverse in the Gulf of Mexico, with a total of at least five species (Gustafson et al. 1998). *Bathymodiolus childressi* occurs at depths ranging from 500 to at least 2,300 m. Two other *Bathymodiolus* species (*B. brooksi* and *B. heckerae*) exclusively occur at deeper sites with depth ranges between 1,080–3,300 and 2,200–3,300 m, respectively (Cordes et al. 2009, see Fig. 10.4 and Plate 24 for images of *B. brooksi*). *Tamu fisheri* has only been collected at shallow sites (500–600 m), and *Idas macdonaldi* has been reported from a single site (GB 386; Gustafson et al. 1998). All of these species are symbiotic with either sulfide-oxidizing, methanotrophic, or both types of bacteria (see chapter 6, this volume). Such a diversity of metabolic types suggests an ability to occupy different ecological niches and avoid competition while co-occurring at the same sites. On the Mid-Atlantic Ridge, Won et al. (2003) showed that co-occurring *Bathymodiolus* species (*B. azoricus* and *B. puteoserpentis*) shared the same sulfide-oxidizing symbiont phylotype, and also harbor very similar methanotrophic symbionts. Based on electron micrographs, the authors suggest that *Bathymodiolus* may continuously acquire symbionts from their environment. Kádár et al. (2005) published data that support this view. These authors showed that the Mid-Atlantic Ridge species *B. azoricus* could lose its symbionts by starving it from sulfide and later re-acquire them when exposed to sulfide again. In the Gulf of Mexico however, when different species of *Bathymodiolus* co-occur, they harbor distinct phylotypes of symbionts. This indicates that, unlike vestimentiferan tubeworms, *Bathymodiolus* species seem to be able to more precisely select their symbionts from the environment.

Bathymodiolus childressi is one of the rare species found at both shallow and deep sites, which prompted interest in the study of the relatedness between shallow and deep populations. An initial study based on allozyme data indicated a clear difference between the two populations separated by 1,700 m of water depth (Craddock et al. 1995). Allozyme markers, however, can be under selection and a specific allelic form may provide a better fitness in deeper populations. Later, Carney et al. (2006) used two mitochondrial and six nuclear markers (including microsatellites, assumed to be neutral), and showed that all the populations studied were interbreeding regardless of depth and distance. A study of spawning and development of *B. childressi* from the shallow water site Brine Pool NR-1 suggests that larvae may spend over a year in the water column (Arellano and Young 2009). This probably provides this species of mussel with a very high dispersive capability, at least throughout the Gulf of Mexico.

Interestingly, genetic comparisons to mussels collected at other sites within the Atlantic Equatorial Belt region (AEB: including the Gulf of Mexico, Blake Ridge, Barbados accretionary prism, Mauritanian margin, and the Gulf of Guinea) revealed the presence of species closely related to *B. childressi* and *B. heckerae* outside of the Gulf of Mexico (Cordes et al. 2007). A genetic study using the mitochondrial COI marker, and the nuclear ITS2 marker confirmed the close relatedness of the different populations, raising questions about possible recent genetic exchange (Olu-Le Roy et al. 2007b). These species complexes are amphi-Atlantic, i.e. found in either side of the Atlantic but not on the Mid-Atlantic Ridge, where two other species of *Bathymodiolus* (*B. azoricus* and *B. puteoserpentis*) inhabit the diffuse areas around hydrothermal vents. This suggests some gene flow in the recent past between the two sides of the Atlantic that may be possible, given the long larval duration estimated for *B. childressi* in the Gulf of Mexico (Arellano and Young 2009).

10.2.3 Iceworms

Another interesting species with a direct tie to seep productivity is the iceworm, *Hesiocaeca methanicola* (see Fig. 10.4c and Plate 24). Iceworms are hesionid polychaetes that inhabit the surface of methane hydrates on the upper Louisiana slope (Desbruyères and Toulmond 1998). They were first discovered living in very high densities (up to 2,500 individuals per square meter) within small depressions carved into an outcropping of methane hydrate (Fisher et al. 2000). The iceworms apparently create these depressions by fanning their parapodia and increasing the water movement over the surface of the clathrate. This may facilitate the sublimation of the hydrate and increase the dissolved methane and sulfide available to free-living microbes from which they likely derive the majority of their nutrition (Fisher et al. 2000). A few small individuals of the same species have occasionally been found in mussel collections at shallow and deep sites of the Gulf of Mexico (Cordes et al. 2007, submitted).

10.2.4 Alvinocaridid Shrimp

The shrimp family Alvinocarididae is endemic to hydrothermal vents and cold seeps worldwide. In the Gulf of Mexico, *Alvinocaris stactophila* is only found at shallow sites (depths around 500–650 m, Carney 1994; Bergquist et al. 2003; Cordes et al. 2006) while a different species, *A. muricola* (see Fig. 10.4a and Plate 24), is encountered at deep sites where they are associated with the tubeworms *Escarpia laminata* and *Lamellibrachia* sp. (Cordes et al. 2007, submitted). A species referred to as *A.* cf. *muricola* was also reported from the Barbados southern seeps at depths of 1,200 and 1,700 m (Olu et al. 1996). Olu-Le Roy et al. (2007a) also report a closely related species in the REGAB pockmark (3,160-m deep) off West Africa, associated with the tubeworms *E. southwardae*. Sequencing of the COI gene for the populations from the Gulf of Mexico, REGAB, and Logatchev (on the Mid-Atlantic Ridge) revealed that they form a monophyletic clade of closely related species (Zelnio and Hourdez 2009). At Blake Ridge cold seeps, however, another species occupies a similar niche (Van Dover et al. 2003). Based on COI, this species is more closely related to another Alvinocarid genus, *Opaepele* (see Zelnio and Hourdez 2009). Similar to the genetic patterns detected in tubeworms and mussels, this suggests either recent or on-going gene flow between the different areas through their larval stage. Interestingly, a recent dive on an experimental wood pile located 300 m away from the Logatchev vent field revealed the presence of *Alvinocaris* sp. on this concentrated biomass source (SH, personal observation). This clearly indicates that habitats other than hydrothermal vents and cold seeps may serve as stepping stones between otherwise isolated populations of *Alvinocaris*.

10.2.5 Tubeworm-Associated Clams

Acesta oophaga (see Fig. 10.4d and Plate 24) was initially thought to be an ecophenotypic variant of *A. bullisi* that grows attached to the tubeworm *Lamellibrachia luymesi*, with a notch in the valves to accommodate the siboglinid's tube (Kohl and Vokes 1994). Later morphological and genetic examination revealed that it was a distinct species, with a shorter and thicker shell, and that the notch in the shell was a normal occurrence for the species (Järnegren et al. 2007). In an earlier study, Järnegren et al. (2005) showed that young *A. oophaga* do not possess a notch, settle near the opening of the tube, with their hinge facing up. At that stage, the bivalves rely on filter-feeding only. Once a little bit larger, the animal turn their shells and eventually grow enclosing the worm's tube opening. The clams are preferentially found on female *Lamellibrachia* (85% of the observations for a tubeworm aggregation with an even sex-ratio), and their stable isotope signatures suggest that as adults a proportion of the diet of the individuals on female worms consists of the embryos released by the tubeworms.

10.2.6 Other Species

The distributional patterns of other seep-associated species also indicate a strong connection between the Gulf of Mexico fauna and other deep-sea areas in the Atlantic. The galatheid *Munidopsis geyeri* is known from the Gulf of Mexico including the Florida Escarpment, the Columbia and Venezuela Basins, and has also been sampled at the seeps of the African margin (MacPherson and Segonzac 2005). The commensal polynoid scale-worm found in *Bathymodiolus* aff. *heckerae* from Africa is genetically similar to *Branchipolynoe seepensis* occurring at the Florida Escarpment (Olu-Le Roy et al. 2007b, D. Jollivet unpublished data). However, the species found in *Tamu fisheri* from the upper slope sites in the Gulf of Mexico is clearly genetically and morphologically distinct (SH, unpublished data). Another polynoid polychaete found at both the shallow and deep seeps of the Gulf of Mexico, *Harmothoe* sp., are genetically indistinguishable based on COI sequences (Cordes et al. submitted).

There are also a few examples of wide distribution in the echinoderms of the seeps in the Equatorial Atlantic Belt. *Ophioctenella acies* is a brittle star that has been found in extremely high abundances in mussel beds from the seeps of the lower slope of the Gulf of Mexico, the Blake Ridge, South Barbados, and also at numerous sites along the mid-Atlantic Ridge (Stöhr and Segonzac 2005). This is one of the first species to be present at all of these habitats and raises the potential for more species to be discovered that cross the seep-vent habitat boundary. The sea cucumber *Chirodota heheva* (see Fig. 10.4b and Plate 24) is considered to be a common resident at seep sites in the Gulf of Mexico and off the southeast coast of the USA, and is also found at anthropogenic habitats (Pawson and Vance 2004), while the same species was reported from the REGAB area off Western Africa (Olu-Le Roy et al. 2007a). *Sarsiaster greigii*, an irregular sea urchin, has been found partially buried in seep sediments, often leaving long trails of apparently reduced sediments behind at the seeps of the Blake Ridge and the deep Gulf of Mexico (van Dover et al 2003; Lessard-Pilon et al. submitted). All of these observations of broad distributions provide evidence for a high dispersal capacity in these seep inhabitants, and allow us to begin to assemble a picture of the biogeography of the cold seep fauna of the world.

10.3 Summary

The seeps of the Gulf of Mexico constitute an incredibly complex but well studied ecosystem that may serve as a model for a variety of other similar settings as they are discovered around the world (Fisher et al. 2007; Cordes et al. 2009). This is particularly relevant given our increasing understanding of the connectivity among deep-water seep ecosystems within the Atlantic basin and world-wide. Even within the familiar setting of the Gulf of Mexico, new discoveries continue to shape our conception of the cold seep ecosystem. New habitat types, new species, and new biogeochemical processes continue to influence our perception of cold seeps, the deep sea, and ecological processes in general. Due to the large volume of existing information on the seeps of the Gulf of Mexico, these new discoveries may be more rapidly interpreted and make the Gulf a relatively accessible and tractable natural laboratory for the further understanding of the complex cold seep ecosystem.

References

Aharon P, Roberts HH, Snelling R (1992) Submarine venting of brines in the deep Gulf of Mexico: observations and geochemistry. Geology 20:483–86

Andersen AC, Hourdez S, Marie B, Jollivet D, Lallier FH, Sibuet M (2004) *Escarpia southwardae* sp. nov., a new species of vestimentiferan tubeworm (Annelida, Siboglinidae) from West African cold seeps. Can J Zool 82:980–999

Arellano SM, Young CM (2009) Spawning, development, and the duration of larval life in a deep-sea cold-seep mussel. Biol Bull 216:149–162

Bergquist DC, Ward T, Cordes EE, McNelis T, Howlett S, Kosoff R, Hourdez S, Carney R, Fisher CR (2003) Community structure of vestimentiferan-generated habitat islands from upper Louisiana slope cold seeps. J Exp Mar Biol Ecol 289:197–222

Bergquist DC, Fleckenstein C, Knisel J, Begley B, MacDonald IR, Fisher CR (2005) Variations in seep mussel bed communities along physical and chemical environmental gradients. Mar Ecol Prog Ser 293:99–108

Brooks JM, Kennicutt MC, Fisher CR, Macko SA, Cole K, Childress JJ, Bidigare RR, Vetter RD (1987) Deep-sea hydrocarbon seep communities: evidence for energy and nutritional carbon sources. Science 238:1138–42

Brooks JM, Wiesenburg DA, Roberts HH, Carney RS, MacDonald IR, Fisher CR, Guinasso NL, Sager WW, McDonald SJ, Burke RA, Aharon P, Bright TJ (1990) Salt, seeps and symbiosis in the Gulf of Mexico. EOS 71:1772–3

Carney RS (1994) Consideration of the oasis analogy for chemosynthetic communities at Gulf of Mexico hydrocarbon vents. Geol Mar Lett 14:149–59

Carney SL, Formica MI, Divatia H, Nelson K, Fisher CR, Schaeffer SW (2006) Population structure of the mussel "*Bathymodiolus*" *childressi* from Gulf of Mexico hydrocarbon seeps. Deep Sea Res I 53:1061–72

Cordes EE, Bergquist DC, Predmore BL, Dienes P, Jones C, Telesnicki G, Fisher CR (2006) Alternate unstable states: convergent paths of succession in hydrocarbon-seep tubeworm-associated communities. J Exp Mar Biol Ecol 339:159–76

Cordes EE, Carney SL, Hourdez S, Carney RS, Brooks JM, Fisher CR (2007) Cold seeps of the deep Gulf of Mexico: community structure and biogeographic comparisons to Atlantic equatorial belt seep communities. Deep Sea Res I 54:637–53

Cordes EE, Bergquist DC, Fisher CR (2009) Macro-ecology of Gulf of Mexico cold seeps. Annu Rev Mar Sci 1:143–168

Cordes EE, Becker EL, Hourdez S, Fisher CR (2010) Biodiversity and biogeography of lower slope vestimentiferan tubeworm and bathymodiolin mussel hosted macrofaunal communities. Deep Sea Res II (in press)

Craddock C, Hoeh WR, Gustafson RG, Lutz RA, Hashimoto J, Vrijenhoek RJ (1995) Evolutionary relationships among deep-sea mytilids (Bivalvia: Mytilidae) from hydrothermal vents and cold-water methane/sulfide seeps. Mar Biol 121:477–85

Desbruyères D, Toulmond A (1998) A new species of hesionid worm, *Hesiocaeca methanicola* sp nov (Polychaeta: Hesionidae), living in ice-like methane hydrates in the deep Gulf of Mexico. Cah Biol Mar 39:93–98

Ding F, Spiess V, Brüning M, Fekete N, Keil H, Bohrmann G (2008) A conceptual model for hydrocarbon accumulation and seepage processes around Chapopote asphalt site, southern Gulf of Mexico: from high resolution seismic point of view. J Geophys Res 113:B08404

Fisher CR, MacDonald IR, Sassen R, Young CM, Macko SA, Hourdez S, Carney RS, Joye S, McMullin E (2000) Methane ice worms: *Hesiocaeca methanicola* colonizing fossil fuel reserves. Naturwissenschaften 87:184–87

Fisher CR, Roberts HH, Cordes EE, Bernard B (2007) Cold seeps and associated communities of the Gulf of Mexico. Oceanography 20:118–29

Gardiner SL, Hourdez S (2003) On the occurrence of the vestimentiferan tube worm *Lamellibrachia luymesi* van der Land and Norrevang, 1975 (Annelida: Pogonophora) in hydrocarbon seep communities in the Gulf of Mexico. Proc Biol Soc Wash 116:380–94

Gustafson RG, Turner RD, Lutz RA, Vrijenhoek RC (1998) A new genus and five new species of mussels (Bivalvia, Mytilidae) from deep-sea sulfide/hydrocarbon seeps in the Gulf of Mexico. Malacologia 40:63–112

Hilario A, Young CM, Tyler PA (2005) Sperm storage, internal fertilization, and embryonic dispersal in vent and seep tubeworms (Polychaeta: Siboglinidae: Vestimentifera). Biol Bull 208:20–28

Hovland M, MacDonald IR, Rueslätten JHK, Naehr T, Bohrmann G (2005) Chapopote asphalt volcano may have been generated by supercritical water. EOS 86:397–402

Humphris CC (1979) Salt movement on continental slope, northern Gulf of Mexico. Am Assoc Petr Geol B 63:782–98

Järnegren J, Tobias CR, Macko SA, Young CM (2005) Egg predation fuels unique species association at deep-sea hydrocarbon seeps. Biol Bull 209:87–93

Järnegren J, Schander C, Sneli JA, Rønningen V, Young CM (2007) Four genes, morphology and ecology: distinguishing a new species of *Acesta* (Mollusca; Bivalvia) from the Gulf of Mexico. Mar Biol 152:43–55

Joye SB, MacDonald IR, Montoya JP, Peccini M (2005) Geophysical and geochemical signatures of Gulf of Mexico seafloor brines. Biogeosciences 2:295–309

Kádár E, Bettencourt R, Costa V, Serrão Santos R, Lobo-da-Cunha A, Dando P (2005) Experimentally induced endosymbiont loss and re-acquirement in the hydrothermal vent bivalve *Bathymodiolus azoricus*. J Exp Mar Biol Ecol 318:99–110

Kennicutt MC, Brooks JM, Bidigare RR, Denoux GJ (1988) Gulf of Mexico hydrocarbon seep communities I. Regional distribution of hydrocarbon seepage and associated fauna. Deep Sea Res A 35:1639–51

Kohl B, Vokes HE (1994) On the living habits of *Acesta bullisi* (Vokes) in chemosynthetic bottom communities, Gulf of Mexico. Nautilus 108:9–14

Lessard-Pilon S, Porter MD, Cordes EE, MacDonald IR, Fisher CR (2010) Community composition and temporal change at deep Gulf of Mexico cold seeps. Deep Sea Res II (submitted)

MacDonald IR, Reilly JF, Guinasso NL, Brooks JM, Carney RS, Bryant WA, Bright TJ (1990) Chemosynthetic mussels at a brine-filled pockmark in the northern Gulf of Mexico. Science 248:1096–99

MacDonald IR, Bohrmann G, Escobar E, Abegg F, Blanchon P, Blinova V, Brükmann W, Drews M, Eisenhauer A, Han X, Heeschen K, Meier F, Mortera C, Naehr T, Orcutt B, Bernard B, Brooks J, de Faragó (2004) Asphalt volcanism and chemosynthetic life in the Campeche Knolls, Gulf of Mexico. Science 304:999–1002

MacPherson E, Segonzac M (2005) Species of the genus *Munidopsis* (Crustacea, Decapoda, Galatheidae) from the deep Atlantic Ocean, including cold-seep and hydrothermal vent areas. Zootaxa 1095:1–60

Mañe-Garzon F, Montero R (1985) Sobre una nueva forma de verme tubicola - *Lamellibrachia victori* n. sp. (Vestimentifera) - Proposicion de un nuevo phylum: mesoneurophora. Rev Biol Uruguay 8:1–28

Marie D, Zhu F, Balagué V, Ras J, Vaulot D (2006) Eukaryotic picoplankton communities of the Mediterranean Sea in summer assessed by molecular approaches (DGGE, TTGE, QPCR). FEMS Microbiol Ecol 55:403–415

McMullin ER (2003) Phylogeography of deep-sea vestimentiferans and a population genetics study of two species, *Lamellibrachia luymesi* and *Seepiophila jonesi*, from the Gulf of Mexico. Ph.D. thesis, Pennsylvania State University, University Park, PA

McMullin ER, Hourdez S, Schaeffer SW, Fisher CR (2003) Phylogeny and biogeography of deep sea vestimentiferan tubeworms and their bacterial symbionts. Symbiosis 34:1–41

Miglietta MP, Hourdez S, Cordes EE, Cowad DA, Fisher CR (2010) Species boundaries of Gulf of Mexico vestimentiferans inferred from mitochondrial genes. Deep Sea Res II (in press)

Nelson K, Fisher CR (2000) Absence of cospeciation in deep-sea vestimentiferan tube worms and their bacterial endosymbionts. Symbiosis 28:1–15

Nix ER, Fisher CR, Vodenichar J, Scott KM (1995) Physiological ecology of a mussel with methanotrophic endosymbionts at three hydrocarbon seep sites in the Gulf of Mexico. Mar Biol 122:605–617

Nussbaumer AD, Fisher CR, Bright M (2006) Horizontal endosymbiont transmission in hydrothermal vent tubeworms. Nature 441:345–348

Olu K, Sibuet M, Harmegnies F, Foucher J-P, Fiala-Medioni A (1996) Spatial distribution of diverse cold seep communities living on various diapiric structures of the southern Barbados prism. Prog Oceanogr 38:347–376

Olu-Le Roy K, Caprais JC, Fifis A, Fabri MC, Galeron J, Budzinski H, Le Menach K, Khripounoff A, Ondreas H, Sibuet M (2007a) Cold-seep assemblages on a giant pockmark off West Africa: spatial patterns and environmental control. Mar Ecol 28:115–30

Olu-Le Roy K, von Cosel R, Hourdez S, Jollivet D (2007b) Do the cold-seep mussels *Bathymodiolus boomerang* and *Bathymodiolus childressi* represent true amphi-Atlantic species? Deep Sea Res I 54:1890–1911

Pawson DL, Vance DJ (2004) *Chiridota heheva*, new species, from Western Atlantic deep-sea cold seeps and anthropogenic habitats (Echinodermata: Holothuroidea: Apodida). Zootaxa 534:1–12

Peel FJ, Travis CJ, Hossack JR (1995) Genetic structural provinces and salt tectonics of the Cenozoic offshore U.S. Gulf of Mexico: a preliminary analysis. In: Jackson MPA, Roberts DG, Snelson S (eds) Salt tectonics: a global perspective. AAPG Mem 65:153–175

Reilly JF, MacDonald IR, Biegert EK, Brooks JM (1996) Geologic controls on the distribution of chemosynthetic communities in the Gulf of Mexico. In: Schumacher D, Abrams MA (eds) Hydrocarbon migration and its near-surface expression. AAPG Mem 66:39–62

Roberts HH (2001) Fluid and gas expulsion on the northern Gulf of Mexico continental slope: mudprone to mineral-prone responses. In: Paull CK, Dillon WP (eds) Natural gas hydrates: occurrence, distribution, and detection, vol 124, Geophysical Monograph. American Geophysical Union, Washington, DC, pp 145–161

Roberts HH, Aharon P (1994) Hydrocarbon-derived carbonate buildups of the northern Gulf of Mexico continental slope: a review of submersible investigations. Geol Mar Lett 14:135–148

Roberts HH, Carney RS (1997) Evidence of episodic fluid, gas, and sediment venting on the northern Gulf of Mexico continental slope. Econ Geol 92:863–79

Roberts HH, Fisher CR, Bernard B, Brooks JM, Bright M, Carney RS, Cordes EE, Goehring L, Hourdez S, Hunt JL Jr, Joye SM, MacDonald IR, Morrison C, Niemann H, Potter J, Samarkin V, Shedd W, Telesnicki G, Becker EL, Bernier M, Bowles M, Kupehik M, Lessard-Pilon S, Petersen C, Petersen J (2007) *ALVIN* explores the deep northern Gulf of Mexico slope. EOS 88:341–42

Salvador A (1987) Late Triassic-Jurassic paleogeography and origin of Gulf of Mexico Basin. Am Assoc Petr Geol B 71:419–451

Sellanes J, Quiroga E, Neira C (2008) Megafaunal community structure and trophic relationships of the recently discovered Concepción Methane Seep Area (Chile, ~36°S). ICES J Mar Sci 65:1102–1111

Smith EB, Scott KM, Nix ER, Korte C, Fisher CR (2000) Growth and condition of seep mussels (*Bathymodiolus childressi*) at a Gulf of Mexico brine pool. Ecology 81:2392–403

Southward EC, Anderson A, Hourdez S (2010) *Lamellibrachia anaximandri* n. sp., a new vestimentiferan tubeworm from the Mediterranean. Zoosystema (submitted)

Stöhr S, Segonzac M (2005) Deep-sea ophiuroids (Echinodermata) from reducing and non-reducing environments in the North Atlantic Ocean. J Mar Biol Ass UK 85:383–402

Van Dover CL, Aharon P, Bernhard JM, Caylor E, Doerries M, Flickinger W, Gilhooly W, Goffredi SK, Knick K, Macko SA, Rapoport S, Raulfs EC, Ruppel C, Salerno JL, Seitz RD, Sen Gupta BK, Shank T, Turnipseed VR (2003) Blake Ridge methane seeps: characterization of a soft-sediment, chemosynthetically-based ecosystem. Deep Sea Res I 50:281–300

Won YJ, Hallam SJ, O'Mullan GD, Pan IL, Buck KR, Vrijenhoek RC (2003) Environmental acquisition of thiotrophic endosymbionts by deep-sea mussels of the genus *Bathymodiolus*. Appl Environ Microbiol 69:6785–6792

Young CM, Vázquez E, Metaxas A, Tyler PA (1996) Embryology of vestimentiferan tube worms from deep-sea methane/sulphide seeps. Nature 381:514–16

Zelnio KA, Hourdez S (2009) A new species of *Alvinocaris* (Crustacea: Decapoda: Caridea: Alvinocarididae) from hydrothermal vents at the Lau Basin, southwest Pacific, and a key to the species of Alvinocarididae. Proc Biol Soc Wash 122(1):52–71

Chapter 11
Biological Communities at Marine Shallow-Water Vent and Seep Sites

Paul R. Dando

Contents

11.1 Introduction

Deep-sea vents and seeps are well known to have communities of animals that are obligate to these environments, where reducing fluids are released from the seafloor and where carbon is fixed by chemosynthesis (Tunnicliffe 1991; Sibuet

P.R. Dando (✉)
Marine Biological Association of the United Kingdom, Citadel Hill, Plymouth PL1 2PB, UK
e-mail: pdando@mba.ac.uk

S. Kiel (ed.), *The Vent and Seep Biota*, Topics in Geobiology 33,
DOI 10.1007/978-90-481-9572-5_11, © Springer Science+Business Media B.V. 2010

and Olu 1998). These habitats are characterized by a high biomass but a low species diversity, mainly consisting of species obligate to such sites (Van Dover 2000). In this respect they differ from 'shallow' sites that have few, if any, obligate species and usually a much lower biomass (Sibuet and Olu 1998; Tarasov et al. 2005). The question therefore arises as to what is the depth cut-off between 'shallow' and 'deep' vent and seep sites. A comparison of vent sites at different depths revealed that this was approximately 200 m, as shown by the degree of obligacy of the fauna (Tarasov et al. 2005). At shallower depths, vent-obligate taxa were either absent or only one or two vent obligate species were present and there were no vent-obligate genera or higher taxa. In contrast, vent-specific genera, families, or higher taxa, occurred at greater depths (Sibuet and Olu 1998; Tunnicliffe 1991; Sahling et al. 2003; Tarasov et al. 2005). This 200 m cut-off may have a physical cause at vent sites in that there is a sharp decrease in the slope of the seawater-pressure boiling curve below this depth (Bischoff and Rosenbauer 1984; Butterfield et al. 1990) at 20 bar.

In their review of the cold-seep communities at 24 sites, Sibuet and Olu (1998) only considered sites between 400 and 6,000 m depth. All of these contained chemosynthesis-based communities. Subsequently a review of the meiofaunal and macrofaunal studies at seep sites ranging from 168 to 5,000 m revealed that species dependent upon chemosynthesis were present at the majority of these sites (Levin 2005). There are a lot of species with chemosynthetic symbionts inhabiting cold-seeps that are also found in non-seep reducing sediments (Sahling et al. 2003) making it difficult to be certain whether many species are seep-obligates. However, in the Gulf of Mexico seep-obligate *Bathymodiolus* sp. and vestimentifera have been reported from a site at 290 m (Roberts et al. 1990) and in the Skagerrak the seep-obligate frenulate with methane-oxidising bacteria, *Siboglinum poseidoni*, occurs as shallow as 280 m (Schmaljohann et al. 1990).

This review brings together, for the first time, details of the shallow-water hydrothermal vents (Table 11.1) and cold seeps (Table 11.2) for which biological data is available. These tables list both the species dependent, or partially dependent, upon chemosynthesis at the different sites and the typical heterotrophic species found at the sites. Species that are believed to be vent or seep obligates or, in some cases, vent and seep obligates, are outlined in bold in these tables. The characteristics and the biological communities at shallow-water and vent sites are compared and contrasted. Comparisons are also made with their deep-sea counterparts. The term 'shallow' has been taken to refer to sites down to 200 m water depth. At such depths organic matter formed by photosynthesis is important for the nutrition of most of the organisms and, unlike deep-sea vent and seep sites, chemosynthesis has a less important, although often significant, role. The locations of the shallow-water vent sites mentioned are shown in Fig. 11.1 and those for the cold seeps are shown in Fig. 11.2. The concentration of sites in the northern hemisphere reflects research activity, rather than a true distribution of shallow vent and seep sites.

Table 11.1 Shallow hydrothermal vent sites for which data on the biota is available – vent obligate species are in bold

Site	Depth	Chemosynthetic spp.	Dominant, non-chemosynthetic spp.	Reference
Various sites, Isle of Ischia, Italy	0–0.5 m	–	Cyanobacteria, nematodes	Meyl (1954)
Castellamare, Bay of Naples, Italy	0–1 m?	*Achromatium volutans*	–	Hinze (1903)
Cape Goryachii, Kunashir Island, Kuriles	0–1 m	**Thermophilic, anaerobic, carboxydotrophic bacteria**	Cyanobacteria, *Haliplannella luciae*	Tarasov and Zhirmunsky (1990); Svetlichny et al. (1991)
Castello Aragonese, Isle of Ischia, Italy	0–3 m	–	*Posidonia oceanica*, non-coraline algae	Hall-Spencer et al. (2008)
Kameni Islands, Santorini, Aegean	1–5 m	Iron-oxidising bacteria	*Malacocerous fuliginosus*, *Capitella* sp.	Holm (1987); Dando et al. (1995c)
Nishino-shima shintoh, Ogasawara Islands	5 m	–	***Xenograpsus novaeinsularis***	Takeda and Kurata (1977)
Akuseki-jima, Tokara Islands	5 m	–	***Xenograpsus novaeinsularis***	Takeda et al. (1993)
Fumarole Bay, Port Foster, Deception Island	0.25–6 m	Monocelid sp.	–	Bright et al (2003)
Vulcano Island, Italy, Tyrrhenian Sea	0.5–8 m	Thiobacilli, ***Ferroglobus placidus***	Heterotrophic (hyper) thermophilic Bacteria and Archaea, cyanobacteria, thermotolerant algae including: *Cystoseira crinita*, *Peyssonnelia harveyana*, Ceramineacea, *Grateloupia dichotoma*	Giaccone (1969); Gugliandolo and Maugeri (1993); Hafenbrandt et al. (1996); Acunto and Rindi (1997); Rusch et al. (2005)
Kodakara-Jima Island, south of Kagoshima, Japan	3–10 m	*Beggiatoa*	Hyperthermophilic, sulphur-dependent heterotrophic bacteria (no macrofauna present)	Hoaki et al. (1995)
Kiriaki Bay, Milos, Aegean	5–10 m	–	*Spio filiforms*, *Capitella* sp., *Microprtous maculata*	Dando et al. (1995c)

(continued)

Table 11.1 (continued)

Site	Depth	Chemosynthetic spp.	Dominant, non-chemosynthetic spp.	Reference
Punta Mita, Mexico	10 m	–	Thecata (Campanularidae), serpulids, spirorbids, ostracods, bryozoans	Canet et al. (2003); P. Dando and A. Howes (1994)
McEwans Bay, Whale Island, Bay of Plenty, New Zealand	3–12 m	*Thiothrix*, stilbotnematids	*Cominella adspersa, Tawera spisa, Perna canaliculus*	Kamenev et al. (1993)
White Island, Bay of Plenty, New Zealand	5–13 m	*Beggiatoa*	No macro-invertebrates, sparse nematodes	Kamenev et al. (1993)
Bahía Concepción, Gulf of California	12–13 m	–	*Holothuria inhabilis*	Forrest (2004); Melwani and Kim (2008)
Lefkos Bay, Nisyros, Aegean	18 m	–	Carnivorous and omnivorous polychaetes	Diapoulis et al. (1994)
White Point, California	Intertidal –20 m	*Thiothrix*	*Haliotis cracherodii, Lottia limatula, Tegula funebralis, Pachygrapsus crassipus, Pagurus samuelis*, spionids	Trager and DeNiro (1990); Melwani and Kim (2008)
Kraternaya Bight, Yankicha Island, Kuriles	2–20 m	*Thiothrix* sp., *Seliberia* sp., **thermophilic, anaerobic, carboxydotrophic bacteria**, *Axinopsida orbiculata orbiculata*	*Mesodinium rubrum, Cerianthus sp., Myxicola infundibulum, Amphitite cirrata*	Tarasov et al. (1990); Kuznetsov and Gebruk (1991); Svetlichny et al. (1991)
Milos brine seeps, Aegean	5–20 m	*Achromatium volutans, Thiomicrospira spp.*	*Cyclope neritea*, ***Capitella* sp. M**, *Oncholaimus campyloceroides*	Dando et al. (1995c); Thiermann et al. (1997); Brinkhoff et al. (1999)
Kueishan Island, Taiwan	8–20 m	–	***Xenograpsus testudinatus***	Jeng et al. (2005)
Bahia de Pozzuoli, Naples, Italy	10–20 m	*Beggiatoa*	*Amphiglena mediterranea*	Gimenez and Marin (1991)
Panarea, Aeolian Islands, Italy	11–21 m	–	Nematodes	Colangelo et al. (1996)
Kito-Iwo-jima, Iwo islands	23 m	–	***Xenograpsus novaeinsularis***	Takeda et al. (1993)
Matupi Harbour, New Britain	0–27 m	*Beggiatoa, Thiovulum, Thyasira* sp.	Sponges (*Petrosia, Scleritoderma, Stromatoderma*), *Sabella* sp., *Hippa* sp., *Macrophalmus* sp.	Tarasov et al. (1999)

Grotta Azzura, Capo Palinuro, Italy	0–30 m	*Beggiatoa, Thiothrix*	*Geodia cydonium, Astroides calycularis, Phyllochaetopterus socialis, Thalassodrilides gurwitschii, Paracypris complanatas*	Morri et al. (1994); Southward et al. (1996); Mattison et al. (1998)
Panarea, Aeolian Islands, Italy,	10–30 m	–	*Posidonia oceanica*	Acunto et al. (1995); Colangelo et al. (1996)
Dom João de Castro Seamount, Azores	19–34 m	*Beggiatoa*	*Codium elisabethae, Sargassum* cf. *vulgare, Enoplus meridonalis, Lumbricillus* sp.	Cardigos et al. (2005)
Volcanic depression, Whale Island, Bay of Plenty, New Zealand	25–45 m	*Thiothrix,* ***Siboglinum*** **sp.**	*Cominella maculosa, Macrothalmus hurtipes, Proharpinia* sp.	Kamenev et al. (1993)
Palaeochori, Milos, Aegean	46 m	*Thioploca* sp.	–	Dando et al. (1998)
Esmeralda Bank, Marianas	63–114 m		***Paraglypturus calderus, Xenograpsus novaeinsularis***	Türkay and Sakai (1995)
Kiriakis Point, Milos, Aegean	60–100 m	–	Sponges, anthozoans, crinoids, serpulids, *Bonellia* cf. *viridis*	Dando et al. (1995c)
Kolbeinsey, Iceland	90–100 m	*Beggiatoa*	*Scypha quadrangulatum, Tethya aurantium, Corymorpha groenlandica*	Fricke et al. (1989)
Milos Bay, Milos, Aegean	60–110 m	–	Bacterial mats, *Bonellia* cf. *viridis*	Dando et al. (1995a, c)
Kagoshima Bay, Japan	82–110 m	***Lamellibrachia satsuma****, Solemya* sp.	Polychaetes, palaeomonid shrimps, galatheids	Hashimoto et al. (1993)
'Calypso' vents, Bay of Plenty, New Zealand	150–200 m	***Siboglinum*** **sp.**	Very diverse background fauna	Kamenev et al. (1993)
Macauley Cone, Kermadec Ridge	200 m	***Bathymodiolus*** **sp., '*Idas*' sp.**	–	Smith et al. (2004)
Steinahóll, Reykanes Ridge, Iceland	200 m	–	*Bonellia viridis*	P.R. Dando (1992)

Table 11.2 Shallow-water cold seep sites for which data on the biota is available

Site	Depth	Chemosynthetic species	Other characteristic species	Reference
Pockmarks with methane seeps				
Elaiona Bay, Gulf of Corinth	29 m	–	Sabellid polychaetes	Christodoulou et al. (2003)
Passamaquoddy Bay, Canada	35–81 m	*Beggiatoa, Thyasira flexuosa, Solemya* sp.	*Yoldia* sp. *Sapotilla* sp., *Euchone incolor*	Wildish et al. (2008)
Tommeliten seepage area, North Sea	65–75 m	–	*Pennatula phosphorea, Funiculina quadrangularis, Flustra foiacea, Sabella* sp., anthozoa	Hovland and Thomsen (1989)
Black Sea, Dneper Paleo-delta	92 m	Bacterial mats, probably*Beggiatoa*	Tunicates	Naudts et al. (2008)
North Sea, Braemar pockmarks	122–125 m	*Beggiatia, Siboglinum fiordicum, Lucinoma borealis, Axinulus croulinensis, Thyasira equalis*	*Paramphione jeffraysi, Paradoneis eliasoni*	Hartley (2005)
Oregon shelf	132 m	–	Bacteria and <10 μmheterotrophic flagellatesmore abundant in bottomwater	Juhl and Taghon (1993)
North Sea, Scanner pockmark	172 m	*Beggiatoa, Astononema southwardorum, Axinulus croulinensis, Thyasira equalis, T. sarsi*	*Uvigerina peregrina,Cassidulina laevigata, Hyalinea balthica, Elphidium clavatum, Retusa truncatula, Spiophanes kröygeri, Pholoe* sp.,*Myriochele* sp.	Dando et al. (1991); Jones (1993); Dando (2001)
Grand Banks, Newfoundland	100–200 m	*'Beggiatoa'*	Ophiuroids, hotothurians	Fader (1991)
Other hydrocarbon seeps				
Kattegat, Denmark	0–2 m	*Beggiatoa, Leptonemella aphanothecae*	*Lanice conchilega, Mytilus edulis, Balanus crenatus, Carcinus maenus, Asterias rubens*	Dando et al. (1994b)
Cape Tarkhankur, Black Sea	2.5 m	–	Nematodes, harpacticoids, ostracods	Schornikova and Syrtlanova (2008)
Hirstholmene 'bubbling reefs', Kattegat, Denmark	10–20 m	*Beggiatoa*	*Metridium senile* and other fouling epifaunal species, *Theristus anoxybiotus*	Jensen et al. (1992); Jensen (1995a)

Laeso, Kattegat, Denmark	15 m	*Leptonemella aphanothecae*	Cnidarians, hydrozoans,bryozoans	Jensen et al. (1992)
Isla Vista oil seeps, California	15 m	*Beggiatoa*	*Limnodriloides verrucosa, L. monothecus*	Davis and Spies (1980)
Californian Shelf	31–53 m	*Beggiatoa*, *Solemya reidi* at one site	*Capitella* sp., *Mediomastus* spp. and *Lumbrineris* sp. at one site	Levin et al. (2000)
Louisiana Shelf	60–75 m	*Beggiatoa*	–	Sassen et al. (2003)
Black Sea, Crimean shelf	60–190 m	Bacterial mats, no invertebrates with symbiotic prokaryotes observed	*Modiolus phaseolinus*, other bivalves, nematodes, polychaetes, oligochaetes	Pimenov et al. (1997); Luth et al. (1999)
Louisiana Shelf	130 m	*Beggiatoa*	Coralline algae, sponges, corals, crinoids, brachiopods	Sassen et al. (1993)
Sakhalin outer shelf	160–250 m	Bacterial mats, *Siboglinum plumosumn*	Asteroids, no infaunal samples taken	Sahling et al. (2003)
Black Sea, Dneper Canyon	182–252 m	Bacterial mats	Nematodes and foraminifera (cladocerans and copepods also important below 200 m)	Sergeeva (2003); Sergeeva and Gulin (2007)
Submarine brine seeps				
East Flower Garden	71 m	Sulphur-oxidising bacteria	Gnathostomulids	Bright et al. (1980); Powell et al. (1986)
Submarine groundwater seeps				
Königshafen, Sylt, North Sea	Intertidal	*Thiovulum majus*	*Nereis diversicolor* and *N. virens*	Zipperle and Reise (2005), P. Dando, unpublished (2007)
Isle of Man, Irish Sea	Intertidal	–	Euryhaline copepods	Moore (1979)
Biscayne Bay, Florida	0–20 m	–	Alga, *Batophora oerstedi*; seagrass, *Diplanthera wrightii*; gastropod *Neritina virginea*; mysid, *Taphromysis bowman*; killifish, *Lucania parva*	Kohout and Kolipinski (1967)
Eckernförde Bay, Baltic Sea	25 m	*Beggiatoa*	–	Bussmann et al. (1999)
Cambridge Fjord, Baffin Island	45 m	–	*Laminaria*, anthozoa, pectinids, decapods	Colbourne and Hay (1990)

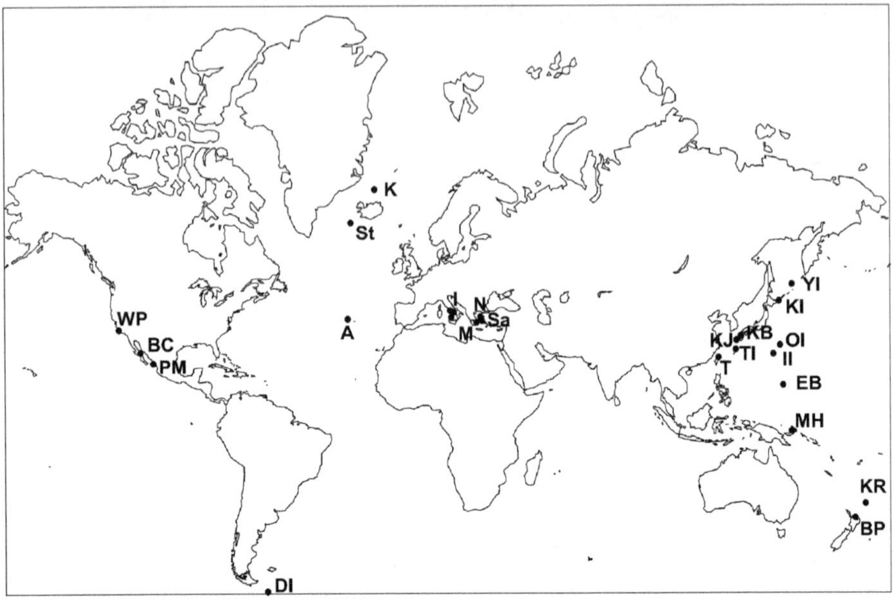

Fig. 11.1 Locations of shallow water hydrothermal vents. Azores (A), Bahía Concepción (BC), Bay of Plenty (BP), Deception Island (DI), Esmeralda Bank (EB), Italy (I), Iwo Islands (II), Kolbeinsey (K), Kagoshima Bay (KB), Kunashir Island (KI), Kodakara-Jima Island (KJ), Kermadec Ridge (KR), Milos (M), Matupi Harbour (MH), Nisyros (N), Ogasawara Islands (OI), Punta Mita (PM), Santorini (Sa), Steinahóll (St), Taiwan (T), Tokara Islands (TI), White Point (WP), Yankicha Island (YI)

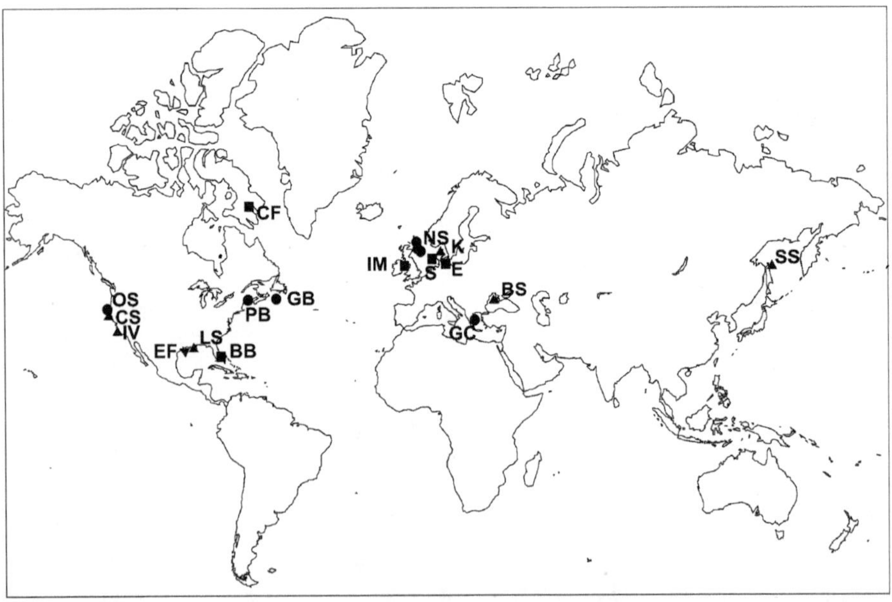

Fig. 11.2 Locations of shallow water cold seeps. Circles: pockmarks; triangles: other hydrocarbon seeps; triangle upside down: brine seep; squares: submarine groundwater seeps. Biscayne Bay (BB), Black Sea seeps (BS), Californian shelf seeps (CS), Cambridge Fjord (CF), Eckernförde Bay (E), East Flower Garden (EF), Grand Banks (GB), Gulf of Corinth (GC), Isle of Man (IM), Isla Vista oil seeps (IV), Kattegat seeps (K), Louisiana shelf seeps (LS), North Sea pockmarks (NS), Oregon Margin (OS), Pasammaquaddy Bay (PB), Sylt (S), Sakhalin shelf (SS)

11.2 Characteristics of Shallow-Water Vent and Seep Sites

11.2.1 Locations

In general, shallow-water vent sites occur near active coastal/island or submarine volcanoes (Gamo and Glasby 2003). Although some occur on mid-ocean ridges, such as on the mid-Atlantic Ridge near Iceland (Fricke et al. 1989; German et al. 1994) and on the ultra-slow spreading Terceira Rift in the Azores (Cardigos et al. 2005). Most shallow vents are found above subducting slabs, especially along volcanic arcs. Examples include: the Kermadec Arc (de Ronde et al. 2001), the Kuriles (Zhirmunsky and Tarasov 1990), the Ryu Kyu Arc (Chen et al. 2005), the Okinawa Arc (Jeng et al. 2005), the Marianas Arc (Türkay and Sakai 1995), the Lesser Antilles Arc (McCarthy et al. 2005), the Aegean Arc (Dando et al. 2000), the Calabrian Arc (Dando et al. 1999). A number of shallow vents sites are located on hot spots, e.g. the Hawaiian (Sansone and Resing 1995; Baker et al. 2002), Icelandic (Benjamínsson 1988), Austral (Cheminee et al. 1991) and Samoan (Staudigel et al. 2004). A few occur in continental margin settings undergoing tectonic extension, as along the east Pacific and Gulf of California coasts of California and Mexico (Vidal et al. 1978; Prol-Ledesma et al. 2000, 2004; Melwani and Kim 2008).

Shallow-water, cold seepage sites are more widely distributed than hydrothermal vent sites, being found on both shelves and slopes, although relatively few have been studied for their biota. Submarine groundwater seeps are common along coasts where fresh water is discharged through marine sediments (Schlüter et al. 2004; Stieglitz 2005). Hydrocarbon seeps are common in sedimentary basins where over-pressured fluids escape along fractures and faults (Hovland and Judd 1988). In contrast, shallow, cold brine seeps, resulting from the dissolution of salt diapers, have only been reported to date from the Gulf of Mexico, although deep-water brine seeps are common in other semi-enclosed seas, such as the Mediterranean.

11.2.2 Physical and Chemical Characteristics of Vents and Seeps

At deep-sea vent sites, physical and chemical factors, such as sulphide concentrations, surface type and topography, affect the distribution of organisms (Sarrazin et al. 1999). An additional factor affecting organisms at many shallow-water vent and seep sites is the release of large quantities of free gas. Although gas venting has been observed from seep sites as deep as 2,080 m (Greinert et al. 2006), the volume of gas released and the frequency of gas venting is markedly higher at shallow sites. This is due to the greatly reduced solubility of gases at lower pressures, leading to bubble formation as gas-saturated water rises through the sediment (Wiesenberg and Guinasso 1979; Duan and Weare 1992). Gas escape causes sediment sorting as finer grain sediment is dispersed (Dando et al. 1995a) and also stimulates a re-circulation of overlying water through the sediment (O'Hara et al. 1995). At hydrothermal sites, the shallow re-charge area surrounding the outlet has a markedly lower temperature.

In shallow-water vents (Dando et al. 2000), as at their deep-sea counterparts (Butterfield et al. 1990), phase separation can occur. This can lead to the discharge of both low and high salinity fluids at different outlets in the same area (Dando et al. 2000). In extreme cases, hot brines can pond on the seabed (Fitzsimons et al. 1997; Dando et al. 2000). At nearby outlets reduced salinity fluids, containing condensates of the boiled seawater, are discharged. Close to shore venting fluids may have reduced salinity due to the input of meteoric water (Stüben and Glasby 1999; Dando et al. 2000). At some sites, the hydrothermal reservoir contains a major input from meteoric water and all discharges are of low salinity fluids (Pichler and Dix 1996; Prol-Ledesma 2003), a situation that differs from that at deep-sea vents. Low salinity discharges are even more pronounced at coastal sites where submarine groundwater springs discharge, e.g. Kohout and Kolipinski (1967), a situation that does not occur at any deep-sea cold seep sites.

Fluid flow rates are important since these affect the geochemical gradients and hence both mineral deposition and the structure of the biological communities (Levin 2005). Particularly in shallow-water settings, these flow rates are tidally influenced and also affected by water pressure (Dando et al. 1994c; Stüben et al. 1996; Aliani et al. 1998a). Abrupt changes due to tectonic activity and seal failure also occur (Dando et al. 1995b; Makropoulos et al. 2000) and these may have dramatic effects on the local ecosystem.

A limiting physical factor at vents is temperature. Steep gradients are found in the sediment with seawater sometimes boiling on the seabed (Dando et al. 1995c; Sievert et al. 2000c). In semi-enclosed embayments venting can raise the overall water temperature, affecting the faunal composition of the entire bay (Aliani et al. 2000). At vents, sulphide is formed by a reaction between basalt and seawater at high temperature (Bischoff and Dickson 1975). Dissolved sulphide and heavy metal concentrations tend to be lower at shallow-water vent sites because of the subsurface precipitation of metals below 300°C (Tivey 2007), rather than forming metallic sulphide deposits around the vent outlets. Thus, shallow vents can be less toxic to organisms than many of their deep-sea counterparts, but see Luther et al. (2001) for details of low-H_2S microhabitats at deep-sea vents. In many shallow-water hydrothermal vent settings most of the sulphide released can be formed by sulphate reduction in the sediment (Miura et al. 2002). Sulphate reduction in hydrothermal vent sediments can occur at temperatures up to 110°C (Jørgensen et al. 1992).

Sulphide at cold seeps is all formed by sulphate reduction. The concentrations of sulphide, at the seeps, are usually higher than those in the surrounding sediments. This is because of the formation of sulphide by a consortium of methane-oxidising archaea, anaerobically oxidizing methane to hydrogen and CO_2, and sulphate reducing bacteria that use the hydrogen to reduce sulphate to sulphide (Boetius et al. 2000).

Most of the methane found at some shallow vents systems, such as those at Milos, is probably formed from CO_2-reduction due to the Fischer-Tropsch reaction (Botz et al. 1996), whereas at other sites thermogenic production within the heated sediment is likely (Welhan and Lupton 1987). Methane production due to methanogenesis by Archaea has a minor role. In contrast, at many shallow cold seeps such as those in the Kattegat, biogenic formation of methane dominates over thermogenesis (Dando et al. 1994c).

Sudden release of methane, due to seal failure or hydrate breakdown, is believed to be the cause of many seabed depressions, pockmarks (McQuillan et al. 1979; Hovland and Judd 1988). High methane concentrations at seeps leads to anaerobic methane oxidation linked to sulphate reduction, with an increase in sulphide, dissolved inorganic caron (DIC) and alkalinity in the sediment. This leads to carbonate precipitation, such that carbonate-cemented sediments are a notable feature of many sites, such as the spectacular carbonate pillars, chimneys and reefs in the Kattegat (Jensen et al. 1992). Carbonate-cemented substrata are also a feature of many methane-seeping pockmarks (Hovland and Judd 1988). Studies on the precipitation of methane-derived carbonate, at a seep in the Kattegat, showed that the most active precipitation zone was at 5–10 cm depth within the sediment (Schuster 1994). Subsequent erosion of the overlying sediment then frequently exposes the resulting lithified deposits on the sea floor (Jørgensen 1992).

Pockmarks that are inactive, or only slightly active, no longer displace sediment and act as sediment traps due to a reduction in bottom currents (Yager et al. 1993). Current reduction, below a critical minimum, means that such depressions can be areas of increased larval settlement (Snelgrove 1994). Pockmark bases might therefore be expected to show an increase in the abundance of deposit feeding organisms compared with the surrounding sediment. Such changes in communities would be dependent on the change in bottom current velocity inside the pockmark compared with outside (Ólafsson et al. 1994). This would be quite unrelated to any factors due to recent or current seepage.

11.3 Species Distribution and Diversity

11.3.1 Hydrothermal Vents

Shallow-water hydrothermal vents are largely colonized by a subset of the background macro- and meio-fauna (Fricke et al. 1989; Tarasov and Zhirmunsky 1989; Dando et al. 1995c; Thiermann 1998; Morri et al. 1999; Tarasov et al. 1999). There is commonly a decline in species numbers along a transect towards the vents due to increasing temperatures and sulphide concentrations and decreasing pH (Fricke et al. 1989; Thiermann et al. 1997; Bodergat et al. 2002; Melwani and Kim 2008). The numbers and the biomass of the dominant species at the vent sites increase towards the centre (Fricke et al. 1989; Kamenev et al. 1993). These dominant species are often relatively scarce, or not present, in the surrounding environment, for example the hydroid *Corymorpha groenlandica* and the solitary sponge *Scypha quarrangulatum* at the Kolbeinsey vents (Fricke et al. 1989) and the gastropod *Cyclope neritea* at the Milos hydrothermal brine seeps (Southward et al. 1997; Plate 27). One of the surprises is that bivalves with sulphur-oxidising bacterial symbionts, found in reducing sediments outside the venting areas, belonging to the Solemyidae, Lucinidae and Thyasiridae, are generally not found in the vicinity of the vents (Dando et al. 1995c; Melwani and Kim 2008). This is probably because they are not tolerant of increasing temperatures, since many of them are deep burrowers. Only one exception has been reported, a species of *Thyasira* was reported from venting areas in Matupi Harbour, New Britain (Tarasov et al. 1999).

A comparison, at Milos, of sessile macroepifauna at three pairs of venting and non-venting sites, at different depths, showed a pattern of increasing species number with depth and distance off-shore at both vent and the equivalent control sites (Morri et al. 1999). At all three depths the venting sites showed an approximately threefold increase in species numbers compared to the equivalent control sites. This difference between vent and control sites applied to all the major groups of epifauna present, poriferans, cnidarians, molluscans, poychaetes, bryozoans, brachiopods and ascidians. This situation was in marked contrast to studies on the infauna, which are more affected by the hot brine (Dando et al. 1995c).

Vent specific species at shallow vent sites are rare and occur at sites in oligotrophic waters (Table 11.1). Most of these are adjacent to oceanic waters. The vestimentiferan, *Lamellibrachia satsuma*, found in Kagoshima Bay is also present at a deep vent on the Nikko Seamount and at a cold seep site at 300 m depth on the Kinsu-no-Se Bank. Genetic studies suggest that the Kagoshima Bay population may have developed from only a few migrating larvae (Kojima et al. 2001). An undescribed frenulate pogonophore, *Siboglinum* sp., has been found only at sedimented hydrothermal sites in the Bay of Plenty, New Zealand (Malakhov et al. 1992; Kamenev et al. 1993) and might be a vent-specific species. The grapsid crab *Xenograpsus novaeinsularis* is found at widely different shallow vent sites in the west Pacific (Table 11.1) and may well be a species specific to shallow vents. The related *Xenograpsus testudinatus* has only been recorded from a single site (Ng et al. 2000) but its unusual habitat and lifestyle (Section 11.7.1), is likely to mean that it is also vent specific. Similarly, the callianassid *Paraglypturus calderus* has only been found at a single site (Esmeralda Bank, in the Marianas Arc) but might be more widely distributed (Türkay and Sakai 1995). The Kermadec Arc has a range of hydrothermal sites at varying depths (de Ronde et al. 2001). At the shallower sites, at 154 m and deeper, vent endemic mytilid bivalves with symbiotic bacteria, occur (Smith et al. 2004). These are vent-obligate species but are probably at the upper depth limit of their range here. At Milos, the polychaete *Capitella* sp. M shows significant adaptations to the vent habitat (Gamenick et al. 1998a, b) and is probably a vent-obligate species.

The high heat input from vents can increase the annual minimum temperature in embayments (Aliani et al. 2000) and in other restricted areas such as submarine caves (Bianchi et al. 1998). This is believed to encourage the settlement of exotic thermophilic species (Giaccone 1969; Acunto et al. 1996a; Koutsoubas and Cinelli 1996; Sartoni and De Biasi 1999; De Biasi and Aliani 2003).

In Palaeochori Bay and the surrounding area of Milos, the seagrasses *Cymodocea nodosa* and *Posidonia oceanica* provide large-scale cover from 4 to 40 m depth (Plate 27). Only *C. nodosa* is found in the parts influenced by hydrothermal venting and forms rings around the brine seeps and the most active vents (Aliani et al. 1998b). This can be seen in an aerial photograph of the Bay in which the pale minerobacterial mats delineate brine seepage through faults and the dark regions are due to seagrass cover (Plate 26). *Cymodocea nodosa* is more tolerant of stress than *P. oceanica* (Mazzella and Buia 1986). Its distribution appears to be temperature limited by the venting fluids (Aliani et al. 1998b) since it has been found growing in the brine, but not in the hot part of the brine seep that can exceed 45°C at 5 cm depth (Fitzsimons et al. 1997). The distribution of *P. oceanica* at Panarea, Italy is also controlled by the distribution of the hydrothermal fluids (Acunto et al. 1995) since this species is known to be very sensitive to sulphide (Garcias-Bonet et al. 2008). In contrast, *P. oceanica* is more productive at the shallow vents at Ischia, Italy, than away from the vent influence (Hall-Spencer et al. 2008). The presence of *P. oceanica* at the latter site may be due to the absence of hydrogen sulphide in the venting fluids (Hall-Spencer et al. 2008), although sediment temperatures were not reported.

Reports on changes in the planktonic species composition at shallow vents are variable. At Milos there were few differences in phytoplankton production, which was low throughout the venting and non-venting areas south of the island (Robinson 2000). The only samples with elevated production appeared to be those containing re-suspended benthic diatoms. No thermophilic phytoplankton species were present, since an increase of 6°C completely inhibited photosynthesis in samples from the water column (Robinson et al. 1997). The high sedimentation rate in the venting area had only a small contribution, of approximately 3%, that could be accounted for by sedimenting phytoplankton: most of the organic sediment was due to zooplankton debris and faecal pellets (Bayona et al. 2002). In Matupi Harbour, New Britain, plankton photosynthetic rates were also greatly reduced over the vent fields but there was high chemosynthetic production in the venting areas that exceeded the photosynthetic rates in the non-venting part of the Harbour (Tarasov et al. 1999). Sulphide, heavy metals and reduced salinity were believed to be responsible for the low photosynthetic rates in areas around the venting sites (Sorokin et al. 1998). There was an increase in cyanobacteria in sulphidic waters, since cyanobacteria are more tolerant to the presence of sulphide (Stanier and Cohen 1977). The complex vertical distribution patterns observed, in photosynthetic and chemosynthetic production and microbial numbers, can be explained by the multi-layered water column plumes observed at shallow-water gasohydrothermal vents (Fig. 11.1). Individual outlets emit fluids at different temperatures and salinities and with different bubble sizes, so that stacked hydrothermal plumes are found.

The distribution of bacterioplankton at Pacific vents was studied by Sorokin et al. (1998). Bacterial densities above the vent sites were uniformly high, with greatly elevated rates of chemosynthesis. Above the Calypso vents, in the Bay of Plenty, the activity of the thiobacilli was greatest in the bottom layers with hydrogen- and methane-oxidizers being more active in the middle to upper water column. However, chemosynthetic production in the water column at this site was not more than 6% of that due to photosynthesis (Sorokin et al. 1998). A similar situation was found at the shallower vents around Whale Island and White Island, with chemosynthetic production in the water column being only a small percentage of total primary production. At these shallower sites, there was a distinct change in the pattern of the grazing community, with amoeba replacing zooflagellates. A similar increase in planktonic amoeba was noted in Kraternaya Bight in the region of the water column with active photosynthesis (Sorokin et al. 1998).

In the algo-bacterial mats over the hot sediment (50–60°C) in the Bay of Plenty, filamentous cyanobacteria dominated the production at 4–6 mm depth while chemosynthetic fixation dominated above this. A similar depth distribution of filamentous cyanobacteria was found at the Milos brine seeps (Wenzhöfer et al. 2000; Plate 27), where the cyanobacteria formed a ring around the main outlets and were, in turn, surrounded by a thick layer of benthic diatoms (Robinson et al. 1997). In Kraternaya Bight, the Kuriles, diatoms were also an important component of the algobacterial mats but cyanobacteria were absent (Fig. 11.3) (Tarasov et al. 1990).

11.3.2 Cold Seeps

Sibuet and Olu (1998) documented data from 24 deep-sea seep sites below 400 m and recorded 211 species, 30% of which were symbiont-containing species. They found that the percentage of symbiont-containing species decreased with

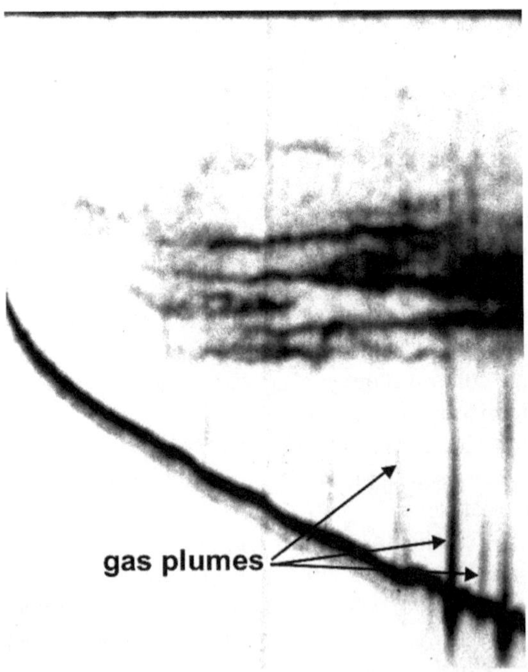

Fig. 11.3 Echo-sounder record showing multi-layered water column plumes above hydrothermal vents at 90–140 m water depth off Milos

increasing water depth, from 37% on the upper Louisiana Slope at 400–1,000 m to 17% in the subduction zones off Japan at 3,800–6,000 m. This trend with depth is reversed at shelf seeps. At three methane-seep sites on the Californian shelf, at 35–50 m depth, only two of 124 species contained symbionts (Levin et al. 2000). In a study of two North Sea seeping pockmarks, at 150–170 m depth, only three of 282 macro-infaunal species contained symbionts (Dando 2001). Similarly a study of three pockmarks in the North Sea Braemar field, at 120 m depth, found that only four of 173 macro-infaunal species had symbionts (Hartley 2005). None of the shallow-water seep symbiont-containing species are seep obligates, all being found in normal reducing sediments. In a comparison of seep sites on the Sakhalin shelf and slope, from 160 to 2,500 m macro-infauna with symbionts were only found at depth below 370 m (Sahling et al. 2003).

At deep-sea seep sites most of the species appear to be seep-obligates, few being shared with hydrothermal vents or organic-fall sites (Sibuet and Olu 1998). However, in a comparison of macrofauna at slope (500 m) and shelf (to 50 m) seeps off northern California (Levin et al. 2000) found that both seep areas were colonized by a sub-set of the background fauna. Similar conclusions were made when comparing foraminifera at the Californian slope site (Rathburn et al. 2000). In their study of the Sakhalin slope seeps, Sahling et al. (2003) suggested that most of the symbiont-containing species, other than the vesicomyids, might not be seep obligates but also inhabit non-seep reducing sediments.

At shallow-water sites, the fauna found at seeps is generally considered a sub-set of the surrounding fauna. The species distributions at seeps differ from those of the surrounding area, because of changes in substratum and because of the presence of higher sulphide concentrations. This was found for foraminifera (Jones 1993), nematodes and macrofauna (Dando et al. 1991) from the Scanner pockmark area in the North Sea. The differences mainly being accounted for by lack of suitable hard substrate outside the pockmarks (Dando et al. 1991, 1994b). In California, at the Isla Vista oil seeps, the same subset of species was found at the seeps and in the surrounding area that were not subject to seepage (Spies and Davis 1979). At methane seeps on the Californian shelf, where neither bacterial mats nor surface carbonate concretions were observed, Levin et al. (2000) found that 55 of the 201 observed taxa were found only at a seep site and 59 were found only at a control site. However, the majority of the species found only at the seep sites were represented by only one or two individuals. Only eight of the 55 species occurred exclusively in seep sediments in at least two of the three sites investigated. These authors considered that only the crustaceans *Cheirimedeia zotea* and *Synidotea angulata* and the polychaete *Capitella* sp. were considered to show a significant seep preference.

In contrast to the, often more abundant and species-rich, epifauna, the infauna at seeps is often a story of species exclusion, compared to the surrounding sediments, due to hydrogen sulphide toxicity, lower oxygen penetration and limited burrowing depth due to sub-bottom concretions (Dando et al. 1993). Spies and Davis (1979) found no change in the vertical distribution of infauna at the Isla Vista oil seeps when compared with control sites. However, at the Kattegat 'Bubbling Reefs', metazoa were confined to the upper 2 cm of muddy sediments surrounding the carbonate concretions. The infauna mainly consisted of nematodes, oligochaetes and polychaetes (Jensen et al. 1992). The polychaete *Scoloplos armiger* and the bivalve *Venus gallina*, dominant species in normal sediments in the area, were absent from the seep sediments. It was assumed that these species were sulphide-intolerant.

Whether any seep-obligate eukaryote species occur at shallow sites is still uncertain. The gutless nematode *Astononema southwardorum*, has only, to date, been found in the Scanner pockmark, where it is common in the sediment on the sides, living just above the zone of peak sulphate reduction (Dando et al. 1991). However, these sediments, and their sulphate reduction rates, were not unusual for reducing sediments and it may be common elsewhere in the North Sea. At a muddy sediment area in the Kattegat 'bubbling reefs' the unusual nematode *Theristus anoxybioticus* was discovered (Jensen 1995b). This species has a juvenile phase that lives at a mean depth of 7 cm in anoxic, sulphidic sediment and dies when exposed to oxic seawater (Jensen 1995a). In contrast, the adults move upwards to the upper sediment on maturity where they feed on diatoms. It is believed that the juveniles are bacteriovores. However, this species is likely to occur in reducing sediments in the area. An unusual ostracod, *Terrestricythere crimaea*, the only known marine member of the genus, was described from the vicinity of methane seeps in the Black Sea (Schornikova and Syrtlanova 2008). Although this may prove to be a seep-obligate species there was uncertainty regarding whether it was actually inhabiting a seep.

Higher abundances of rotaliid foraminifera and lower abundances of buliminids foraminifera were found at seeps (Jones 1996). In addition, there was an overall reduction of in the diversity and abundance of foraminifera at seeps, but a higher relative abundance of epifaunal species. This agreed with the presence of more abundant macro-epifauna and macro-epifaunal species in the Scanner pockmark (Dando 2001).

At the Isla Vista oil seeps the total abundance of macro-infauna and the species abundance was higher at the seep sites (Davis and Spies 1980). This is believed to be due to the higher abundance and biomass of bacteria, growing on the hydrocarbons, supporting a larger population of deposit feeders that in turn supports more carnivores (Davis and Spies 1980; Montagna et al. 1989). Sediment disturbance by larger predators helped to restructure the community and increase diversity (Davis and Spies 1980). In particular, oligochaetes were nearly eight times more abundant at seep stations than at a comparable control site.

Substrate types (e.g. rocks or concreted sediment, soft substrates of different grain size distribution or organic matter content) can vary over short differences at seeps, causing higher variability in diversity values at seep stations when compared with control stations (Dando 2001). In the study of the Scanner pockmark and surrounding sediments both the maxima and minima in organism densities and in biomass were obtained at stations in the base of the pockmark, although there were no differences between mean values of biomass between pockmark stations and control stations (Dando 2001). Closely spaced cores in the pockmark base showed very different species compositions, reflecting the great heterogeneity in substrate type and geochemistry.

Submarine groundwater seeps flush solutes, including methane, sulphide and nutrients, out of the sediments (Johannes 1980; Capone and Bautista 1985; Bussmann et al. 1999; Schlüter et al. 2004). Thus, bacterial mats are frequently found above such seepage sites (Table 11.2). However, the effects of seepage on the fauna are mainly due to the reduced salinity and sediment instability. Under suitable hydrological head conditions the water flow can give rise to fluid sediment (Schlüter et al. 2004; Karpen et al. 2006), making the discharge sites unsuitable habitats for macrobenthos. The salinity gradient away from the seeps controls the faunal composition, euryhaline species being favoured closer to the seeps (Kohout and Kolipinski 1967; Moore 1979) and an increase in numbers of species and biomass is found with decreasing influence of the seepage (Zipperle and Reise 2005).

11.4 Biological Communities

11.4.1 Hydrothermal Vents

11.4.1.1 Prokaryote Communities

The thermophilic/hyperthermophilic Bacteria and Archaea are the two groups of organisms that have numerous obligate taxa at shallow-water vents. Shallow vents, especially those at Vulcano, the Bay of Naples, Milos and Iceland have been productive sites for the isolation of new taxa in these groups. Dando et al. (1999) listed 28 taxa, mainly Crenarchaeota, that had been isolated from Mediterranean vents by 1998. In many cases similar taxa have also been isolated from deep-sea vents (Gugliandolo and Maugeri 1993). Of particular note was the discovery of the first representative, *Nanoarchaeum equitans*, of a new group of Archaea, the Nanoarchaeota, with one of the smallest cell and genome sizes known, at the Kolbeinsey vents (Huber et al. 2002). This species grows at 90°C in symbiotic association with a species of the archaeum *Ignicococcus* but little is, as yet, known of its nutrition and ecology.

As at the deep-sea vents, most of the thermophilic and hyperthermophilic taxa isolated have been heterotrophs. The autotrophs, or facultative autotrophs, include methanogens such as *Methanococcus thermolithotrophicus* (Huber et al. 1982); hydrogen oxidisers such as *Methanopyrus kandleri* (Kurr et al. 1991) and *Archaeoglobus fulgidus* (Stetter 1988); an anaerobic iron, sulphide and hydrogen oxidiser, *Ferroglobus placidus* (Hafenbrandt et al. 1996): sulphur oxidisers such as *Halothiobacillus kellyi* (Sievert et al. 2000a) and sulphur reducers including *Pyrodictium occultum* and *P. brockii* (Stetter et al. 1983). High densities of *Pyrodictium, Pyrococcus, Thermococcus* and *Archaeoglobus* were found in the active crater of the MacDonald Seamount and viable cells were collected 1 km away (Huber et al. 1990).

A high density of hyperthermophilic sulphur-dependent heterotrophs were studied in hot sediments in an inlet of Kodakara-Jima Island, Kagoshima (Hoaki et al. 1995). The heterotrophs were probably supplied with organic matter, from *Beggiatoa* and cyanobacteria, and created a strong gradient of amino acids in the sediment, which declined sharply from surface concentrations. Similarly, Sievert et al. (2000b) found that heterotrophs dominated the benthic prokaryote population at the Milos brine seeps. The main organic inputs were from photosynthetic sources (benthic diatoms, benthic cyanobacteria and debris from surrounding seagrass), with a smaller input from sulphur-oxidising bacteria.

11.4.1.2 Eukaryote Communities

Shallow-water hydrothermal vents are largely colonized by a subset of the background fauna and flora (Fricke et al. 1989; Tarasov and Zhirmunsky 1989; Dando et al. 1995c; Morri et al. 1999; Tarasov et al. 1999; Forrest 2004). There is commonly a decline in the numbers of macrofaunal species along a transect towards the vents due to increasing temperatures and sulphide concentrations (Fricke et al. 1989; Thiermann et al. 1997; Forrest 2004). Thus, particularly with infauna few species are present at the venting sites. Major differences in infaunal communities can be found over the scale of 1 m or less (Dando et al. 1995c; Thiermann et al. 1997; Forrest 2004). Tube-dwelling organisms and motile swimmers became more abundant in the hot sediments at Bahía Concepción, suggesting that tubes gave some protection to infaunal species from the hot sediments (Forrest 2004; Melwani and Kim 2008). Crustaceans dominated the vent and surrounding sediments here with molluscs and sipunculids being more abundant in control areas (Melwani and Kim 2008). The two zones, at vents and their surroundings sediments and in sediments outside the venting area, had very different macrofaunal communities. In highly sulphidic sediments, such as those at White Point, California and at Milos, tube-dwellers were not common (Thiermann et al. 1997; Melwani and Kim 2008), whereas mobile polychaetes dominated in the vent sediments at White Point. Mobile carnivorous and omnivorous polychaetes also dominated the benthic fauna in the geothermally heated Lefkos Bay, Nysiros, in the Aegean (Diapoulis et al. 1994). However, no data was provided of the chemistry or sediment temperatures at this site.

The numbers of meiofauna in sediments at Vulcano was found to be heavily depleted in areas where colloidal sulphur was deposited (Colangelo et al. 1996). This depletion in numbers of organisms occurred, to a lesser extent, in sediments disturbed by venting gas emissions. However, an increase in nematode numbers was found in the sediments disrupted by gas release.

The epifaunal community tended to be less affected than that of the infaunal by venting. Epifaunal communities, both close to the vents and away from vent influence, at Milos were similar in structure (De Biasi et al. 2004). Only under rock overhangs, where hydrothermal fluids could accumulate, were small-scale differences found. The macro-algal communities were not unusual for the region, with the exception that 14 species with warm-water affinities occurred, including the Lessepsian migrant, *Stypopdium schimperi* (Sartoni and De Biasi 1999; De Biasi and Aliani 2003). At Vulcano Island, Italy, large changes in the macro-algal communities were recorded, with species of *Cystoseira* often becoming dominant near the vents (Giaccone 1969; Acunto et al. 1996b). Major changes in the algal communities were also recorded at vents at Ischia, the cover of coralline algae was reduced from >60% outside the venting area to zero close within the venting area, due to reduced pH (Hall-Spencer et al. 2008). Exotic species and those with warm-water affinities were found close to the vents. Near the vents in Porto di Levante, a 200 m wide bed of the sub-tropical seagrass *Halophila stipulacea* has become established (Acunto et al. 1996a) and this is bringing changes to the benthic community as a result of its different algal epiflora. This seagrass is not native to the Mediterranean but is common in the Red Sea and on the east coast of Africa. It was not reported during detailed surveys made in 1968 (Giaccone 1969).

De Biasi and Aliani (2003) showed that Palaeochori Bay, Milos, had higher winter minimum temperatures than a nearby bay without venting influence and suggested that the increase in minimum temperature was responsible for the survival of warmer water species. This effect might be expected to be even more important for infauna and soft-bottom epifauna, than for epifauna living above the seabed, since much of the seabed in Palaeochori Bay is geothermally heated (Dando et al. 1995c). A similar situation is likely to be found in other semi-enclosed sites such as some of those at Vulcano (Giaccone 1969) and in the Kuriles (Tarasov et al. 1990).

11.4.2 Cold Seeps

11.4.2.1 Prokaryote Communities

The surfaces of the carbonate concretions and of the epifauna at methane-seeps are often coated with methane-oxidizing bacteria. Methane-oxidation rates by microbes on the concretions were tenfold to 100-fold higher than those on similar substrates away from the seeps (Jensen et al. 1992). Hydrogen sulphide is also commonly present in the escaping gas, although it is rapidly oxidized, and leads to strands of filamentous sulphur-oxidizing bacteria where the gas bubbles contact the concreted sediment, especially under overhangs (Dando et al. 1991, 1994b; Jensen et al. 1992). Indeed, mats of sulphur-oxidizing bacteria are a conspicuous feature of many seep sites and the presence of mats of *Beggiatoa* have been reported from most of the described sites listed in Table 11.2.

Only one study of bacterioplankton in a pockmark has been reported. In a seeping pockmark on the Oregon shelf there was an approximately twofold increase in heterotrophic bacteria both in the bottom water and again at the surface (Juhl and Taghon 1993). However, since the control and pockmark samples were taken 12 h apart, the difference may not be significant. In contrast to shallow-water vents, few studies have been carried out on microbial communities at shallow seep sites. The majority of these have concentrated on rates of sulphate reduction, methane oxidation and the breakdown of higher hydrocarbons, e.g. Montagna et al. (1986), Dando et al. (1994b), Bussmann et al. (1999), that were found to be enhanced at the seeps.

Microbial activity, measured as ATP levels, in the East Flower Garden brine were equivalent to those found in surface waters, with sulphur-oxidizing bacteria accumulating at the brine-seawater interface (Bright et al. 1980). Counts of prokaryote cell numbers and calculations on microbial productivity at the Isla Vista oil seeps showed mixed results, with the seep site not having consistently higher values (Montagna et al. 1989). This study estimated that two-thirds of the aerobic benthic metabolism was due to microbes. More studies on the microbial communities at shallow seeps are badly needed.

11.4.2.2 Eukaryote Communities

Methane accumulates in sedimentary basins where fine, organic-rich sediments settle (Martens and Klump 1984). Outgassing leads to sediment sorting (the finer sediments being dispersed), formation of carbonate cemented sediments and often pockmark depressions (Dando and Hovland 1992). These changes in bottom topography in areas on soft seabed lead to changes in the epifauna with colonization of hard substrata by a fouling community. Thus, at the Kattegat seeps, species of macroalgae, poriferans, hydrozoans, anthozoans, cirripedes, bryozoans and ascidians are common at the seeps but seldom seen on the surrounding soft bottoms (Jensen et al. 1992). The carbonates are bored into by sponges, such as *Cliona celata*, polychaetes, such as *Dodocaceria concharum*, and the bivalve *Hiatella arctica* (Jensen et al. 1992). A rich and varied fauna, including nematodes, polychaetes, bivalves and echinoderms, in turn, inhabits these borings. On the Sakhalin shelf Sahling et al. (2003) also found that the elevated abundances of epifauna at seep sites was caused primarily by the presence of suitable substrata, rather than an increased food supply.

The shelter provided by the carbonates and the seabed depressions also leads to an accumulation of large crustaceans and demersal fish at seep sites and in both active and inactive pockmarks (Hovland and Sommerville 1985; Dando et al. 1991; Fader 1991; Jensen et al. 1992; Dando 2001). Fish concentrations can be so high that the bottom of pockmarks can be almost completely covered with fish otoliths (Fig. 11.4) such that polychaete worms in the base of the Scanner pockmark coated their tubes with fish otoliths (Dando 2001). Since the mean otolith density outside the pockmarks was 85 m^{-3} and the maximum depth of the pockmarks was 25 m, then the pockmark densities are much higher than could be explained by simple sediment sorting, suggesting that they have been a fish attractant since their formation.

Thus, in a similar manner to hydrothermal vents in the deep-sea, shallow-water cold seeps can appear like oases of life in an otherwise featureless soft-bottomed seabed. Most of the epifauna are passive or active filter-feeders (Dando 2001). However, the accumulation of holothurians in dense clusters in many Canadian pockmarks (Fader 1991; Wildish et al 2008) may be due to their deposit feeding on the bacterial mats growing over the seeps.

The high densities of epifaunal communities on authigenic carbonates, led to the suggestion, when they were first observed, that they were substantially supported by chemosynthetic bacterial production, by analogy to the situation at deep-sea

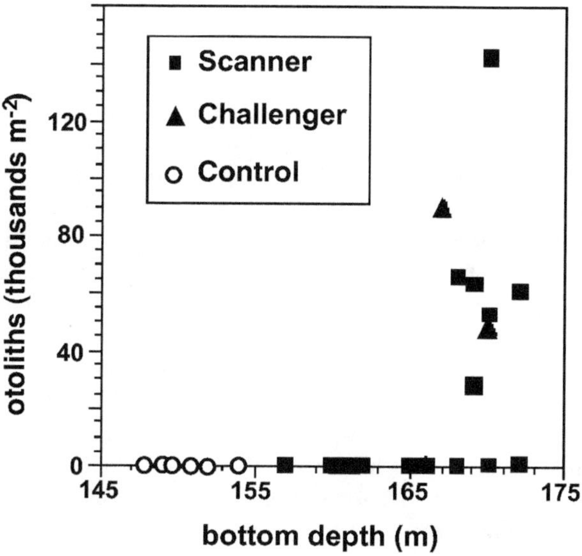

Fig. 11.4 Otolith densities, in thousands per square meter, in two North Sea pockmarks and in nearby control sediments. Water depths outside the pockmarks are 148–154 m

vents (Hovland and Judd 1988). Stable isotope and fatty acid biomarker studies have not supported this hypothesis (Section 11.7.2).

A similar pattern of an oasis of epifauna is seen at the freshwater seeps in Cambridge Fjord, Canada, where the force of the freshwater flow has removed the fine-grained muds, exposing cobbles and rocks (Colbourne and Hay 1990). The community at these seeps was very different from that on the surrounding mud, since the rocks were able to support a typical epifaunal community and associated species. In addition, the freshwater flow injected nutrients into the seawater, thus stimulating photosynthetic primary production. Since the seeps were at the head of the fjord there was less dilution of the freshwater than would have occurred at a more exposed coastal site.

Predatory epifauna and fish were observed to be much more prevalent at the Isle of Vista oil seep (Spies and Davis 1979). Fish were also attracted to the outflow of the East Flower Garden Bank brine seep in the Gulf of Mexico (Bright et al. 1980). Crabs, *Carcinus maenus*, were frequently found directly above gas bubble outlets at the shallow Frederikshavn seeps (Dando et al. 1994b) and mysids were found congregating around gas bubble outlets on the Calfornian shelf (Levin et al. 2000). The bubbles may aid camouflage and the water circulation produced by the rising bubbles will carry small organisms into the site.

The mats of sulphur-oxidizing bacteria that overly many seeps are inhabited by their own community of grazers and predators. Ciliates and nematodes are common (Spies and Davis 1979), as in mats at hydrothermal sites. A distinct meiofaunal community was found associated with the thick microbial mats at methane seeps in the Black Sea. Foraminifera, especially *Ammonia compacta, Cribroelphidium bartletti* and *Glomospira gordiales*, became abundant in sulphide-rich samples and nematodes were present in all samples. Juvenile cladocera and copepods were especially common in samples below 200 m water depth but were absent above this (Sergeeva and Gulin 2007). At very shallow sites, nematodes, harpacticoid copepods and ostracods dominated the infauna, which had the highest biomass and species diversity when seepage was at a minimum (Schornikova and Syrtlanova 2008).

At sites subject to storm surges bacterial mats can be important in exporting organic matter to the wider community (Grant and Bathman 1987). They also structure the community by preventing most of the oxygen penetrating below the mat, hence reducing the abundance of the infaunal community by orders of magnitude (Vetter 1998). Conversely, they act as a barrier, often preventing the escape of H_2S to the surface and thus making the habitat more suitable for epifauna.

The macro-infauna, when found, at the seep sites is a sub-set of that found in the surrounding areas (Davis and Spies 1980; Gittings et al. 1984; Dando et al. 1991; Levin et al. 2000). Changes in the composition of the community can include an increase in the number of filter-feeders and a higher abundance closer to the seeps (Gittings et al. 1984). At particularly sulphide-rich seeps sulphide toxicity also affects the composition of the community (Gittings et al. 1984). In contrast to the epifauna, infauna show an even greater effect of species exclusion, compared to the surrounding sediments, due to hydrogen sulphide toxicity, lower oxygen penetration and limited burrowing depth due to sub-bottom concretions (Dando et al. 1993). Nematode communities change to a dominance of deposit feeders in sulphidic sediments (Jensen 1986).

11.5 Adaptations to Life at Vents and Seeps

One of the major factors limiting colonization of vent and seep environments is the high levels of hydrogen sulphide, H_2S being the toxic species (Vismann 1991). This can be compounded by changes in salinity and, at vents, by high temperatures and concentrations of heavy metal. Only a limited number of experimental studies have been made on the tolerance of species dominating the fauna at shallow sites, in contrast to numerous studies on their deep-sea counterparts (Childress and Fisher 1992). One of the best-studied shallow vent ecosystems is that of the hydrothermal brine seeps at Milos and the three dominant species there will be considered in more detail.

Cyclope neritea (L.), the gastropod that dominates the mineral-algal-bacterial mats above the hydrothermal brine seeps at Milos, was not affected when placed in seawater containing 1 mM sulphide, showed periods of torpidity when exposed to 10 mM and could even recover from brief periods in 50 mM sulphide (Southward et al. 1997). *Cyclope neritea* appeared to be able to exclude most of the sulphide by some unknown mechanism, since tissue concentrations increased little with time and there was no increase in oxidation products. In situ this species can minimize the effects of sulphide by using its extensible siphon to draw water from above the brine over its gills. *Cyclope neritea* is widespread in coastal lagoons and salt-marshes elsewhere in the Mediterranean and can tolerate varying temperatures and salinities. The species has an upper lethal temperature of approximately 33°C (Massé et al. 1978), higher than that of the upper 2 cm of the mat-covered sediment at Milos (Fitzsimons et al. 1997). However, it is unknown whether the populations elsewhere are so resistant to sulphide. The Milos population is also unusual in that, instead of depositing its eggs on hard substratum, such as bivalve shells, on the seabed (Gomoiu 1964), it crawls onto the leaves of the seagrass *Cymodocea nodosa* and deposits its eggs there, thus the young hatch in water with a low sulphide content (Dando et al. 1995c).

Another inhabitant of the brine seep area that also showed unusual tolerance to sulphide was a polychaete in the *Capitella capitata* species complex, *Capitella* sp. M (Gamenick et al. 1998b). This species reached maximum densities on the edge of the brine seeps in sediment containing up to 710 μM dissolved sulphide, close to its tolerance limit (Gamenick et al. 1998a). It could survive anaerobically even in the presence of high sulphide (Gamenick et al. 1998b). In contrast the nematode *Oncholaimus campyloceroides* has reduced tolerance to sulphide in the presence of anoxia and high salinities (Thiermann 1998) and may only enter the seep area for short periods, possibly to feed on organisms killed by the sulphide. However, it has an efficient sulphide detoxification system, depositing oily inclusions of elemental sulphur and polysulphide chains, in the epidermis, that are rapidly re-oxidized on exposure to oxygen (Thiermann et al. 2000). Similar detoxification mechanisms have been found in the body walls of meiofauna at seeps (Powell et al. 1980), resulting in the formation of thiosulphate, elemental sulphur, sulphite or sulphate, depending on the species.

Echiurans burrow deep into muddy sediment, or hide their bodies under rocks, and most species are surface deposit feeders using an extensible proboscis. They are often found feeding on bacterial mats at shallow vent and seep sites. Although no seep or vent echiurans have been investigated for sulphide detoxification mechanisms, a species living in sulphidic intertidal sediments, *Urechis caupo*, was found to use a variety of mechanisms against sulphide toxicity (Arp et al. 1995). These included mucus secretion, sulphide binding followed by oxidation to thiosulphate, oxidation by symbiotic bacteria and/or specialized organelles and sloughing off epidermal cells.

Chaetopterid polychaetes are frequently found at high densities at both shallow and deep seeps and vents (Olu et al. 1996; Abbiati et al. 1994; Desbruyères et al. 2006; P. Dando 2007). Most chaetopterid species live partly buried in the sediment and pump water through their tubes, releasing it into the sediment through the posterior end of the tube (Barnes 1965; Sendall et al. 1995). While this life-style applies to those species found at seeps, some species at vents have been observed lying mainly exposed on the substratum (Fouquet et al. 2008) or attached to rock with their tubes projecting into the water (Abbiati et al. 1994). Chaetopterids probably thrive at vent and seep sites, given suitable substrate, because of the high bacterial biomass that they can collect for food by filtering water through the mucosal sacs they form in their tubes (Barnes 1965; Sendall et al. 1995). In addition, for species living in the sediment, the flux of seawater through their tubes would prevent their exposure to high sulphide concentrations. The chaetopterids are also able to reverse the water flow, thus preventing toxic fluids from entering the tube (Barnes 1965).

A response to the lack of oxygen, that is common under sulphidic conditions, is for organisms to increase their surface area to volume ratio. This is particularly evident among nematodes of the thiobios (Jensen 1986). A similar strategy is used by frenulate pogonophores to obtain sulphide for their endosymbionts in a low sulphide environment (Dando et al. 2008). Other species shuttle between the deeper layers with high sulphide concentrations and the upper sediment where nitrate or oxygen is available. The stilbotnematids with sulphur-oxidizing epibacteria are believed to do this, their bacteria partially oxidizing sulphide to sulphur deeper in the sediment (Schiemer et al. 1990). The nematodes then move towards the surface layers for their bacteria to oxidize the elemental sulphur. This is a strategy used by the giant sulphur-oxidizing bacterium *Thioploca* (Hüttel et al. 1996). Although common at deeper seep sites (Robinson et al. 2004), it has been found at only one shallow seep or vent site, the 40 m vent site at Milos (Fig. 11.5). *Beggiatoa* spp. and *Thiovulum* spp. at sedimented sites are able to glide to optimum positions in the sulphide-oxygen gradient, allowing them to take up oxygen or nitrate as well as sulphide (Jørgensen and Revsbech 1983). *Achromatium volutans*, the dominant giant sulphur bacterium at the Milos vents similarly positions itself in

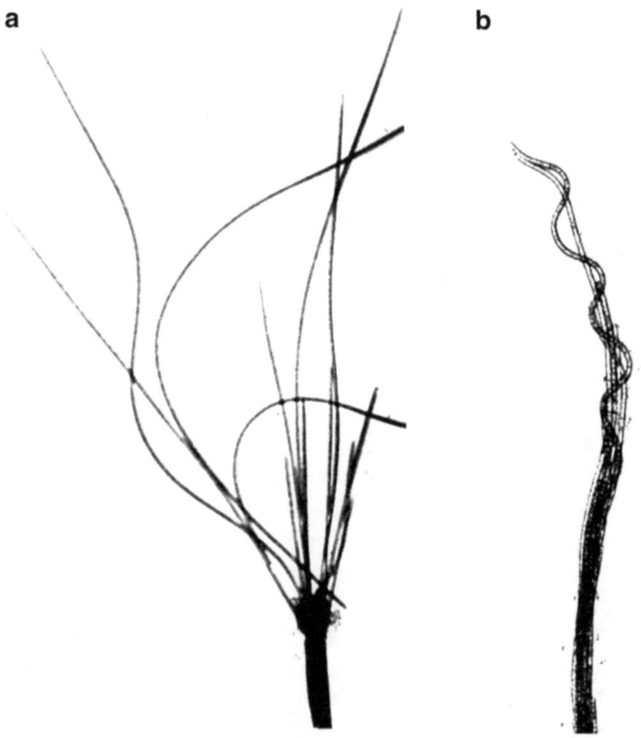

Fig. 11.5 *Thioploca* trichomes, from bacterial mats at a 40 m vent site at Milos, showing their ability to move vertically within their sheath. (**a**) Trichomes leaving the sheath in the upper sediment. (**b**) Trichomes withdrawing into the sheath and moving vertically downwards

a narrow band in similar gradients (Dando et al. 1998). In contrast, *Beggiatoa* sp. on consolidated surfaces form a 'holdfast'-type end to the filament to attach to the substratum, or to the shells or epidermis of other organisms (Mattison et al. 1998), presumably to prevent the filaments from being carried away by currents.

11.6 Fine-Scale Distribution of Organisms at Seeps and Vents

11.6.1 Prokaryotes

Archaeal and bacterial populations were studied along 2.0 and 2.35 m-long transects from the centre of a vent in the Milos brine seeps (Brinkhoff et al. 1999; Sievert et al. 1999, 2000c). The number of taxa decreased sharply with increasing sediment temperatures towards the vents (Sievert et al. 1999), although on one sampling occasion storms had mixed the sediments and disrupted both the vertical and horizontal gradients in bacterial distribution. Total cell numbers also showed a trend of decreasing numbers towards the vents at all sediment depths. In contrast, the surface samples, including the algo-minero-bacterial mat overlying the sediment, showed a sharp peak in cell numbers 1.2 m from the vent, where the precipitates were greatest (Sievert et al. 1999). *Thiomicrospira* numbers peaked 2 m away from the vent centre where oxygen penetrated further into the sediment and none were found 30 cm from the vent centre where temperatures were 30–60°C (Brinkhoff et al. 1999). Unexpectedly, Archaea made up less than 25% of the total, reaching maximum densities 1.2 m from the vent centre at 27 mm depth where the temperature was 41°C. There was no evidence for them in the hotter sediments below this and they only made up 12% of the population at the same depth in the vent centre, where the temperature was 82°C. Together with cyanobacteria, diatoms were found to be active even at the centre of the venting area where surface temperatures were 30–40°C (Sievert et al. 2000b).

At the vents in the inlet of Kodakara-Jima Island, cell numbers of both sulphur-dependent hyperthermophilic autotrophs and heterotrophs increased with sediment depth to at least 30 cm and 105°C (Hoaki et al. 1995). *Thermococcus* spp. appeared to be dominant at this site.

The structure of the 5 mm thick algal-bacterial–mineral mats, of the vents at 0.7 m depth in Kraternaya Bight, has proved to be complex (Tarasov et al. 1990). At the surface was a white film containing elemental sulphur, thiobacilli and *Thiothrix* and below this was a loose layer containing diatom species with some purple sulphur-oxidising bacteria and non-sulphur purple bacteria. Below these layers was a region containing green sulphur-oxidising bacteria, sulphate-reducing bacteria, methanogens and anaerobic saprophytes. Cyanobacteria were absent from these mats (Tarasov et al. 1990). No similar studies have been made on prokaryote distribution at shallow-water cold seep sites.

11.6.2 Meiofauna

The vertical and lateral distribution of nematodes was studied along 5 m transects into hydrothermal brine seeps at Milos (Thiermann et al. 1997). Three species, *Oncholaimus camplyoceroides*, a species of *Chromdorina* and a species of *Sabatieria* were responsible for over 50% of the abundance. *Oncholaimus* and *Chromdorina* were mainly restricted to the upper 1 cm of sediment, with *Sabatieria* living below this. These three species were particularly abundant in the zone between the end of the seagrass beds and the start of the brine seep, i.e. the region just before a sharp rise in the salinity and dissolved sulphide concentration in the sediment.

In the Scanner pockmark the gutless, endosymbiont-containing nematode *Astononema southwardorum* accumulated, just above the zone of sulphate reduction, in a zone of elemental sulphur accumulation (Dando et al. 1991). This suggests that it migrates to a zone where it has access to both traces of oxygen and sulphide, since the endosymbionts are probably sulphur-oxidizing bacteria (Giere et al. 1995). Fine-scale sampling at the shallow Kattegat methane seeps showed that the highest nematode densities and biomass were in sediment collected 5 cm away from the gas outlet (Dando et al. 1994b), which is probably due to the shallow re-charge flow (O'Hara et al. 1995) carrying bacteria and small plankton down into the sand for the mainly microbial feeders. At the seep itself, the major nematode biomass was due to adults of the large predatory nematode *Enoplolaimoides brunetti*. This species was probably able to resist displacement by the gas bubbles and capture smaller meiofauna carried in the fluid flow.

11.6.3 Macro-infauna

In a study along 5 m long transects into the Milos hydrothermal brine seeps at 10 m water depth, Thiermann et al. (1997) noted a decline in species numbers with increasing salinity and dissolved sulphide concentrations. Only the gastropod *Cyclope neritea* was found at the end of the transect where salinity and sulphide were at a maximum. With a slight fall in these parameters, the polychaete *Microspio mecznikowianus* made up 10–20% of individuals and just outside this zone, in the region where salinity and sulphide had started to decrease, the polychaete *Capitella capitata* started to become dominant. Along the transect, grain size, ATP concentrations and chlorophyll concentrations were similar. It was therefore concluded that factors such as salinity, temperature and sulphide controlled the fine-scale distribution of the fauna.

A similar study was made at a site at 5 m depth with an 11.5 m long transect across an area of variable sediment temperatures and gas vents (Dando et al. 1995c). In sediment with the highest temperatures, 39–93°C at 10 cm depth, only a few crustaceans were found. The burrows of the decapod *Callianassa truncata* only occurred where the sediment temperature was below 40°C. The dominant species along this transect was the amphipod *Perioculodes longimanus*. Few annelids or gastropods were present, unlike the 10 m site, and it was concluded that the higher temperatures and sediment disturbance caused by the outgassing favoured a highly mobile fauna. At vents in the Bay of Plenty, New Zealand, it was also noted that, at sediment temperatures above 40°C, motile crustaceans dominated the fauna (Kamenev et al. 1993).

As noted by Levin (2005), there is extensive heterogeneity of distribution at seeps. This is especially true at shallow seeps where heterogeneity of bottom structure and sediment geochemistry create a large number of microhabitats (Dando and Hovland 1992). For this reason it is frequently difficult to obtain replicate cores since no two samples are alike. There can be large-scale heterogeneity within a single box core with differences in chemical concentrations and faunal densities being two orders of magnitude or more across a seepage boundary (Dando et al. 1994a). Steep geochemical gradients at both seeps and vents are always found where there is release of free gas (Dando et al. 1994b; O'Hara et al. 1995; Robinson et al. 1997). The distribution of species with methane or sulphide-oxidizing symbionts will obviously be affected by such gradients. In addition, gas pathways will also affect macrofaunal distribution due to differences in grain size due to sediment sorting and because of the formation of hard ground due to methane oxidation and carbonate deposition.

11.7 Food Web Studies

11.7.1 Hydrothermal Vents

At deep-sea vents most of the community depends for a substantial amount of its nutrition on carbon fixed by chemosynthesis (Van Dover 2000). Since shallow vent sites are in the euphotic zone a major input from photosynthesis would be expected. The mixture of food sources and microhabitats has been shown, at the Milos vents, to give rise to a higher epifaunal species biodiversity in the immediate area of the vents than in comparable control areas (Morri et al. 1999). Determining the food webs in such situations is difficult and attempts have been made using fatty acids (Kharlamenko et al. 1995) and stable isotopes (Trager and DeNiro 1990; Southward et al. 1996; Forrest 2004) as nutritional indicators.

In Bahía Concepción, Gulf of California, most of the organisms at the vent site, with the exception of the sponge *Porites californica*, were slightly more depleted in ^{13}C than those living away from the vents, implying that they derived some nutrition from carbon fixed by chemosynthetic prokaryotes (Forrest 2004). In a study of the Capo Palinuro caves with hot sulphur springs, the sponges again showed no significant input to their nutrition of carbon fixed by chemosynthesis. However, other organisms, including the polychaete *Phyllochaetopterus socialis*, a serpulid, the oligochaete *Thalassodrilides gurwitschii*, the ostracod *Paracypris complanata*, a vermetid gastropod and the echinoderms *Amphiura* sp. and *Cucumaria* sp., all showed a nutritional input from sulphur-oxidizing bacteria (Southward et al. 1996). The difficulty in obtaining comparable samples of species from normal sea caves made overall calculations of the importance of chemosynthesis to the ecosystem impossible. Sediment trap studies suggest that, even in the inner caves, most of the particulate organic carbon (POC) was derived from photosynthesis, with only 31% of the suspended POC above the geothermal springs being derived from chemosynthesis (Airoldi and Cinelli 1997). The carbon isotope ratio data indicated diverse feeding habits for different species of both filter feeders and deposit feeders (Southward et al. 1996). Large isotopic differences were found for individuals of the same species living in different microhabitats. For example, the filter-feeding polychaete *Phyllochaetopterus socialis* on the cave floor had a tissue $\delta^{13}C$ of –21.8‰, those living at the chemocline in Grotta Azzurra were more depleted (–25‰) and those in a similar situation in Grotta Sulfurea were even more depleted in ^{13}C (–30.9‰ $\delta^{13}C$). This probably reflected the composition of suspended particles in the different situations.

At another semi-enclosed hydrothermal site, the caldera of Kraternaya Bight in the Kurile Islands, cyanobacteria and diatoms form an important component of the algo-bacterial mats at the vents (Starynin et al. 1989). All the species of macrobenthos studied had higher concentrations of bacterial fatty acids than similar species in normal ecosystems. This suggested that the high bacterial production at the site contributed to the productivity (Kharlamenko et al. 1995). However, most animals also had high concentrations of diatom-specific fatty acids and it was concluded that photosynthesis was the most important source for POC supply to the benthos. Cluster analysis of the fatty acid components revealed seven groups of organisms, based on nutritional similarity. Bacteria were particularly important in the nutrition of the polychaete *Pectinaria hyperborea* and the holothurians *Eupentacta pseudoquinqesemita* and *Psolus* sp., although diatoms also formed a major component of the diets of the holothurians. Brown algae, which were abundant in the Bight, were only a food source for the gastropods *Littorina kurile* and *Nassa freycinetti* (Kharlamenko et al. 1995).

Although grazing on the algo-bacterial mats in Kraternaya Bight was not considered to be important in the food web there (Kharlamenko et al. 1995), it is important in other vent ecosystems. At the vents off White Point, California, the abalone, *Haliotus cracheradii*, and the limpet, *Lottia limatula*, both obtain substantial nutrition from the mats of *Thiothrix*; in contrast, *Mytilus* sp. at the vents obtained little carbon from chemoautotrophic bacteria (Stein 1984; Trager and DeNiro 1990). At the Milos vents, the minero-bacterial mats over the hot brine seeps were depleted in ^{13}C, $\delta^{13}C$ –23.9‰, compared with values of –19.2‰ to –20.9‰ for sediment POC from control areas. However, the most abundant grazer on the mats, the gastropod *Cyclope neritea*, had much heavier tissue carbon, from –11.2‰ to –16.0‰, that varied with the season (Southward et al. 1997). Stomach content studies showed that *Cyclope neritea* ingested sediment, containing diatoms, the sulphur-oxidizing bacterium *Achromatium volutans* and elemental sulphur, but not seagrass leaves (that had a $\delta^{13}C$ of −7.4 to −10.0). The gastropod spends most of the year on the mats, only moving onto the seagrass to deposit eggs. It thus appears that *Cyclope neritea* is an extremely selective feeder, selecting components of the algo-bacterial mats, e.g. those bacteria using form II ribulose*bis*phosphate carboxylase or the reverse TCA cycle for fixing CO_2, and may not be digesting the ingested diatoms.

Cyclope neritea is also a scavenger and additionally feeds on organisms that are killed by sulphide if they get too close to the hydrothermal brine seeps (Dando et al. 1995c). This scavenging strategy is utilized by the crab *Xenograpsus testudinatus* at the Kueishan Island vents where molten sulphur erupts from the seabed (Jeng et al. 2005). At slack water, a plume of colloidal elemental sulphur and sulphide rises vertically from the seabed, killing, or narcotizing, zooplankton in the water above. The crabs leave the shelter of crevices in the solidified sulphur on the edges of the vent to feed on the zooplankton that fall to the bottom at this time. Scavenging on dead planktonic organisms is likely to be important in the food webs of other sites where high concentrations of hydrogen sulphide or colloidal sulphur have been recorded (Powell et al. 1986; Colangelo et al. 1996). Even fish can be killed by swimming into sulphide-rich water (MacDonald 1992).

Fish appear to make use of the entrainment of water by rising gas bubble plumes, stationing themselves near the base of a plume so that the entrained water, with food organisms, is drawn towards them (Cardigos et al. 2005).

There is an especially wide variety of primary producers at shallow-water vent sites. Benthic photosynthesis is stimulated by the elevated temperatures and nutrient levels due to the vents and by the downward supply of seawater in the re-charge zones. Primary producers include chemoautotrophic Bacteria and Archaea, cyanobacteria, photosynthetic sulphur bacteria, macroalgae, phanerograns, benthic diatoms and a variety of planktonic photosynthesisers, as well as significant inputs of terrestrial organic matter at some sites. This makes the task of unraveling food webs at shallow-water sites particularly challenging when compared to their deep-sea counterparts below the euphotic zone. The diet of some species changes seasonally, e.g. *Cyclope neritea*. In other species, different individuals in an area can consume very different food items, due to microscale differences in conditions at vent and seep sites.

11.7.2 Cold Seeps

Nutrient input from the sediment is significant at shallow-water cold seep settings. Algal species indicative of eutrophication, such as *Enteromorpha*, may be found at such sites (Johannes 1980; Dando et al. 1994b). Similar nutrient enrichment has been shown at shallow vents (Dando et al. 1995b). As with shallow-water hydrothermal vents, cold seeps supply the surrounding ecosystem with a diverse mix of nutritional sources that may include chemoautotrophic bacteria, benthic cyanobacteria, benthic diatoms and seagrasses as well as phytoplankton. Near fluid outlets, the shallow sediment water re-circulation causes organic enrichment of a zone around the fluid outlet (O'Hara et al. 1995), allowing a more normal community to exist close to the outlet in a zone free of sulphides and, in the case of vents, lower temperatures (Dando et al. 1994b; Fitzsimons et al. 1997).

Underwater photographs and videos of many seeps give the impression that they are rich oases of life on an otherwise fairly barren sea floor. This has led to the hypothesis that the rich fauna surrounding the seeps is largely dependent upon a food chain based on hydrocarbon-oxidizing bacteria (Spies and Davis 1979; Hovland and Judd 1988; Hovland and Thomsen 1989). Food web studies have been made at a number of sites, using stable isotopes as indicators, especially of methane input, to test this hypothesis (Spies and DesMarais 1983; Juhl and Taghon 1993; Dando et al. 1994b; Levin et al. 2000; Dando 2001).

Isotope studies on organisms living along a transect downstream from the East Flower Garden Brine Pool showed that they become more depleted in ^{13}C the closer they live to the brine lake outlet, implying a significant input from carbon fixed by chemosynthesis, that is diluted with distance from the Lake (Powell et al. 1986). In contrast to studies at hydrothermal sites, the sponges close to the pool were highly depleted in ^{13}C. The interactions between meiofauna and macrofauna at this complex site appear to depend mainly on the food supply but there is little direct evidence of macrofauna having a significant predation effect on the meiofauna (Powell et al. 1986). This was supported by studies at the Isla Vista oil seeps using ^{14}C as a tracer of fossil carbon from the seep (Bauer et al. 1990). The meiofauna and macrofauna showed quite different patterns of fossil carbon incorporation with distance from the seep, implying no linkage between the two groups and a different feeding selectivity for sediment particulate organic carbon.

At the Isla Vista oil seeps, the isotope data indicated that the nematodes within the bacterial mats were feeding on the mats. The majority of the larger infauna were only slightly depleted in ^{13}C at the seeps while the maldanid *Praxillella affinus* showed a contribution of chemosynthetic carbon to its nutrition of 15.6% at the seep (Spies and DesMarais 1983). Thus, fossil carbon from the seeps made, overall, a minimal contribution to the biomass of the macrofauna. A similar finding was made by Juhl and Taghon (1993): fauna at a pockmark on the Oregon shelf were very slightly more depleted in ^{13}C than at control sites. Similar conclusions, that the majority of seep fauna derive the vast majority of their nutrition through photosynthetically fixed carbon, were reached in studies of methane seeps on the Californian shelf (Levin et al. 2000), the Scanner pockmark in the North Sea (Dando et al. 1991), the bubbling reefs in the Kattegat (Jensen et al. 1992) and the shallower Kattegat seeps (Dando et al. 1994b). This contrasts with studies at deep-sea seeps where it has been shown that carbon fixed by chemosynthesis makes a major contribution to the nutrition of seep organisms (Levin and Michener 2002; MacAvoy et al. 2003). The difference may be due to the availability and preferred use of phytoplankton and/or benthic diatoms at shallow sites, e.g. Kharlamenko et al. (1995).

11.7.3 Comparison with Deep-Sea Sites and Other Reducing Environments

In general, the food webs are similar at both shallow-water vent and seep sites. One exception is the input of organic carbon from thermophilic and hyperthermophilic Bacteria and Archaea at vents sites, although little is know about their contribution to the nutrition of organisms. A second difference is in the role of pockmarks that act as sediment traps, so that even those pockmarks that are no longer active will have a higher organic matter content in the sediment and a finer-grained deposit, thus altering the structure of the benthic and epibenthic communities. At both shallow vent and seep sites there is evidence for a large degree of heterogeneity in diets, both between species and within a given species. The latter is due to microhabitat differences causing large differences in the mixture of food available (Southward et al. 1996; Morri et al. 1999; De Biasi et al. 2004).

The complexity in food webs at shallow-water vents and seeps is also found in biological communities inhabiting reducing sediments rich in organic matter. In this latter habitat, which may be common in areas close to vent and seep sites, POC inputs also come from chemosynthetic bacteria, heterotrophic prokaryotes, cyanobacteria, seagrasses, benthic diatoms and photosynthetic plankton as well as terrestrial organic matter. Related species utilize different components of the above, or different mixtures thereof. For example, in a study of fatty acids in four bivalve species from Vostock Bay, Sea of Japan, Zhukova et al. (1992) showed that one species obtained most of its nutrition from endosymbiotic sulphur-oxidising bacteria, and another from heterotrophic bacteria in the sediment. The remaining two species appeared to utilize different proportions of phytoplankton and heterotrophic bacteria. Even bivalves with symbiotic, sulphur-oxidising bacteria from reducing sediments can show changes in nutrition between years (Dando and Spiro 1993). Microhabitat differences in the sediment can affect the supply of sulphide to thyasirid bivalves, and hence the degree of mixotrophy that individuals exhibit (Dando and Spiro 1993).

A difference for the situation in reducing sediments is that major fluid flows from both vents and seeps create seawater downflows into the sediment (O'Hara et al. 1995; Klein 2006). Thus, the fluid outlets are surrounded by a zone of sediment acting as a filter and are enriched with POC. Deposit and filter-feeding organisms can thrive in this narrow zone, especially since the downflow will dilute the effect of the vent and seep fluids. A related situation occurs close to gas bubble outlets where the rising bubble stream entrains water as it rises above the seabed (McDougall 1978), with the entrained water spreading out horizontally at a certain level in the water column (Fig. 11.3). This will affect both the nekton, which can use the rising plume as a feeding station (Cardigos et al. 2005), and the food supply to plankton in the neutrally buoyant plume.

Although many epifaunal species can live close to vent and seep outlets in shallow-water, because of the rapid dilution of fluids rising from the seabed, this does not apply to infaunal species. In both vent and seep situations, high soluble sulphide concentrations or cementation of sediment can prevent animals burrowing deeply. At vents, high temperatures in the sediment will also limit the distribution of the biota.

11.8 Discussion

Shallow-water hydrothermal vents and cold seeps differ from their deep-sea counterparts in the absence of a general vent- or seep-obligate fauna, although a few individual species appear to be exceptions. Only thermophilic and hyperthermophilic Bacteria and Archaea can be considered, as a group, to be obligate species at hydrothermal sites. In contrast to the prokaryotes, the general mix of species of eukaryotes at the shallow sites is a subset of those found in the surrounding region, although there is evidence, in semi-enclosed areas, of exotic thermophilic species establishing themselves in hydrothermal areas, or becoming more predominant in such areas. Benthic fauna with chemoautotrophic or methylotrophic symbionts are comparatively rare at these shallow sites, only eight species, four of which are considered vent obligates, are recorded for the vents listed in Table 11.1. In addition, only four eukaryote species, without symbionts, recorded at shallow-water vent sites can be considered vent obligates. For the shallow seeps, there are 11 eukaryote species listed with chemoautotrophic symbionts in Table 11.2, but no clear seep-obligate species were found among the eukaryotes. These figures compare with a mean of 82% of vent-obligate species at deep-sea vent sites (Tunnicliffe et al. 1998). For deep-water cold seeps Sibuet and Olu (1998) recorded 64 eukaryotes with chemoautotrophic or methylotrophic symbionts, 30% of the total species, with most of them being considered seep-obligates. Most of the remaining species were also considered seep-obligates.

In contrast to the lack of animals with symbionts at shallow-water vents, nematodes and bivalves with chemoautotrophic symbionts may be quite common in shallow-water reducing sediments near to vents (e.g. Dando et al. 1995c). Mironov et al. (2002) have suggested that this might be because there is no energetic advantage in developing such complex symbiotic systems in the photic zone. However, such species are often abundant in shallow-water sediments, for example within, or on the edge of, seagrass beds (Dando et al. 1986a; Gros et al. 1998) or in organic-rich fjord sediments (Dando et al. 1986b). Given the excess CO_2 released at vent sites, it is also a surprise that there have been no reports, to date, of some algal species adapting to these conditions and becoming vent obligates.

The central question is: why are there so few obligate species at the shallow sites compared with their deeper counterparts? Tarasov et al. (2005) discussed this issue at length in their review of the communities at shallow and deep vent sites. At shallow sites, the lower pressures and temperatures can result in the fluids being less toxic and thus more accessible to the background fauna so that there is a greater predatory pressure than at comparable deep-sea sites. In addition, the food web data indicates that most species living at the shallow vent and seep sites depend principally for their nutrition on carbon fixed by photosynthesis. This may be because there has not been the oligotrophic pressure in shallow-water, compared with the deep-sea, for animals to adapt to bacterial food sources, although there are situations, as discussed in the previous paragraph, where eukaryotes have developed symbiotic associations with chemoautotrophs in shallow-water settings. These species are able to invade shallow cold seeps but, with a few exceptions, not usually hydrothermal vents.

Although many epifaunal species can live close to vent and seep outlets in shallow-water, because of the rapid dilution of fluids rising from the seabed, this does not apply to infaunal species. In both vent and seep situations, high, soluble sulphide concentrations or cementation of sediment can prevent animals burrowing deeply. At vents, high temperatures and/or heavy metal concentrations in the sediment will also limit the distribution of the biota.

The 200 m boundary approximately coincides with the transition from the euphotic to the aphotic zone and with a faunistic depth boundary for non-vent fauna (Tarasov et al. 2005). These authors suggested that there might be a threshold in the ratio between organic matter derived from chemosynthesis and that from photosynthesis that controls the evolution of vent-obligate taxa. If this hypothesis is correct, we should find more shallow vent- or seep-obligate species in oligotrophic areas, where the critical threshold will be shallower. There is some evidence for this at shallow-water vent sites. Most of the sites in the oligotrophic western Pacific have at least one vent-obligate species; compare Table 11.1 and Fig. 11.1. The four exceptions in this region are at Matupi Harbour, the inlet of Kodakara-Jima Island and at Yankicha and Kunashir Islands in the Kuriles. At these four sites, there are significant photosynthetic inputs, from planktonic cyanobacteria and photosynthetic eukaryotes and from benthic cyanobacteria and diatoms. The remaining site with a vent-obligate species is at Milos, which is also in an oligotrophic area (Robinson 2000). No eukaryote species have been reported from the vents at Kodakara-Jima Island.

Shallow-water cold seeps, in contrast to many of their vent counterparts, are mainly found close to the coast along continental margins, in regions where primary productivity by photosynthesis is high. There would thus be less pressure on species to adopt alternative nutritional strategies. Nearly all the shallow seep species with symbionts are found in reducing sediments in the same region as the seeps. The one possible exception is the gutless nematode, *Astononema southwardorum*. Although the microniche occupied by this species in the Scanner pockmark is of the type common in reducing sediments and it is probably common in non-seep sediments in the area. Unfortunately, few meiofaunal studies have been made in reducing sediments at this depth in the North Sea.

In their study of seeps on the Sakhalin shelf and slope, Sahling et al. (2003) proposed that predatory pressure restricted the number of macrofaunal species with symbionts found at the shallower seeps. Carnivorous gastropods, asteroids and a crab, *Hyas* sp., were abundant at sites shallower than 250 m water depth and the reduction in the numbers of these species with depth coincided with a rise in shelled megafauna. This pressure of predators would be expected to especially affect the distribution of species exposed at the surface, but to have a lesser effect on the deep-burrowing bivalves with symbionts, such as some of the thyasirids and lucinids that are found in reducing coastal sediments. Predation pressure is likely to affect the structure of the shallow-water vent and seep faunal communities and may explain the absence, or low numbers of, species of mytilids, vesicomyids and vestimentiferans at shallow sites.

The range of microniches at both vent and seep sites, together with the multifarious food supplies, enhanced fluid circulation and regular habitat disturbance exposing areas for settlement, means that a larger number of species are often found living at and around such sites than in the nearby area. Depending on the variety of POC inputs, substratum types, bottom topography and water flow patterns; such sites can be hotspots of species diversity. Examples of such hot-spots include: the epifauna at the venting areas off Palaeochori, Milos (Morri et al. 1999), the biota at the 'Bubbling Reefs' in the Kattegat (Jensen et al. 1992) and the biota in the region of the brine seeps on East Flower Garden Bank, Texas (Bright et al. 1980). In the Scanner pockmark, Hurlebert rarefaction analysis (Hurlebert 1971) on the infauna alone, showed that the greatest species diversity was in samples from the pockmark's base.

The unusual nature and species richness at these sites has resulted in several of them being given, or considered for, protected area status. The shallow-water hydrothermal venting area of Kraternaya Bight was made a Biological Reserve in 1988 by the Sakhalin District Council and the Dom João de Castro Seamount was declared a Special Area for Conservation and a Site of European Community Importance by the Portuguese government under the 1992 EC Habitat Directive. Some of the most-studied cold seep areas have been similarly protected: the Flower Garden Banks are now a U.S. National Marine Sanctuary and the 'Bubbling Reefs' in the Kattegat are a Special Area of Conservation under the EU Habitats Directive. The UK Joint Nature Conservation Committee is currently proposing that the Scanner and Braemar pockmarks in the North Sea are made offshore marine Special Areas of Conservation.

11.9 Future Directions

There is currently a lack of detailed studies of the fauna at both vent and seep sites at depths between 200 and 500 m. This depth zone is likely to show the transition between the faunal characteristics of shallow sites, described in this review, and their deep-sea counterparts where most of the fauna are vent- or seep-obligates. Comprehensive studies are needed in regions where vent sites occupy a depth gradient. For hydrothermal vents the best areas to study would be seamounts such as those on the Kermadec Ridge (de Ronde et al. 2001), or the Dom João de Castro Seamount in the Atlantic (Cardigos et al. 2005) where venting is known to occur at a range of depths. For seep studies, faulted sedimentary basins such as the Gulf of Mexico, where most studies to date are at sites below 500 m (MacAvoy et al. 2008) and the Gulf of Cadiz (Pinheiro et al. 2003), where mud volcanoes exist at a range of depths from 200 to >3,000 m, would be good areas for more detailed studies.

The evidence that some shallow-water hydrothermal vents and cold seeps are areas of high species biodiversity suggests that wider studies are needed at geographically diverse sites, especially those in areas sheltered from storm disturbance, to identify sites that should be considered for protection. There have been few studies on the existence of sibling species occurring at vent and seep sites and more comparisons are needed, similar to those carried out by Gamenick et al. (1998a, b), into the 'background' biota found in the most extreme conditions.

Both vent and seep sites are exposed to reduced pH and elevated dissolved inorganic carbon, so that organisms at such sites are those tolerant of these conditions. While recent studies have concentrated on species exclusion, especially with respect to reducing pH (Hall-Spencer et al. 2008; Martin et al. 2008), there has been little effort made to examine the adaptation of biota inhabiting such sites to these conditions. Genetic studies are needed to see if significant selection for tolerance to low pH has occurred. Such information would help in modeling the response of the marine ecosystem to increasing acidification (Doney et al. 2009).

11.10 Conclusions

The depth boundary between shallow-water vents and cold-seeps and their 'deep-sea' counterparts is at approximately 200 m. This faunistic boundary is probably controlled by the amount of POC input from photosynthetic production, since this will decrease below the photic zone, and the greater number of predators that are able to invade the shallow sites. In the deep-sea POC derived from photosynthesis becomes limiting and there is a greater pressure on organisms to adapt and use carbon produced by chemosynthesis in environments with high sulphide concentrations. Shallow-water vents and seeps differ from their deep-sea counterparts in the low numbers of obligate species, no definite seep-obligates being reported. Both shallow vents and seeps are colonized by a subset of the local fauna that are tolerant of sulphide, and in the case of vents, higher temperature conditions. The multifarious food sources, repeated disturbance and habitat types at many vent and seep sites create a large number of microniches for fauna so that the fluid outlets and their surroundings often support a greater species diversity.

Acknowledgments I thank Steffen Kiel for the invitation to write this review, Nike Bianchi, Carla Morri, Alan Southward and Eve Southward for discussions on vent and seep ecosystems over the years and Andrey Gebruk and Lisa Levin for reviews of the manuscript and helpful comments. The studies at Milos and in Denmark were funded by EC MAST Programme projects and I thank my numerous collaborators for their efforts in unraveling the complex ecosytems there. The acquisition of the airborne data by the NERC Airborne Remote Sensing Facility under Project MC04/10 is gratefully acknowledged.

References

Abbiati M, Airoldi L, Castelli A, Cinelli F, Southward AJ (1994) Preliminary observations on a dense population of *Phyllochaetopterus socialis* Claparede at the sulphurous water boundary in a Mediterranean submarine cave. Mem Mus Nat Hist Nat (France) 162:323–329

Acunto S, Rindi F (1997) Variabilità spaziale di popolamenti fiobentici in relazione ad attivà idrotermali nella Baia di levante dell'Isolla di Vulcano (Isole Eolie): studio preliminare. Biol Mar Med 4:351–352

Acunto S, Maltagliati F, Cinelli F (1995) Osservazioni sui popolamenti bentonici di un'area interessata da attivitá idrotermali nei pressi dell'Isola di Panarea (Isole Eolie). Biol Mar Med 3:434–436

Acunto S, Maltagliati F, Rindi F, Rossi F, Lardicci C, Cinelli F, Cognetti G (1996a) *Halophila stipulacea* (Forssk.) Aschers (Hydrocharitacea) nel Porto di Levante dell'Isola di Vulcano: fenologia e popolamenti zoobentonici associati. In: Faranda FM, Povero P (eds) Caratterizzazione ambientale marine del sistema Eolie e dei bacini limitrofi di Cefalù e gioia (EOCUMM95) Data Report. Consorzio Nazionale Interuniversitario per le Scienze del Mare, Italy, pp 327–332

Acunto S, Rindi F, Cinelli F (1996b) Observazioni sui popolamenti fitobentici della Baia di Levante dell'Isola di Vulcano (isole Eolie): variabilità spaziale in relazione ad attività idrotermali. In: Faranda FM, Povero P (eds) Caratterizzazione ambientale marine del sistema Eolie e dei bacini limitrofi di Cefalù e gioia (EOCUMM95) Data Report. Consorzio Nazionale Interuniversitario per le Scienze del Mare, Italy, pp 323–326

Airoldi L, Cinelli F (1997) Sources and biochemical composition of suspended particulate material in a submarine cave with sulphur water springs. Mar Biol 128:537–545

Aliani S, Amici L, Dando PR, Meloni R (1998a) Time series and bottom temperature in a marine shallow water hydrothermal vent off Milos Island (Aegean Volcanic Arc): preliminary results. Rapp P-V Reun (Monaco) 35:46–47

Aliani S, Bianchi CN, Cocito S, Dando PR, Meloni R, Morri C, Niemeyer A, Peirano A, Ziebis W (1998b) A map of seagrass meadows in Palaeochori Bay (Milos Island, Greece), a marine area with hydrothermal activity. Rapp P-V Reun (Monaco) 35:512–513

Aliani S, De Biasi AM, Meloni R (2000) Seasonal variations of sea-water temperature at shallow-water hydrothermal vents in Milos Island (Aegean Sea): faunistic remarks. Atti Assoc Ital Oceanol Limnol 13:217–226

Arp AJ, Menon JG, Julian D (1995) Multiple mechanisms provide tolerance to environmental sulfide in *Urechis caupo*. Am Zool 35:132–144

Baker ET, Massoth GJ, de Ronde CEJ, Lupton JE, McInnes BIA (2002) Observations and sampling of an ongoing subsurface eruption of Kavachi volcano, Solomon Islands, May 2000. Geology 30:975–978

Barnes RD (1965) Tube-building and feeding in chaetopterid polychaetes. Biol Bull 129: 217–233

Bauer JE, Spies RB, Vogel JS, Nelson DE, Southon JR (1990) Radiocarbon evidence of fossil-carbon cycling in sediments of a nearshore hydrocarbon seep. Nature 348:230–232

Bayona JM, Monjonell A, Miquel JC, Fowler SW, Albaigés J (2002) Biogeochemical characterization of particulate organic matter from a coastal hydrothermal vent zone in the Aegean Sea. Org Geochem 33:1609–1620

Benjamínsson J (1988) Jarðhiti í sjó of flaeðamáli við Ìsland. Natturufraedingurinn 58:153–169

Bianchi CN, Abbiati M, Airoldi L, Alvisi M, Benedetti-Cecchi L, Cappelletti A, Cinelli F, Colantoni P, Dando PR, Morri C, Niccolai I, Picco P, Southward AJ, Southward E (1998) Hydrology and water budget of a submarine cave with sulphidic springs: the "Grotta Azzurra" of Capo Palinuro (Southern Italy). In: Piccazzo M (ed) Atti del 12° Congresso dell'Associazone Italiana di Oceanologia e Limnologia, Isola di Volcano, Settembre 1996. Associazone Italiana di Oceanologia e Limnologia, Genova, pp 285–301

Bischoff JL, Dickson FW (1975) Seawater-basalt interaction at 200°C and 500 bars: implications for origin of seafloor heavy–metal deposits and regulation of seawater chemistry. Earth Planet Sci Lett 25:385–397

Bischoff JL, Rosenbauer RJ (1984) The critical point and phase boundary of seawater 200–500 °C. Earth Planet Sci Lett 68:172–180

Bodergat AM, Oki K, Ishizaki K, Rio M (2002) Impact of volcanism, human activities, and water mass circulation on the distribution of ostracod populations in Kagoshima Bay (Kyushu Island, southern Japan). CR Acad Sci Paris 334:1053–1059

Boetius A, Ravenschlag K, Schubert CJ, Rickert D, Widdel F, Gieseke A, Amann R, Jørgensen BB, Witte U, Pfannkuche O (2000) A marine microbial consortium apparently mediating anaerobic oxidation of methane. Nature 407:623–626

Botz R, Stüben D, Winckler G, Bayer R, Schmitt M, Faber E (1996) Hydrothermal gases from offshore Milos Island, Greece. Chem Geol 130:161–173

Bright TJ, LaRock PA, Lauer RD, Brooks JM (1980) A brine seep at the East Flower Garden Bank, northwestern Gulf of Mexico. Int Rev Gesamten Hydrobiol 65:535–549

Bright M, Arndt C, Keckeis H, Felbeck H (2003) A temperature-tolerant interstitial worm with associated epibiotic bacteria from the shallow water fumaroles of Deception Island, Antarctica. Deep Sea Res II 50:1859–1871

Brinkhoff T, Sievert SM, Kuever J, Muyzer G (1999) Distribution and diversity of sulfur-oxidizing Thiomicrospira spp. at a shallow-water hydrothermal vent in the Aegean Sea (Milos, Greece). Appl Environ Microbiol 65:3843–3849

Bussmann I, Dando PR, Niven SJ, Suess E (1999) Groundwater seepage in the marine environment: role for mass flux and bacterial activity. Mar Ecol Prog Ser 178:169–177

Butterfield A, Massoth GJ, McDuff RE, Lupton JE, Lilley MD (1990) Geochemistry of hydrothermal fluids from Axial Seamount hydrothermal emissions study vent field, Juan de Fuca Ridge: subseafloor boiling and subsequent fluid-rock interaction. J Geophys Res 95: 12895–12921

Canet C, Prol-Ledesma MR, Melgarejo J-C, Reyes A (2003) Methane-related carbonates found at submarine hydrothermal springs: a new setting for microbially-derived carbonates? Mar Geol 199:245–261

Capone DG, Bautista MF (1985) A groundwater source of nitrate in nearshore marine sediments. Nature 313:214–216

Cardigos F, Colaço A, Dando PR, Ávila SP, Sarradin P-M, Tempera F, Conceição P, Pascoal A, Serrão Santos R (2005) Characterization of the shallow water hydrothermal vent field communities of the D. João de Castro Seamount (Azores). Chem Geol 224:153–168

Cheminee J-L, Stoffers P, McMurtry G, Richnow H, Puteanua D, Sedwick P (1991) Gas-rich submarine exhalations during the 1989 eruption of MacDonald Seamount. Earth Planet Sci Lett 107:318–327

Chen C-TA, Zeng Z, Kuo F-W, Yang TF, Wang B-J, Tu Y-Y (2005) Tide-influenced hydrothermal system offshore NE Taiwan. Chem Geol 224:69–81

Childress JJ, Fisher CR (1992) The biology of hydrothermal vent animals: physiology, biochemistry and autotrophic symbioses. Oceanogr Mar Biol Annu Rev 30:337–441

Christodoulou D, Papatheodorou G, Ferentinos G, Masson M (2003) Active seepage in two contrasting pockmark fields in the Patras and Corinth gulfs, Greece. Geol Mar Lett 23:194–199

Colangelo MA, Dall'Olio P, Ponti M, Ceccherelli VU (1996) Prime osservavazioni sul meiobenthos associato a fondali interessati da fenomeni vulcanismo scondario nell'Isola di Panarea (Me). In: Faranda FM, Povero P (eds) Caratterizzazione ambientale marine del sistema Eolie e dei bacini limitrofi di Cefalù e gioia (EOCUMM95) Data Report. Consorzio Nazionale Interuniversitario per le Scienze del Mare, Italy, pp 369–374

Colbourne EB, Hay AE (1990) An acoustic remote sensing and submersible study of an Arctic submarine spring plume. J Geophys Res 95:13219–13234

Dando PR (2001) A review of pockmarks in the UK part of the North Sea, with particular respect to their biology. Department of Trade and Industry, Technical Report TR 001. Strategic Environmental Assessment – SEA2, London

Dando PR, Hovland M (1992) Environmental effects of submarine seeping natural gas. Cont Shelf Res 12:1197–1207

Dando PR, Spiro B (1993) Varying nutritional dependence of the thyasirid bivalves *Thyasira sarsi* and *T. equalis* on chemoautotrophic symbiotic bacteria, demonstrated by isotope ratios of tissue carbon and shell carbonate. Mar Ecol Prog Ser 92:151–158

Dando PR, Southward AJ, Southward EC (1986a) Chemoautotrophic symbionts in the gills of the bivalve mollusc *Lucinoma borealis* and the sediment chemistry of its habitat. Proc R Soc B 227:227–247

Dando PR, Southward AJ, Southward EC, Barrett RL (1986b) Possible energy sources for chemoautotrophic prokaryotes symbiotic with invertebrates from a Norwegian fjord. Ophelia 26:135–150

Dando PR, Austen MC, Burke RJ, Kendall MA, Kennicutt MC, Judd AG, Moore DC, O'Hara SCM, Schmaljohann R, Southward AJ (1991) Ecology of a North Sea pockmark with an active methane seep. Mar Ecol Prog Ser 70:49–63

Dando PR, Fenchel T, Jensen P, O'Hara SCM, Niven SJ, Schuster U (1993) Ecology of gassy, organic rich sediment in a shallow subtidal area on the Kattegat coast of Denmark. Mar Ecol Prog Ser 100:265–271

Dando PR, Bussmann I, Niven SJ, O'Hara SCM, Schmaljohann R, Taylor LJ (1994a) A methane seep area in the Skagerrak, the habitat of the Pogonophore, *Siboglinum poseidoni*, and the bivalve mollusc *Thyasira sarsi*. Mar Ecol Prog Ser 107:157–167

Dando PR, Jensen P, O'Hara SCM, Niven SJ, Schmaljohann R, Schuster U, Taylor LJ (1994b) The effects of methane seepage at an intertidal/shallow subtidal site on the shore of the Kattegat, Vendsyssel, Denmark. Bull Geol Soc Denmark 41:65–79

Dando PR, Jensen P, O'Hara SCM, Taylor LJ, Clayton C, Baylis S, Schuster U, Laier T (1994c) Gas seepage from a carbonate-cemented sandstone reef on the Kattegat coast of Denmark. Mar Petr Geol 11:182–189

Dando PR, Hughes JA, Leahy Y, Niven SJ, Taylor LJ, Smith C (1995a) Gas venting rates from submarine hydrothermal areas around the island of Milos, Hellenic Volcanic Arc. Cont Shelf Res 15:913–929

Dando PR, Hughes JA, Leahy Y, Taylor LJ, Zivanovic S (1995b) Earthquakes increase hydrothermal venting and nutrient inputs into the Aegean. Cont Shelf Res 15:655–662

Dando PR, Hughes JA, Thiermann F (1995c) Preliminary observations on biological communities at shallow hydrothermal vents in the Aegean Sea. In: Parson LM, Walker CL, Dixon DR (eds) Hydrothermal vents and processes. Geological Society (Special Publication), London, pp 303–317

Dando PR, Thomm M, Arab H, Brehmer M, Hooper L, Jochimsen B, Schlesner H, Stöhr R, Miquel J-C, Fowler S (1998) Microbiology of shallow hydrothermal sites off Palaeochori Bay, Milos (Hellenic Volcanic Arc). Cah Biol Mar 39:369–372

Dando PR, Stüben D, Varnavas SP (1999) Hydrothermalism in the Mediterranean Sea. Prog Oceanogr 44:333–367

Dando PR, Aliani S, Arab H, Bianchi CN, Brehmer M, Cocito S, Fowler SW, Gundersen J, Hooper LE, Kölbl R, Kuever J, Linke P, Makropoulos KC, Meloni R, Miquel J-C, Morri C, Müller S, Robinson C, Schlesner H, Sievert S, Stöhr R, Stüben D, Thomm M, Varnavas SP, Ziebis W (2000) Hydrothermal studies in the Aegean Sea. Phys Chem Earth (B) 25:1–8

Dando PR, Southward AJ, Southward EC, Lamont P, Harvey R (2008) Interactions between sediment chemistry and frenulate pogonophores (Annelida) in the north-east Atlantic. Deep Sea Res I 55:966–996

Davis PH, Spies RB (1980) Infaunal benthos of a natural petroleum seep: study of community structure. Mar Biol 59:31–41

De Biasi AM, Aliani S (2003) Shallow-water hydrothermal vents in the Mediterranean Sea: stepping stones for Lessepsian migration? Hydrobiologia 503:37–44

De Biasi AM, Bianchi CM, Alliani S, Cocito S, Peirano A, Dando PR, Morri C (2004) Epibenthic communities in a marine shallow area with hydrothermal vents (Milos Island, Aegean Sea). Chem Ecol 20(Suppl1):S89–S105

de Ronde CEJ, Baker ET, Massoth GJ, Lupton JE, Wright IC, Feely RA, Greene RG (2001) Intra-oceanic subduction related hydrothermal venting, Kermadec volcanic arc, New Zealand. Earth Planet Sci Lett 193:359–369

Desbruyères D, Segonzac M, Bright M (2006) Handbook of deep-sea hydrothermal vent fauna. Second completely revised version. Denisia 18:1–544

Diapoulis A, Koussouris T, Boganos C, Frotis G, Berthas I (1994) Distribution of benthic organisms in a volcanic sea area, Aegean Sea, Greece. Fresenius Environ Bull 3:16–23

Doney SC, Fabry VJ, Feely RA, Kleypas JA (2009) Ocean acidification: the other CO_2 problem. Annu Rev Mar Sci 1:169–192

Duan Z, Weare JH (1992) An equation of state for the CH4 -CO2-H2O system: I pure systems from 0 to 1000°C and 0 to 8000bar. Geochim Cosmochim Acta 56:1451–1460

Fader GBJ (1991) Gas-related sedimentary features from the eastern Canadian continental shelf. Cont Shelf Res 11:1123–1153

Fitzsimons MF, Dando PR, Hughes JA, Thiermann F, Akoumianaki I, Pratt SM (1997) Submarine hydrothermal brine seeps off Milos, Greece: observations and geochemistry. Mar Chem 57:325–340

Forrest MJ (2004) The geology, geochemistry and ecology of a shallow water submarine hydrothermal vent in Bahía Concepcíón, Baia California Sur, Mexico. M. Sc. thesis, University of Southern California, Monterey Bay, CA

Fouquet Y, Cherkashev GA, Charlou JL, Ondréas H, Birot D, Cannat M, Bortnikov NC, Silantyev S, Sudarikov T, Cambon-Bonavita MA, Desbruyères D, Fabri M-C, Querellou J, Hourdez S, Gebruk AV, Sokolova T, Hoise E, Marcier E, Kohn C, Donval JP, Etoubleau J, Normand A, Stephan M, Briand P, Crozon J, Fernagu P, Buffier E (2008) Serpentine cruise - ultramafic hosted hydrothermal deposits on the Mid-Atlantic Ridge: first submersible studies on Ashadze 1 and 2, Logatchev 2 and Krasnov vent fields. InterRidge News 17: 16–21, online supplement

Fricke H, Giere O, Stetter K, Alfredsson GA, Kristjansson JK, Stoffers P, Svavarsson J (1989) Hydrothermal vent communities at the shallow subpolar Mid-Atlantic Ridge. Mar Biol 102:425–429

Gamenick I, Abbiati M, Giere O (1998a) Field distribution and sulphide tolerance of *Capitella capitata* (Annelida: Polychaeta) around shallow water hydrothermal vents off Milos (Aegean Sea). A new sibling species? Mar Biol 130:447–453

Gamenick I, Vismann B, Grieshaber MK, Giere O (1998b) Ecophysiological differentiation of *Capitella capitata* (Polychaeta). Sibling species from different sulfidic habitats. Mar Ecol Prog Ser 175:155–166

Gamo T, Glasby CJ (2003) Submarine hydrothermal activity in coastal zones. In: Taniguchi M, Wang K, Gamo T (eds) Land and marine hydrogeology. Elsevier, Tokyo, pp 151–163

Garcias-Bonet N, Marbà N, Holmer M, Duarte CM (2008) Effects of sediment sulfides on seagrass *Posidonia oceanic* meristematic activity. Mar Ecol Prog Ser 372:1–6

German CR, Briem J, Chin C, Danielsen M, Holland S, James R, Jónsdóttir A, Ludford E, Moser C, ólafsson J, Palmer MR, Rudnicki MD (1994) Hydrothermal activity on the Rejkjanes Ridge: the Steinahóll vent-field at 63°06'N. Earth Planet Sci Lett 121:647–654

Giaccone G (1969) Associazioni algali e fenomeni secondari di vulcanismo nelle acque marine di Vulcano (Mar Tirreno). Giorn Bot Ital 103:353–366

Giere O, Windoffer R, Southward EC (1995) The bacterial endosymbiosis of the gutless nematode *Astomonema southwardorum*: ultrastructural aspects. J Mar Biol Assoc UK 75:153–164

Gimenez F, Marin A (1991) Los anélidos poliquetos de una solfataria submarina en el Golfo de Napoles. Ann Biol Murcia 17:143–151

Gittings SR, BT J, Powell EN (1984) Hard-bottom macrofauna of the East Flower Garden brine seep: impact of a long-term, sulfurous brine discharge. Contrib Mar Sci 27:105–125

Gomoiu MT (1964) Biologisches Studium der Arten *Nassa reticulata* L. und *Cyclonassa neritea* (L.) im Schwarzen Meer (rumänischer Küstenbereich). Rev Roum Biol Zool 9:39–49

Grant J, Bathman UV (1987) Swept away: resuspension of bacterial mats regulates benthic-pelagic exchange of sulphur. Science 236:1472–1474

Greinert J, Artemov Y, Egorov V, De Batist M, McGinnis D (2006) 1300–m-high rising bubbles from mud volcanoes at 2080 m in the Black Sea: hydroacoustic characteristics and temporal variability. Earth Planet Sci Lett 244:1–15

Gros O, Durand P, Frenkile L, Mouëza M (1998) Putative environmental transmission of sulfur-oxidising gill endosymbionts in four tropical lucinid bivalves. FEMS Microbiol Lett 160:257–262

Gugliandolo C, Maugeri TL (1993) Chemolithotrophic, sulfur-oxidizing bacteria from a marine, shallow hydrothermal vent of Vulcano (Italy). Geomicrobiol J 11:109–120

Hafenbrandt D, Keller M, Dirmeier R, R R, Rossnagel P, Burggraf S, Huber H, Stetter KO (1996) *Ferroglobus placidus* gen. nov., sp. nov., a novel hyperthermophilic archaeum that oxidises Fe^{2+} at neutral pH under anoxic conditions. Arch Microbiol 166:308–314

Hall-Spencer JM, Rodolfo-Metalpa R, Martin S, Ransome E, Fine M, Turner SM, Rowley SJ, Tedesco D, Buia M-C (2008) Volcanic carbon dioxide vents show ecosystem effects of ocean acidification. Nature 454:94–99

Hartley JP (2005) Seabed investigations of pockmark features in UKCS Block 16/3. Report to Joint Nature Conservation Committee. Hartley Anderson, Ellon

Hashimoto J, Miura T, Fujikura K, Ossaka J (1993) Discovery of vestimentiferan tube-worms in the euphotic zone. Zool Sci 10:1063–1067

Hinze G (1903) *Thiophysa volutans*, ein neues Schwefelbakterium. Ber Dtsch Bot Ges 21: 309–316

Hoaki T, Nishijima M, Miyashita H, Maruyama T (1995) Dense community of hyperthermophilic sulfur-dependent heterotrophs in a geothermally heated shallow submarine biotope near Kodakara-Jima Island, Kagoshima, Japan. Appl Environ Microbiol 61: 1931–1937

Holm NG (1987) Biogenic influences on the geochemistry of certain ferruginous sediments of hydrothermal origin. Chem Geol 63:45–57

Hovland M, Judd AG (1988) Seabed pockmarks and seepages. Graham & Trotman, London

Hovland M, Sommerville JH (1985) Characteristics of two natural gas seepages in the North Sea. Mar Petr Geol 2:319–326

Hovland M, Thomsen E (1989) Hydrocarbon-based communities in the North Sea? Sarsia 74:29–42

Huber H, Thomm M, König H, Thies G, Stetter KO (1982) *Methanococcus thermolithotrophicus*, a novel thermophilic lithotropic methanogen. Arch Microbiol 132:47–50

Huber R, Stoffers P, Cheminee JL, Richnow HH, Stetter KO (1990) Hyperthermophilic archaebacteria within the crater and open-sea plume of erupting Macdonald Seamount. Nature 345:179–182

Huber H, Hohn MJ, Rachel R, Fuchs T, Wimmer VC, Stetter KO (2002) A new phylum of Archaea represented by a nanosized hyperthermophilic symbiont. Nature 417:63–67

Hurlbert SH (1971) The non-concept of species diversity: a critique and alternative parameters. Ecology 52:577–586

Hüttel M, Forster S, Klöser S, Fossing HA (1996) Vertical migration in the sediment-dwelling sulphur bacteria *Thioploca* spp. in overcoming diffusion limitations. Appl Environ Microbiol 62:1863–1872

Jeng M-S, Ng NK, Ng PKL (2005) Hydrothermal vent crabs feast on sea 'snow'. Nature 432:969

Jensen P (1986) Nematode fauna in the sulfide–rich brine seep and adjacent bottoms of the East Fower Garden, NW Gulf of Mexico IV ecological aspects. Mar Biol 92:489–503

Jensen P (1995a) Life history of the nematode *Theristus anoxybioticus* from sublittoral muddy sediment at methane seepages in the northern Kattegat, Denmark. Mar Biol 123:131–136

Jensen P (1995b) *Theristus (Penzancia) anoxybioticus* n. sp. (Nematoda: Xylidae) from sublittoral methane seepages in the northern Kattegat, Denmark. J Nematol 27:231–236

Jensen P, Aagaard I, Burke RA, Dando PR, Jørgensen NO, Kuijpers A, Laier T, O'Hara SCM, Schmaljohann R (1992) 'Bubbling reefs' in the Kattegat: submarine landscapes of carbonate-cemented rocks support a diverse ecosystem at methane seeps. Mar Ecol Prog Ser 83:103–112

Johannes RE (1980) The ecological significance of the submarine discharge of groundwater. Mar Ecol Prog Ser 3:365–373

Jones RW (1993) Preliminary observations on benthonic foraminifera associated with biogenic gas seep in the North Sea. In: Jenkins I, Graham D (eds) Micropalaeontology. Kluwer, Dordrecht, pp 69–91

Jones RW (1996) Micropalaeontology in petroleum exploration. Clarendon, Oxford

Jørgensen NO (1992) Methane-derived carbonate cementation of marine sediments from the Kattegat, Denmark: geochemical and geological evidence. Mar Geol 103:1–13

Jørgensen BB, Revsbech NP (1983) Colorless sulfur bacteria, *Beggiatoa* spp. and *Thiovulm* spp. in O_2 and H_2S microgradients. Appl Environ Microbiol 45:1261–1270

Jørgensen BB, Isaksen MF, Jannasch HW (1992) Bacterial sulfate reduction above 100°C in deep-sea hydrothermal vent sediments. Science 258:1756–1757

Juhl AR, Taghon GL (1993) Biology of an active methane seep on the Oregon continental shelf. Mar Ecol Prog Ser 102:287–294

Kamenev GM, Fadeev VI, Selin NI, Tarasov VG, Malakhov VV (1993) Composition and distribution of macro- and meiobenthos around sublittoral hydrothermal vents in the Bay of Plenty, New Zealand. New Zeal J Mar Fresh Res 27:407–418

Karpen V, Thomsen L, Suess E (2006) Groundwater discharges in the Baltic Sea: survey and quantification using a schlieren technique application. Geofluids 6:241–250

Kharlamenko VI, Zhukova NV, Khotimchenko SV, Svetashev VI, Kamenev GM (1995) Fatty acids as markers of food sources in a shallow-water hydrothermal ecosystem (Kraternaya Bight, Yankich Island, Kurile Islands). Mar Ecol Prog Ser 120:231–241

Klein S (2006) Sediment porewater exchange and solute release during ebullition. Mar Chem 102:60–61

Kohout FA, Kolipinski MC (1967) Biological zonation related to groundwater discharge along the shore of Biscayne Bay, Miami, Florida. In: Lauff GH (ed) Estuaries. American Association for the Advancement of Science, Washington, DC, pp 488–508

Kojima S, Ohta S, Yamamoto T, Miura T, Fujiwara Y, Hashimoto J (2001) Molecular taxonomy of vestimentiferans of the western Pacific and their phylogenetic relationship to species of the eastern Pacific. 1. Family Lamellibrachiidae. Mar Biol 139:211–219

Koutsoubas D, Cinelli F (1996) Indo-Pacific origin gastropod species in the Aegean Sea. *Melibe fimbrata* Alder & Hancock, 1864. A new invader. B Malacol 32:35–40

Kurr M, Huber R, König H, Jannasch HW, Fricke H, Trincone A, Kristjansson JK, Stetter KO (1991) *Methanopyrus kandleri*, gen. and sp. nov. represents a novel group of hyperthermophilic methanogens, growing at 110°C. Arch Microbiol 156:239–247

Kuznetsov AP, Gebruk AV (1991) Bacterial symbiotrophy and structure of the respiration-feeding system in *Axinopsida orbiculata* (Thyasiridae, Bivalvia) from the Craternaya Bay (Kurile Islands). Okeanologiya 31:162–169

Levin LA (2005) Ecology of cold seep sediments: interactions of fauna with flow, chemistry and microbes. Oceanogr Mar Biol Annu Rev 43:1–46

Levin LA, Michener R (2002) Isotopic evidence for chemosynthesis-based nutrition of macrobenthos. The lightness of being at Pacific methane seeps. Limnol Oceanogr 47:1336–1345

Levin LA, James DW, Martin CM, Rathburn AE, Harris LH, Michener RH (2000) Do methane seeps support distinct macrofaunal assemblages? Observations on community structure and nutrition from the northern California slope and shelf. Mar Ecol Prog Ser 208:21–39

Luth U, Luth C, Gebruk AV, Thiel H (1999) Methane gas seeps along the oxic/anoxic gradient in the Black Sea: manifestations, biogenic sediment compounds and preliminary results on benthic ecology. Mar Ecol Prog Ser 20:221–249

Luther GW, Rozan TF, Taillefert M, Nuzzio DB, Di Meo C, Shank TM, Lutz RA, Cary SG (2001) Chemical speciation drives hydrothermal vent ecology. Nature 410:813–816

MacAvoy SE, Macko SA, Carney RS (2003) Links between chemosynthetic production and mobile predators on the Louisina continental slope: stable carbon isotopes of specific fatty acids. Chem Geol 201:229–237

MacAvoy SE, Morgan E, Carney RS, Macko SA (2008) Chemoautotrophic production incorporated by heterotrophs in Gulf of Mexico hydrocarbon seeps: an examination of mobile benthic predators and seep residents. J Shellfish Res 27:153–162

MacDonald IR (1992) Sea-floor brine pools affect behavior, mortality, and preservation of fishes in the Gulf of Mexico: Lagerstätten in the making? Palaios 7:383–387

Makropoulos K, Kouskouna V, Karnassopoulou A, Dando P, Varnavas SP (2000) Seismicity in the Hellenic Volcanic Arc Hydrothermal System in relation to geochemical parameters. Phys Chem Earth B 25:19–23

Malakhov VV, Obzhirov AI, Tarasov VG (1992) Pogonophoran genus *Siboglinum* in relation to zones of high concentrations of methane. Dokl Akad Nauk 135:195–197

Martens CS, Klump JV (1984) Biogeochemical cycling in an organic-rich coastal marine basin - 4 An organic carbon budget for sediments dominated by sulphate reduction and methanogenesis. Geochim Cosmochim Acta 44:471–490

Martin S, Rodolfo-Metalpa R, Sonia Rowley E, Buia M-C, Gattuso1 J-P, Hall-Spencer J (2008) Effects of naturally acidified seawater on seagrass calcareous epibionts. Biol Lett 4:689–692

Massé H, Nodot C, Macé A-M (1978) Influence de la temperature sur la reproduction et la survie de quelques Nassariidae (Molusca, Gasteropoda). In: McClusky DS, Berry AJ (eds) Proceedings of the 12th European marine biological symposium. Permagon, London, pp 367–374

Mattison RG, Abbiati M, Dando PR, Fitzsimons MF, Pratt SM, Southward AJ, Southward EC (1998) Chemoautotrophic microbial mats in submarine caves with hydrothermal sulphidic springs at Cape Palinuro, Italy. Microb Ecol 35:58–71

Mazzella L, Buia MC (1986) Strategie evolutive nelle fanerogame marine del Mediterraneo. Nova Thalassia 8(Suppl 3):651

McCarthy KT, Pichler T, Price RE (2005) Geochemistry of Champagne Hot Springs shallow hydrothermal vent field and associated sediments, Dominica, Lesser Antilles. Chem Geol 224:55–68

McDougall T (1978) Bubble plumes in stratified environments. J Fluid Mech 85:65–672

McQuillan R, Fannin NGT, Judd JR (1979) I. G. S. Pockmark investigations 1974–1978. Report no. 98. Institute of Geological Science, Marine Geophysics Unit, Edinburgh

Melwani AR, Kim SL (2008) Benthic infaunal distributions in shallow hydrothermal vent sediments. Acta Oecol 33:162–175

Meyl AH (1954) Beiträge zur Kenntnis der Nematodenfauna vulkanisch erhitzter Biotope. III. Mitteilung. Nematoden aus der Mischungszone strandnaher, heisser Süsswasserquellen mit dem Meerwasser auf der Insel Ischia. Z Morphol Ökol Tiere 42:421–458

Mironov AN, Gebruk AV, Moskalev LI (2002) Biogeography of hydrothermal vent communities and obligate hydrothermal taxa. In: Gebruk AV (ed) Biology of hydrothermal systems. KMK Press, Moscow, pp 410–455

Miura T, Nedachi M, Hashimoto A (2002) Sulphur sources for chemoautotrophic nutrition of shallow water vestimentiferan tubeworms in Kagoshima Bay. J Mar Biol Assoc UK 82: 537–540

Montagna PA, Bauer JE, Prieto MC, Hardin D, Spies RB (1986) Benthic metabolism in a natural coastal petroleum seep. Mar Ecol Prog Ser 34:31–40

Montagna PA, Bauer JE, Hardin D, Spies RB (1989) Vertical distribution of microbial and meiofaunal populations in sediments of a natural coastal hydrocarbon seep. J Mar Res 47:657–680

Moore CG (1979) The zonation of psammolittoral harpacticoid copepods around the Isle of Man. J Mar Biol Assoc UK 59:711–724

Morri C, Cinelli F, Bianchi CM (1994) Sessile epifauna gigantism in a submarine cave with sulphur springs. Cave Diving 6:4–9

Morri C, Bianchi CN, Cocito S, Peirano A, De Biasi AM, Aliani S, Pansini M, Boyer M, Ferdeghini F, Pestarino M, Dando P (1999) Biodiversity of marine sessile epifauna at an Aegean island subject to hydrothermal activity: Milos, Eastern Mediterranean Sea. Mar Biol 135:729–739

Naudts L, Greinert J, Artemov Y, Beaubien SE, Borowski C, De Batist M (2008) Anomalous sea-floor backscatter patterns in methane venting areas, Dnepr paleo-delta, NW Black Sea. Mar Geol 251:253–267

Ng NK, Huang J-F, Ho P-H (2000) Description of a new species of hydrothermal crab, *Xenograpsus testudinatus* (Crustacea: Decapoda: Brachyura: Grapsidae) from Taiwan. Natl Taiwan Mus (Spec Publ Ser) 10:191–199

O'Hara SCM, Dando PR, Schuster U, Bennis A, Boyle JD, Chiu FTW, Hatherell TVJ, Niven SJ, Taylor LJ (1995) Gas seep induced interstitial water circulation: observations and environmental implications. Cont Shelf Res 15:931–948

Ólafsson EB, Petersen OH, Ambrose WG (1994) Does recruitment limitation structure populations and communities of macroinvertebrates in marine soft sediments: the relative significance of pre- and post-settlement processes. Oceanogr Mar Biol Annu Rev 32:65–109

Olu K, Sibuet M, Harmegnies F, Foucher J-P, Fiala-Médoni A (1996) Spatial distribution of diverse cold seep communities living on various diapiric structures of the southern Barbados prism. Prog Oceanogr 38:347–376

Pichler T, Dix GR (1996) Hydrothermal venting within a coral reef ecosystem, Ambitle Island, Papua New Guinea. Geology 24:435–438

Pimenov NV, Rusanov II, Poglazova MN, Mityushina LL, Sorokin DY, Khmelenina VN, Trotsenko YA (1997) Bacterial mats on coral-shaped carbonate structures in methane-seep areas of the Black Sea. Mikrobiologiia 66:354–360

Pinheiro LM, Ivanov MK, Sautkin A, Akhmanov G, Magalhaes VH, Volkonskaya A, Monteiro J H, Somoza L, Monteiro J J, Hamouni N, Cunha MR (2003) Mud volcanism in the Gulf of Cadiz: results from the TTR-10 cruise. Mar Geol 195:131–151

Powell EN, Crenshaw MA, Rieger RM (1980) Adaptations to sulfide in sulfide-system meiofauna. End-products of sulfide detoxification in three turbellarians and a gastrotrich. Mar Ecol Prog Ser 2:169–177

Powell EN, Bright TJ, Brooks JM (1986) The effect of sulfide and an increased food supply on the meiofauna and macrofauna at the East Flower Garden brine seep. Helgoländer Meeresun 40:57–82

Prol-Ledesma RM (2003) Similarities in the chemistry of shallow submarine hydrothermal vents. Geothermics 32:639–644

Prol-Ledesma RM, Canet C, Melgarejo JC, Núñez-Cornú F (2000) Mineralogy of a shallow hydrothermal submarine center near Punta Mita (Western Mexico). In: World geothermal congress 2000, Kyushu-Tohoku, Japan, pp 1599–1604

Prol-Ledesma RM, Canet C, Torres-Vera MA, Forrest MJ, Armienta MA (2004) Vent fluid chemistry in Bahia Concepción coastal submarine hydrothermal system, Baja California Sur, Mexico. J Volcanol Geotherm Res 137:311–328

Rathburn AE, Levin LA, Held Z, Lohmann KC (2000) Benthic formainifera associated with cold methane seeps on the northern California margin: ecology and stable isotope composition. Mar Micropaleontol 38:247–266

Roberts HH, Aharon P, Carney R, Larkin J, Sassen R (1990) Sea floor response to hydrocarbon seeps, Louisiana Continental Slope. Geol Mar Lett 10:232–243

Robinson C (2000) Plankton gross production and respiration in the shallow water hydrothermal systems off Milos, Aegean Sea. J Plank Res 22:887–906

Robinson C, Ziebis W, Müller S, Eichstadt K, Dando P, Linke P, Varnavas S, Megalovasilis P, Panagiotaras D (1997) *In situ* investigations of shallow water hydrothermal vent systems, Palaeochori Bay, Milos, Aegean Sea. In: Fourth underwater science symposium. Society for Underwater Technology, Newcastle upon Tyne, pp 85–100

Robinson CA, Bernhard JM, Levin LA, LA MGF, Blanks JK (2004) Surficial hydrocarbon seep infauna from the Blake Ridge (Atlantic Ocean, 2150 m) and the Gulf of Mexico (690–2240 m). PSZNI: Mar Ecol 25:313–316

Rusch A, Walpersdorf E, de Beer D, Gurrieri S, Amend JP (2005) Microbial communities near the oxic/anoxic interface in the hydrothermal system of Vulcano Island, Italy. Chem Geol 224:169–182

Sahling H, Galkin SV, Salyuk A, Greinert J, Foerstel H, Piepenburg D, Suess E (2003) Depth-related structure and ecological significance of cold-seep communities—a case study from the Sea of Okhotsk. Deep Sea Res I 50:1391–1409

Sansone FJ, Resing JA (1995) Hydrology and geochemistry of sea surface hydrothermal plumes resulting from Hawaiian coastal volcanism. J Geophys Res 100:13555–13569

Sarrazin J, Juniper SK, Massoth G, Legendre P (1999) Physical and chemical factors influencing species distributions on hydrothermal sulfide edifices of the Juan de Fuca Ridge, northeast Pacific. Mar Ecol Prog Ser 190:89–112

Sartoni G, De Biasi AM (1999) A survey of the marine algae of Milos Island, Greece. Cryptogamie Algol 20:271–283

Sassen R, Roberts HH, Aharon P, Larkin J, Chinn EW, Carney R (1993) Chemosynthetic bacterial mats at cold hydrocarbon seeps, Gulf of Mexico continental slope. Org Geochem 20:77–89

Sassen R, Milkov AV, Roberts HH, Sweet ST, DeFreitas DA (2003) Geochemical evidence of rapid hydrocarbon venting from a seafloor-piercing mud diapir, Gulf of Mexico continental shelf. Mar Geol 198:319–339

Schiemer F, Novak R, Ott J (1990) Metabolic studies on thiobiotic free-living nematodes and their symbiotic microorganisms. Mar Biol 106:129–137

Schlüter M, Sauter EJ, Andersen CE, Dahlgaard H, Dando PR (2004) Spatial distribution and budget for submarine groundwater discharge in Eckernförde Bay (W-Baltic Sea). Limnol Oceanogr 49:157–167

Schmaljohann R, Faber E, Whiticar MJ, Dando PR (1990) Co-existence of methane- and sulphur-based endosymbioses between bacteria and invertebrates at a site in the Skagerrak. Mar Ecol Prog Ser 61:119–124

Schornikova EI, Syrtlanova NM (2008) A new species of *Terrestricythere* from the Black Sea, in zones of gas seepage. Senckenb Lethaea 88:121–126

Schuster U (1994) Microbial methane oxidation in a marine sediment around intertidal gas seeps. Ph. D. thesis, University of Essex, Essex

Sendall KA, Fontaine AR, O'Foighil D (1995) Tube morphology, and activity patterns related to feeding and tube-building in the polychaete *Mesochaetopterus taylori* Potts. Can J Zool 73:509–517

Sergeeva NG (2003) Meiobenthos in the region with methane gas seeps. In: Eremeev VN, Gaevskaya AV (eds) Present conditions of biological diversity in the near-shore zone of the Crimea Peninsular (the Black Sea sector). Ecosi-Gidrophizika, Sevastopol, Ukraine, pp 258–267

Sergeeva NG, Gulin MB (2007) Meiobenthos from an active methane seepage area in the NW Black Sea. PSZN I: Mar Ecol 28:152–159

Sibuet M, Olu K (1998) Biogeography, biodiversity and fluid dependence of deep-sea cold-seep communities at active and passive margins. Deep Sea Res 45:517–567

Sievert SM, Brinkhoff T, Muyzer G, Ziebis W, Kuever J (1999) Spatial heterogeneity of bacterial populations along an environmental gradient at a shallow submarine hydrothermal vent near Milos Island (Greece). Appl Environ Biol 65:3834–3842

Sievert SM, Heidorn T, Kuever J (2000a) *Halothiobacillus kellyi* sp. nov., a mesophilic obligately chemolithoautotrophic sulfur-oxidizing bacterium isolated from a shallow-water hydrothermal vent in the Aegean Sea and emended description of the genus *Halothiobacillus*. Int J Evol Microbiol 50:129–137

Sievert SM, Muyzer G, Kuever J (2000b) Identification of 16S ribosomal DNA-defined bacterial populations at a shallow submarine hydrothermal vent near Milos Island (Greece). Appl Environ Microbiol 66:3102–3109

Sievert SM, Ziebis W, Kuever J, Sahm K (2000c) Relative abundance of Archaea and Bacteria along a thermal gradient of a shallow-water hydrothermal vent quantified by rRNA slot-blot hybridisation. Microbiology 146:1287–1293

Smith PJ, McVeagh SM, Won Y, Vrijenhoek RC (2004) Genetic heterogeneity among New Zealand species of hydrothermal vent mussels (Mytilidae: Bathymodiolus). Mar Biol 144:537–545

Snelgrove PVR (1994) Hydrodynamic enhancement of invertebrate larval settlement in microdepositional environments – colonization tray experiments in muddy habitat. J Exp Mar Biol Ecol 176:149–166

Sorokin Y, Sorokin PY, Zakuskina OY (1998) Microplankton and its functional activity in zones of shallow hydrotherms in the Western Pacific. J Plank Res 20:1015–1031

Southward AJ, Kennicutt MC, Alcalà-Herrara J, Abbiati M, Airoldi L, Cinelli F, Bianchi CN, Morri C, Southward EC (1996) On the biology of submarine caves with sulphur springs: $^{13}C/^{12}C$ ratios as a guide to trophic relations. J Mar Biol Assoc UK 76:265–285

Southward AJ, Southward EC, Dando PR, Hughes JA, Kennicutt MC, Alcala-Herrera J, Leahy Y (1997) Behaviour and feeding of the nassarid gastropod *Cyclope nerita* abundant at hydrothermal brine seeps off Milos (Aegean Sea). J Mar Biol Assoc UK 77:753–771

Spies RB, Davis PH (1979) The macrofaunal benthos of a natural oil seep in the Santa Barbara Channel. Mar Biol 50:227–237

Spies RB, DesMarais DJ (1983) Natural isotope study of trophic enrichment of marine benthic communities by petroleum seepage. Mar Biol 73:67–71

Stanier RY, Cohen G (1977) Phototrophic prokaryotes: the cyanobacteria. Annu Rev Microbiol 31:225–274

Starynin DA, Gorlenko VM, Ivanov MV, Karnachuk OV, Namsaraev BB (1989) Algobacterial mats of Kraternaya Bight. Biol Morya 3:70–77

Staudigel H, Hart SR, Koppers AAP, Constable C, Workman R, Kurz M, BE T (2004) Hydrothermal venting at Vailulúu Seamount: the smoking end of the Samoan Chain. Geochem Geophys Geosys 5:Q02003. doi:02010.01029/02003GC000626

Stein JL (1984) Subtidal gastropods consume sulphur-oxidising bacteria: evidence from coastal hydrothermal vents. Science 223:696–698

Stetter KO (1988) *Archaeoglobus fulgidus* gen. nov., sp. nov.: a new taxon of extremely thermophilic Archaebacteria. Syst Appl Microbiol 10:172–173

Stetter KO, König H, Stackebrandt E (1983) *Pyrodictium* gen. nov., a new genus of submarine disc-shaped sulphur reducing archaebacteria growing optimally at 105°C. Syst Appl Microbiol 4:535–551

Stieglitz T (2005) Submarine groundwater discharge into the near-shore zone of the Great Barrier Reef, Australia. Mar Pollut Bull 51:51–59

Stüben D, Glasby GP (1999) Geochemistry of shallow water hydrothermal fluids from Paleohori Bay, Milos, Aegean Sea. Explor Mining Geol 8:273–287

Stüben D, Sedwick P, Colantoni P (1996) Geochemistry of submarine warm springs in the limestone cavern of Grotta Azzurra, Capo Palinuro, Italy: evidence for mixing-zone dolomitisation. Chem Geol 131:113–125

Svetlichny VA, Sokolova TG, Gerhardt M, Kostrikina NA, Zavarzin GA (1991) Anaerobic extremely thermophilic carboxydotrophic bacteria in hydrotherms of Kuril Islands. Microb Ecol 21:1–10

Takeda M, Kurata Y (1977) Crabs of the Ogasawara Islands IV. A collection made at the new volcanic island, Nishino-shima-shinto, in 1975. Bull Natl Sci Mus (Jpn) A 3:91–111

Takeda M, Takeuchi H, Suganuma H (1993) Occurence of Xenograpsus novaeinsularis Takeda et Kurata (Crustacea: Decapoda: Brachyura) in the Tokara and Iwo Islands. Nat Environ Sci Res 6:59–64

Tarasov VG, Zhirmunsky AV (1989) Investigation of the ecosystem of Kraternaya Bight (Kurile Islands). Sov J Mar Biol 15:139–148

Tarasov VG, Zhirmunsky AV (1990) Investigation of the ecosystem of Kraternaya Bight (Kurile Islands). Biol Morya 3:4–12

Tarasov VG, Propp MV, Propp LN, Zhirmunsky AV, Namsaraev BB, Gorlenko VM, Starynin DA (1990) Shallow-water gasohydrothermal vents of Ushishir Volcano and the ecosystem of Kraternaya Bight (The Kurile Islands). PSZN I: Mar Ecol 11:1–23

Tarasov VG, Gebruk AV, Shulkin VM, Kamenev GM, Fadeev VI, Kosmynin VN, Malakhov VV, Starynin DA, Obzhirov AI (1999) Effect of shallow-water hydrothermal venting on the biota of Matupi Harbour (Rabaul Caldera, New Britain Island, Papua New Guinea). Cont Shelf Res 19:79–161

Tarasov VG, Gebruk AV, Mirononov AN, Moskalev LL (2005) Deep-sea and shallow-water hydrothermal vent communities: two different phenomena? Chem Geol 224:5–39

Thiermann F (1998) Zur Ökologie der Nematodenfauna besonders der dominanten Art *Oncholaimus camplyoceroides* (De Coninck & Stekhoven, 1933), an sublitoralen Hydrothermalquellen der Insel Milos. Shaker Verlag, Aachen, Deutschland

Thiermann F, Akoumianaki I, Hughes JA, Giere O (1997) Benthic fauna of a shallow-water gaseohydrothermal vent area in the Aegean Sea (Greece). Mar Biol 128:149–159

Thiermann F, Vismann B, Giere O (2000) Sulphide tolerance of the marine nematode *Oncholaimus campuloceroides* – a result of internal sulphur formation. Mar Ecol Prog Ser 193:251–259

Tivey MK (2007) Generation of seafloor hydrothermal vent fluids and associated mineral deposits. Oceanography 20:50–66

Trager GC, DeNiro MJ (1990) Chemoautotrophic sulphur bacteria as a food source for mollusks at intertidal hydrothermal vents: evidence from stable isotopes. Veliger 33:359–362

Tunnicliffe V (1991) The biology of hydrothermal vents: ecology and evolution. Oceanogr Mar Biol Annu Rev 29:3119–3207

Tunnicliffe V, McArthur AG, McHugh D (1998) A biogeographical perspective of the deep-sea hydrothermal vent fauna. Adv Mar Biol 34:353–442
Türkay M, Sakai K (1995) Decapod crustaceans from a volcanic hot spring in the Marianas. Senckenb Marit 26:25–35
Van Dover CL (2000) The ecology of deep-sea hydrothermal vents. Princeton University Press, Princeton, NJ
Vetter EW (1998) Population dynamics of a dense assemblage of marine detritivores. J Exp Mar Biol Ecol 226:131–161
Vidal VMV, Vidal FV, Isaacs JD (1978) Coastal submarine hydrothermal activity off Northern Baja California. J Geophys Res 83:1757–1774
Vismann B (1991) Sulfide tolerance: physiological mechanisms and ecological implications. Ophelia 34:1–27
Welhan JA, Lupton JE (1987) Light hydrocarbon gases in Guaymas Basin hydrothermal fluids: thermogenic versus abiogenic origin. Am Assoc Petr Geol B 71:215–223
Wenzhöfer F, Holby O, Glud RN, Nielsen HK, Gundersen JK (2000) In situ microsensor studies of a hot vent at Milos (Greece). Mar Chem 69:43–54
Wiesenberg DA, Guinasso NL (1979) Equilibrium solubilities of methane, carbon dioxide and hydrogen in water and seawater. J Chem Eng Data 24:356–360
Wildish DJ, Akagi HM, McKeown DL, Pohle GW (2008) Pockmarks influence benthic communities in Passamaquoddy Bay, Bay of Fundy, Canada. Mar Ecol Prog Ser 357:51–66
Yager PL, Nowell ARM, Jumars PA (1993) Enhanced deposition to pits – a local food source for benthos. J Mar Res 51:209–236
Zhirmunsky AV, Tarasov VG (1990) Unusual marine ecosystem in the flooded crater of Ushisher volcano. Mar Ecol Prog Ser 65:95–102
Zhukova NV, Kharlamenko VI, Svetashev VI, Rodionov IA (1992) Fatty acids as markers of bacterial symbionts of marine bivalve molluscs. J Exp Mar Biol Ecol 162:253–263
Zipperle A, Reise K (2005) Freshwater springs on intertidal sand flats cause a switch in dominance among polychaete worms. J Sea Res 54:143–150

Chapter 12
Japan: Vents and Seeps in Close Proximity

Hiromi Watanabe, Katsunori Fujikura, Shigeaki Kojima, Jun-Ichi Miyazaki, and Yoshihiro Fujiwara

Contents

12.1 Introduction: Vent and Seep Communities Around Japan

Since the discovery of dense animal communities associated with deep-sea hydrothermal venting (Lonsdale 1977), biological knowledge of those animals has accumulated (Van Dover 2000). Some unique animals associated with vent fields were found to depend on chemosynthetic primary production (Corliss et al. 1979). Subsequently, similar chemosynthetic animal assemblages were also discovered associated with deep-sea methane-seep areas, whale falls, and sunken wood (Paull et al. 1984; Smith et al. 1989). To understand the pathways of adaptation to these environments, species shared between different habitats are of particular interest (Distel et al. 2000; Lorion et al. 2008). On a global scale, the number of species shared between vents and seeps

H. Watanabe (✉), K. Fujikura, and Y. Fujiwara
Japan Agency for Marine-Earth Science and Technology, 2-15 Natsushima-cho, Yokosuka 237-0061, Japan
e-mail: hwatanabe@jamstec.go.jp

S. Kojima
Atmosphere and Ocean Research Institute, University of Tokyo, 5-1-5 Kashiwanoha, Kashiwa, Chiba 277-8564, Japan

J.-I. Miyazaki
Yamanashi University, 4-4-37 Takeda, Kofu, Yamanashi 400-8510, Japan

S. Kiel (ed.), *The Vent and Seep Biota*, Topics in Geobiology 33,
DOI 10.1007/978-90-481-9572-5_12,

is less than 10% of the total recorded vent or seep species (e.g. Tunnicliffe et al. 1998, 2003; Sibuet and Olu 1998). In the vent and seep communities around Japan, however, this figure exceeds 20% (based on a faunal list provided by Fujikura et al. 2008), although the identification of species is still in progress. This relatively high abundance of both vent- and seep-inhabiting species suggests close relationships between vent and seep communities around Japan. A high similarity between megafaunal communities at vents and seeps around Japan was already noted by Fujikura et al. (1995); however, that study was based on species abundances investigated at only a single vent and two methane-seep communities. To date, at least 55 vent and seep communities have been discovered around Japan (Fujikura et al. 2008), and further analyses are required to elucidate the nature of this similarity.

In this chapter, we focus on similarities between megafaunal communities inhabiting vents and seeps. As Kojima (2002) has already provided an review with an almost complete list of studies on vent and seep animals around Japan as of the time of publication, here we only provide brief, essential information on their geologic settings and ecologic characteristics. Then, we summarize the species distributional records with statistical analyses based on previous studies of those communities around Japan and discuss their proximity. In addition, we also summarize recent genetic studies of both vent- and seep-inhabiting species, i.e., three *Calyptogena* clams, *Lamellibrachia* and *Paraescarpia* tubeworms, and *Bathymodiolus* mussels, and discuss the connectivity among populations of these species. Finally, we discuss whether the similarity of communities is accompanied by population connectivity.

12.2 Tectonic Background and Faunal Characteristics of Vents and Seeps Around Japan

Generally, hydrothermal vents and methane seeps are distributed along plate boundaries. Four plates converge around Japan: the Eurasian and North American continental plates and the Pacific and Philippine Sea oceanic plates (Fig. 12.1). Trench–Arc–Back-arc systems are formed on the continental margin, and about 75% of all back-arc basins worldwide are found in the western Pacific (Ishibashi and Urabe 1994). Hydrothermal vent systems have developed associated with the volcanic-arc and back-arc basins in the western Pacific, and methane-seep sites are associated with faults, cracks, and permeable layers in subduction zones. Around Japan, hydrothermal vent and methane-seep communities are aggregated in eight areas: the Japan Trench, Sagami Bay, Suruga Bay, Nankai Trough, Kagoshima Bay, Nansei-Shoto Trench, Izu-Ogasawara Arc, and Okinawa Trough (Fig. 12.1, Table 12.1). Although they are aggregated in a narrow area horizontally, they are scattered in a wide depth range, from 80 m in Kagoshima Bay to 7,434 m in the Japan Trench. The geologic background and megafaunal characteristics of the eight areas are briefly described below. Recently, a hydrothermal vent in the Wakamiko Protocaldera in Kagoshima Bay and some methane seeps in the Japan Sea have been discovered (Yamanaka et al. 2000; Matsumoto et al. 2009), but

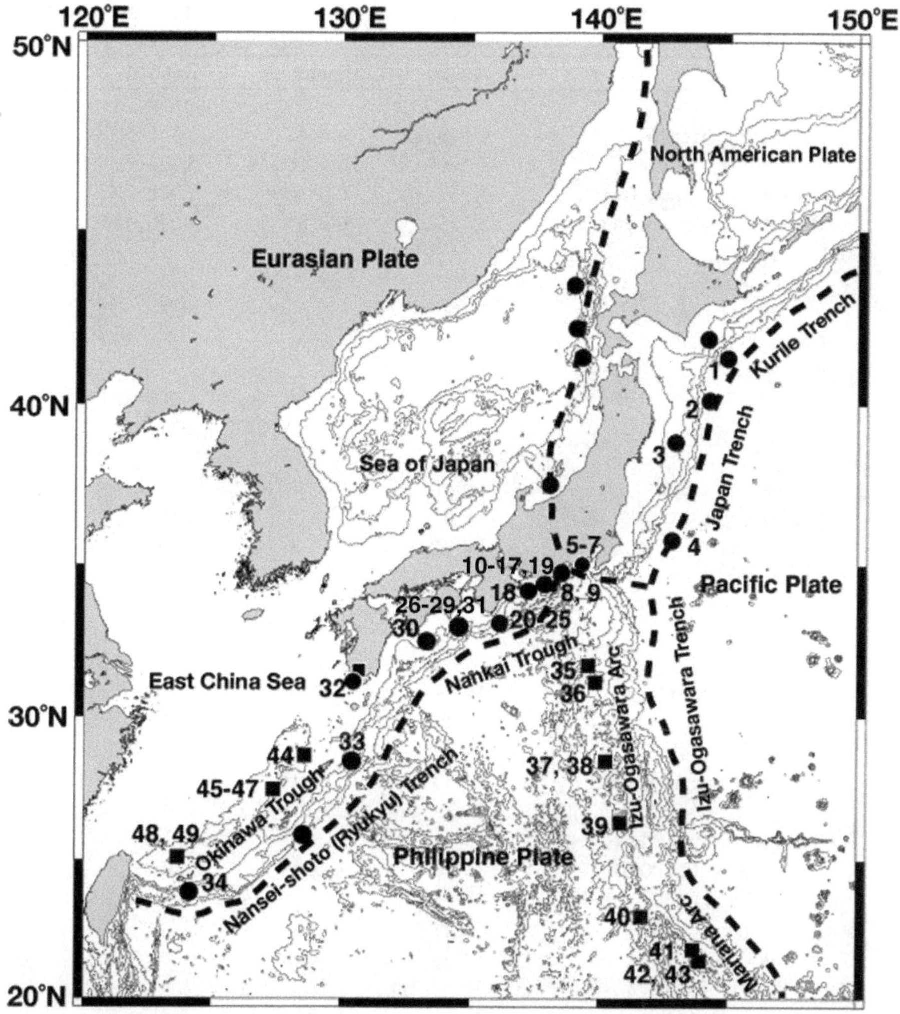

Fig. 12.1 Distribution map of vent and seep communities around Japan (Modified from Fujikura et al. 2008). The numbers on the map are correlated with those listed in Table 12.1. Solid circles indicate distribution of seep communities and solid squares indicate distribution of vent communities

biological data from these areas are still very limited. Therefore, they are not included in the discussion here.

The Japan Trench results from the subduction of the Pacific plate under the North American plate. Sediment porewater seeps with animal communities were first discovered by the Japanese-French KAIKO project (Juniper and Sibuet 1987; Ohta and Laubier 1987) and animal communities associated with sediment-covered methane seeps are known from various depths ranging from 5,343 to 7,434 m (Sibuet et al. 1988; Fujikura et al. 1999). Seep communities in 5,343–6,809 m depth

Table 12.1 List of hydrothermal vents and methane seeps around Japan (Modified from Fujikura et al. 2008). ID, identification number correlated with the numbering of symbols in Fig. 12.4

Area	Site	Latitude	Longitude	Depth	ID
Kurile and Japan Trenches					
	West Kurile Trench	41°19′	144°48′	5,131–5,785	1
	Sanriku Escarpment	40°06′–08′	144°10′–14′	5,343–6,809	2
	Near Axis of Japan Trench	40°03′–04′	144°16′–17′	7,326–7,434	3
	Dai-ichi Kashima Seamount	35°54′	142°31′	5,640–5,695	4
Sagami Bay					
	Sagami Knoll	35°06′	139°20′	1,400–1,500	5
	Off Hatsushima	35°00′	139°14′	800–1,300	6
	Okinoyama Bank	34°59′	139°31′	750–1,300	7
Suruga Bay					
	Off Toi	34°55′	138°39′	1,500–1,600	8
	Senoumi Bank	34°35′–38′	138°24′–26′	200–400	9
Nankai Trough					
	Kanesu-no-se Bank	34°17′–18′	138°15′	270–300	10
	Omaezaki Spur	34°15′	138°02′	1,100–1,200	11
	Ryuyo Canyon	34°12′	137°46′	1,000–1,100	12
	Tokai Thrust	34°06′–07′	138°07′–08′	2,120–2,360	13
	Dai-ni Tenryu Knoll	34°04′	137°48′	500–900	14
	Dai-ni Atsumi Knoll	35°51′	137°23′	1,042–1,100	15
	Yukie Ridge	33°50′	137°54′–55′	1,940–2,180	16
	Dai-san Tenryu Canyon	33°39′	137°55′	3,700–3,800	17
	Tenryu Canyon	33°36′–37′	137°32′	3,654–3,855	18
	Zenisu Ridge	33°33′	138°26′	3,300	19
	Dai-ichi Kumano Knoll	33°43′	137°05′	1,900	20
	Dai-yon Kumano Knoll	33°39′	136°38′	2,000	21
	Dai-roku Kumano Knoll	33°41′	136°34′	2,000	22
	Off Kumano (2,100 m)	33°16′	136°43′	2,100	23
	Off Kumano (2,500–2,900 m)	33°07′–08′	136°28′–29′	2,500–2,900	24
	Off Kumano (3,250 m)	33°08′	136°34′	3,250	25
	Muroto Knoll	32°58′	134°14′	600	26
	Off Muroto (3,200–3,700 m)	32°32′–35′	134°41′–43′	3,260–3,706	27
	Off Muroto (4,600 m)	32°21′	134°51′	4,600	28
	Off Muroto (4,800 m)	32°21′	134°56′	4,800	29
	Ashizuri Knoll	32°29′	133°38′	600	30
	Dai-ichi Minami-Muroto Knoll	32°21′	134°32′	3,540–3,620	31
Kagoshima Bay					
	Tubeworm site	31°40′	130°48′	80–130	32
Nansei-ShotoTrench					
	Off Kikaijima Island	28°27′	130°19′	1,400–1,500	33
	Kuroshima Knoll	24°07′–08′	124°11′–12′	636–812	34
Izu-Ogasawara Islands					
	Myojin Knoll	32°06′	139°53′	1,200–1,397	35
	Sumisu Caldera	31°29′	140°02′	920	36
	Suiyo Seamount	28°35′	140°39′	1,300–1,400	37
	Mokuyo Seamount	28°19′	140°34′	1,200–1,300	38
	Kaikata Seamount	26°42′	141°05′	300–500	39

(continued)

Table 12.1 (continued)

Area	Site	Latitude	Longitude	Depth	ID
Northern Mariana Islands					
	Nikko Seamount	23°05′	142°20′	400–500	40
	Dai-ni Kasuga Seamount	21°36′	143°39′	300–400	41
	Eifuku Seamount	21°29′	144°02′	1,580	42
	Daikoku Seamount	21°25′	144°10′	350–450	43
Okinawa Trough					
	Minami Ensei Knoll	28°24′	127°39′	600–800	44
	North Knoll of Iheya Ridge	27°48′	126°54′	900–1,000	45
	Iheya Ridge	27°33′	126°58′	1,300–1,500	46
	Izena Hole	27°13′–16′	127°03′–06′	1,300–1,600	47
	Hatoma Knoll	24°52′	123°51′	1,400–1,530	48
	Dai-yon Yonaguni Knoll	24°51′	122°42′	1,320–1,387	49

along thrust fault-connected fractures are dominated by the vesicomyid bivalve *Calyptogena phaseoliformis* (Ogawa et al. 1996; Fig. 12.2a, Plate 28). The biomass at these seeps was estimated to reach 51 kg/m^2 using deep-tow camera observations (Ohta and Laubier 1987). In deeper areas (7,326–7,434 m), communities dominated by the thyasirid clam *Axinulus hadalis* with two types of endosymbionts are patchily distributed in sediments, and no *Calyptogena* clams have yet been found in this area (Fujikura et al. 1999; Fujiwara et al. 2001).

The plate boundary of the North American and Philippine Sea plates is located in Sagami Bay, where at least nine seep sites with animal communities have been discovered (Okutani and Egawa 1985; Hashimoto et al. 1989; Fujikura et al. 2008). Most seep communities are characterized by dense aggregations of *Calyptogena* clams and initially these clams were all identified as *Calyptogena soyoae*. However, in 1997, molecular analysis and detailed morphologic observations revealed two species, *Calyptogena okutanii* and *C. soyoae* (Kojima and Ohta 1997). Both species harbor sulfur-oxidizing bacteria in their gills (Barry et al. 2002; Kuwahara et al. 2007) and partially bury themselves in the sediment to supply hydrogen sulfide to their symbionts (Fig. 12.2b, Plate 28); these two species are difficult to identify at the species level in situ. The Off Hatsushima Island site is the most thoroughly investigated methane seep site around Japan and was found to have flourishing communities (Fujikura et al. 1995, 2002). Therefore many ecologic studies of *Calyptogena* clams have been carried out here, for example, on the correlation of *Calyptogena* distribution with the concentration of hydrogen sulfide in sediment (Hashimoto et al. 1995) and in situ observations of spawning of *Calyptogena* clams (Fujiwara et al. 1998; Fujikura et al. 2007). As the two *Calyptogena* species live together, it remains unclear which species of *Calyptogena* show these ecologic characteristics. Other than *Calyptogena* clams, the mussel *Bathymodiolus platifrons* with methanotrophic symbionts (Fujiwara et al. 2000; Barry et al. 2002) has been observed on patchily distributed hard substrata. Large colonies of *Lamellibrachia* and *Alaysia* tubeworms (Fig. 12.2c, Plate 28) and the gastropod *Oenopota sagamiana* have also been found, but these colonies were spatially separated from the *Calyptogena* and mussel beds.

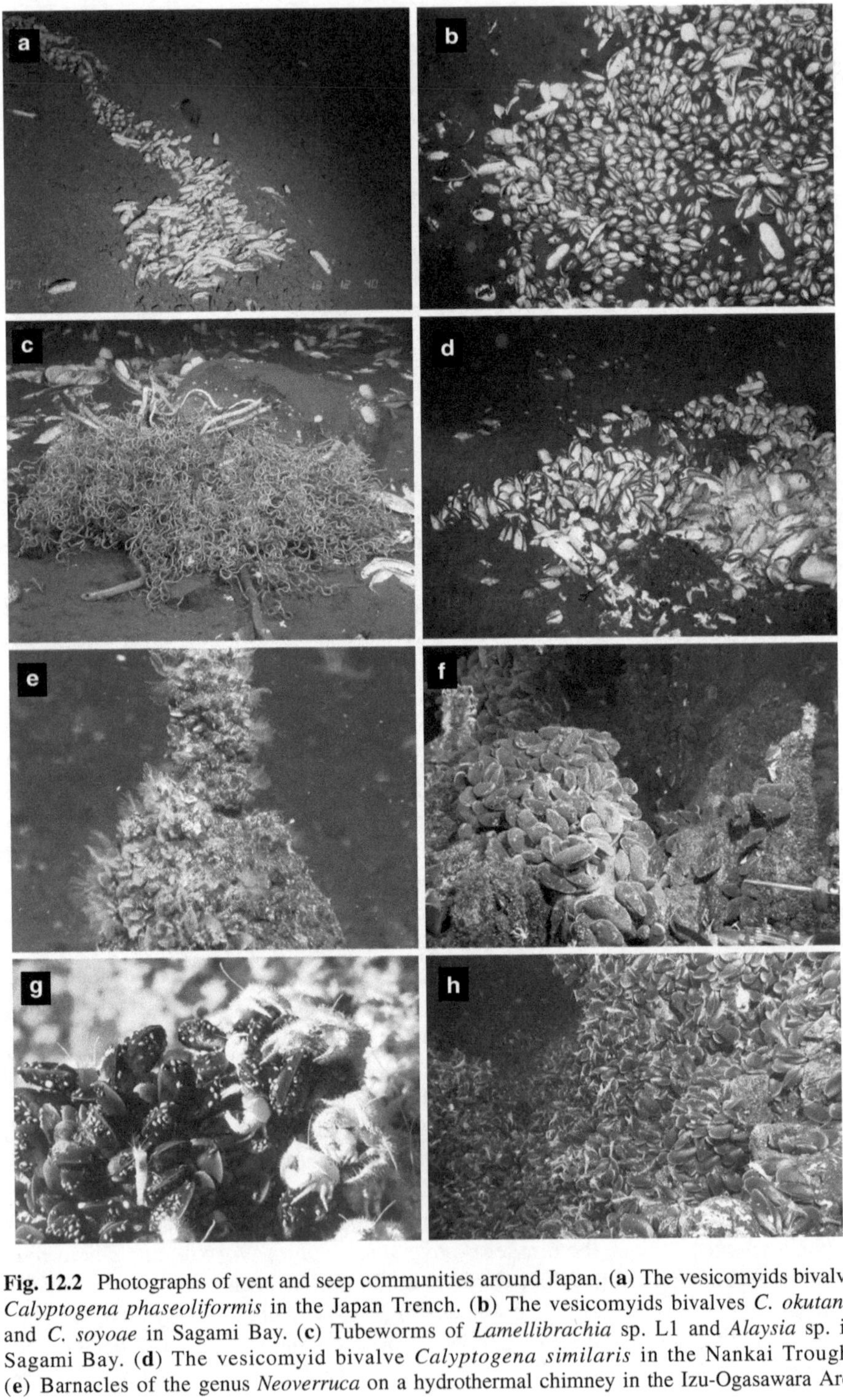

Fig. 12.2 Photographs of vent and seep communities around Japan. (**a**) The vesicomyids bivalve *Calyptogena phaseoliformis* in the Japan Trench. (**b**) The vesicomyids bivalves *C. okutanii* and *C. soyoae* in Sagami Bay. (**c**) Tubeworms of *Lamellibrachia* sp. L1 and *Alaysia* sp. in Sagami Bay. (**d**) The vesicomyid bivalve *Calyptogena similaris* in the Nankai Trough. (**e**) Barnacles of the genus *Neoverruca* on a hydrothermal chimney in the Izu-Ogasawara Arc.

At the Off Hatsushima Island site, the seeping fluid contains hydrogen sulfide as well as methane, and its temperature reaches up to 10°C (Tsunogai et al. 1994). The chemical composition suggested that the lateral transfer of fresh water may be a source of the seepage in the orange-colored sediment in the Off-Hatsushima Island site (Tsunogai et al. 1994).

In Suruga Bay, some seep communities have been discovered along the boundary of the Eurasian and Philippine Sea plates. Most of the seeps are located in steep valleys with high-speed currents and high conductivity. Therefore investigations in this area are limited. *Calyptogena* clams such as *Calyptogena fausta* have been collected in this area (Fujikura et al. 1999).

The Nankai Trough is located along the convergent boundary between the Philippine Sea and Eurasian plates, and the Japanese-French KAIKO project also targeted this area (Sibuet et al. 1988). Here, methane seeps have been recognized in three different settings: at thrust faults, mud volcanoes, and permeable layers (Ashi 2003). Animal communities dominated by vesicomyid clams and *Alaysia, Escarpia, Paraescarpia*, and *Lamellibrachia* tubeworms (Ohta and Laubier 1987) are patchily distributed where fluid seeps through thrust faults (Colwell et al. 2004) in a wide depth range from 270 to 4,800 m (Lallemand et al. 1992; Fujikura et al. 2008; Fig. 12.2e, Plate 29). At least 12 species of vesicomyid clams have been reported from this area, and the distribution of each species appears to be determined by depth (Fujikura et al. 2000). The biomass at these seeps was estimated to reach 16 kg/m^2 (Ohta and Laubier 1987).

The shallowest (seep) site with a dense aggregation of the tubeworm *Lamellibrachia satsuma* was discovered in 80 m depth in Kagoshima Bay (Hashimoto et al. 1993). Interestingly, the symbionts of *L. satsuma* appear to utilize hydrogen sulfide generated by bacteria in the sediment (Mizota and Yamanaka 2003), even though the Sakurajima volcano is located near the seep site.

The Nansei-Shoto Trench results from the subduction of the Philippine Sea plate beneath the Eurasian plate. The seep communities in this area consist of dense aggregations of *Bathymodiolus hirtus* associated with carbonate concretions, and *Bathymodiolus securiformis*, *Calyptogena* and *Vesicomya* clams, and *Lamellibrachia* tubeworms are associated with sediment (Fujikura et al. 2003).

On the continental side of the Izu-Ogasawara Trench, which results from the subduction of the Pacific plate under the Philippine Sea plate, at least five hydrothermal vents were found on the frontal volcanoes along the Izu-Ogasawara Island arc (Ishibashi and Urabe 1994). In this paper, the vents in the northern area of the Mariana Arc are also included in this area, as the species compositions of the communities are similar. (The Mariana Arc is the southern continuation of the Izu-Ogasawara Arc and is a different region than the Mariana Trough, which is the back-arc basin of the Mariana Arc. See below for details.) As this area is located offshore, no sediment

Fig. 12.2 (continued) (**f**) The mussel *Bathymodiolus septemdierum* in the Izu-Ogasawara Arc. (**g**) The mussel *Bathymodiolus platifrons* and the galatheid crab *Shinkaia crosnieri* in the Okinawa Trough, (**h**) *Bathymodiolus platifrons* in the Okinawa Trough. Color plates of these images can be found in Appendix (Plates 28, 29) except for volcanic dust or precipitation of vent fluid is available, and infaunal animals are rare in this area. Aggregations of *Bathymodiolus septemdierum* harboring sulfur-oxidizing bacteria (Fujiwara et al. 2000; Miyazaki et al. 2004) and barnacles of the genus *Neoverruca* (Watanabe et al. 2005) were both found on the walls of chimneys (Fig. 12.2f, Plate 29). The flatfish *Symphurus thermophylus* appears to be endemic to the shallow sites (300–450 m) in this area (Munroe and Hashimoto 2008).

The Okinawa Trough is a back-arc basin formed as a result of the subduction of the Philippine Sea plate beneath the Eurasian plate at the Nansei-Shoto Trench, and hydrothermal activities are divided into two areas, the middle and south (Glasby and Notsu 2003). Thick sediment covers the Okinawa Trough, and hydrothermal vents are characterized by carbon dioxide enrichment (Sakai et al. 1990). In contrast to other hydrothermal vent communities in the West Pacific, those in the southern area of the Okinawa Trough lack brachyuran crabs; instead, large *Paralomis* crabs (probably non-vent-endemic species) have been observed at several sites. The crustaceans *Shinkaia crosnieri*, *Alvinocaris longirostris*, and the mussel *Bathymodiolus platifrons* are abundant (Figs. 12.2g and h, Plate 29), and the tube-building polychaete worm *Paralvinella hessleri* was found on vent chimneys. In addition, the vesicomyid clams *Calyptogena nankaiensis* and *C. okutanii* are associated with sediment-hosted vents, and scalpellid barnacles (probably *Ashinkailepas seepiophila* and *Leucolepas* sp.) are found in the Iheya Ridge in the middle area (Ohta and Kim 2001), while *Alaysia* tubeworms are aggregated on the Dai-yon Yonaguni Knoll in the southern area (Fujikura et al. 2001).

In summary, some species of vesicomyid clams, *Bathymodiolus* mussels, and some vestimentiferan tubeworms are common to many of the vent and seep areas around Japan. Other taxa such as the thyasirid *Axinulus hadalis* in the Japan Trench, *Calyptogena soyoae* in Sagami Bay, and the galatheid crab *Shinkaia crosnieri* in the Okinawa Trough characterize specific sites and show endemicity in each vent and seep area. However, the community composition of each locality has not been examined statistically. Fujikura et al. (2008) summarized species distributional data on the regional scale. Here, we summarize the megafaunal data on the site scale based on previous reports (partially shown in Table 12.2; the complete table will be available on the Web site of the Deep-Sea Ecosystem Research Team, Institute of Biogeosciences in JAMSTEC: http://www.jamstec.go.jp/e/about/research/biogeosciences). The similarities between species composition of vent and seep communities and populations are discussed below, with statistical analyses based on the species composition of animal communities and genetic data.

12.3 Similarity Among Chemosynthetic Communities Around Japan

The first step in recognizing a pattern of communities is to compare their species compositions. The similarity coefficient is the easiest and simplest index to evaluate variation in species composition among areas (Magurran 1988). Species compositions of chemosynthetic communities have been compared using the

Table 12.2 Species composition of vent and seep sites around Japan. The numbers on this table are correlated with those listed in Table 12.1. (the complete data set will be available on the website of the Deep-Sea Ecosystem Research Team, JAMSTEC; http://www.jamstec.ac.jp/e/about/research/biogeosciences). Asterisks indicate animals with symbionts

	1	2	3	4	5	6	7	8	9	10	11	12	13	14	15	16	17	18	19	20	21	22	23	24	25	26	27	28	29	30	31	32	33	34	35	36	37	38	39	40	41	42	43	44	45	46	47	48	49
Gastropoda																																																	
Bathyacmaea nipponica						+	+																																										
Bathyacmaea secunda																																												+	+	+	+	+	+
Bathyacmaea tertia																																													+			+	
Puncturella parvinobilis																																												+	+		+		
Lepetodrilus nux																																													+	+	+	+	+
Margarites ryukyuensis																																													+	+	+	+	+
Margarites shinkai						+	+																																								+		
Cantrainea jamsteci																																		+										+	+				
Provanna glabra					+	+	+																																										
Provanna aff. *gabra*																																												+	+	+	+	+	+
Bivalvia																																																	
Acharax johnsoni		+				+	+																																										
Neilonella profunda		+	+																																														
Bathymodiolus japonicus *						+	+																																					+	+	+			

(continued)

Table 12.2 (continued)

	1	2	3	4	5	6	7	8	9	10	11	12	13	14	15	16	17	18	19	20	21	22	23	24	25	26	27	28	29	30	31	32	33	34	35	36	37	38	39	40	41	42	43	44	45	46	47	48	49
Bathymodiolus platifrons *						+	+																																						+	+	+	+	+
Bathymodiolus septemdierum *																																			+	+	+	+										+	
Bathymodiolus aduloides *							+			+	+	+																					+											+		+			
Bathymodiolus securiformis *															+																			+															
Lucinoma yoshidai						+	+																																										
Conchocele biecta					+	+	+																																										
Vesicomya kaikoae																	+	+									+				+																		
Vesicomya crenulomarginata																					+	+		+																									
Calyptogena fausta *								+								+				+																													
Calyptogena soyoae *					+	+	+																																										
Calyptogena okutanii *						+	+					+			+																														+	+			
Calyptogena kawamurai *									+					+												+				+				+										+					
Calyptogena laubieri *																	+	+													+																		
Calyptogena nankaiensis *												+																																	+				

Calyptogena similaris *											+			+					+	+	+										+																
Calyptogena tsubasa *															+	+	+												+																		
Calyptogena magnocultellus *														+							+	+																									
Calyptogena nautilei *																+	+					+	+		+				+																		
Calyptogena kaikoi *															+	+										+	+		+																		
Calyptogena phaseoliformis *	+	+		+																																											
Calyptogena fossajaponica *	+	+																																													
Polychaeta																																															
Lamellibrachia satsuma *									+																					+							+										
Lamellibrachia sp. L2 *														+								+			+																						
Lamellibrachia sp. L1 *					+	+			+	++		+																				+		+									+	+			
Paraescrpia echinospica *										++																					+													+			
*Escarpia*sp. E1 *									+																									+													
Alaysia sp. A1 *				+	+	+								+																				+													
Alaysia sp. A2 *																																												+	+		
Paralvinella hessleri																																	+									+	+	+	+	+	+
Nicomache ohtai					+	+																																									

(continued)

Table 12.2 (continued)

										1	1	1	1	1	1	1	1	1	2	2	2	2	2	2	2	2	2	2	3	3	3	3	3	3	3	3	3	3	4	4	4	4	4	4	4	4	4	4
	1	2	3	4	5	6	7	8	9	0	1	2	3	4	5	6	7	8	9	0	1	2	3	4	5	6	7	8	9	0	1	2	3	4	5	6	7	8	9	0	1	2	3	4	5	6	7	8	9
Brachipolynoe pettiboneae																																							+								+		
Shinkai longipedata																																													+	+			
Mytilidiphila okinawaensis																																												+		+			
Crustacea																																																	
Ashinkailepas seepiophila						+																													+										+	+			
Leucolepas longa																																													+	+			
Neoverruca sp.																																			+				+		+		+		+	+	+	+	+
Alvinocais brevitelsonis																																			+		+	+				+		+					
Alvinocais longirostris						+																																							+	+	+	+	+
Opaepele loihi																																								+		+							
Shinkaia crosnieri																																												+	+	+	+	+	+
Austinograea yunohana																																			+		+	+	+	+	+	+	+	+					
Xanthidae gen et sp.																																							+	+	+		+						

similarity coefficient and other species diversity indices at the worldwide level (Tunnicliffe and Fowler 1996; Kiel and Goedert 2006). At smaller geographic levels around Japan, the community composition and structure of a hydrothermal vent field on the Minami-Ensei Knoll in the Okinawa Trough and of two methane-seep sites, Off Hatsushima Island and the Okinoyama Bank in Sagami Bay, were compared (Fujikura et al. 1995). Although more than 50 vent and seep sites have been discovered so far, the similarity among most has not yet been analyzed. Here, we summarize presence/absence data from previous reports (Table 12.2) and estimate Jaccard's similarity coefficient among the species compositions of 49 vent and seep sites listed in Table 12.2 and original data shown on the JAMSTEC web site mentioned above. Based on the similarity coefficient, the communities were plotted using nonmetric multidimensional scaling (MDS) and group average clustering was carried out. The similarities among communities as well as between vent and seep were also analyzed statistically by analysis of similarity (ANOSIM). All of the above procedures were conducted with the computer software PRIMER ver. 6.0 (PRIMER-E Ltd, UK, Clarke and Warwick 2001).

As shown in Table 12.2, the original data contain many absence or zero values, and the similarity coefficient was estimated at quite low levels. These results may show high endemicity of megafauna in each vent and seep area. The MDS plot is shown in Fig. 12.3, with a low stress value (<0.05) inferring "an excellent representation with prospects for misinterpretation" and better representation than when using cluster analysis (Clarke and Warwick 2001). The

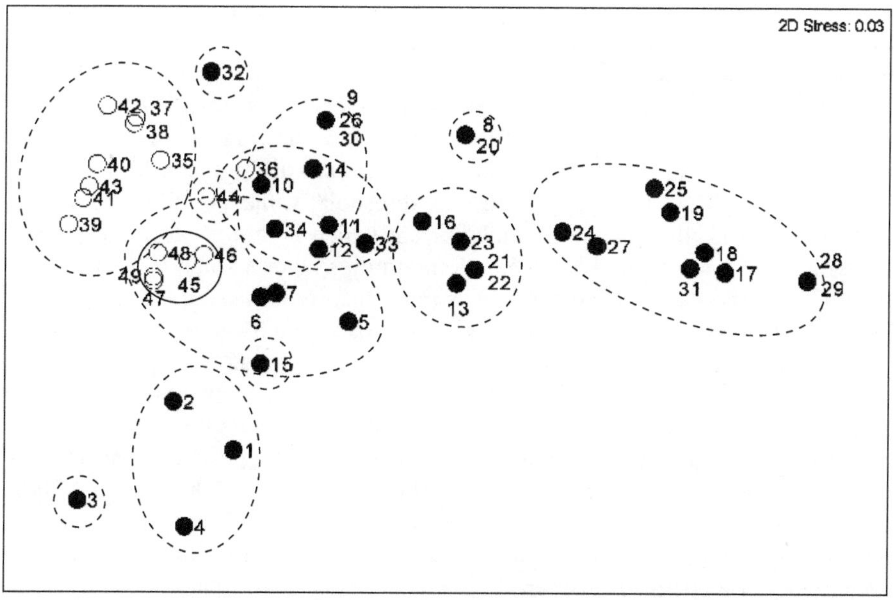

Fig. 12.3 MDS plot of similarity of vent and seep communities in Japan. The numbers beside the circles indicate the communities listed in Table 12.1. The filled circles indicate seep communities, and the open circles indicate vent communities. The results of group-average cluster in 5% similarity were imposed as dashes lines

MDS plot shows a gradual change in community composition, rather than discrete grouping. For the group-average clustering, methane seep communities in Sagami Bay and the Okinawa Trough hydrothermal vents except for the Minami-Ensei Knoll are classified in the same group (Fig. 12.3). On the other hand, communities in the Sumisu Caldera in the Izu-Ogasawara Arc, at 1,000–1,400-m depth in the Nankai Trough, and on the Kuroshima Knoll in the Nansei-Shoto Trench are classified in the other group (Fig. 12.3). The results of ANOSIM showed that community composition between vents and seeps were found to be significantly different (Global $R = 0.169$, $P = 0.001$) although not all pairwise tests between community compositions of vent and seep were significantly different. Most of the vent fields in the Okinawa Trough were associated with sediment, as mentioned above. Sedimented vent sites have been thought to emulate seep conditions better than rocky ridge crest sites (Tunnicliffe et al. 1998). Due to the sedimentation in the Okinawa Trough, the vent communities there may show similarity to the seeps in Sagami Bay. Therefore, no clear distinction could be found between vent and seep communities. Another community in the vent field in the Sumisu Caldera also showed greater similarity to communities in some methane seeps than to those in other hydrothermal vents. The Sumisu Caldera is located offshore, and it seems unlikely that the similarity between these areas was generated by sedimentation in the Sumisu Caldera. There may thus be other factors resulting in similarities between vent and seep communities.

The relationship between similarities (Jaccard's coefficients) of megafaunal communities and horizontal difference in depth or distance are shown in Fig. 12.4. The correlation between similarity and the difference in the depth and horizontal distance were examined, but the intensity of both depth ($R = 0.429$, $P < 0.01$) and distance ($R = 0.216$, $P < 0.01$) associations were not found to be high. However, apparently most of the similarity values greater than zero were obtained for pairs of communities located where the difference in depth is less than 1,320 m (Fig. 12.4a), whereas no threshold value could be found for distance (Fig. 12.4b). As the oceanic current which is thought to largely influence on the larval dispersal between communities, shows some oscillation and the linear horizontal distance may not as important in discussing the relationship between two communities. As mentioned previously, methane seep communities in the Nankai Trough are found associated in three different geologic settings at various depths ranges (Ashi 2003). The present analyses show that the similarities among communities are to a greater extent related to depth than to tectonic background. A bathymetric zonation has already been suggested for vesicomyid clams and vestimentiferan tubeworms, (Fujikura et al. 2000; Kojima 2002; Goffredi et al. 2003). Clams and tubeworms are prominent species among vent and seep communities and they appear to cause the depth-related similarity of community composition between vents and seeps around Japan.

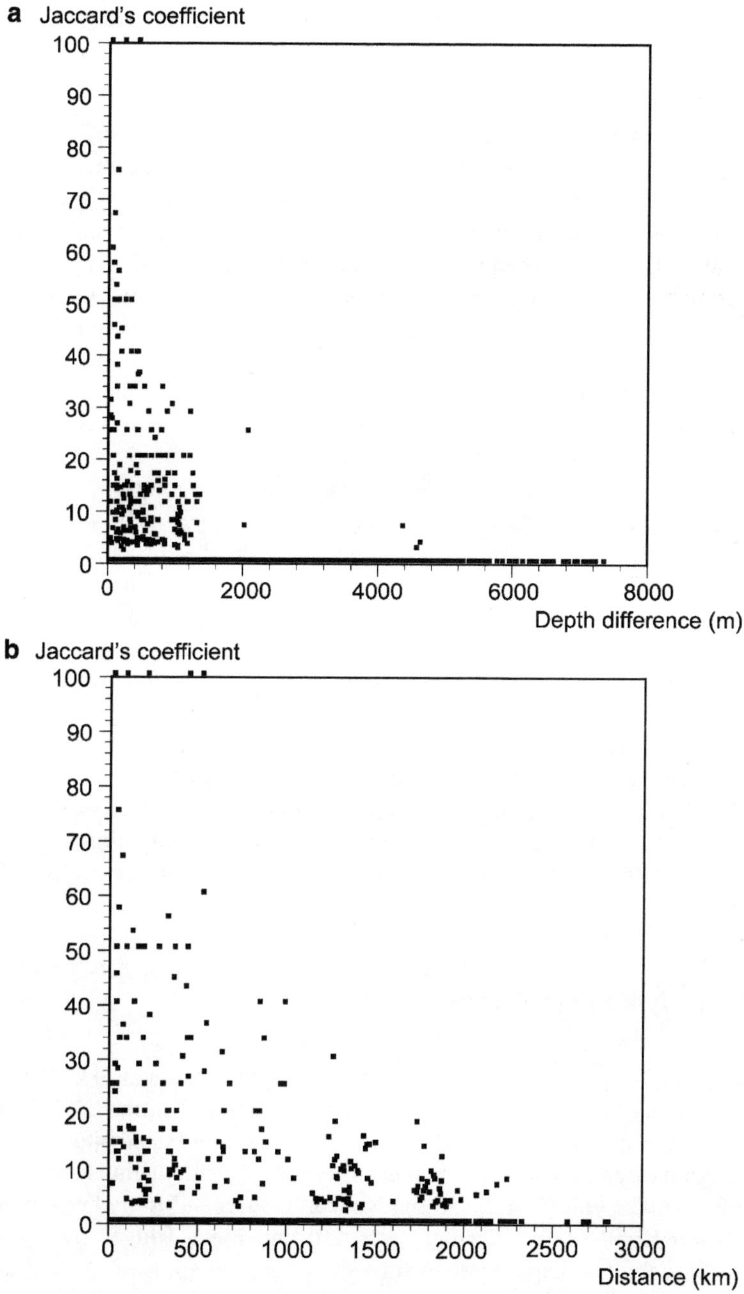

Fig. 12.4 Relationships between Jaccard's similarity coefficient and depth difference (**a**) and horizontal distance (**b**) for each pairwise community analyzed in the present study

12.4 Connectivity Among Vent and Seep Communities Around Japan

How is the similarity of species composition generated? One of the plausible processes is larval dispersal and supply among populations. Determining the larval supply process between deep-sea vents and seeps is very difficult, and population genetic analyses have been used to determine gene flow or connectivity resulting from larval supply and subsequent development and reproductive success. As the species compositions are similar between some vents and seeps, there are some species common to both vents and seeps. In general, not all populations of the same species belong to a single panmictic population. Sometimes populations are isolated and accumulate sufficient genetic differences to be treated as genetically different populations. In the western Pacific, there are many potential barriers to the dispersal of vent- and seep-obligate species (Tunnicliffe et al. 1998). *Calyptogena* clams, *Lamellibrachia* tubeworms, and *Bathymodiolus* mussels are good model groups to examine genetic differentiation for the evaluation of dispersal between sites with different geologic settings, because they inhabit several vents and seeps around Japan. Molecular phylogenetic studies of these animals around Japan have been carried out (Kojima et al. 2001, 2002, 2005; Miyazaki et al. 2004; Iwasaki et al. 2006; Kyuno et al. 2009). Here, we summarize the published data of the mitochondrial COI sequences of the DNA barcoding region and analyze them further to estimate gene flow among vent and seep populations. On the basis of the published datasets, parsimony networks of haplotypes were constructed using the program TCS 1.21 (Clement et al. 2000) with a connection limit of 95% and modified in Adobe Illustrator 10.0.3. In addition, population differences were examined using the pairwise Φ_{ST} with Jukes-Cantor distance with 1,000 permutation and exact test with the computer software package Arlequin ver. 3.11 (Excoffier et al. 2005).

12.4.1 *Calyptogena* Clams

Sequence data for *C. okutanii* (1,101 bp of the partial mitochondrial COI sequence) were obtained and phylogenetically analyzed previously (Kojima et al. 2001); the specimens are from methane seeps at the Off Hatsushima Island site ($N = 13$) and the Okinoyama Bank ($N = 11$) in Sagami Bay, the Dai-ni Atsumi Knoll ($N = 13$), and the Ryuyo Canyon ($N = 22$) in the Nankai Trough and in hydrothermal vents on the North Knoll of Iheya Ridge ($N = 20$) and Iheya Ridges ($N = 6$) in the Okinawa Trough. The haplotype network shows a geographically biased distribution of haplotypes, that is, a single haplotype dominated in the Okinawa Trough while another haplotype, which is also observed at low frequency in the Okinawa Trough, dominated in Sagami Bay and the Nankai Trough (Fig. 12.5a). The estimated pairwise Φ_{ST} values and P value in the exact test among six populations of *C. okutanii* indicated that populations in the Okinawa Trough differed significantly

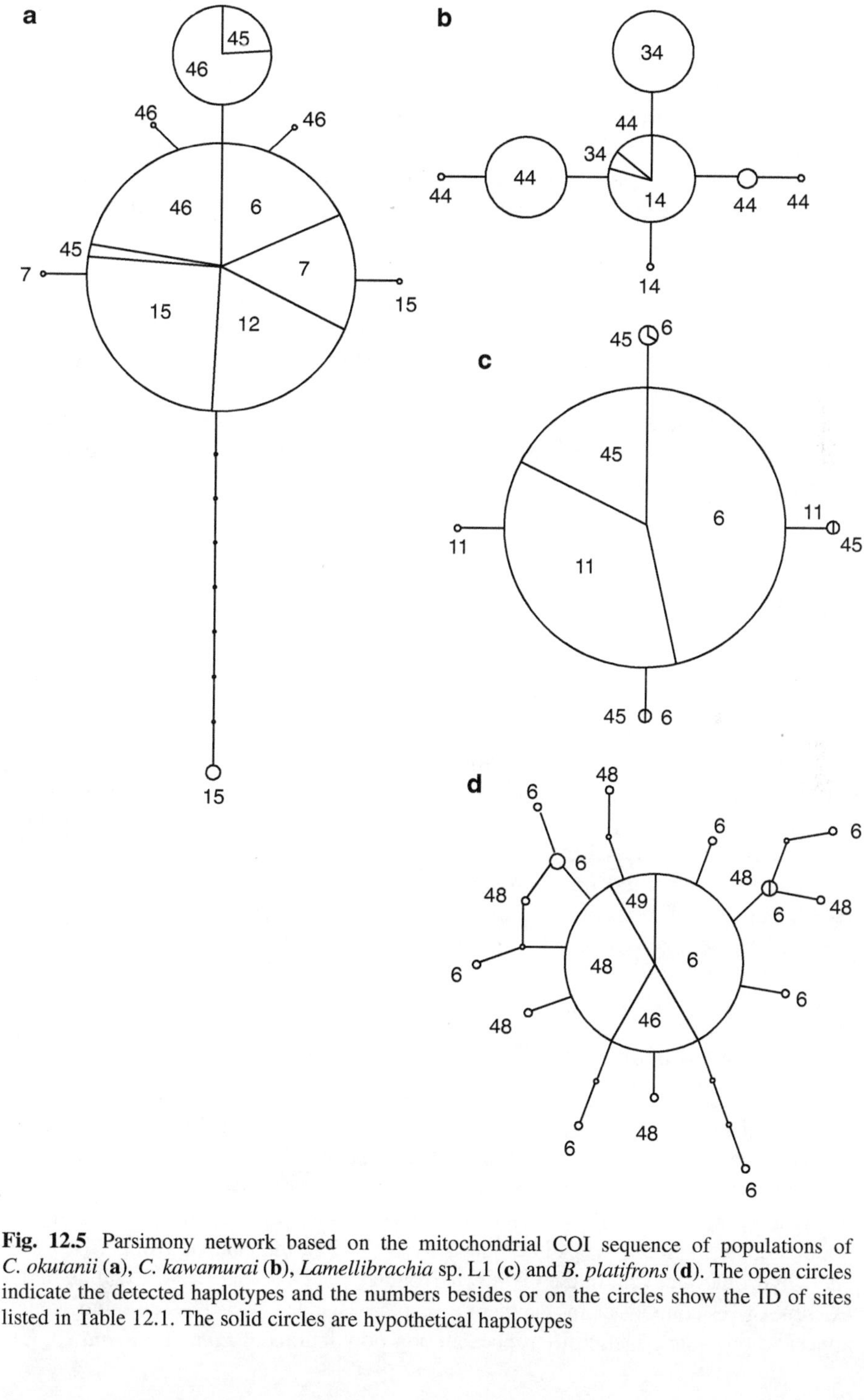

Fig. 12.5 Parsimony network based on the mitochondrial COI sequence of populations of *C. okutanii* (**a**), *C. kawamurai* (**b**), *Lamellibrachia* sp. L1 (**c**) and *B. platifrons* (**d**). The open circles indicate the detected haplotypes and the numbers besides or on the circles show the ID of sites listed in Table 12.1. The solid circles are hypothetical haplotypes

from those in Sagami Bay and the Nankai Trough. Three individuals with a markedly different sequence (8-bp difference from the major haplotype) were obtained from the Dai-ni Atsumi Knoll (Fig. 12.5a). The communities in the Ryuyo Canyon and on the Dai-ni Atsumi Knoll in the Nankai Trough were classified in a different cluster in the group-average clustering on the MDS plot (Fig. 12.3) and significant difference could not be detected between these populations (Table 12.3A), although they were located at the same depth range (1,000–1,100 m). As there was also no significant genetic difference between the populations in the Ryuyo Canyon and Sagami Bay, the subduction zone of the Philippine Sea plate appears not to act as a dispersal barrier to *C. okutanii*.

Specimens of the second analyzed vesicomyid clam, *Calyptogena kawamurai* (*Calyptogena solidissima* is regarded as a junior synonym of this species), were collected from a hydrothermal vent field on the Minami Ensei Knoll ($N = 20$) in the Okinawa Trough, methane seeps on the Dai-ni Tenryu Knoll ($N = 12$) in the Nankai Trough, and on the Kuroshima Knoll ($N = 14$) in the Nansei-Shoto Trench (Kojima et al. 2006). The haplotype network based on 1,101 bp of the partial mitochondrial COI sequence (Fig. 12.5b) showed that the dominant haplotypes differed among the three populations, and therefore, the haplotypic compositions of populations were statistically different (Table 12.3B). Dispersal barriers, probably the Nansei-Shoto Trench and Ryukyu Islands, must exist among the three populations of *C. kawamurai*.

In the third vesicomyid, *C. nankaiensis*, no common haplotype was detected for 1,101 bp of the partial mitochondrial COI sequences among specimens collected from the Ryuyo Canyon methane-seep area in the Nankai Trough and the North Knoll of the Iheya Ridge hydrothermal vent area in the Okinawa Trough (Kojima et al. 2005), although the number of available specimens (11 in total) was not sufficient for statistical analysis. None of the three *Calyptogena* clams showed population connectivity associated with the similarity of community composition between vent and seep sites around Japan. Among vent animals, the genetic differences in barnacles of the genus *Neoverruca* have been suggested to be caused by limitations in larval dispersal related to thermal tolerance (Watanabe et al. 2005, 2006). Detailed analyses of larval dispersal and settlement mechanisms are required to elucidate those differences.

12.4.2 **Lamellibrachia** ***Tubeworms***

In case of the undescribed vestimentiferan tubeworm *Lamellibrachia* sp. L1 (informal Japanese name *Sagami-haorimushi*; see Kojima et al. 2001), the distribution of haplotypes based on 1,029 bp of the partial mitochondrial COI sequence was not biased geographically (Fig. 12.5c), and there was no significant difference among the populations at methane seeps in Sagami Bay ($N = 23$), the Nankai Trough ($N = 18$), and a hydrothermal vent in the Okinawa Trough ($N = 12$; Table 12.3C). Unlike in the example of vesicomyids clams, neither the Ryukyu Islands nor the Nansei-Shoto Trench appear to promote genetic differences among populations of *Lamellibrachia* sp. L1.

Table 12.3 Genetic differentiation indices for seven populations of *C. okutanii* (A), *C. kawamurai* (B), and three populations of *Lamellibrachia* sp. (C). The bold numbers in labels show the ID of vent and seep sites in Table 12.1. The numbers in parentheses indicate the numbers of individuals used in the analyses. Upper right: exact test; lower left: pairwise Φ_{ST}

A	**6** (13)	**7** (11)	**12** (13)	**15** (22)	**46** (20)	**45** (6)
6		0.463	−1.000	0.376	0.000	0.000
7	0.016		0.458	0.362	0.000	0.001
12	0.000	0.016		0.365	0.000	0.001
15	0.051	0.039	0.051		0.000	0.001
46	0.770**	0.716**	0.770**	0.416**		1.000
45	0.874**	0.743**	0.874**	0.276**	−0.090	

B	**14** (12)	**34** (14)	**44** (20)
14		0.000**	0.000**
34	0.848**		0.000**
44	0.421**	0.680**	

C	**6** (23)	**11** (18)	**45** (12)
6		0.588	0.1381
11	0.002		0.1795
45	0.009	0.030	

$*P < 0.05$; $**P < 0.01$

The larval characteristics of this species are not known, and how connectivity has been established may not be discussed here. A congeneric species, *L. satsuma*, was found on a whale-fall carcass as well as in vents and seeps (Fujiwara et al. 2007). It seems thus possible that *Lamellibrachia* sp. L1 may also be able to use such steppingstones, and thereby maintaining a large larval pool in deep water, enabling populations to be highly connected among spatially distant vent and seep communities.

Another tubeworm species, *Paraescarpia echinospica*, is also distributed in a methane seep in the Nankai Trough and a hydrothermal vent in the Okinawa Trough, as well as in vent and seep areas in Papua New Guinean waters (Kojima et al. 2002). Molecular phylogenetic analyses showed that the dominant haplotype in *P. echinospica* was shared among populations in all these habitats (Kojima et al. 2002). It appears that dispersal barriers such as islands, subduction zones, and interposing plates that result in distinct populations among vesicomyid clams do not effect the connectivity of tubeworm species inhabiting vents and seeps.

12.4.3 Bathymodiolus *Mussels*

The two bathymodiolin mussels with methanotrophic symbionts, *Bathymodiolus japonicus* and *B. platifrons*, are both found in methane-seep areas in Sagami Bay and in the vent field in the Okinawa Trough, whereas most other *Bathymodiolus* mussels have been found either at vents or at seeps. Therefore, *B. japonicus* and *B. platifrons* are suitable species to examine the connectivity between vent and seep sites. Phylogenetic analyses showed that some haplotypes were shared by the populations in Sagami Bay and the Okinawa Trough (Miyazaki et al. 2004; Iwasaki et al. 2006; Kyuno et al. 2009), suggesting the existence of connectivity between the two populations of *Bathymodiolus* mussels (Fig. 12.5d shows parsimony network of *B. platifrons*). However, the numbers of individuals analyzed were not sufficient to elucidate connectivity in the populations in the Okinawa Trough, as *Bathymodiolus* mussels are one of the most dominant animals there. Further statistical analyses with more sequence data are required for elucidation. For the shrimp *A. longirostris*, a single haplotype was shared between the populations in the methane seep in Sagami Bay and hydrothermal vents in the Okinawa Trough (Tokuda et al. 2006), the same as found for *Bathymodiolus* mussels. There may be some common features which determine their distributions or dispersal, between *Alvinocaris* shrimps and *Bathymodiolus* mussels around Japan.

12.5 Conclusions

More than 50 animal communities associated with deep-sea hydrothermal vents and methane seeps around Japan are distributed in a wide depth range (80–7,434 m) in various geologic settings such as volcanic arcs, back-arc basins, subduction zones, and other plate boundaries. As explained in this chapter, these vent and seep communities share 20% of their species, a much higher figure than seen on a global scale. Although significant distinction in the community composition between vents and seeps was suggested, similarities were still detected for some vent and seep communities, probably due to sedimentation at the vent fields in the Okinawa Trough as well as in the similarity in their distributional depth rather than horizontal distance and tectonic setting. Genetic connectivity between populations at vents and seeps was shown for *Lamellibrachia* tubeworms and *Bathymodiolus* mussels, whereas populations of *Calyptogena* clams were genetically different between vent and seep sites. These results suggest that similarities in community composition are not necessarily accompanied by population connectivity of the constituting species. Therefore the similarity between vent and seep communities may not be generated by larval supply. Rather than larval supply, environmental similarity between vents and seeps appears to be a plausible factor that leads to the similarity of animal communities in distant sites. Bathymetric zonation may be one possible factor generating dissimilarity among communities (Figs. 12.3 and 12.4). However, quantitative data are still limited except for those reported by Fujikura et al. (1995) and improved quantitative datasets are required to elucidate the factor controlling the distribution of species in vent and seep communities around Japan. Due to the close proximity of vent and seep communities in different settings and depths, Japanese waters and the adjacent northwestern Pacific area provides a good model for resolving the question of how similarity and connectivity between vent and seep communities is generated.

Acknowledgements We appreciate Drs. Motohiro Shimanaga and Dhugal J. Lindsay for their useful advices on analyzing similarities among vent and seep communities. We also thank the three reviewers and the editor Dr. Steffen Kiel, who gave us helpful comments to improve this paper.

References

Ashi J (2003) Characteristics of Nankai Trough and origin of seeping water. Kaiyo Mon 35:296–300 (in Japanese)

Barry JP, Buck KR, Kochevar RK et al (2002) Methane-based symbiosis in a mussel, *Bathymodiolus platifrons*, from cold seeps in Sagami Bay, Japan. Invertebr Biol 121:47–54

Clarke KR, Warwick RM (2001) Change in Marine Communities: an approach to statistical analysis and interpretation, 2nd edn. PRIMER-E, Plymouth, UK

Clement M, Posada D, Crandall KA (2000) TCS: a computer program to estimate gene genealogies. Mol Ecol 9:1657–1659

Colwell F, Matsumoto R, Reed D (2004) Review of the gas hydrates, geology, and biology of the Nankai Trough. Chem Geol 205:391–404

Corliss JB, Dymond J, Gordon LI et al (1979) Submarine thermal springs on the Galapagos Rift. Science 203:1073–1083

Distel DL, Baco AR, Chuang E et al (2000) Do mussels take wooden steps to deep-sea vents? Nature 403:725–726

Excoffier L, Laval G, Schneider S (2005) Arlequin ver. 3.0: an integrated software package for population genetics data analysis. Evol Bioinform (Online) 1:47–50

Fujikura K, Hashimoto J, Fujiwara Y et al (1995) Community ecology of the chemosynthetic community at Off Hatsushima site, Sagami Bay, Japan. JAMSTEC Deep Sea Res 11:227–241 (in Japanese with English figure captions)

Fujikura K, Kojima S, Tamaki K et al (1999) The deepest chemosynthesis-based community yet discovered from the hadal zone, 7326 m deep, in the Japan Trench. Mar Ecol Prog Ser 190:17–26

Fujikura K, Kojima S, Fujiwara Y et al (2000) New distribution records of vesicomyid bivalves from deep-sea chemosynthesis-based communities in Japanese waters. Venus 59:103–121

Fujikura K, Fujiwara Y, Ishibashi J et al (2001) Report on investigation of hydrothermal vent ecosystems by the crewed submersible 'Shinkai 2000' on the Dai-yon (No. 4) Yonaguni Knoll and the Hatoma Knoll, the Okinawa Trough. JAMSTEC Deep Sea Res 19:141–154 (in Japanese with English abstract and figure captions)

Fujikura K, Hashimoto J, Okutani T (2002) Estimated population densities of megafauna in two chemosynthesis-based communities: a cold seep in Sagami Bay and a hydrothermal vent in the Okinawa Trough. Benthos Res 57:21–30

Fujikura K, Aoki M, Fujiwara Y et al (2003) Report on investigation of vent and methane seep ecosystems by the crewed submersible 'Shinkai 2000' and the ROV 'Dolphin 3K' on the Hatoma and the Kuroshima Knolls, the Nansei-shoto area. JAMSTEC Deep Sea Res 22:21–30 (in Japanese with English abstract and figure captions)

Fujikura K, Amaki K, Barry J et al (2007) Long-term *in situ* monitoring of spawning behavior and fecundity in *Calyptogena* spp. Mar Ecol Prog Ser 333:185–193

Fujikura K, Okutani T, Maruyama T (eds) (2008) Deep-sea life – biological observations using research submersibles. Tokai University Press, Kanagawa, Japan (in Japanese with English figure captions)

Fujiwara Y, Tsukahara J, Hashimoto J et al (1998) *In situ* spawning of a deep-sea vesicomyid clam: evidence for environmental cue. Deep Sea Res I 45:1881–1889

Fujiwara Y, Takai K, Uematsu K et al (2000) Phylogenetic characterization of endosymbionts in three hydrothermal vent mussels: influence on host distributions. Mar Ecol Prog Ser 208:147–155

Fujiwara Y, Kato C, Masui N et al (2001) Dual symbiosis in the cold-seep thyasirid clam *Maorithyas hadalis* from the hadal zone in the Japan Trench, western Pacific. Mar Ecol Prog Ser 214:151–159

Fujiwara Y, Kawato M, Yamamoto T et al (2007) Three-year investigations into sperm whale-fall ecosystems in Japan. Mar Ecol 28:219–232

Glasby GP, Notsu K (2003) Submarine hydrothermal mineralization in the Okinawa Trough, SW of Japan: an overview. Ore Geol Rev 23:299–233

Goffredi SK, Hurtado LA, Hallam SJ, Vrijenhoek RC (2003) Evolutionary relationships of deep-sea vent and cold seep clams (Mollusca: Vesicomyidae) of the "*pacifica/lepta*" species complex. Mar Biol 142:311–320

Hashimoto J, Ohta S, Tanaka T et al (1989) Deep-sea communities dominated by the giant clam, *Calyptogena soyoae*, along the slope foot of Hatsushima Island, Sagami Bay, Central Japan. Palaeogeogr Palaeoclimatol Palaeoecol 71:179–192

Hashimoto J, Fujikura K, Miura T et al (1993) Discovery of vestimentiferan tube-worms in the euphotic zone. Proc JAMSTEC Symp Deep Sea Res 9:321–326

Hashimoto J, Ohta S, Fujikura K et al (1995) Life habit of vesicomyid clam, *Calyptogena soyoae*, and hydrogen sulfide concentration in interstitial waters in Sagami Bay, Japan. J Oceanogr 51:341–350

Ishibashi J, Urabe T (1994) Hydrothermal activity related to arc-backarc magmatism in the Western Pacific. In: Taylor B (ed) Backarc basins, tectonics and magmatism. Plenum, New York/London, pp 451–495

Iwasaki H, Kyuno A, Shintaku M et al (2006) Evolutionary relationships of deep-sea mussels inferred by mitochondrial DNA sequences. Mar Biol 149:1111–1122

Juniper K, Sibuet M (1987) Cold seep benthic communities in Japan subduction zones: spatial organization, trophic strategies and evidence for temporal evolution. Mar Ecol Prog Ser 40:115–126

Kiel S, Goedert JL (2006) Deep-sea food bonanzas: early Cenozoic whale-fall communities resemble wood-fall rather than seep communities. Proc R Soc B 273:2625–2631

Kojima S (2002) Deep-sea chemoautosynthesis-based communities in the Northwestern Pacific. J Oceanogr 58:343–363

Kojima S, Ohta S (1997) *Calyptogena okutanii* n. sp. a sibling species of *Calyptogena soyoae* Okutani, 1957 (Bivalvia: Vesicomyidae). Venus 56:189–195

Kojima S, Ohta S, Yamamoto T et al (2001) Molecular taxonomy of vestimentiferans of the western Pacific and their phylogenetic relationship to species of the eastern Pacific. I. Family Lamellibrachiidae. Mar Biol 139:211–219

Kojima S, Ohta S, Yamamoto T et al (2002) Molecular taxonomy of vestimentiferans of the western Pacific and their phylogenetic relationship to species of the eastern Pacific II. Families Escarpiidae and Arcovestiidae. Mar Biol 141:57–64

Kojima S, Fujikura K, Okutani T (2005) Genetic differentiation of two vesicomyid bivalves, *Calyptogena okutanii* and *Calyptogena nankaiensis*, between seep areas off the central Honshu and hydrothermal vent fields in the Okinawa Trough. Venus 64:45–53

Kojima S, Tsuchida E, Numanami H et al (2006) Synonymy of *Calyptogena solidissima* with *Calyptogena kawamurai* (Bivalvia: Vesicomyidae) and its population structure revealed by mitochondrial DNA sequences. Zool Sci 23:835–842

Kuwahara H, Yoshida T, Takaki Y et al (2007) Reduced genome of the thioautotrophic intracellular symbiont in a deep-sea clam, *Calyptogena okutanii*. Curr Biol 17:881–886

Kyuno A, Shintaku M, Fujita Y et al (2009) Dispersal and differentiation of deep-sea mussels of the genus *Bathymodiolus* (Mytilidae, Bathymodiolinae). J Mar Biol Article ID 625672, pp 1–15

Lallemand SE, Glaçon G, Lauriat-Rage A et al (1992) Seafloor manifestations of fluid seepage at the top of a 2000-metre-deep ridge in the eastern Nankai accretionary wedge: long-lived venting and tectonic implications. Earth Planet Sci Lett 109:333–346

Lonsdale P (1977) Clustering of suspension-feeding macrobenthos near abyssal hydrothermal vents at oceanic spreading centers. Deep Sea Res 24:857–863

Lorion J, Duperron S, Gros O et al (2008) Several deep-sea mussels and their associated symbionts are able to live both on wood and on whale falls. Proc R Soc B 276:177–185

Magurran AE (1988) Ecological diversity and its measurement. Princeton University Press, Princeton, NJ

Matsumoto R, Okuda Y, Hiruta A et al (2009) Formation and collapse of gas hydrate deposits in high methane flux area of the Joetsu Basin, eastern margin of Japan Sea. J Geogr 118:43–71 (in Japanese with English abstract and figure captions)

Miyazaki J, Shintaku M, Kyuno A et al (2004) Phylogenetic relationships of deep-sea mussels of the genus *Bathymodiolus* (Bivalvia: Mytiidae). Mar Biol 144:527–535

Mizota C, Yamanaka T (2003) Strategic adaptation of a deep-sea, chemosynthesis-based animal community: an evaluation based on soft body part carbon, nitrogen, and sulfur isotopic signatures. Jpn J Benthol 58:56–59 (in Japanese with English abstract and figure captions)

Munroe TA, Hashimoto J (2008) A new western Pacific tonguefish (Pleuronectiformes: Cynoglossidae): the first pleuronectiform discovered at active hydrothermal vents. Zootaxa 1839:43–59

Ogawa Y, Fujioka K, Fujikura K et al (1996) En echelon patterns of *Calyptogena* colonies in the Japan Trench. Geology 24:807–810

Ohta S, Kim DS (2001) Submersible observations of the hydrothermal vent communities on the Iheya Ridge, Mid Okinawa Trough. Jpn J Oceanogr 57:663–677

Ohta S, Laubier L (1987) Deep biological communities in the subduction zone of Japan from bottom photographs taken during "Nautile" dives in the Kaiko project. Earth Planet Sci Lett 83:329–342

Okutani T, Egawa K (1985) The first underwater observation on living habitat and thanatocenoses of *Calyptogena soyoae* in bathyal depth of Sagami Bay. Venus 44:285–289

Paull CK, Hecker B, Commeau R et al (1984) Biological communities at the Florida Escarpment resemble hydrothermal vent taxa. Science 226:965–967

Sakai H, Gamo T, Kim ES et al (1990) Venting of carbon dioxide-rich fluid and hydrate formation in Mid-Okinawa Trough backarc basin. Science 248:1093–1096

Sibuet M, Olu K (1998) Biogeography, biodiversity and fluid dependence of deep-sea cold-seep communities at active and passive margins. Deep Sea Res II 45:517–567

Sibuet M, Juniper K, Pautot G (1988) Cold-seep benthic communities in the Japan subduction zones: geological control of community development. J Mar Res 46:333–348

Smith CR, Kukert H, Wheatcroft RA et al (1989) Vent fauna on whale remains. Science 341:27–28

Tokuda G, Yamada A, Nakano K et al (2006) Occurrence and recent long-distance dispersal of deep-sea hydrothermal vent shrimp. Biol Lett 2:257–260

Tsunogai U, Ishibashi J, Wakita H et al (1994) Source of "cold seepage" at the off Hatsushima Island, Sagami Bay: a preliminary report of the Dive 721 of "Shinkai 2000". JAMSTEC Deep Sea Res 10:395–403 (in Japanese with English abstract and figure captions)

Tunnicliffe V, Fowler MR (1996) Influence of sea-floor spreading on the global hydrothermal vent fauna. Nature 379:531–533

Tunnicliffe V, McArthur AG, McHugh D (1998) A biological perspective of the deep-sea hydrothermal vent fauna. Adv Mar Biol 34:353–442

Tunnicliffe V, Juniper K, Sibuet M (2003) Reducing environments of the deep-sea floor. In: Tyler PA (ed) Ecosystems of the deep oceans, vol 28, Ecosystems of the world. Elsevier, Amsterdam, The Netherlands, pp 81–110

Van Dover CL (2000) The ecology of deep-sea hydrothermal vent. Princeton University Press, Princeton, NJ, p 424

Watanabe H, Tsuchida S, Fujikura K et al (2005) Population history associated with hydrothermal vent activity inferred from genetic structure of neoverrucid barnacles around Japan. Mar Ecol Prog Ser 288:233–240

Watanabe H, Kado R, Kaida M et al (2006) Dispersal of vent-barnacle (genus *Neoverruca*) in the Western Pacific. Cah Biol Mar 47:353–357

Yamanaka T, Ishibashi J, Hashimoto J (2000) Organic geochemistry of hydrothermal petroleum generated in the submarine Wakamiko caldera, southern Kyushu, Japan. Org Geochem 31:1117–1132

Chapter 13
Shaping Vent and Seep Communities: Habitat Provision and Modification by Foundation Species

Breea Govenar

Contents

13.1 Introduction

Ecologists strive to identify mechanisms that determine the structure or composition of biological communities. Until the past decade, the role of positive species interactions in regulating community structure had been relatively overlooked, compared to negative species interactions, such as competition and predation. While mutualism and commensalism have long been recognized as critically important in pairwise species interactions, the impact of facilitation and habitat provision or modification on community dynamics and species diversity has received increased attention only more recently (Bertness and Callaway 1994; Hacker and Gaines 1997).

B. Govenar (✉)
Biology Department, Woods Hole Oceanographic Institution, Woods Hole, MA 02543, USA
e-mail: bgovenar@whoi.edu

S. Kiel (ed.), *The Vent and Seep Biota*, Topics in Geobiology 33,
DOI 10.1007/978-90-481-9572-5_13, © Springer Science+Business Media B.V. 2010

Part of the reason for the oversight is that positive species interactions are more prevalent in habitats with harsh environmental conditions (Bertness et al. 1999; Crain and Bertness 2006) and are more often detectable at regional scales (Bertness and Leonard 1997; van de Koppel et al. 2006). Communities may be structured by facilitation cascades, where foundation species create the habitat or modify the environment that facilitates the settlement of other species, which subsequently interact with one another (Bruno 2000; Altieri et al 2007). Foundation species are large or spatially dominant organisms that create or provide habitats, colonized by other species (Bruno and Bertness 2001; for review).

The definition of foundation species overlaps with the definitions of ecosystem engineers and keystone species. However, each classification is unique and has a different emphasis for explaining the cause of variation in biological communities. Ecosystem engineers are broadly defined as organisms that modify, maintain, create, or destroy habitats (Jones et al. 1994; Wright and Jones 2004, 2006). The primary role of ecosystem engineers is the alteration of the availability of resources, such as water, light, and nutrients. The effect of ecosystem engineers on the associated biological community may be either large or small, and either positive or negative (Jones et al. 1997; Wright et al. 2002; Wright and Jones 2004). Beavers are common examples that modify the local environment by building dams that change the hydrology of the watershed at regional scales and create additional habitats that are colonized by different species, resulting in greater local and regional species diversity than rivers and streams without beaver dams (Wright et al. 2002). In contrast, keystone species can have either positive or negative effects on other species, but are expressly defined by the large impact that they have on the associated community, relative to their abundance (Paine 1974; Mills et al. 1993; Power et al. 1996; Hacker and Gaines 1997). The most common examples are keystone predators (e.g. Paine 1969) that exert top-down control on biological community structure by removing a competitively dominant species and thus allow for greater species coexistence. Accordingly, the term "keystone" may be better used as a modifier of a particular biological interaction that emphasizes the relative importance or the strength of the interaction, whereas ecosystem engineers are characterized by the action rather than their impact of the interactions. The distinction between foundation species and either ecosystem engineers or keystone species is that habitat provision or modification by the presence of foundation species is a positive biological interaction that has a large impact on the structure of the associated community. Thus, it could be said that foundation species are keystone physical ecosystem engineers. Foundation species tend to be sessile, large, spatially dominant, or have complex physical structures, and their interaction with other species is defined by their presence, rather than their actions. Although foundation species may compete for resources with other species, serve as a source of food for some consumers, or consume potential colonists (Crooks 1998; Bustamante et al. 1995), these trophic interactions are not directly related to the role of habitat provision or modification.

At deep-sea hydrothermal vents and hydrocarbon seeps, the density and diversity of macrofaunal invertebrates are greater in association with foundation species than the adjacent seafloor. At vents and seeps, the most widely studied foundation species are siboglinid tubeworms and bathymodiolin mussels. Other tube-dwelling polychaetes, including frenulate tubeworms and alvinellid tube-dwelling worms, and bivalves such as vesicomyid clams, are also important foundation species in some vent and seep habitats. Many, if not most, foundation species at vents and seeps (e.g. tubeworms, mussels, and clams) have bacterial symbionts that use chemical rather than solar energy to fix inorganic carbon and are the primary source of nutrition to the animal. Consequently, tubeworms, mussels, and clams live in microhabitats where they are exposed to the mixing of reduced chemicals in either vent or seep fluids and relatively well-oxygenated seawater. Reduced chemicals, particularly hydrogen sulfide, can be lethal to animals by reducing the availability of oxygen or by inhibiting cellular respiration. Thus, some species may benefit from the positive biological interactions with foundation species but are also exposed to physiologically stressful conditions, where the source of energy to the ecosystem can also be lethal to the resident species.

This chapter examines the ecological role of habitat provision and modification by foundation species in deep-sea hydrothermal vent and hydrocarbon seep communities and how foundation species contribute to the maintenance of species diversity at local and regional scales in chemosynthesis-based ecosystems. At the local scale, the presence of foundation species coincides with access to high local productivity habitats; the physical structure of foundation species provides additional area available for colonization, increases environmental heterogeneity, or serves as a refuge from predation; and the physiological function of the foundation species can modify the physical or chemical environment by altering fluid flow or composition. At the regional scale, aggregations of foundation species form biogenic "islands" that connect biological communities through migration or dispersal. Changes to immigration and extinction rates are caused by the size and distance of islands (McArthur and Wilson 1967; Hanski 1998), the rate of production of colonists within an island, and the rate by which the resources to the foundation species are generated or depleted (Wright et al. 2004). Therefore, ecological factors that affect the distribution and abundance of foundation species can have either direct or indirect effects on associated species and impact the community structure at local and regional scales.

The first section of this chapter presents an overview of the habitats of vent and seep communities in the context of the underlying geology and fluid chemistry. The second section introduces how evolutionary and large-scale ecological processes affect global patterns of biogeography and regional species diversity at vents and seeps. The next section outlines local-scale patterns in the distribution of foundation species and the associated epifaunal communities. The fourth section discusses a range of the primary abiotic factors and biological interactions that have been shown to regulate the structure of vent and seep communities. The last section specifically focuses on the ecological role of foundation species, through a discussion of the mechanisms of (1) providing access to high productivity habitat,

(2) increasing habitat complexity, and (3) modifying the physical or chemical environment. This chapter does not provide an exhaustive characterization of the ecology or evolution of either vent or seep communities, which can be found in comprehensive reviews elsewhere (e.g. Tunnicliffe 1991 for vents, and Sibuet and Olu 1998; Levin 2005 for seeps). In fact, the examples in this chapter tend to be heavily weighted to the community types that have been studied with the greatest uniformity across habitats and study sites. To illustrate how vent and seep communities are shaped by foundation species, the most useful examples are siboglinid tubeworms and bathymodiolin mussels at hydrothermal vents in the eastern Pacific Ocean and the Atlantic Ocean and hydrocarbon seeps on the North American continental margins. Tubeworms and mussels have been found in other localities, and other organisms may act as foundation species (e.g. clams and serpulid tube-dwelling polychaetes), but they are not discussed here, because the associated communities have not been adequately sampled to examine the relationship between the foundation species and the associated community or to compare to other vent and seep communities. Lastly, some species (e.g. bacterial mats and highly mobile species, such as rimicarid shrimp) may be characteristic of a particular habitat and have unique associated species assemblages, but are not considered foundation species in the strict sense, if they do not provide or modify habitats through the physical architecture of their presence.

13.2 Overview of Vent and Seep Communities

Hydrothermal vents and hydrocarbon seeps are characterized by high concentrations of reduced chemicals that are used as an energy source to drive local primary production through chemolithoautotrophy. The most common reduced chemicals utilized to support chemosynthetic ecosystems at vents and seeps are sulfide and methane (Fisher 1990; McCollom and Shock 1997; Dubilier et al. 2008). Oxygen or other oxidants available in the seawater react with the reduced chemicals in the vent and seep fluids and provide the chemical energy needed to fix inorganic carbon into simple sugars. Despite the benefits of sulfide-oxidation for primary production, sulfide exposure has physiological costs to aerobic organisms. As a result, most organisms at vents and seeps are patchily distributed along the seafloor with respect to the concentration or rate of the fluid flow.

At mid-ocean ridges, back-arc basins, and other volcanically and tectonically active areas of the seafloor, hydrothermal vents are formed by seafloor spreading, where seawater is entrained into cracks in the deformed oceanic crust and transformed under high heat and pressure into hydrothermal fluids. End-member hydrothermal fluids are characterized by high temperatures and high concentrations of dissolved metals and other reduced chemicals. The highest recorded temperatures reach above 400°C, and concentrations of metals and reduced chemicals can be as high as several hundred millimolar (Von Damm and Lilley 2004). Mineralization, resulting from the mixing hydrothermal fluids with seawater, forms high-temperature vents or "chimneys" that are often characterized by the visible "black smoke" of dissolved metals precipitating from the rising hydrothermal fluids, at the opening of a chimney. The average maximum size of an individual chimney or chimney complex varies among oceanic spreading centers, from heights of ~20 m at the East Pacific Rise to 45 m at the Juan de Fuca Ridge (Haymon et al. 1991; Robigou et al. 1993). Vent fluids are also diluted with seawater and further modified in cracks and conduits below the surface of the seafloor to form diffuse-flow or low-temperature vents, where maximum temperatures are usually less than 50°C (Von Damm and Lilley 2004). Diffuse-flow hydrothermal vents are found at the base of chimneys or develop in patches along the seafloor and can extend up to several hundred square-meters. Hydrothermal vent habitats are considered to be ephemeral and patchy, because of the interconnected relationship between volcanism, tectonism, and hydrothermal circulation. Eruptions or dike injections can destroy and create hydrothermal vent habitats on decadal scales, as observed at the East Pacific Rise (Haymon et al. 1993; Fornari et al. 1998; Tolstoy et al. 2006). However, there are also hydrothermal vents, such as the TAG hydrothermal mound, that appear to have been continuously active for tens of thousands of years, if not longer (Rona et al. 1984; You and Bickle 1998).

Cold seeps are found in pockmarks, mud volcanoes, and other seafloor features on active and passive continental margins, where hydrocarbons leak from reservoirs buried under the sediment as dissolved methane, petroleum, and gas hydrates. As petroleum is degraded and gas hydrates are disturbed or exposed to seawater, additional methane is produced (Ergorov et al. 1999; Aitken et al. 2004). The cause of hydrocarbon release at seeps varies among localities, as a result of sediment compaction, slumping, or the movement of buried salt layers (Sibuet and Olu 1998). Anaerobic methane oxidation coupled with sulfate reduction produces sulfide, and in the presence carbon dioxide, carbonate crusts and deposits develop (Boetius et al. 2000). Concentrations of methane and sulfide can reach several millimolar at the sediment-seawater interface, but may be only a few micromolar to non-detectable concentrations among tubeworms and mussels (Julian et al. 1999; Smith et al. 2000; Bergquist et al. 2003a). While some seeps in the Gulf of Mexico and the Blake Ridge extend over several hundred square-meters and seem to persist for centuries (Cordes et al. 2003; Van Dover et al. 2003), other seeps on the northern California slope are smaller and more transient (Levin et al. 2000).

Compared to other deep-sea habitats, invertebrate communities at hydrothermal vents and hydrocarbon seeps are characterized by high biomass and low species diversity (Tunnicliffe 1991; Carney 1994). These ecological characteristics are ultimately related to the source of primary production in vent and seep ecosystems. The base of the food web is formed by chemolithoautotrophic bacteria, which utilize the chemical energy from the oxidation of reduced chemicals, present in vent and seep fluids, to fix inorganic carbon. The most common electron-donors are reduced sulfur species (H_2S, HS^-, S^{-2}), but other reduced chemicals, such as hydrogen and dissolved metals, can also be used to support chemolithoautotrophy- or "chemosynthesis" (Jannasch and Wirsen 1979; Karl 1995).

Another form of chemosynthetic primary production is methanotrophy, in which methane is used as both the carbon and the energy source. Some animals at vents and seeps have symbiotic relationships with chemoautotrophic, methanotrophic, or both types of bacteria, which can provide the bulk of the animal's nutrition or serve to detoxify the microhabitat of the animal (Fisher 1990, for review). Animals with chemoautotrophic or methanotrophic symbionts tend to grow to large individual sizes and/or high densities (reviewed in Tunnicliffe 1991; Sibuet and Olu 1998), and can contribute 75–90% of the biomass in vent and seep communities (Hessler and Smithey 1983; Govenar et al. 2005; Bergquist et al. 2003b). A greater number of endemic vent and seep animals do not have symbionts, but exhibit behavioral, physiological, and biochemical adaptations to tolerate different degrees of mixing between relatively homogeneous seawater and variable concentrations of potentially toxic, reduced chemicals in vent and seep fluids (Grieshaber and Volkel 1998).

The steep environmental gradients caused by the mixing of vent and seep fluids with background seawater may limit species diversity in chemosynthetic ecosystems (Tunnicliffe 1992; Carney 1994). Simply, the high concentrations of reduced chemicals could inhibit colonization by deep-sea fauna from neighboring habitats. The most abundant taxa are bivalve and gastropod molluscs, tube-dwelling and scale-covered (polynoid) polychaetes, and crustaceans; all of which posses protective coverings to shield their soft tissues from exposure to sulfide and heavy metals (Tunnicliffe 1992). Intrusions from deep-sea predators also appear to be reduced at vents and seeps, either because of evolutionary adaptations to exploit the soft-bodied animals that are more common in the deep-sea, or because of the chemical gradients associated with vent and seep fluids that deter opportunistic foraging (Carney 1994; Sahling et al. 2002). In the absence or reduction of predators, competition could further lead to the numerical dominance of one or a few species and a decrease in the species diversity.

13.3 Biogeographic Patterns and Regional Species Diversity

Evolutionary and ecological processes influence global patterns in biogeography and regional species diversity at hydrothermal vents and hydrocarbon seeps. Historical effects, such as plate tectonics, physical oceanographic processes, or other vicariance events open and close dispersal pathways, which lead to shifts in the species composition of different localities (Tunnicliffe et al. 1996; Van Dover et al. 2002). Ecological factors can also act on patterns caused by historical effects and either increase or decrease species diversity though immigration and extinction processes (McArthur and Wilson 1967; Ricklefs 1987). While often viewed separately, the evolutionary history of vents and seeps are intrinsically linked, as taxonomic and phylogenetic relationships among some species reveal many transitions between vent and seep habitats over geologic time (Tunnicliffe et al. 1998; Van Dover et al. 2002).

Among deep-sea hydrothermal vents, there are at least four biogeographic provinces, (1) the East Pacific Rise and Galapagos Rift, (2) mid-ocean ridges of the northeast Pacific Ocean, (3) the Mid-Atlantic Ridge, and (4) the spreading centers of the Southwest Pacific Ocean (Tunnicliffe et al. 1998; Van Dover et al. 2002; Bachraty et al. 2009). The distribution of the dominant foundation species and the regional species diversity varies among the four provinces. Different species of siboglinid tubeworms are found at the East Pacific Rise (and Galapagos Rift), the mid-ocean ridges of the northeast Pacific, and some vent sites in the southwest Pacific Ocean; but siboglinid tubeworms have not been reported from the Mid-Atlantic Ridge. Bathymodiolin mussels, on the other hand, form dense beds at the East Pacific Rise (and Galapagos Rift), throughout the southwest Pacific Ocean, and along the Mid-Atlantic Ridge. A few, small bathymodiolin mussels have been found at the Juan de Fuca Ridge (McKinness et al. 2005), but mussel beds have not been located. To date, the greatest regional species diversity is associated with the East Pacific Rise (Tunnicliffe et al. 1998; Van Dover et al. 2002). Large-scale ecological factors associated with seafloor spreading may contribute to this pattern, including the older geologic age, fast spreading rate, high to intermediate disturbance frequency and intensity, and large areas of diffuse hydrothermal flow at the East Pacific Rise (Tunnicliffe 1988; Tunnicliffe et al. 1996; Juniper and Tunnicliffe 1997; Van Dover et al. 2002; Van Dover 2003).

Rather than geographic differences, depth seems to play an important role in large-scale distribution patterns among seep communities (McMullin et al. 2003; Sahling et al. 2002; Cordes et al. 2007). In a comparison among deep methane seeps (3,200–4,500 m) along the Aleutian Margin and the Florida Escarpment, infaunal communities have lower densities but similar community structure to shallower seeps along the northern California margin (Levin and Mendoza 2007). However, in epifaunal communities associated with siboglinid tubeworm aggregations, densities are not significantly different between shallow and deep sites, but the species diversity decreases with depth (Cordes et al. 2007). Furthermore, the composition of seep communities in the deep Gulf of Mexico (>1,000 m) are more similar to deeper seep communities at the Blake Ridge and the Barbados accretionary prism than to communities that at closer, but shallower seeps on the upper continental slope of the Gulf of Mexico (Cordes et al. 2007).

In different habitats associated with bathymodiolin mussel beds, species diversity is greater in seep than vent communities (Fig. 13.1; Turnipseed et al. 2003, 2004; Levin 2005). Hydrocarbon seeps are embedded with the continental margin, either on the continental shelf or slope, where the species diversity background deep-sea fauna is much higher and increases with depth (Grassle and Maciolek 1992; Rex et al. 1993; Levin et al. 2001). In contrast, hydrothermal vents are more isolated on mid-ocean ridges and have greater endemicity (Tunnicliffe 1991, 1992). Although seep communities may have greater species diversity, seeps also have lower species endemicity, which may reflect the proximity of seep habitats to source populations (Carney 1994).

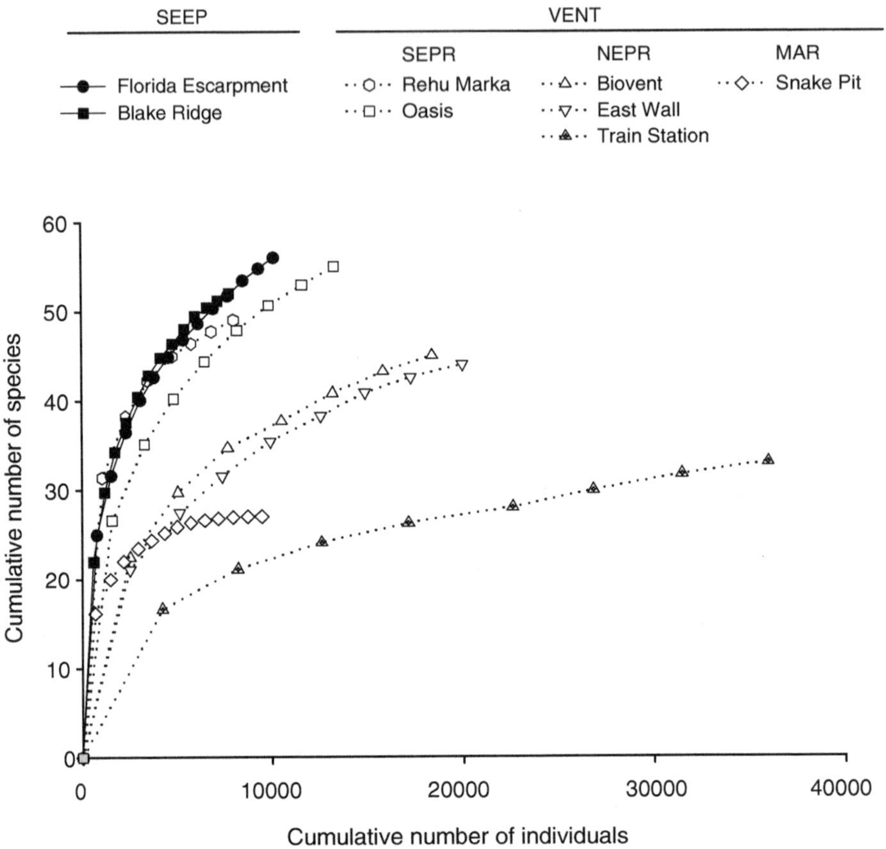

Fig. 13.1 Comparisons of species diversity between analogous vent and seep habitats (From Turnipseed et al. 2003)

13.4 Local-Scale Patterns in Vent and Seep Communities

Large-scale ecological factors play an important role in the distribution of organisms, but patterns at local scales provide critical insights for identifying processes that drive community dynamics and regulate community structure. A vent or a seep site generally refers to the semi-continuous area of visible fluid flow, bacterial mats, or benthic macrofauna. Several sites often comprise a locality (e.g. Hydrate Ridge, Cascadia Margin; East Pacific Rise, 9°50′N). Another important scale for observing local ecological processes is at the level of the foundation species (e.g. tubeworm aggregation, mussel bed). Abiotic factors and biological interactions that affect the distribution of foundation species may operate at scales that are different from other species within the epifaunal community. From another perspective, factors that determine the distribution of foundation species would thus have indirect effects on the epifaunal community structure. Therefore, it is important to first characterize the patterns before analyzing the processes that regulate community structure at vents and seeps (Underwood et al. 2000).

13.4.1 Spatial Patterns

The distribution of symbiont-containing foundation species at the East Pacific Rise follows a gradient in hydrothermal fluid flux. On the surfaces of high-temperature vents, the dominant megafauna are alvinellid polychaetes, which harbor episymbiotic chemoautotrophic bacteria, form colonies of parchment-like tubes (Desbruyères et al. 1998). In areas of vigorous diffuse-flow, siboglinid tubeworms with endocellular chemoautotrophic symbionts form dense aggregations where maximum temperatures are ~30–50°C (Hessler et al. 1988; Govenar et al. 2005; Le Bris et al. 2006a). Bathymodiolin mussels harbor chemoautotrophic symbionts and maintain the ability to filter-feed, which allows them to form extensive beds and occupy a wide range of environmental conditions, from ~15°C with tubeworms to close to ambient temperatures (~2°C) with or without vesicomyid clams (Fisher et al. 1988a; Childress and Fisher 1992). Clams also have functional feeding structures but rely more on chemoautotrophic symbionts for their nutrition, and although temperatures around the exposed (posterior) end of the clams may be close to 2°C, they may stick their foot (at the anterior end) into cracks and form small clusters, where temperatures can be as high as 7°C (Fisher 1988b; Childress and Fisher 1992).

While individual species at the East Pacific Rise also occupy different microhabitats along gradients of fluid flux, the structure of epifaunal communities associated with foundation species appears to vary at different scales. The numerically dominant species are similar in epifaunal communities associated with tubeworms and with mussels, but some species, including gastropods *Rhyncopelta concentrica* and *Cyathermia naticoides*, have greater relative abundances with tubeworms, and other gastropods *Bathymargarites symplector* and *Eulepetopsis vitrea* are relatively more abundant in mussel beds (Van Dover 2002, 2003; Dreyer et al. 2005; Govenar et al. 2005). Similarly, gastropod assemblages associated with alvinellids and siboglinid tubeworms are numerically dominated by one species, *Lepetodrilus elevatus*, but peltospirid gastropod species are relatively more abundant with alvinellids than with tubeworms (Matabos et al. 2008). In contrast, comparisons of epifaunal communities with the same foundation species in different habitats do not appear to vary significantly over local scales. The density and biomass in epifaunal communities associated with aggregations of the siboglinid tubeworm *Rifitia pachyptila* in high- and low-sulfide sites were not significantly different, and the community composition was remarkably similar between sites (Govenar et al. 2005). Epifaunal communities associated with *R. pachyptila* aggregations were also similar before and after a shift in the habitat chemistry from high sulfide concentrations and no apparent iron to low sulfide and high iron concentrations (Govenar et al. 2004, 2005). Epifaunal communities also had high similarity among mussel beds, geographically separated by ~30° of latitude (Van Dover 2002, 2003; Dreyer et al. 2005).

At the Juan de Fuca Ridge, there are two extreme morphotypes of the single species of sibolginid tubeworm, *Ridgeia piscesae*, which occur in different habitats (Carney et al. 2002), but the structure and composition of the epifaunal communities also differs between these two habitat types. A "short-fat" morphotype occurs on the surfaces of high-temperature chimneys, and a "long-skinny" morphotype occurs on the basalt, where temperatures and concentrations of hydrothermal fluids are considerably lower (Govenar et al. 2002; Urcuyo et al. 2003, 2007). Both types of *R. piscesae* harbor chemoautotrophic symbionts but appear to exhibit different physiological adaptations to acquire sulfide in the habitats where they occur (Anderson et al. 2006; Carney et al. 2007). The species diversity is considerably lower with the short-fat morphotype of *R. piscease*, where the high temperatures and concentrations of reduced chemicals or chronic disturbance may limit colonization or opportunist species (Sarrazin et al. 1999; Govenar et al. 2002). Communities associated with the long-skinny *R. picessae* have greater species diversity (Govenar et al. 2004; Bergquist et al. 2007), and the species composition among basalt-dwelling *R. piscesae* aggregations does not vary significantly with location, substrate type, or vent fluid temperature or chemistry (Tsurumi and Tunniclffe 2001, 2003).

Along the Mid-Atlantic Ridge, the dominant megafauna are bathymodiolin mussels and rimicarid shrimp, which occur in different microhabitats. Mussels have both chemoautotrophic and methanotrophic symbionts in varying proportions, depending on the vent fluid chemistry (Duperron et al. 2006), and the shrimp *Rimicaris exoculata* has episymbiotic bacteria in its gill chamber that use either sulfide-oxidation or iron-oxidation as an energy source for chemoautotrophy (Schmidt et al. 2008). Accordingly, mussels occur in habitats with lower temperature and hydrothermal fluid concentration than rimicarid shrimp, which form dense swarms around the openings of the high-temperature vents (Desbruyères et al. 2000; Cuvelier et al. 2009). Neither mussels nor shrimp are present at every site along the Mid-Atlantic Ridge, which may be explained by the lack of suitable habitat for one species or the other, or large-scale ecological factors, such as distance or difference in depth (Desbruyères et al. 2000). Water depth can affect phase separation and water-rock reactions, resulting in different fluid chemistry among hydrothermal vents (Von Damm and Lilley 2004). A comparison among the epifaunal communities associated with mussel beds at three sites, "Lucky Strike" (1,700 m, northern MAR), "Snake Pit" (3,480 m, 'southern' MAR), and "Logatchev" (3,050 m, 'southern' MAR) revealed high dissimilarity (<40%) across geographic and bathymetric differences, and relatively low dissimilarity across environmental differences and closely located sites (Gebruk et al. 2000; Van Dover and Doerries 2005).

The distribution of megfauna along environmental gradients is also common at hydrocarbon seeps. The patchiness in seep habitats is due to the deposition of organic material, which is reduced and migrates at different rates, given the permeability in the sediment (Sibuet and Olu 1998). Anaerobic methane oxidation and sulfate reduction can cause further heterogeneity by forming carbonates and areas of high sulfide concentrations (Levin et al. 2001). Like their vent relatives, siboglinid tubeworms and clams at seeps have sulfur-oxidizing chemoautotrophic symbionts and both occur in sulfidic sediments (Fisher 1990; Cordes et al. 2009 for review). Clams are able to burrow in the sediment, but tubeworms require a point of attachment for colonization and settle on exposed carbonate crusts (Bergquist et al. 2002; Olu-le Roy et al. 2004). Seep mussels have chemoautotrophic, methanotrophic or both types of bacterial symbionts (Fisher 1990; Cordes et al. 2009) and are found in areas with higher concentrations of methane (MacDonald et al. 1989; Bergquist et al. 2005, Olu-le Roy et al. 2007). Corals, such as the reef-forming *Lophelia pertusa* have been found in the vicinity of hydrocarbon seeps, but do not appear to derive their nutrition from chemosynthetic primary production (Becker et al. 2009) and host unique epifaunal communities that include background deep-sea fauna, seep endemic fauna, and coral specific associates (Cordes et al. 2008).

13.4.2 Temporal Patterns

Large-scale disturbances associated with seafloor volcanism and tectonism can create or destroy habitats at hydrothermal vents. Following a seafloor eruption at the East Pacific Rise (9°50′N), nascent areas of diffuse hydrothermal flow were colonized by the siboglinid species *Tevnia jerichonana* within 11 months (Shank et al. 1998). Colonization by a larger siboglinid *Riftia pachyptila* followed 20 months later, and within 2.5 years after the eruption had formed dense aggregations that spatially dominated the diffuse flow habitat. Six months later, small mussels (*Bathymodilus thermophilus*) appeared in cracks on the basalt, over the course of the following year, the density of mussels on the basalt increased and mussels were found among *R. pachyptila*. Another seafloor eruption occurred in the same area in 2005–2006, and similar patterns have been observed (Shank et al. 2006). The mechanisms of community succession is not fully understood, but the transition from *T. jerichonana* to *R. pachyptila* to *B. thermophilus* is likely to a combination of pre-settlement factors, such as geochemical changes (Shank et al. 1998) or biogenic cues (Mullineaux et al. 2000; Mullineaux et al. 2003), and post-settlement factors, including physical overgrowth (Hessler et al. 1988) and resource competition (Johnson et al. 1994). Eruptions have also been observed and documented on the Juan de Fuca Ridge, leading to rapid colonization and community assembly within 2 years (Tunnicliffe et al. 1997). However, predictable changes in the community structure have not been observed on annual scales and directional succession has not been described for vent communities on the JdFR (Sarrazin et al. 1997).

Small-scale disturbances on shorter timescales also influence community structure at hydrothermal vents. A mosaic of different faunal assemblages occurs on the surfaces of geologically young, high-temperature vents on the Juan de Fuca Ridge (Sarrazin et al. 1997). These structures are characterized by multiple channels of hydrothermal fluids that vary by fluid intensity and chemical composition. Exposed surfaces, caused by the opening of a fluid channel or a rock collapse, are colonized by the temperature-tolerant alvinellid polychaete *Paralvinella sulfincola* that secretes a mucus tube, which forms a marcasite crust and plugs conduits of fluid flow through the permeable substrate (Juniper et al. 1992). As fluid flow decreases, the next assemblage that develops is dominated by another alvinellid polychaete *P. palmiformis*. Over time, densities of the gastropods *Lepetodrilus fucensis* and *Depressigyra globulus* increase in density, and eventually the tubeworm *Ridgeia piscease* settles and forms dense aggregations (Sarrazin et al. 1997, 1999). The species diversity is greatest in the assemblages associated with the foundation species *R. piscesae* (Sarrazin et al. 1999), and species diversity is positively correlated with the tube surface area of *R. piscesae* aggregations (Govenar et al. 2002). Depending on the degree of reactivation of fluid channels or other small-scale disturbances, the faunal assemblage will revert to earlier stages, as demonstrated by the dynamic succession model for high-temperature vent communities at the JdFR (Sarrazin et al. 1997, 1999).

Environmental changes and ecological succession occur over much longer timescales at hydrocarbon seeps. The best example of the changes in fluid chemistry and community composition overtime is from seep habitats on the Upper Louisiana Slope of the Gulf of Mexico (reviewed in Cordes et al. 2009). Areas of active methane seepage are colonized by the mussel *Bathymodiolus childressi*, and the tubeworms *Lamellibrachia luymesi* and *Seepiophila jonesi* later settle on exposed carbonates, formed by anaerobic oxidation of methane (Bergquist et al. 2003a). Individuals of *B. childressi* may reach ages of 100–150 years old, but intact beds of mussel shells may persist for much longer (Nix et al. 1995). In a 10- to 30-year window during this period, there are high concentrations of methane and sulfide in seep fluids, and high settlement rates of *L. luymesi* and *S. jonsei* juveniles (Bergquist et al. 2002) coupled with low juvenile mortality result in fast population growth rates (Cordes et al. 2003). As aggregations of *L. luymesi* develop, the epifaunal community is dominated by high abundances of a small number of primary consumers that can profit from the physical structure of the foundation species, in areas of chemosynthetic primary production (Cordes et al. 2005). After 30–50 years, sulfide concentrations among the *L. luymesi* aggregations begin to decline, as a result of increasing carbonated precipitation and consumption by growing aggregations of *L. luymesi*. Once the sulfide concentrations are no longer detectable at the sediment-seawater interface, the epfiaunal community in the oldest (>200 years) includes more non-endemics and is dominated in terms biomass by higher-order consumers (Fig. 13.2; Bergquist et al. 2003; Cordes et al. 2005).

13.5 From Patterns to Processes: Local Ecological Processes

Variation in community structure over space and time is driven by a combination of abiotic factors and biological interactions. Although the reliance on chemical energy in chemosynthetic ecosystems underlines the importance of abiotic factors, such as reduced chemical concentrations, biological interactions also play an important role in the regulation of vent and seep communities. One of the important lessons learned from shallow-water marine habitats is that the relative strength or importance of negative or positive species interactions can vary over gradients in abiotic factors (Menge and Sutherland 1976; Bertness et al. 1999). A distinction of chemosynthetic ecosystems is that the relationship between productivity and physiological stress may be positively correlated along environmental gradients (Micheli et al. 2002).

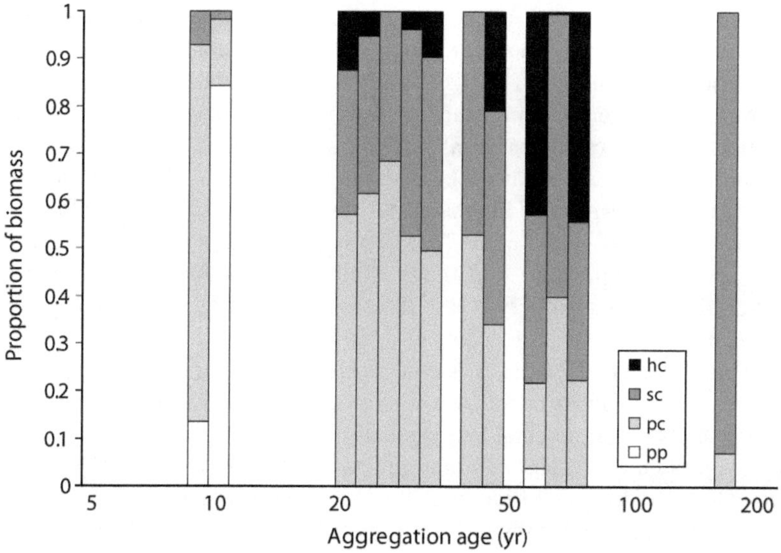

Fig. 13.2 Change over time in proportion of biomass contributed by primary producers (pp), primary consumers (pc), secondary consumers (sc), and higher consumers (hc) in seep communities associated with aggregations of *Lamellibrachia luymesi* (From Cordes et al. 2005)

13.5.1 Abiotic Factors

In chemosynthetic ecosystems, the primary abiotic factors are associated with variable mixing between seawater and either vent or seep fluids. In general, seawater in the deep sea is cold and relatively well oxygenated. In contrast, vent and seep fluids can have high concentrations of reduced chemicals, including sulfide, hydrocarbons, and dissolved metals. All reduced chemicals have essentially the same dual effect on vent and seep communities: as electron donors, reduced chemicals can both support chemosynthetic primary production and reduce local oxygen concentrations. Methane is important for chemosynthesis but is non-toxic. Sulfide and metals, on the other hand, can have further detrimental effects on animal physiology. High metal concentrations and temperatures are characteristic of most vent habitats, but seep fluids are typically not associated with elevated temperatures (Sibuet and Olu 1998). At vents, temperatures and chemical concentrations can vary over orders of magnitude within seconds and centimeters (Johnson et al. 1988; Scheirer et al. 2006). Patterns in temperature variation can also vary over larger spatial and temporal scales, as a result of volcanic and tectonic processes (Scheirer et al. 2006). Although seep fluid concentrations are considered to be more temporally stable, steep environmental gradients that can be found across the width of a single mussel bed demonstrate the high spatial variability in chemical concentrations in seep habitats (Smith et al. 2000; Van Dover et al. 2003).

The correlation between temperature and sulfide concentrations over time (Johnson et al. 1988) led to the question of whether temperature is a good proxy for chemical concentrations (e.g. Le Bris et al. 2006b). The relationship between sulfide concentrations and temperature can be linear within a site, but the sulfide to temperature ratio can vary substantially among sites (Le Bris et al. 2006a; Nees et al. 2008). The variation may be caused by water-rock reactions and biological processes that alters the composition of hydrothermal fluids, beyond a simple dilution of the source end-member (Le Bris et al. 2003, 2006a; Von Damm and Lilley 2004). For example, in high-temperature vent fluids, the formation of FeS reduces the overall concentration of free sulfide H_2S/HS^- (Luther et al. 2001), and in the absence of significant concentrations of iron, low pH shifts the equilibrium of sulfide species to a greater proportion of H_2S (Le Bris et al. 2005). Thus, point measurements of one or a few physicochemical parameters may not fully describe the physicochemical characteristics or environmental heterogeneity in vent or seep habitats.

Arguably, the most important abiotic factor in the distribution of vent and seep communities is hydrogen sulfide, which is a source of energy for chemosynthetic primary production and inhibits aerobic cellular respiration (Vismann 1991; Grieshaber and Volkel 1998). Many vent and seep animals exhibit physiological adaptations to cope with sulfide exposure, including trapping sulfide before diffusion across cell membranes, molecular detoxification or resistance to sulfide, or harboring sulfur-oxidizing symbionts (Vismann 1991; Childress and Fisher 1992). Sulfide can also serve as a chemical cue for colonization (Renninger et al. 1995). Consequently, the zonation pattern in visually dominant organisms at vents and seeps often correlates to gradients in sulfide exposure. At hydrothermal vents on the East Pacific Rise, the siboglind tubeworms *Riftia pachyptila* and *Tevnia jerichonana* occur in diffuse-flow habitats with vigorous mixing between hydrothermal fluids and the ambient seawater, where sulfide concentrations can be as high as several hundred micromolar. In contrast, the clam *Calpyptogena magnifica* lives in cracks where there is no detectable sulfide at their siphon, but low concentrations of sulfide near their foot (Arp et al. 1984; Fisher 1988b). On the Juan de Fuca Ridge, two genetically indistinguishable morphotypes of the tube-worm *Ridgeia piscesae* live in different habitats of high- and low-sulfide exposure (Carney et al. 2002). A "short-fat" morphotype with shorter diffusion distance across the gill epidermis and higher hemoglobin gene expression lives on the surfaces of high-temperature chimneys, where sulfide concentrations around the plume (anterior gas-exchange organ) can be over an order of magnitude higher than at the plume-level of a "long-skinny" morphotype of the same species, which occurs in cracks in the seafloor basalt (Andersen et al. 2006; Carney et al. 2007; Urcuyo et al. 2007). At seeps, the relative abundance of five vesicomyid clams varied with sulfide concentrations in Monterey Bay (Barry et al. 1997), and the distribution of microbial mats, vesicomyid clams, and solemyid clams followed a trend of decreasing sulfide concentrations within a seep at the Hydrate Ridge on the Cascadia Margin (Sahling et al. 2002). The distribution and abundance of individual vent and seep species is also strongly related to sulfide exposure, and the diversity and composition of macrofaunal communities varies with the visually dominant species along environmental sulfide gradients (e.g. Robinson et al. 2004; Matabos et al. 2008; Levin et al. 2009). However, sulfide concentrations do not necessarily explain the variation in macrofaunal communities associated with the same foundation species in different sulfidic conditions, which suggests that other factors contribute to the regulation of biological community structure at vents and seeps (Tsurumi and Tunnicliffe 2001; Govenar et al. 2005; Bergquist et al. 2003b).

Dissolved metals also have direct effects on species in hydrothermal vent communities. Heavy metals can cause oxidative stress and damage in aerobic organisms, but the specific impacts on vent fauna have been difficult to assess (e.g. Colaço et al. 2006; Company et al. 2008). Rather than simply exposure to high metal concentrations, the ratio of dissolved to particulate metal concentrations can serve as an important abiotic factor in determining the distribution of some vent species (Sarradin et al. 2007). Among hydrothermal vent fields on the Mid-Atlantic Ridge, the relative abundance of visually dominant species appears to be correlated with the metal concentrations in the fluids (Desbruyères et al. 2000). The mussels *Bathymodiolus azoricus* and *B. puteoserpentis* dominate vent habitats with lower mineral particle fluxes and metal concentrations (Desbruyères et al. 2000). High metal content in hydrothermal fluids may foul the filtering mechanism of mussels (Desbruyères et al. 2000). Nevertheless, low metal accumulation in the shells (Cravo et al. 2007) and high metal storage in the gills (Cosson et al. 2008) also suggests that mussels are able to tolerate elevated metal concentrations (Kadar et al. 2005a; Company et al. 2006, 2008). On the other hand, the highly mobile shrimp *Rimicaris exoculata* dominates habitats with higher metal contents (Desbruyères et al. 2000). Somewhat contrary to expectation, tissues of *R. exoculata* had lower metal concentrations than the same tissues in another shrimp, *Mirocaris fortunata* that occurs on the surfaces of high-temperature vents, but not exposed to the most vigorous hydrothermal fluid flow (Kadar et al. 2006). The lower metal accumulation in *R. exoculata* may reflect a physiological adaptation to the high metal content in hydrothermal fluids (Kadar et al. 2006). Metals, particularly iron, may also have an important indirect effect on biological communities by binding to sulfide and potentially ameliorating the habitat (Luther et al. 2001). The role of metals as an abiotic factor that influences biological community structure is still unclear and requires further study.

Although elevated temperatures are often coincident with high concentrations of reduced chemicals, temperature itself may play an important role in niche partitioning and species coexistence of closely related hydrothermal vent species. Gastropods at the Juan de Fuca Ridge and the East Pacific Rise appear to occupy distinct microhabitats, based on temperature (Bates et al. 2005; Mills et al. 2007). Similarly, physiological tolerances to temperature determine the habitat ranges of alvinellid polychaetes (Girguis and Lee 2006; Cottin et al. 2008). However, species do not necessarily respond in the same way to temperature fluctuations (Boutet et al. 2009), making it difficult to predict species distribution patterns based solely on temperature measurements. Like sulfide and metal concentrations, temperature can determine the distribution and abundance of single species or groups of closely related species, but the effect of these abiotic factors on the diversity and composition of hydrothermal vent communities may be confounded with other factors or only be observable at larger spatial and temporal scales.

13.5.2 Biological Interactions

In addition to steep environmental gradients and high heterogeneity associated with vent and seep habitats, biological interactions also play an important role in shaping vent and seep communities. A challenge in studying biological interactions in the remote deep-sea habitats of vents and seeps is identifying the limited resources that underlie negative or positive interactions and the consequences on the individual or population growth rates of the species. In the absence of frequent sampling or a sustained presence on the seafloor, manipulative field experiments have been invaluable for examining patterns and developing models of how biological interactions can shape vent and seep communities (Van Dover and Lutz 2004).

Competition for shared resources has been demonstrated among species in low- and high-temperature hydrothermal vent habitats. In the diffuse-flow communities at the Galapagos Rift and the East Pacific Rise, aggregations of the tubeworm *Riftia pachyptila* are commonly overgrown by beds of the mussel *Bathymodiolus thermophilus* (Hessler and Smithey 1983; Hessler et al. 1988). Both *R. pachyptila* and *B. thermophilus* harbor sulfur-oxidizing chemoautotrophic symbionts and must acquire sulfide from hydrothermal fluids (Fisher 1990, for review). Thus, the thick mat created by mussel beds may redirect the flow of hydrothermal fluids emanating from the seafloor, away from the gas-exchange organ or "plume" at the anterior end of the tubeworms. A clearance experiment demonstrated that removing mussels increased the concentrations of sulfide at the plumes of the tubeworms (Johnson et al. 1994). The formation of the bed would then have a negative effect on tubeworms, as well as a positive effect for conspecific mussels within the bed that may have greater access to the laterally dispersed vent fluids.

On the surfaces of high-temperature chimneys, the roles of resource partitioning and competition were examined through stable carbon and nitrogen isotope analyses (Levesque et al. 2003). Two species, *Paralvinella sulfincola* and *P. palmiformis* that occur together in high densities were found to have significantly different diets. However, a shared diet was found for *P. palmiformis* and *P. pandorae*. When the two species occurred together, individuals of *P. pandorae* were smaller than when *P. palmiformis* was absent, which suggests that *P. pandorae* is negatively affected by intra- or interspecific competition (Levesque et al. 2003). A similar approach has been used to examine the mechanisms of species coexistence in hydrothermal vent gastropods. Results of molecular identification of gut contents and stable isotope analysis reveal that three congeneric species may compete for a shared resource, which have a distinctly different diet than another species of a different genus (Govenar and Shank, in preparation). Additional work is needed to both reveal patterns of differences in relative abundance and to identify shared resources, including nutrients and food sources, to uncover other examples of competition that influence the structure of vent and seep communities.

While the intrusion of predators from the background deep-sea are relatively rare, vent endemic predators, such as zoarcid fish, can exert top-down control on vent and seep communities. In a manipulative field experiment on the East Pacific Rise, the effect of predation was strongest in the most vigorously mixing diffuse-flow habitat, where siboglinid tubeworms were the dominant foundation species (Micheli et al. 2002). The removal of predators from caged colonization blocks among siboglind tubeworm aggregations led to an increase in the abundance of gastropod limpets and amphipods. In agreement with the results of the caged block experiment, the gut contents of the zoarcid fish *Thermarces cerebus* contained mostly gastropod limpets and amphipods (Sancho et al. 2005). Increased limpet abundances, in the absence of predation, could contribute to decreased rates of colonization by siboglinid tubeworms and other sessile species (Micheli et al. 2002). Other predators and scavengers that prey on invertebrates at hydrothermal vents include octopods, bythograeid crabs, amphinomid and polynoid polychaetes and turrid snails, but their role in regulating community structure is less clearly defined (Micheli et al. 2002; Voight 2005; Voight and Sigwart 2007). Stable isotope analyses have identified background deep-sea and endemic predators in seep communities (MacAvoy et al. 2002, 2008), but role of predation in shaping seep communities has not been studied directly. Some vagrant deep-sea predators at hydrocarbon seeps, such as the hagfish *Eptatretus* sp., the giant isopod *Bathynomus giganteus* and the spider crab *Rochina crasse* make intrusions into seep habitats but feed mostly on background deep-sea fauna (MacAvoy et al. 2008), while others, including individuals the rat tail fish *Nezumia* sp. and the eels *Synaphobranchus* sp., *Ophicthus cruentifer* and *Dysomma rugos* derive their nutrition from food sources dependent on chemosynthetic primary production (MacAvoy et al. 2002) and may reduce faunal densities in seep communities, but the identity of the prey is not clear.

Positive species interactions are also prevalent in vents and seep communities. The most well-studied examples are invertebrates that harbor internal or external bacterial symbionts. In addition to the symbiont-containing foundation species, other vent and seep species have unique associations with bacteria that can serve as a source of nutrition or as a buffer from reduced chemicals (reviewed in Fisher 1990; Ott et al. 2005; Stewart et al. 2005). Another symbiont that occurs with bathymodiolin mussels at vents and seeps is a commensal polynoid polychaete that resides in the mantle of some individuals, where it presumably benefits from the protection of the mussel shell and the increase in food availability from the psuedofeces of the mussel (Fisher 1994; Chevaldonné et al. 1998). There does not appear to be any relationship between the size of the mussel and the size or the abundance of the commensal polychaetes, and rarely is there any evidence of a negative impact to the mussel (Fisher 1994).

Facilitation has also been shown to play an important role in larval settlement and community assembly at hydrothermal vents on the East Pacific Rise and the Mid-Atlantic Ridge. The tubeworm *Tevnia jerichonana* is the first sessile metazoan to colonize new diffuse-flow vents, created by seafloor eruptions, and is later replaced by the tubeworm *Riftia pachyptila* as the biomass dominant species (Shank et al. 1998). On colonization blocks deployed in different diffuse-flow habitats, *R. pachyptila* only colonized blocks where *T. jerichonana* had settled, which suggests that *T. jerichonana* may provide a biochemical cue that facilitates the colonization of *R. pachyptila* (Mullineaux et al. 2000). In a subsequent experiment, there was no relationship between the settlement of either *T. jerichonana* or *R. pachyptila* with the tubes of either species as colonization substrates, which led to the revised hypothesis that *T. jerichionana* may serve as a biogenic cue in the early stages of community development after an eruption (Hunt et al. 2004). A different experiment found that sessile colonists, such as serpulid polychaetes and foraminifers also facilitate the colonization of sessile colonists in the periphery of hydrothermal vents (Mullineaux et al. 2002). At the Juan de Fuca Ridge, the colonization of four out of five gastropod species appeared to be facilitated by gregarious settlement (Kelly and Metaxas 2007). The role of facilitation on colonization and community organization is likely to be an important factor in isolated habitats of hydrothermal vents on mid-ocean ridges and the patchy distribution of hydrocarbon seeps, embedded in the sedimented deep-sea.

13.6 Habitat Provision and Modification by Foundation Species

13.6.1 Access to Highly Productive Habitats

At vents and seeps, most foundation species have chemosynthetic symbionts and require access to both reduced chemicals and oxygen for carbon-fixation. Although foundation species may contribute the greatest proportion of the biomass in vent and seep communities (Hessler et al. 1988; Bergquist et al. 2003b; Govenar et al. 2005), very few endemic species base their nutrition on the consumption of the tubeworms and mussels (Fisher 1994; Colaço et al. 2002; Bergquist et al. 2007; Levin 2005). Instead, the presence of foundation species may coincide with availability of high productivity habitats, where heterotrophic macrofauna feed on particulate and dissolved organic matter, free-living bacteria, protists, and other small invertebrates.

As both free-living bacterial primary production and growth rates of symbiont-containing foundation species are dependent on chemoautotrophy, it cannot be determined whether increased species diversity with foundation species is due to higher local productivity or larger habitat area provided by the biogenic structure. To tease apart the variables of chemosynthetic primary production and the size of the area provided by foundation species, artificial tubeworm aggregations were constructed from flexible hose in a similar size frequency and density as natural tubeworm aggregations and were deployed along a gradient in hydrothermal fluid flux, representing three ranges of chemosynthetic primary production (Govenar and Fisher 2007; Plate 30). After 1 year, artificial aggregations in the low productivity treatment, ~20 m away from any visible source of hydrothermal fluid flow, had the lowest species density and species diversity. Aggregations in high productivity treatment, embedded in a natural

Table. 13.1 Univariate measures of community structure for artificial aggregations in low-, intermediate- and high-productivity zones and natural aggregations of *Riftia pachyptila*, collected at the same site (Tica) in 2001 and 2002

	Density (individuals m^{-2})	*S*	*J′*	*H′*
Low (n = 3)	14 (8)	10 (4)	0.94 (0.02)	2.09 (0.31)
Intermediate (n = 3)	617 (195)	28 (1)	0.71 (0.05)	2.38 (0.17)
High (n = 3)	942 (654)	25 (6)	0.65 (0.19)	2.09 (0.66)
Natural (n = 4)	4,301 (1,612)	26 (6)	0.51 (0.08)	1.65 (0.31)

Average density, species richness (S), Pielou's evenness (J′), Shannon–Wiener diversity (H′) of macrofaunal species; standard deviation in parentheses (From Govenar and Fisher 2007)

aggregation of tubeworms, had the greatest density. However, aggregations in the high productivity treatment had the same or lower species diversity than aggregations in the intermediate productivity treatment, <0.5 m away from natural tubeworm aggregations, on the bare basalt substrate. Furthermore, species diversity was similar between artificial aggregations, in high and intermediate productivity treatments, to natural aggregations collected at the same site in the previous year (Table 13.1; Govenar and Fisher 2007). The results of this experiment demonstrate the importance of the physical structure provided by foundation species in the regulation of the epifaunal community structure in areas of chemosynthetic primary production.

13.6.2 Physical Structure and Habitat Complexity

Added to the positive effects of increased access to high productivity habitats, epifaunal communities associated with foundation species at vents and seeps benefit from increased area and habitat complexity provided by the physical architecture of the foundation species. In the deep-sea, hard substrate and upright structures are limited resources that are often necessary for colonization and filter-feeding. In addition, the interstitial spaces among intertwined tubes of siboglinid tubeworm aggregations and among bivalve shells in mussel beds can create greater environmental variability and refuge from predation that can increase survivorship, promote species coexistence, and lead to increased species diversity, relative to the unmodified seafloor (Bruno and Bertness 2001).

The relative importance of the habitat provided by the foundation species may also outweigh the effects of the environmental chemistry in regulating the structure of the epifaunal community at vents and seeps. On the East Pacific Rise, epifaunal communities associated with mussels were similar despite the differences in the apparent age of the foundation species and geographic distance among sites (Van Dover 2002, 2003; Dreyer et al. 2005), with exception of epifaunal communities sampled from beds where most (~75%) of the mussels were dead (Van Dover 2002). Furthermore, the structure of the epifaunal communities associated with aggregations of *Riftia pachyptila* aggregations were remarkably similar between low- and high-sulfide sites, and species richness was positively correlated with the tube surface area of the *R. pachyptila* aggregations at both sites (Govenar et al. 2005). On the Juan de Fuca Ridge, epifaunal communities associated with aggregations of *Ridgeia piscease* did not vary significantly with the location, year of collection, temperature of vent fluids, or substratum type, but species richness was correlated with the structural complexity of the *R. piscesae* aggregations (Tsurumi and Tunniclffe 2003). The species diversity was greater in experimental treatments with complex structure than treatments that were a mimic of the seafloor (Kelly and Metaxas 2007). In the Gulf of Mexico, the species richness of epifaunal communities associated with aggregations of the tubeworms *Lamellibrachia luymesi* and *Seepiophila jonesi* tended to increase with the areal coverage and tube surface area of the tubeworm aggregations, and the epifaunal density increased with habitat complexity (tube surface area/areal coverage) (Bergquist et al. 2003).

13.6.3 Modification

In addition to providing access to high productivity habitats, adding area for colonization and increasing environmental heterogeneity, foundation species can also modify physical and chemical characteristics of vent and seep habitats. Foundation species with chemosynthetic symbionts must acquire sulfide and methane and subsequently alter the chemical composition of the surrounding fluids. In seep habitats associated with pockmarks in the North Atlantic and mud volcanoes in the Gulf of Cadiz, small frenulates can alter the distribution and composition of seep fluids (Dando et al. 2008; Sommer et al. 2009), by decreasing the rate of sulfide and methane release. The impact of the habitats created or modified by frenulates on other macrofaunal invertebrates is not well understood and warrants further investigation.

As individuals and aggregations of the siboglind tubeworm *Lamellibrachia luymesi* grow, interstitial sulfide concentrations are expected to similarly decrease. In fact, theoretical models suggest that sulfide reservoirs in seep habitats would not be sufficient to sustain the growth estimates for *L. luymesi* (Bergquist et al. 2000; Cordes et al. 2003, 2005). However, sulfate release from the posterior extension of *L. luymesi* tubes in the sediment (Dattagupta et al. 2006, 2008), stimulates sulfide production by sulfate-reducing bacteria and can allow tubeworm aggregations to persist for centuries (Bergquist et al. 2000, Cordes et al. 2003). As tubeworm aggregations age, less sulfide is present above the sediment-seawater interface and the composition of the epifaunal community changes (Bergquist et al. 2003a; Cordes et al. 2005).

13.6.4 Effect of Foundation Species on Regional Species Diversity

Differences in the epifaunal communities associated with biogenic islands, created by foundation species, may also contribute to higher regional species diversity at vents and seeps (Govenar and Fisher 2007; Cordes et al. 2008). Among artificial tubeworm aggregations deployed in a diffuse-flow vent habitat at the East Pacific Rise, there were significant differences in the species assemblages that provided insights into the relationship between foundation species and the associated epifaunal communities (Govenar and Fisher 2007). Although tubeworms have been shown to colonize plastic tubing (Hunt et al. 2004), no tubeworms were found on the surfaces of the artificial aggregations. In addition, numerically dominant species found in natural tubeworm aggregations were either absent or had lower relative abundance artificial aggregations. In contrast, mussels settled on artificial aggregations in the high and intermediate productivity treatments, and numerically dominant species in mussel beds at a nearby site had greater relative abundance in artificial than natural tubeworm aggregations. The similarity in the composition of the faunal assemblages fell between the epifaunal communities associated with either natural aggregations of tubeworms or mussels (Fig. 13.3). These patterns highlight the importance of the identity or functional traits of the foundation species (e.g. McGill et al. 2006). The tendency for some species to occur either with tubeworms or with mussels suggests that foundation species may not be interchangeable within vent communities (Machicote et al. 2004; Berkenbusch and Rowden 2007).

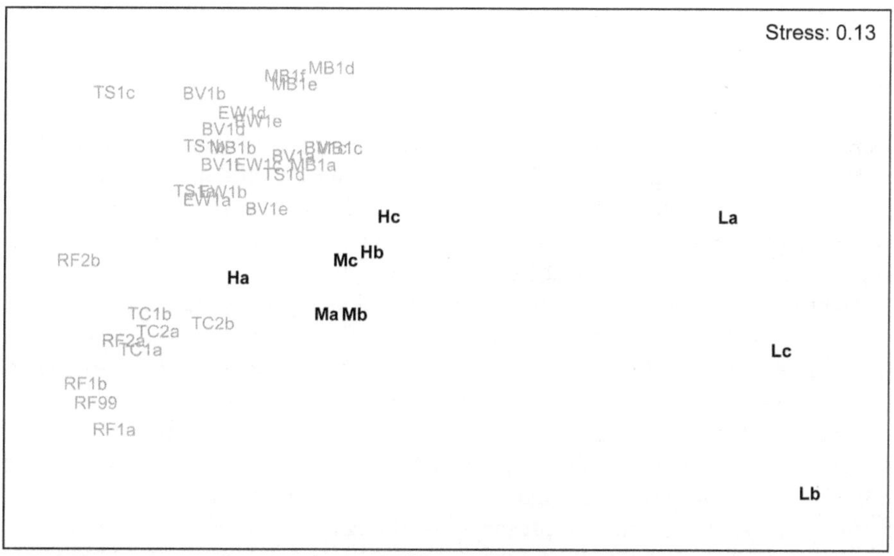

Fig. 13.3 Non-metric dimensional scaling plot of the Bray–Curtis similarity values (based on presence-absence data) between artificial aggregations in low (L), intermediate (M), and high (H) productivity zones, and natural aggregations of *Riftia pachyptila* from Riftia Field (RF) and Tica (TC) and mussel beds from Biovent (BV), Mussel Bed (MB), East Wall (EW), and Train Station (TS), on the East Pacific Rise (9°50′N). The number corresponds to the year 2001 (1) or 2002 (2), and the lowercase letter represents a unique replicate (Data from Dreyer et al. 2005; Govenar et al. 2004, 2005; Govenar and Fisher 2007)

Source–sink dynamics between communities associated with foundation species and species turnover associated with the succession from tubeworms dominance to mussel dominance may increase species diversity at the regional scale of the East Pacific Rise (EPR), relative to other mid-ocean ridges. High regional species diversity at the EPR has been attributed to large-scale ecological factors such as older geologic age, fast spreading rate, high to intermediate disturbance frequency and intensity, and large areas of diffuse hydrothermal flow (Tunnicliffe 1988; Tunnicliffe et al. 1996; Juniper and Tunnicliffe 1997; Van Dover et al. 2002; Van Dover 2003). Thus, habitat provision or modification at local scales by different foundation species may also contribute to the high regional species diversity at the EPR. At the global scale, the EPR and associated ridges is the only mid-ocean ridge system where both tubeworm aggregations and mussel beds are present. This may result in higher regional species diversity relative to the Juan de Fuca Ridge, which does not have dense mussel beds, and the Mid-Atlantic Ridge, where tubeworms have never been reported (Van Dover et al. 2002).

13.7 Conclusion

On the backdrop of historical events that lead to global biogeographic patterns, local ecological processes affect the composition of hydrothermal vent and hydrocarbon seep communities at local and regional scales. Plate tectonics and other vicariance events open and close pathways to dispersal and contribute to evolutionary processes that determine the composition of species in different localities. Geologic processes, including volcanism, tectonism, organic deposition, and sedimentation, regulate the flow and composition of vent and seep fluids that determine the distribution of species at large spatial and temporal scales. Abiotic factors and biological interactions further refine the structure of vent and seep communities and the maintenance of local and regional species diversity.

Habitat provision and modification by foundation species can facilitate species coexistence and increase species diversity relative to the surrounding seafloor. At vents and seeps, most foundation species have sulfur-oxidizing or methanotrophic bacterial symbionts and require access to both reduced chemicals and oxygen. Thus, epifaunal communities associated with foundation species are exposed to potentially harsh environmental conditions. However, the positive effects of habitat provision may outweigh the relative importance of the environmental chemistry. The presence of foundation species can increase rates of colonization or survivorship and local species diversity in epifaunal communities through the various ecological mechanisms, including (1) providing access to high productivity habitats, (2) adding area available for colonization, increasing environmental variability, and providing refuge from predation, and (3) ameliorating habitat conditions. Furthermore, habitats created or modified by foundation species form biogenic islands that can increase migration and dispersal and increase regional species diversity.

References

Aitken CM, Jones DM, Larter SR (2004) Anaerobic hydrocarbon biodegradation in deep subsurface oil reservoirs. Nature 431:291–294

Altieri AH, Silliman BR, Bertness MD (2007) Hierarchical organization via facilitation cascade in intertidal cordgrass bed communities. Am Nat 160(2):195–206

Andersen AC, Flores JF, Hourdez S (2006) Comparative branchial plume biometry between two extreme ecotypes of the hydrothermal vent tubeworm *Ridgeia piscesae*. Can J Zool 84(12):1810–1822

Arp AJ, Childress JJ, Fisher CR (1984) Metabolic and blood-gas transport characteristics of the hydrothermal vent bivalve *Calyptogena magnifica*. Physiol Zool 57(6):648–662

Bachraty C, Legendre P, Desbruyères D (2009) Biogeographic relationships among deep-sea hydrothermal vent faunas at global scale. Deep Sea Res I 56(8):1371–1378

Barry JP, Kochevar RE, Baxter CH (1997) The influence of pore-water chemistry and physiology on the distribution of vesicomyid clams at cold seeps in Monterey Bay: Implications for patterns of chemosynthetic community organization. Limnol Oceanogr 42(2):318–328

Bates AE, Tunnicliffe V, Lee RW (2005) Role of thermal conditions in habitat selection by hydrothermal vent gastropods. Mar Ecol Prog Ser 305:1–15

Becker EL, Cordes EE, Macko SA, Fisher CR (2009) Importance of seep primary production to *Lophelia pertusa* and associated fauna in the Gulf of Mexico. Deep Sea Res I 56(5):786–800

Bergquist DC, Williams FM, Fisher CR (2000) Longevity record for deep-sea invertebrate – the growth rate of a marine tubeworm is tailored to different environments. Nature 403:4999–5000

Bergquist DC, Urcuyo IA, Fisher CR (2002) Establishment and persistence of seep vestimentiferan aggregations on the upper Louisiana slope of the Gulf of Mexico. Mar Ecol Prog Ser 241:89–98

Bergquist DC, Andras JP, McNelis T, Howlett S, van Horn MJ, Fisher CR (2003a) Succession in Gulf of Mexico cold seep vestimentiferan aggregations: The importance of spatial variability. Mar Ecol 24(1):31–44

Bergquist DC, Ward T, Cordes EE, NcNelis T, Howlett S, Kosoff R, Hourdez S, Carney R, Fisher CR (2003b) Community structure of vestimentiferan-generated habitat islands from Gulf of Mexico cold seeps. J Exp Mar Biol Ecol 289(2):197–222

Bergquist DC, Fleckenstein C, Knisel J, Begley B, MacDonald IR, Fisher CR (2005) Variations in seep mussel bed communities along physical and chemical environmental gradients. Mar Ecol Prog Ser 293:99–108

Bergquist DC, Eckner JT, Urcuyo IA, Cordes EE, Hourdez S, Macko SA, Fisher CR (2007) Using stable isotopes and quantitative community characteristics to determine a local hydrothermal vent food web. Mar Ecol Prog Ser 330:49–65

Berkenbusch K, Rowden AA (2007) An examination of the spatial and temporal generality of the influence of ecosystem engineers on the composition of associated assemblages. Aquat Ecol 41:129–147

Bertness MD, Callaway R (1994) Positive interactions in communities. Trends Ecol Evol 9(5):191–193

Bertness MD, Leonard GH (1997) The role of positive interactions in communities: lessons from intertidal habitats. Ecology 78:1976–1989

Bertness MD, Leonard GH, Levine JM, Schmidt PR, Ingraham AO (1999) Testing the relative contribution of positive and negative interactions in rocky intertidal communities. Ecology 80(8):2711–2726

Boetius A, Kavenschlag K, Schubert CJ, Rickert D, Widdel F, Gleseke A, Amann R, Jorgensen BB, Witte U, Pfannkuche O (2000) A marine microbial consortium apparently mediating anaerobic oxidation of methane. Nature 407:623–626

Boutet I, Jollivet D, Shillito B, Moraga D, Tanguy A (2009) Molecular identification of differentially regulated genes in the hydrothermal-vent species *Bathymodiolus thermophilus* and *Paralvinella pandorae* in response to temperature. BMC Genomics 10:222–239

Bruno JF (2000) Facilitation of cobble beach plant communities through habitat modification by *Spartina alterniflora*. Ecology 81(5):1179–1192
Bruno JF, Bertness MD (2001) Habitat modification and facilitation in benthic marine communities. In: Bertness MD, Gaines SD, Hay ME (eds) Marine community ecology. Sinauer, Sunderland, MA
Bustamante RH, Branch GM, Eekhout S, Robertson B, Zoutendyk P, Schleyer M, Dye A, Hanekom N, Keats D, Jurd M, McQuaid C (1995) Gradients of intertidal primary productivity around the coast of South Africa and their relationships with consumer biomass. Oecologia 102:189–201
Carney RS (1994) Consideration of the oasis analogy for chemosynthetic communities at Gulf of Mexico hydrocarbon vents. Geo Mar Lett 14:149–159
Carney SL, Peoples JR, Fisher CR, Schaeffer SW (2002) AFLP analyses of genomic DNA reveal no differentiation between two phenotypes of the vestimentiferan tubeworm *Ridgeia piscesae*. Cah Biol Mar 43(3–4):363–366
Carney SL, Flores JF, Orobona KM, Butterfield DA, Fisher CR, Schaeffer SW (2007) Environmental differences in hemoglobin gene expression in the hydrothermal vent tubeworm, *Ridgeia piscesae*. Comp Biochem Phys B 146(3):326–337
Chevaldonné P, Jollivet D, Feldman RA, Desbruyères D, Lutz RA, Vrijenhoek RC (1998) Commensal scale-worms of the genus *Branchipolynoe* (Polychaeta: Polynoidae) at deep-sea hydrothermal vents and cold seeps. Cah Biol Mar 39(3–4):347–350
Childress JJ, Fisher CR (1992) The biology of hydrothermal vent animals: physiology, biochemistry and autotrophic symbioses. Oceanogr Mar Biol Annu Rev 30:337–441
Colaço A, Deshairs F, Desbruyères D (2002) Nutritional relations of deep-sea hydrothermal fields at the Mid-Atlantic Ridge: a stable isotope approach. Deep Sea Res I 49(2):395–412
Colaço A, Bustamente P, Fouquet Y, Sarradin PM, Serrao-Santos R (2006) Bioaccumulation of Hg, Cu, Zn, in the Azores triple junction hydrothermal vent fields food web. Chemosphere 65:2260–2267
Company R, Serafim A, Cosson R, Camus L, Shillito B, Fiala-Médioni A, Bebianno MJ (2006) The effect of cadmium on antioxidant responses and the susceptibility to oxidative stress in the hydrothermal vent mussel *Bathymodiolus azoricus*. Mar Biol 148(4):817–825
Company R, Serafim A, Cosson RP, Fiala-Médioni A, Camus L, Colaço A, Serrao-Santos R, Benianno MJ (2008) Antioxidant biochemical responses to long-term copper exposure in *Bathymodiolus azoricus* from Menez-Gwen hydrothermal vent. Sci Tot Environ 289:407–417
Cordes EE, Bergquist DC, Shea K, Fisher CR (2003) Hydrogen sulphide demand of long-lived vestimentiferan tube worm aggregations modifies the chemical environment at deep-sea hydrocarbon seeps. Ecol Lett 6(3):212–219
Cordes EE, Hourdez S, Predmore BL, Redding ML, Fisher CR (2005) Succession of hydrocarbon seep communities associated with the long-lived foundation species *Lamellibrachia luymesi*. Mar Ecol Prog Ser 305:17–29
Cordes EE, Carney SL, Hourdez S, Carney RS, Brooks JM, Fisher CR (2007) Cold seeps of the deep Gulf of Mexico: community structure and biogeographic comparisons to Atlantic equatorial belt seep communities. Deep Sea Res I 54:637–653
Cordes EE, McGinley MP, Podowski EL, Becker EL, Lessard-Pilon S, Viada ST, Fisher CR (2008) Coral communities of the deep Gulf of Mexico. Deep Sea Res I 55:777–787
Cordes EE, Bergquist DC, Fisher CR (2009) Macro-ecology of Gulf of Mexico cold seeps. Annu Rev Mar Sci 1:143–168
Cosson RP, Thiébaut E, Company R, Castrec-Rouelle M, Colaço A, Martins I, Sarradin P-M, Bebianno MJ (2008) Spatial variation of metal bioaccumulation in the hydrothermal vent mussel *Bathymodiolus azoricus*. Mar Environ Res 65:405–415
Cottin D, Ravaux J, Léger N, Halary S, Toullec J-Y, Sarradin P-M, Gaill F, Shillito B (2008) Thermal biology of the deep-sea vent annelid *Paralvinella grasslei*: in vivo studies. J Exp Biol 211:2196–2204
Crain CM, Bertness MD (2006) Ecosystem engineering across environmental gradients: implications for conservation and management. Bioscience 56(3):211–218
Cravo A, Foster P, Almeida C, Company R, Cosson RP, Bebianno MJ (2007) Metals in the shell of *Bathymodiolus azoricus* from a hydrothermal vent site on the Mid-Atlantic Ridge. Environ Intl 33:609–615

Crooks JA (1998) Habitat alteration and community-level effects of an exotic mussel, *Musculista senhousia*. Mar Ecol Prog Ser 162:137–152

Cuvelier D, Sarrazin J, Colaço A, Copley J, Desbruyères D, Glover AG, Tyler P, Santos RS (2009) Distribution and spatial variation of hydrothermal faunal assemblages at Lucky Strike (Mid-Atlantic Ridge) revealed by high-resolution video image analysis. Deep Sea Res I 56:2026–2040

Dando PR, Southward AJ, Southward EC, Lamont P, Harvey R (2008) Interactions between sediment chemistry and frenulate pogonophores (Annelida) in the north-east Atlantic. Deep Sea Res I 55:966–996

Dattagupta S, Miles LL, Barnabei MS, Fisher CR (2006) The hydrocarbon seep tubeworm *Lamellibrachia luymesi* primarily eliminates sulfate and hydrogen ions across its roots to conserve energy and ensure sulfide supply. J Exp Biol 209(19):3795–3805

Dattagupta S, Arthur MA, Fisher CR (2008) Modification of sediment geochemistry by the hydrocarbon seep tubeworm *Lamellibrachia luymesi*: A combined empirical and modeling approach. Geochem Cosmochem Acta 72(9):2298–2315

Desbruyères D, Chevaldonné P, Alayse A-M, Jollivet D, Lallier FH, Jouin-Toulmond C, Zal F, Sarradin P-M, Cosson R, Caprais J-C, Arndt C, O'Brien J, Guezennec J, Hourdez S, Riso R, Gaill F, Laubier L, Toulmond A (1998) Biology and ecology of the "Pompeii worm" (*Alvinella pompejana* Desbruyères and Laubier) a normal dweller of an extreme deep-sea environment: A synthesis of current knowledge and recent developments. Deep Sea Res II 45:383–422

Desbruyères D, Almeida A, Biscoito M, Comtet T, Khripounoff A, Le Bris N, Sarradin P-M, Segonzac M (2000) A review of the distribution of hydrothermal vent communities along the northern Mid-Atlantic Ridge: dispersal vs. environmental controls. Hydrobiologia 440:201–216

Dreyer JC, Knick KE, Flickinger WB, Van Dover CL (2005) Development of macrofaunal community structure in mussel beds on the northern East Pacific Rise. Mar Ecol Prog Ser 302:121–134

Dubilier N, Bergin C, Lott C (2008) Symbiotic diversity in marine animals: the art of harnessing chemosynthesis. Nat Rev Microbiol 6:725–740

Duperron S, Bergin C, Zielinski F, Blazejak A, Pernthaler A, McKiness ZP, DeChaine E, Cavanaugh CM, Dublier N (2006) A dual symbiosis shared by two mussel species, *Bathymodiolus azoricus* and *Bathymodiolus puteoserpentis* (Bivalvia: Mytilidae), from hydrothermal vents along the northern Mid-Atlantic Ridge. Environ Microbiol 8(8):1441–1447

Ergorov AV, Crane K, Vogt PR, Rozhkov AN (1999) Gas hydrates that outcrop the sea floor: stability models. Geo Mar Lett 19:89–96

Fisher CR (1990) Chemoautotrophic and methanotrophic symbioses in marine invertebrates. Rev Aquat Sci 2:399–436

Fisher CR, Childress JJ, Arp JJ, Brooks JM, Distel D, Favuzzi JA, Felbeck H, Hessler RR, Johnson KS, Kennicutt MC, Macko SA, Newton A, Powell MA, Somero GN, Soto T (1988a) Microhabitat variation in the hydrothermal vent mussel *Bathymodiolus thermophilus*, at Rose Garden vent on the Galapagos rift. Deep Sea Res A 35:1769–1792

Fisher CR, Childress JJ, Arp JJ, Brooks JM, Distel D, Dugan JA, Felbeck H, Fritz LW, Hessler RR, Johnson CR, Kennicut MC II, Lutz RA, Macko SA, Newton A, Powell MA, Somero GN, Soto T (1988b) Variation in the hydrothermal vent clam, *Calyptogena magnifica*, at the Rose Garden vent on the Galapagos spreading center. Deep Sea Res A 35:1811–1831

Fisher CR, Childress JJ, Macko SA, Brooks JM (1994) Nutritional interactions in Galapagos Rift hydrothermal vent communities: inferences from stable carbon and nitrogen isotope analyses. Mar Ecol Prog Ser 103:45–55

Fornari DJ, Shank TM, Von Damm KL, Gregg TKP, Lilley M, Levai G, Bray A, Haymon RM, Perfit MR, Lutz RA (1998) Time-series temperature measurements at high-temperature hydrothermal vents, East Pacific Rise (9°49-51'N): evidence for monitoring a crustal cracking event. Earth Planet Sci Lett 160(3–4):419–431

Gebruk AV, Chevaldonné P, Shank TM, Lutz RA, Vrijenhoek RC (2000) Deep-sea hydrothermal vent communities of the Logatchev area (14°45'N, Mid-Atlantic Ridge): diverse biotopes and high biomass. J Mar Biol Assoc UK 80(3):383–393

Girguis PR, Lee RW (2006) Thermal preference and tolerance of alvinellids. Science 312:231

Govenar B, Fisher CR (2007) Experimental evidence of habitat provision by aggregations of *Riftia pachyptila* at hydrothermal vents on the East Pacific Rise. Mar Ecol 28:3–14

Govenar B, Bergquist DC, Urcuyo IA, Eckner JT, Fisher CR (2002) Three *Ridgeia piscesae* assemblages from a single Juan de Fuca Ridge sulphide edifice: structurally different and functionally similar. Cah Mar Biol 43(3–4):247–252

Govenar B, Freeman M, Bergquist DC, Johnson GA, Fisher CR (2004) Composition of a one-year-old *Riftia pachyptila* community following a clearance experiment: Insight to succession patterns at deep-sea hydrothermal vents. Biol Bull 207(3):177–182

Govenar B, Le Bris N, Gollner S, Glanville J, Aperghis AB, Hourdez S, Fisher CR (2005) Epifaunal community structure associated with *Riftia pachyptila* in chemically different hydrothermal vent habitats. Mar Ecol Prog Ser 305:67–77

Grassle JF, Maciolek NJ (1992) Deep-sea species richness-regional and local diversity estimates from quantitative bottom samples. Am Nat 139(2):313–341

Grieshaber MK, Volkel S (1998) Animal adaptations for tolerance and exploitation of poisonous sulfide. Annu Rev Physiol 60:33–53

Hacker SD, Gaines SD (1997) Some implications of direct positive interactions for community diversity. Ecology 78(7):1990–2003

Hanski I (1998) Metapopulation dynamics. Nature 396:41–49

Haymon RM, Fornari DJ, Edwards MH, Carbotte S, Wright D, Macdonald KC (1991) Hydrothermal vent distribution along the East Pacific Rise crest (9°09'-54'N) and its relationship to magmatic and tectonic processes on fast-spreading mid-ocean ridges. Earth Planet Sci Lett 104:513–534

Haymon RM, Fornari DJ, Von Damm KL, Lilley MD, Perfit MR, Edmond JM, Shanks WC, Lutz RA, Grebmeier JM, Carbotte S, Wright D, McLaughlin E, Smith M, Beedle N, Olson E (1993) Volcanic eruption of the midocean ridge along the East Pacific Rise crest at 9°45-52'N: Direct submersible observations of sea-floor phenomena associated with an eruption event in April, 1991. Earth Planet Sci Lett 119(1–2):85–101

Hessler RR, Smithey WM Jr (1983) The distribution and community structure of megafauna at the Galapagos Rift hydrothermal vents. In: Rona PA, Bostrom K, Laubier L, Smith KL Jr (eds) Hydrothermal processes at seafloor spreading centers. Plenum, New York

Hessler RR, Smithey WM, Boudrias MA, Keller CH, Lutz RA, Childress JJ (1988) Temporal change in megafauna at the Rose Garden hydrothermal vent (Galapagos Rift; eastern tropical Pacific). Deep Sea Res A 35:1681–1710

Hunt HL, Metaxas A, Jennings RM, Halanych KM, Mullineaux LS (2004) Testing biological control of colonization by vestimentiferan tubeworms at deep-sea hydrothermal vents (East Pacific Rise, 9°50'N). Deep Sea Res I 51(2):225–234

Jannasch HW, Wirsen CO (1979) Chemosynthetic primary production at East Pacific Rise seafloor spreading centers. Biosci 29(10):592–598

Johnson KS, Childress JJ, Beehler CL (1988) Short-term temperature variability in the Rose Garden hydrothermal vent field: an unstable deep-sea environment. Deep Sea Res A 35: 1711–1721

Johnson KS, Childress JJ, Beehler CL, Sakamoto CM (1994) Biogeochemistry of hydrothermal vent mussel communities: the deep-sea analogue to the intertidal zone. Deep Sea Res I 41:993–1011

Jones CG, Lawton JH, Shachak M (1994) Organisms as ecosystem engineers. Oikos 69:373–386

Jones CG, Lawton JH, Shachak M (1997) Positive and negative effects of organisms as physical ecosystem engineers. Ecology 78(7):1946–1957

Julian D, Gaill F, Wood E, Arp AJ, Fisher CR (1999) Roots as a site of hydrogen sulphide uptake in the hydrocarbon seep vestimentiferan *Lamellibrachia* sp. J Exp Biol 202:2245–2257

Juniper SK, Tunnicliffe V (1997) Crustal accretion and the hot vent ecosystem. Phil Soc Roy Soc Lond A 355:459–474

Juniper SK, Johnasson IR, Tunnicliffe V, Southward AJ (1992) Influence of a tube-building polychaete on hydrothermal chimney mineralization. Geology 20(10):895–898

Kadar E, Costa V, Martins I, Serrao-Santos R, Powell JJ (2005a) Enrichment in trace metals (Al, Mn, Co, Cu, Mo, Cd, Fe, Zn, Pb, and Hg) of macro-invertebrate habitats at hydrothermal vents along the Mid-Atlantic Ridge. Hydrobiologia 548:191–205

Kadar E, Costa V, Segonzac M (2005b) Trophic influences of metal accumulation in natural pollution laboratories at deep-sea hydrothermal vents of the Mid-Atlantic Ridge. Sci Tot Environ 373:464–472

Kadar E, Costa V, Santos RS (2006) Distribution of micro-essential (Fe, Cu, Zn) and toxic (Hg) metals in tissues of two nutritionally distinct hydrothermal shrimps. Sci Tot Environ 358:143–150

Karl DM (1995) Ecology of free-living, hydrothermal vent microbial communities. In: Karl DM (ed) The microbiology of deep-sea hydrothermal vents. CRC Press, Boca Raton, FL

Kelly N, Metaxas A (2007) Diversity of invertebrate colonists on simple and complex substrates at hydrothermal vents on the Juan de Fuca Ridge. Aquat Biol 3:271–281

Le Bris N, Sarradin PM, Caprias JC (2003) Contrasted sulphide chemistries in the environment of 13° N EPR vent fauna. Deep Sea Res I 50:737–747

Le Bris N, Zbinden M, Gaill F (2005) Processes controlling the physico-chemcial micro-environments associated with Pompeii worms. Deep-Sea Res I 52:1071–1083

Le Bris N, Govenar B, Le Gall C, Fisher CR (2006a) Variability of physico-chemical conditions in 9°50'N EPR diffuse flow vent habitats. Mar Chem 98:167–182

Le Bris N, Rodier P, Sarradin P-M, Le Gall C (2006b) Is temperature a good proxy for sulfide in hydrothermal vent habitats? Cah Biol Mar 47(4):465–470

Levesque C, Juniper SK, Marcus J (2003) Food resource partitioning and competition among alvinellid polychaetes of Juan de Fuca Ridge hydrothermal vents. Mar Ecol Prog Ser 246:173–182

Levin LA (2005) Ecology of cold seep sediments: interactions of fauna with flow, chemistry, and microbes. Oceanogr Mar Biol 43:1–46

Levin LA, Mendoza GF (2007) Community structure and nutrition of deep methane-seep macrobenthos from the North Pacific (Aleutian) Margin and the Gulf of Mexico (Florida Escarpment). Mar Ecol 28:131–151

Levin LA, James DW, Martin CM, Rathburn AE, Harris LH, Michener RH (2000) Do methane seeps support distinct macrofaunal assemblages? Observations on community structure and nutrition from the northern California slope and shelf. Mar Ecol Prog Ser 208:21–39

Levin LA, Etter RJ, Rex MA, Gooday AJ, Smith CR, Pineda J, Stuart CT, Hessler RR, Pawson D (2001) Environmental influences on regional deep-sea species diversity. Annu Rev Ecol Syst 32:51–93

Levin LA, Mendoza GF, Konotchick T, Lee R (2009) Macrobenthos community structure and trophic relationships within active and inactive Pacific hydrothermal sediments. Deep Sea Res II 56:1632–1648

Luther GW III, Rozan TF, Talliefert M, Nuzzio DB, Di Meo CA, Shank TM, Lutz RA, Cary SC (2001) Chemical speciation drives hydrothermal vent ecology. Nature 410:813–816

MacAvoy SE, Carney RS, Fisher CR, Macko SA (2002) Use of chemosynthetic biomass by large, mobile, benthic predators in the Gulf of Mexico. Mar Ecol Prog Ser 225:65–78

MacAvoy SE, Morgan E, Carney RS, Macko SA (2008) Chemoautotrophic production incorporated by heterotrophs in Gulf of Mexico hydrocarbon seeps: An examination of mobile benthic predators and seep residents. J Shell Res 27(1):153–161

MacDonald IR, Boland GS, Baker JS, Brooks JM, Kennicutt MC, Bidigare RR (1989) Gulf of Mexico hydrocarbon seep communities, 2: Spatial distribution of seep organisms and hydrocarbons at Bush Hill. Mar Biol 101(2):235–247

Machicote M, Branch LC, Villarreal D (2004) Burrowing owls and burrowing mammals: are ecosystem engineers interchangeable as facilitators? Oikos 106:527–535

Matabos M, Le Bris N, Pendlebury S, Thiébaut E (2008) Role of physico-chemical environment on gastropod assemblages at hydrothermal vents on the East Pacific Rise (13°N/ EPR). J Mar Biol Assoc UK 88(5):995–1008

McArthur RA, Wilson EO (1967) The theory of island biogeography. Princeton University Press, Princeton, NJ

McCollom TM, Shock EL (1997) Geochemical constraints on chemolithoautotrophic metabolism by microorganisms in seafloor hydrothermal systems. Geochem Cosmochim Acta 61:4375–4391

McGill BJ, Enquist BJ, Weiher E, Westoby M (2006) Rebuilding community ecology from functional traits. Trends Ecol Evol 21(4):178–185

McKinness ZP, McMullin ER, Fisher CR, Cavanaugh CM (2005) A new bathymodioline mussel symbiosis at the Juan de Fuca hydrothermal vents. Mar Biol 348:109–116

McMullin ER, Hourdez S, Schaeffer SW, Fisher CR (2003) Phylogeny and biogeography of deep sea vestimentiferan tubeworms and their bacterial symbionts. Symbiosis 34(1):1–41

Menge BA, Sutherland JP (1976) Species diversity gradients, synthesis of the roles of predation, competition, and temporal heterogeneity. Am Nat 110:351–369

Micheli F, Peterson CH, Mullineaux LS, Fisher CR, Mills SW, Sancho G, Johnson GA, Lenihan HS (2002) Predation structures communities at deep-sea hydrothermal vents. Ecol Mongr 72:365–382

Mills LS, Soulé ME, Doak DF (1993) The keystone-species concept in ecology and conversation. Bioscience 43(4):219–224

Mills SW, Mullineaux LS, Tyler PA (2007) Habitat associations in gastropod species at East Pacific Rise hydrothermal vents. Biol Bull 212:185–194

Mullineaux LS, Fisher CR, Peterson CH, Schaeffer SW (2000) Tubeworm succession at hydrothermal vents: use of biogenic cues to reduce habitat selection error? Oecologia 123:275–284

Mullineaux LS, Peterson CH, Micheli F, Mills SW (2003) Successional mechanism varies along a gradient in hydrothermal fluid flux at deep-sea vents. Ecol Mongr 73:523–542

Nees HA, Moore TS, Mullaugh KM, Holyoke RR, Janzen CP, Ma S, Metzger E, Waite TJ, Yucel M, Lutz RA, Shank TM, Vetriani C, Nuzzio DB, Luther GW III (2008) Hydrothermal vent mussel habitat chemistry, pre- and post-eruption at 9°50'N on the East Pacific Rise. J Shell Res 27(1):169–175

Nix ER, Fisher CR, Vodenichar J, Scott KM (1995) Physiological ecology of a mussel with methanotrophic endosymbionts a three hydrocarbon seep sites in the Gulf of Mexico. Mar Biol 122(4):605–617

Olu-le Roy K, Sibuet M, Fiala-Médioni A, Gofas S, Salas C, Mariotti A, Foucher J-P, Woodside J (2004) Cold seep communities in the deep eastern Mediterranean Sea: composition, symbiosis and spatial distribution on mud volcanoes. Deep Sea Res I 51:1915–1936

Olu-le Roy K, Caprais J-C, Fifis A, Fabri M-C, Galéron J, Budzinsky H, Le Ménach K, Khripounoff A, Ondréas H, Sibuet M (2007) Cold-seep assemblages on a giant pockmark off West Africa: spatial patterns and environmental control. Mar Ecol 28:115–130

Ott J, Bright M, Bulgheresi S (2005) Marine microbial thiotrophic ectosymbioses. Ocean Mar Biol 42:95–118

Paine RT (1974) Intertidal community structure: experimental studies on the relationship between a dominant competitor and its principle competitor. Oecologia 15:93–120

Power ME, Tilman D, Estes JA, Menge BA, Bond WJ, Mills LS, Daily G, Castilla JC, Lubchenco J, Paine RT (1996) Challenges in the quest for keystones. Bioscience 46(8):609–620

Renninger GH, Kass L, Gleeson RA, Van Dover CL, Battelle B-A, Jinks RN, Herzog ED, Chamberlain SC (1995) Sulfide as a chemical stimulus for deep-sea hydrothermal vent shrimp. Biol Bull 189:69–76

Rex MA, Stuart CT, Hessler RR, Allen JA, Sanders HL, Wilson GDF (1993) Global-scale latitudinal patterns of species diversity in the deep-sea benthos. Nature 365:636–639

Ricklefs RE (1987) Community diversity: Relative roles of local and regional processes. Science 235:167–171

Robigou V, Delaney JR, Stakes DS (1993) Large massive sulphide deposits in a newly discovered active hydrothermal system, the High Rise Field, Endeavour Segment, Juan de Fuca Ridge. Geophys Res Lett 20(17):1887–1890

Robinson CA, Bernhard JM, Levin LA, Mendoza GF, Blanks JK (2004) Surficial hydrocarbon seep infauna from the Blake Ridge (Atlantic Ocean, 2150 m) and the Gulf of Mexico (690–2240 m). PSZN Mar Ecol 25(4):313–336

Rona PA, Thompson G, Mottl MJ, Karson JA, Jenkins WJ, Graham D, Mallette M, Von Damm K, Edmond JM (1984) Hydrothermal activity at the Trans-Atlantic Geotraverse Hydrothermal Field, Mid-Atlantic Ridge crest at 26°N. J Geophys Res 89:1365–1377

Sahling H, Rickert D, Lee RW, Linke P, Suess E (2002) Macrofaunal community structure and sulfide flux at gas hydrate deposits from the Cascadia convergent margin, NE Pacific. Mar Ecol Prog Ser 231:121–138

Sancho G, Fisher CR, Mills SW, Micheli F, Johnson GA, Lenihan HS, Peterson CH, Mullineaux LS (2005) Selective predation by the zoarcid fish *Thermarces cerberus* at hydrothermal vents. Deep Sea Res I 52:837–844

Sarradin P-M, Lannuzel D, Waeles M, Crassous P, Le Bris N, Caprais JC, Fouqet Y, Fabri MC, Riso R (2007) Dissolved and particulate metals (Fe, Zn, Cu, Cd, Pb) in two habitats from an active hydrothermal field on the EPR at 13°N. Sci Tot Environ 392:119–129

Sarrazin J, Robigou V, Juniper SK, Delaney JR (1997) Biological and geological dynamics over four years on a high-temperature sulfide structure at the Juan de Fuca Ridge hydrothermal observatory. Mar Ecol Prog Ser 153:5–24

Sarrazin J, Juniper SK, Massoth G, Legendre P (1999) Physical and chemical factors influencing species distributions on hydrothermal sulfide edifices of the Juan de Fuca Ridge, northeast Pacific. Mar Ecol Prog Ser 190:89–112

Scheirer DS, Shank TM, Fornari DJ (2006) Temperature variations at diffuse and focused flow hydrothermal vent sites along the northern East Pacific Rise. Geochem Geophys Geosyst 7:Q03002

Schmidt C, Vuillemin R, Le Gall C, Gaill F, Le Bris N (2008) Geochemical energy sources for microbial primary production in the environment of hydrothermal vent shrimps. Mar Chem 108:18–31

Shank TM, Fornari DJ, Von Damm KL, Lilley MD, Haymon RM, Lutz RA (1998) Temporal and spatial patterns of biological community development at nascent deep-sea hydrothermal vents (9°50' N, East Pacific Rise). Deep Sea Res II 45:465–515

Shank TM, Govenar B, Buckman K, Fornari DJ, Soule SA, Luther GW III, Lutz RA, Vetriani C, Tolstoy M, Cowen JP, Von Damm KL (2006) Initial biological, chemical, and geological observations after the 2005-6 volcanic eruption on the East Pacific Rise. Eos Trans. AGU, 87(52), Fall Meet. Suppl., Abstract V13C–04

Sibuet M, Olu K (1998) Biogeography, biodiversity and fluid dependence of deep-sea cold-seep communities at active and passive margins. Deep Sea Res II 45(1–3):517–567

Smith EB, Scott KM, Nix ER, Korte C, Fisher CR (2000) Growth and condition of seep mussels (*Bathymodiolus childressi*) at a Gulf of Mexico Brine Pool. Ecology 81(9):2392–2403

Sommer S, Linke P, Pfannkuche O, Schleicher T, Schneider von Deimling J, Reitz A, Haeckle M, Flogel S, Hensen C (2009) Seabed methane emissions and the habitat of frenulate tubeworms on the Captain Arutyunov mud volcano (Gulf of Cadiz). Mar Ecol Prog Ser 382:69–86

Stewart FJ, Newton ILG, Cavanaugh CM (2005) Chemosynthetic endosymbioses: adaptations to oxic–anoxic interfaces. Trends Microbiol 13(9):439–448

Tolstoy M, Cowen JP, Baker ET, Fornari DJ, Rubin KH, Shank TM, Waldhauser F, Bohnenstiehl DR, Forsyth DW, Holmes RC, Love B, Perfit MR, Weekly RT, Soule SA, Glazer B (2006) A sea-floor spreading event captured by seismometers. Science 314:1920–1922

Tsurumi M, Tunniclffe V (2001) Characteristics of a hydrothermal vent assemblage on a volcanically active segment of Juan de Fuca Ridge, northeast Pacific. Can J Fish Aquat Sci 58(3):530–542

Tsurumi M, Tunniclffe V (2003) Tubeworm-associated communities at hydrothermal vents on the Juan de Fuca Ridge, northeast Pacific. Deep Sea Res I 50(5):611–629

Tunnicliffe V (1988) Biogeography and evolution of hydrothermal-vent fauna in the eastern Pacific Ocean. Proc R Soc Lond B 233:347–366

Tunnicliffe V (1991) The biology of hydrothermal vents – ecology and evolution. Oceanogr Mar Biol 29:319–407

Tunnicliffe V (1992) The nature and origin of the modern hydrothermal vent fauna. Palaios 7:338–350

Tunnicliffe V, Fowler CMR, McArthur AG (1996) Plate tectonic history and hot vent biogeography. In: Rona PA, MacLeod CJ, Tyler PA, Walker CL (eds) Tectonic, magmatic, hydrothermal

and biological segmentation of Mid-Ocean Ridges. Geological Society Special Publication No. 118. Geological Society of London
Tunnicliffe V, Embley RW, Holden JF, Butterfield DA, Massoth GJ, Juniper SK (1997) Biological colonization of new hydrothermal vents following an eruption on Juan de Fuca Ridge. Deep Sea Res I 44(9–10):1627–1644
Tunnicliffe V, McArthur AG, McHugh D (1998) A biogeographical perspective of the deep-sea hydrothermal vent fauna. Adv Mar Biol 34:353–442
Turnipseed M, Knick KI, Lipcius RN, Dreyer J, Van Dover CL (2003) Diversity in mussel beds at deep-sea hydrothermal vents and cold seeps. Ecol Lett 6:518–523
Turnipseed M, Jenkins CD, Van Dover CL (2004) Community structure in Florida Escarpment seep and Snake Pit (Mid-Atlantic Ridge) vent mussel bed. Mar Biol 145:121–132
Underwood AJ, Chapman MG, Connell SD (2000) Observations in ecology: you can't make progress on processes without understanding the patterns. *J Exp Mar Biol Ecol* 250 (1-2):97–115
Urcuyo IA, Massoth GJ, Julian D, Fisher CR (2003) Habitat, growth, and physiological ecology of a basaltic community of *Ridgeia piscesae* from the Juan de Fuca Ridge. *Deep Sea Res I* 50 (6):763–780
Urcuyo IA, Bergquist DC, MacDonald IR, Van Horn MJ, Fisher CR (2007) Growth and longevity of the tubeworm *Ridgeia piscesae* in the variable diffuse flow habitats of the Juan de Fuca Ridge. Mar Ecol Prog Ser 344:143–157
van de Koppel J, Altieri AH, Silliman BR, Bruno JF, Bertness MD (2006) Scale-dependent interactions and community structure on cobble beaches. Ecol Lett 9:45–50
Van Dover CL (2002) Community structure of mussel beds at deep-sea hydrothermal vents. Mar Ecol Prog Ser 230:137–158
Van Dover CL (2003) Variation in community structure within hydrothermal vent mussel beds of the East Pacific Rise. Mar Ecol Prog Ser 253:55–66
Van Dover CL, Doerries MB (2005) Community structure in mussel beds at Logatchev hydrothermal vents and a comparison of macrofaunal species richness on slow- and fast-spreading mid-ocean ridges. Mar Ecol 26:110–120
Van Dover CL, Lutz RA (2004) Experimental ecology at deep-sea hydrothermal vents: a perspective. J Exp Mar Biol Ecol 300:273–307
Van Dover CL, German CR, Speer KG, Parson LM, Vrijenhoek RC (2002) Evolution and biogeography of deep-sea vent and seep invertebrates. Science 295:1253–1257
Van Dover CL, Aharon P, Bernhard JM, Caylor E, Doerries M, Flickinger W, Gilhooly W, Goffredi SK, Knick KE, Macko SA, Rapoport S, Raulfs EC, Ruppel C, Salerno JL, Seitz RD, Sen Gupta BK, Shank T, Turnipseed M, Vrijenhoek R (2003) Blake Ridge methane seeps: characterization of a soft-sediment, chemosynthetically based ecosystem. Deep Sea Res I 50(2):281–300
Vismann B (1991) Sulfide tolerance: Physiological mechanisms and ecological implications. Ophelia 34(1):1–27
Voight JR (2005) Hydrothermal vent octopuses of *Vulcanoctopus hydrothermalis*, feed on bathypelagic amphipods of *Halice hesmonectes*. J Mar Biol Assoc UK 85(4):985–988
Voight JR, Sigwart JD (2007) Scarred limpets at hydrothermal vents: evidence of predation by deep-sea whelks. Mar Biol 152(1):129–133
Von Damm KL, Lilley MD (2004) Diffuse flow hydrothermal fluids from 9°50'N East Pacific Rise: origin, evolution and biogeochemical controls. AGU Monogr 144:245–268
Wright JP, Jones CG (2004) Predicting effects of ecosystem engineers on patch-scale species richness from primary productivity. Ecology 85(8):2071–2081
Wright JP, Jones CG (2006) The concept of organisms as ecosystem engineers ten years on: progress, limitations, and challenges. Bioscience 56(3):203–236
Wright JP, Jones CG, Flecker AS (2002) An ecosystem engineer, the beaver, increases species richness and the landscape scale. Oecologia 132(1):96–101
Wright JP, Gurney WSC, Jones CG (2004) Patch dynamics in a landscape modified by ecosystem engineers. Oikos 105:336–348
You CF, Bickle MJ (1998) Evolution of an active sea-floor massive sulphide deposit. Nature 394:668–671

Chapter 14
An Eldorado for Paleontologists: The Cenozoic Seeps of Western Washington State, USA

Steffen Kiel

Contents

14.1 Introduction

Most seep communities occur in deep water and it requires certain geologic processes – particularly their preservation within carbonates and their uplift above sea-level – before paleontologists can study them. Just like their modern analogs fossil seeps are highly localized and finding them requires walking through endless meters of strata that are usually barren of megafossils. The outcrop situation in western Washington is far from being ideal because most of the area is covered in thick forest. Many seep deposits are exposed along, sometimes even in, river beds (Plate 31) and can be sampled only at certain times of the year when water levels are low. Sites at coastal outcrops may only be reached by kayak or canoe, during low tide or only early in the year before they are covered by algal growth.

S. Kiel (✉)
Institut für Geowissenschaften, Christian-Albrechts-Universität, Ludewig-Meyn-Str. 10, 24118 Kiel, Germany
e-mail: steffen.kiel@gmx.de

S. Kiel (ed.), *The Vent and Seep Biota*, Topics in Geobiology 33,
DOI 10.1007/978-90-481-9572-5_14, © Springer Science+Business Media B.V. 2010

Nevertheless, extensive searching over the past 20 years produced seep fossils that are unrivaled world-wide in their diversity and the quality of their preservation (Plate 32), and they are now probably the best-studied fossil seep faunas on Earth. A particular appeal is the fact that the seep-bearing sediment also revealed diverse whale-fall (Squires et al. 1991; Goedert et al. 1995) and wood-fall communities (Lindberg and Hedegaard 1996; Kiel and Goedert 2006b) so that evolutionary interactions between these ecosystems can be traced through nearly 45 million years of time (e.g., Kiel and Goedert 2006a). Furthermore, the seep carbonates preserve a wide range of molecular fossils (biomarkers) that reveal past fluid compositions, and the biochemical processes and microbial consortia involved in the precipitation of the carbonates, even on very small spatial scales (Peckmann et al. 2002, 2003; Goedert et al. 2003; Hoffmann 2006).

The history of their discovery began in 1979 when Jim and Gail Goedert made fossil collections from a limestone deposit in the Bear River area in Pacific County, Washington, and sent the specimens to the Los Angeles County Museum (LACM). Nothing was done until someone at LACM mentioned the large bivalves to Richard L. Squires, thinking that he might be interested in them because of his work on some vesicomyids from Dead Man's Island in California. This was about 1990, and Richard was the first to recognize that these might represent a fossil cold-seep fauna. Using a report on the limestone resources of western Washington (Danner 1966) Jim Goedert checked limestones found in similar depositional settings in the Lincoln Creek Formation (the Menlo deposit) and rocks now called the Humptulips Formation. Not surprisingly, the recovered fossils were similar to the Bear River limestone fauna. The resulting publication (Goedert and Squires 1990) was the starting shot to extensive and continuing research on fossil deep-water ecosystems in this area.

This chapter aims to give an overview of the fossil seep deposits in western Washington, their fauna, paleoecology, and mode of occurrence, to outline how these sites are recognized, and why they are exposed on land.

14.2 Fossil Seeps on Land: A Brief Tectonic History of Western Washington State

Most fossil seeps in western Washington are found within a tectonic unit called the Coast Range terrane (Brandon and Calderwood 1990; Stewart and Brandon 2004; Fig. 14.1). This unit consists of early Eocene basalts at the base with a thick cover of middle Eocene to early Miocene, seep-bearing marine sediments on top. There are different opinions in the literature as to where this terrane formed (either in situ, or as seamounts on the oceanic Farallon plate just to the southwest of their present-day position), and when it was accreted to the North American plate (50 versus 38 million years; see Snavely and MacLeod 1974; Babcock et al. 1992; Brandon and Vance 1992). What is important here is that the seep-bearing sediments of the Coast Range terrane were deposited before and after the subduction of the Juan de Fuca

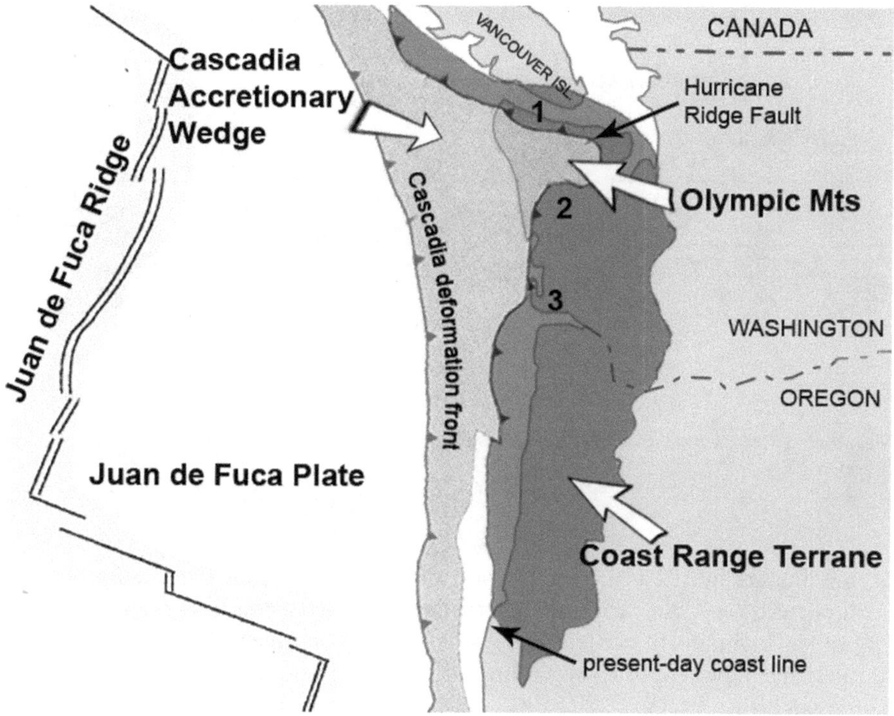

Fig. 14.1 Sketch of the tectonic setting of the Pacific Northwest. The major areas with fossil seeps are the Makah and Pysht formations along the north coast of the Olympic Peninsula (1), the Humptulips Formation and northern part of the Lincoln Creek Formation just south of the Olympic Mountains (2), and the Astoria and Lincoln Creek formations along the shore of the Columbia River (3) (Redrawn from Stewart and Brandon 2004)

plate beneath the North American plate began around 35 million years ago. This implies that the oldest seeps in western Washington (those of the Humptulips Fm) were deposited on a passive margin, whereas all younger seeps formed in an active, subduction-related setting.

Exhumation of the Coast Range terrane began in the middle Miocene about 15 million years ago, when the formerly straight, NNW trending Cascadia subduction zone acquired its modern, arched shape. While most subduction zones are concave toward the arc (the Aleutian and Java trenches are fine examples), the Cascadia subduction zone is concave seaward. The hinge point of this curved margin is located in the Olympic Mountains area. As a result, the subducting slab of the Juan de Fuca plate is forced to arch upward as it is subducted, which causes the uplift of the Olympic Mountains (Brandon and Calderwood 1990). Most of the Olympic Mountains belong to a tectonic unit called the Olympic Structural Complex, which is part of the Cascadia accretionary wedge. The Olympic Structural Complex is underplating the Coast Range terrane, along a major fault called the Hurricane Ridge fault (Fig. 14.2). Thus while the Olympic

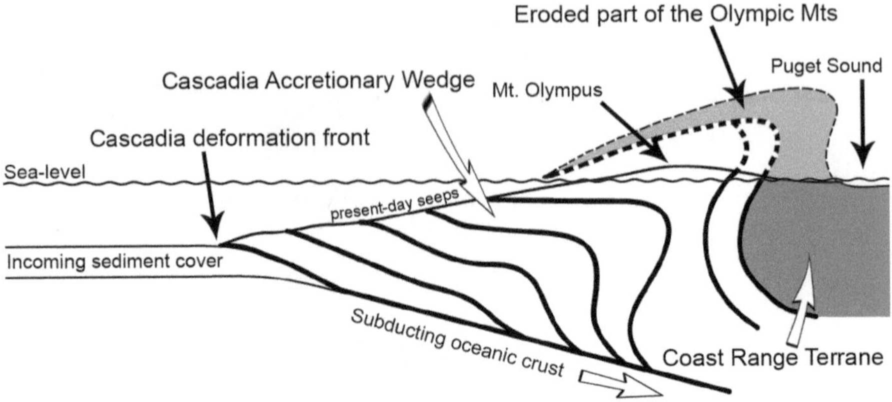

Fig. 14.2 Cross section through the Cascadia subduction zone (Redrawn from Stewart and Brandon 2004)

Mountains are uplifted, the seep-bearing sediments of the Coast Range terrane are uplifted, too. Due to ongoing erosion in western Washington, and of the Olympic Mountains in particular, the seep-bearing sediments of the Coast Range terrane are now exposed in a horseshoe-like fashion around the Olympic Mountains (Fig. 14.1).

14.3 Recognition of Fossil Seeps

The recognition of seep deposits in the fossil record is based on several lines of evidence. In western Washington, where most of the exposed sediments have been deposited in cold and deep water (as evident from foraminiferal assemblages and the abundance of glendonites), the occurrence of fossil-rich limestones within deep-water mud- and siltstones is a good first proxy for a seep deposit. This is particularly true when the limestones are dominated by taxa that are also known from modern seeps sites. This line of evidence was used to identify the first fossil seep deposits in western Washington (Goedert and Squires 1990).

Seep carbonates have several petrographic peculiarities including wavy laminations, banded and botryoidal cement, clotted micrite, and so-called yellow calcite (cf. Greinert et al. 2001; Peckmann et al. 2001) which can be recognized in fossil examples (see Goedert and Campbell [1995] and Peckmann et al. [2002] for examples from Washington State). These characteristics are often obvious even in the field, making the difference between seep carbonates and non-seep related carbonates such as concretions readily apparent. Such petrographic evidence for ancient seep deposits is often used in concert with the geochemical and biogeochemical methods described below.

Probably the most commonly used line of evidence is the carbon isotope signature of the limestone, especially of those carbonate facies that are associated with microbial methane oxidation. Methane is isotopically very light and carbonates that precipitate due to methane oxidation inherit (to some extent) this isotopic signature. This signature is usually given as the deviation of the $^{12}C/^{13}C$-ratio from the standard Vienna PeeDee belemnite (in ‰$\delta^{13}C$ PDB). Typical methane seep carbonates have $\delta^{13}C$ values of −35‰ to −60‰, those with crude oil as the main hydrocarbon in the seeping fluid have $\delta^{13}C$ values of around −20‰ to −30‰. In western Washington, carbonate $\delta^{13}C$ values were for example used in the recognition of seep deposits in the Mio-Pliocene Quinault Formation (Campbell 1992), the early Oligocene part of the Makah Formation (Goedert and Campbell 1995), and in the late Eocene to late Oligocene Lincoln Creek Formation (Peckmann et al. 2002).

A powerful tool to prove that methane oxidation occurred at putative ancient seep sites is the use of biomarkers (molecular fossils) of methane-oxidizing archaea. This method is outlined in detail in Chapter 4 of this book, and has been used in studies of seep deposits in the Lincoln Creek and Pysht Formations in western Washington (Peckmann et al. 2002; Goedert et al. 2003).

14.4 The Bear River Deposit

Prior to the discovery of Recent chemosymbiotic communities associated with vents and seeps in the deep sea (Corliss et al. 1979; Paull et al. 1984), the paleoecology of fossil bivalve-dominated invertebrate assemblages preserved in carbonate deposits in deep-marine strata was difficult to interpret. An excellent example of this is the ‘Bear River’ site, a localized, relatively small-sized limestone deposit within nearly barren Late Eocene deep-water strata, which preserved an anomalous and diverse invertebrate assemblage (Goedert and Benham 2003). The Bear River limestone deposit was reported as early as 1916 (Danner 1966 states this but does not cite the report) and was first used locally for agricultural purposes. The first detailed study of this site was an analysis of the limestone by Danner (1966), and he mentions the fossils but simply referred to the deposit as a ‘reef’. A few other reports show the Bear River locality on maps (Weaver 1937) or in lists of fossil localities (Weaver 1942) although for reasons unknown there is no mention of its fossil content. Sponge fossils from the Bear River site were described by Rigby and Jenkins (1983), but they did not discuss the associated fauna or speculate as to the paleoecology of this site. Finally, the Bear River assemblage was interpreted as an ancient cold-seep community (Goedert and Squires 1990; Squires and Goedert 1991), and this was largely on the basis of the vesicomyid and mytilid bivalve taxa that are present.

Probably the most abundant macrofossils are specimens of the mytilid *Bathymodiolus willapaensis* (Squires and Goedert 1991); they appear to make up much of the rock in hand samples, most are closed-valved and appear to be in random orientations. Specimens of the vesicomyid *Adulomya chinookensis* (Squires and Goedert 1991) are fairly common, up to 9 cm long, and in many cases they are found in clusters. Solemyids are relatively rare and most are articulated and seem to be both randomly oriented and distributed. Some large individuals are up to 13 cm long and appear to be *Acharax dalli*, a species found at many outcrops of Eocene and Oligocene strata in the Pacific Northwest. Lucinids are few but large (possibly *Cryptolucina*, up to 9 cm long) and only one small thyasirid has been found so far (Goedert and Benham 2003). Gastropods are mostly small and inconspicuous, although they are numerous and include typical extant vent and seep taxa such as *Provanna*, *Hyalogyrina*, *Depressigyra*, *Lurifax*, and two trochoids (Kiel 2006). Gastropod limpets are rare, while worm tubes are abundant.

An unusual feature of the Bear River deposit is the abundance of uncrushed fossils of the hexactinellid sponge *Aphrocallistes polytretos* Rigby and Jenkins, 1983. Hexactinellid sponges were recently reported from recent seeps offshore New Zealand where they were found in close association with seep carbonates and *Lamellibrachia* tubeworms (Baco et al. 2010). The mode of life of these sponges, however, is unclear. The hexactinellids at the Bear River deposit must at least have been able to tolerate the hydrocarbon discharge because they are abundant, large, branching and in most cases intertwined with mussels, vesicomyids, solemyids, and worm tubes. Symbioses between sponges and methanotrophic bacteria have to date only been documented for the poecilosclerid *Cladorhiza* (Vacelet et al. 1995) and a new species of demosponge (*Pseudosuberites* sp.; Thurber et al. 2009).

14.5 Humptulips Formation

The seeps of the Humptulips Formation are the oldest in the area, being of middle Eocene age (cf., Prothero 2001). Compared to the younger seep deposits in Washington they have their own distinctive character. Many of the seep carbonates were found in situ and extend laterally for several tens of meters. Some of these deposits are dominated by large, smooth lucinids of the genera *Cryptolucina* and *Nipponothracia*, which gives them a slight 'Mesozoic' character. Nevertheless, at the same seeps the earliest vesicomyids and bathymodiolins were found, though they are rare compared to the Late Eocene sites in other formations such as the Bear River deposit.

The seeps of the Humptulips Formation show distinctive faunal assemblages within individual seep deposits. Large, inflated thyasirids are commonly found in clusters of five or six; other parts of the deposit contain loosely scattered solemyids and nothing else. Tube worm clusters are usually associated with a diverse and relatively abundant gastropod fauna of neomphalids, limpets, and trochoids. In most deposits, large portions consist of wavy, laminated limestone that is entirely fossil-free. These faunal assemblages within one seep site resemble those described from modern seeps (Sahling et al. 2002; Levin and Mendoza 2007; Olu-Le Roy et al. 2007), indicating that the present-day adaptations to different chemical microenvironments at seeps already existed more than 40 million years ago. At least two seep sites in the Humptulips Formation contain numerous specimens of a large and slender gastropod, up to 90 mm high, that is an unusual genus named *Humptulipsia*

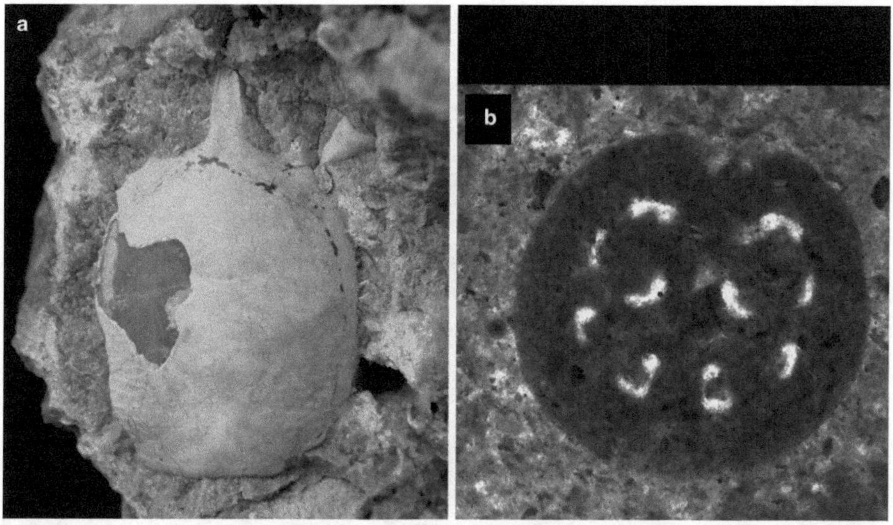

Fig. 14.3 Crustacean fossils and fecal pellets. (**a**) Carapace of the galatheid crustacean *Shinkaia katapsyxis* Schweitzer and Feldmann, 2008 from a seep deposit in the Humptulips Formation. (**b**) The trace fossil *Palaxius habanensis*, a fecal pellet probably produced by callianassid shrimp, from an early Miocene seep deposit of the Astoria Formation

that may be distantly related to provannids (Kiel 2008b). Another member of this genus has recently been identified from an early Cretaceous seep site in southern France (Kiel et al. 2010). One of these sites also preserves a neritid-like gastropod called *Thalassonerita eocenica* by Squires and Goedert (1996), which was later considered as belonging to the enigmatic vetigastropod genus *Sahlingia* (Warén and Bouchet 2001). Despite extensive subsequent sampling at this site, no new specimens that could clarify its identity have been found.

Another modern aspect of a seep in the Humptulips Formation is the recent discovery of abundant and well-preserved specimens of the galatheid crustacean *Shinkaia katapsyxis* Schweitzer and Feldmann, 2008. Its unusual mass occurrence suggests that *Shinkaia katapsyxis* (Fig. 14.3a) lived gregariously in swarms with large numbers of individuals of its own species, like its modern congener at vent sites in the western Pacific Ocean (Schweitzer and Feldmann 2008). Galatheids are not the only crustaceans at seeps in the Humptulips Formation. Peckmann et al. (2007) reported a seep deposit with several callianassid chelipeds which they were able to link to perforated fecal pellets known as *Palaxius* (Fig. 14.3b).

14.6 Lincoln Creek Formation

Sediments of the Lincoln Creek Formation (LCF) are geographically widespread, ranging from the southern slopes of the Olympic Mountains to the north shore of the Columbia River, and they contain a wide diversity of seep communities.

Fig. 14.4 Silicified fossils from a late Oligocene seep deposit of the Lincoln Creek Formation, found float near Knappton on the bank of the Columbia River. (**a**) Straight, segmented worm tube. (**b, c**) The protobranch bivalve *Tindaria*. (**d, e**) *Benthomangelia*, a predatory turrid gastropod, note the fine sculpture on the larval shell in c. (**f**) The trochid gastropod *Solariella*. (**g**) The seguenziid gastropod *Halystina*. (**h**) The turrid gastropod *Ptychosyrinx* (Images from Kiel 2010)

Many of the fossils found in seep carbonates of the LCF must have been enclosed in the carbonate very rapidly, because they often preserve the original mineralogy of molluskan shells, articulated specimens of polyplacophorans, as well as very delicate structures such as ornamentations of molluskan larval shells (Peckmann et al. 2002; Kiel 2006; Fig. 14.4). The preservation of original shell mineralogy is confined to seep deposits found in the area just south of the Olympic Mountains, in the Canyon and Satsop River area (Fig. 14.5; Plate 31). A seep deposit near Menlo, along the Willapa River, is poorly exposed (Goedert and Squires 1990; Campbell and Bottjer 1993) and not studied in detail but does contain abundant specimens of *Bathymodiolus willapaensis* and other bivalves. Other localities along the Willapa River are suspected to be seep deposits (Nesbitt et al. 1994; Goedert and Peckmann 2005) but have yet to be studied in detail and more field-work in the Willapa River area would undoubtedly be productive. Further to the south, along the shore of the Columbia River, fossils are mostly silicified, but the preservation of fine details is just as good as in the northern part of the Formation. Most of the deposits in the Lincoln Creek Formation are small in size, in some cases consisting of less than one cubic meter of limestone. One notable exception

Fig. 14.5 Seep deposit in the Lincoln Creek Formation, cropping out along the bank of the Satsop River; hammer for scale (site SR2 of Peckmann et al. 2002). This image was taken in 1997, the deposit has now (2009) been completely eroded away. An extended color plate of this figure can be found in Appendix (Plate 32)

to this is a deposit termed the "West Fork Satsop locality" (Campbell and Bottjer 1993) which is exposed in a cliff above the river for distance of one kilometer. The "West Fork Satsop locality" also preserves bathymodiolin mussels in large numbers, and a number of other molluskan taxa, but the deposit and fauna have yet to receive any detailed study.

Using material from the LCF (Goedert et al. 2000) proposed a model of worm tube preservation. Botryoidal aragonite precipitates on the inner and outer surface of the tube, is then engulfed by clotted micrite and finally the aragonitic material (and also the original tube wall) is replaced by silica (Goedert et al. 2000). A few sponges and corals were also found at seeps in the Lincoln Creek Formation (Peckmann et al. 2002; Goedert and Peckmann 2005). One LCF seep site yielded four different species of beautifully preserved hexactinellid sponges, two of which have yet to be reported elsewhere and were possibly endemic (Rigby and Goedert 1996). As in the Bear River deposit, these sponges must have been able to tolerate hydrocarbon discharge, because they are numerous and intertwined with worm tubes and abundant provannid gastropods. The relationship between sponges, corals and seeps in modern oceans is debated and not fully understood, thus it is notable that at least one small coral, *Deltocyathus insperatus* Goedert and Peckmann, 2005 appears to have been endemic to one seep deposit (SR4 of Peckmann et al. 2002) in the Lincoln Creek Formation (Goedert and Peckmann 2005).

Chemosymbiotic bivalves are diverse but generally neither as abundant nor as large as in the Bear River site or seeps in the Humptulips Formation. The most common vesicomyid is *Archivesica knapptonensis*, which is much smaller than the slightly older *Adulomya chinookensis* (Amano and Kiel 2007); another vesicomyid is a small (~1 cm) and rare *Pliocardia* spp. Large lucinids such as *Cryptolucina* or *Nipponothracia* are rare, most lucinids in the Lincoln Creek Formation belong to the genus *Lucinoma*. Thyasirids, solemyids and bathymodiolins are moderately common and are not different from those found elsewhere in western Washington. One small seep deposit along the Canyon River preserves a few brachiopods, possibly *Hemithyris astoriana*, along with thyasirids, solemyids, and *Lucinoma* sp.; few Cenozoic seeps worldwide are known to preserve brachiopods (Campbell 2006; Chapter 9 in this book).

Perhaps thanks to the excellent preservation, there is a high diversity of small gastropods in the seep deposits of the Lincoln Creek Formation (Squires 1995; Peckmann et al. 2002; Kiel 2006). These gastropod faunas show a marked ecologic difference between those from the northern part of the Lincoln Creek Formation and those found along the Columbia River. Whereas bacteria-grazers such as *Provanna*, *Retiskenea*, *Lurifax*, *Depressigyra*, *Hyalogyrina* and *Pyropelta* are common in the north, these groups are virtually absent from the seep deposits found along the Columbia River. Gastropods from the southern localities all belong to known deep-water genera and may be members of the background, deep-water fauna, although their ecology is not yet fully understood. Also their mode of occurrence differs: the bacteria-grazers in the northern seeps are usually enclosed in the micritic carbonate matrix, whilst gastropods in the seep deposits from the Columbia River are most abundant in patches of fine sediment trapped within the seep carbonate.

Apart from seep deposits, the LCF produced many invertebrate communities associated with sunken wood (wood-fall communities). The two most common fossils in these communities are the bathymodiolin *Idas*? *olympicus* and the patellogastropod limpet *Pectinodonta palaeoxylodia*; extant members of these two genera are commonly found on wood-falls. The overall diversity of wood-inhabiting taxa is high in the LCF, the most interesting taxa include the neomphalid *Leptogyra squiresi*, provannids, skeneids, cocculinids, two different polyplacophorans, and an ostracod belonging to *Xylocythere*, a genus that is today endemic to vents and wood-falls (Kiel and Goedert 2006a, b, 2007). Very few taxa are found in both wood-fall and seep communities in the LCF, despite the high diversity in both types of communities. Shared taxa include *Provanna antiqua*, *Leptochiton*, and general deep-sea browsers such as nuculanid bivalves and turrid gastropods (Kiel and Goedert 2006a). Further overlap might be found among the small 'skeneiform' gastropods, but their poor preservation so far precludes a detailed investigation. The two oldest whale-falls known to date (latest Eocene) are from the LCF and were found in the Satsop and Canyon River area. Associated taxa include modioliform mussels which might perhaps be bathymodiolins, and the buccinid gastropod *Colus*, which is a general predator and scavenger (Kiel and Goedert 2006a; Kiel 2008a). Modiolin mussels were also found associated with a fish skeleton in the LCF (Kiel 2008a).

14.7 Makah and Pysht Formations

These two formations are late Eocene to early Miocene in age, consist mainly of bathyal deposits, and are best exposed along the shore of the Juan de Fuca Strait on the north side of the Olympic peninsula. Seep deposits are known from late Eocene to early Oligocene strata and can be found as erosional lag materials along beach exposures, as well as in situ within deep water siltstones, especially east of the mouth of the East Fork of Twin River, in Clallam County. Only two sites have been investigated in detail so far, but several more have been sampled and await description. Goedert and Campbell (1995) reported an early Oligocene seep deposit from Shipwreck Point in the Makah Formation. It included a diverse assemblage (25 taxa) including *Bathymodiolus willapaensis*, vesicomyids, lucinids, *Acharax* sp. of up to 75 mm length, *Provanna antiqua*, small, globular gastropods resembling hyalogyrinids or *Retiskenea*, several turrid and trochid gastropods, plates of the polyplacophoran *Leptochiton*, as well as worm tubes, shrimp remains (cf. *Callianassa*) and scaphopods (Goedert and Campbell 1995). The other well-studied seep deposit is from the late Eocene part of the Pysht Formation at the mouth of Whiskey Creek (Goedert et al. 2003; Peckmann et al. 2003). The fauna includes large lucinids, vesicomyids, bathymodiolins, thyasirids, *Acharax*, and a few poorly preserved gastropods and worm tubes. Biomarker studies revealed the presence of crocetane, biphytane and squalane strongly depleted in ^{13}C ($\delta^{13}C$ values as low as −101‰), indicating that methane oxidation by archaea was a significant carbon source at this site (Goedert et al. 2003). Pyrite with ^{34}S-depleted sulfur indicates that bacterial sulfate reduction was the sulfide-generating process at this site, supplying the abundant bivalves with sulfophilic symbionts with nutrients (Peckmann et al. 2003).

In the Murdock Creek area (Pysht Formation) seep limestone contains thyasirid, solemyid, modiolid, and vesicomyid bivalves (Goedert and Squires 1993). Other invertebrate taxa have been found as well, including gastropods and echinoids, crinoid fragments, as well as small blocks of seep limestone with abundant, relatively large (up to 20 mm length) nuculanid bivalves. In outcrops near the Twin Rivers seep limestones are exposed as small pods, up to 1.5 m across, many containing specimens of thyasirid bivalves up to 75 mm long and 50 mm wide (J.L. Goedert and S. Kiel 2010).

The Makah and Pysht formations are renowned for their whale- and wood-fall communities (Squires et al. 1991; Goedert et al. 1995; Lindberg and Hedegaard 1996; Kiel and Goedert 2006a). Taxa associated with wood-falls are essentially the same as described for the Lincoln Creek Formation (Kiel and Goedert 2006a). The whale-fall communities of the Makah and Pysht formations are mostly of upper early Oligocene age and thus stratigraphically younger than those of the LCF. The species diversity in the Makah and Pysht whale falls is higher than in the LCF. There are two species of bathymodiolin mussels, thyasirid and lucinid bivalves occur, as well as predatory gastropods such as naticids, buccinids, and cephalaspideans.

14.8 Astoria Formation

Deep-water sediments of the early to middle Miocene Astoria Formation crop out along the Columbia River, to the east of the Lincoln Creek Formation (Wolfe and McKee 1968, 1972; Wells 1989). Seep carbonates found so far are mostly small float blocks from the foot of landslides, but some thin (up to 10 cm thick) lens-like seep carbonates have been found in situ. There are no whale- or wood-fall communities yet. Two types of seep deposits have been found so far. In one of them, thin sections show few if any typical seep carbonate structures but are of the 'mudstone' type sensu (Greinert et al. 2001). The fauna of these deposits is of low diversity, but the fossils are silicified and often remarkably well preserved. They were first mentioned when Amano and Kiel (2007) described the small vesicomyid bivalve *Isorropodon frankfortensis* from them. The remaining fauna includes a few worm tubes, a possible member of the heterobranch gastropod *Hyalogyrina*, and a few neogastropods and opisthobranchs. The other type of seep carbonate contains cements and pyrite, and a diverse mollusk fauna whose shells are recrystallized to calcite. There are several chemosymbiotic bivalves including an undescribed vesicomyid of moderate size, resembling *Pliocardia*, a common thyasirid, and few solemyids (see Plate 32d for example). Mytilids and lucinids have not been found so far. Gastropods include many predatory neogastropods and opisthobranchs, a few small provannids, but no skeneiform bacteria-grazers (Kiel 2010).

14.9 Further Seeps

Other seep sites are known in other rock units, but have not been studied in detail as of yet. Some seep carbonates have been recognized in Eocene deep-water rocks mapped as "Unit B" (cf., Wolfe and McKee 1968) on the West Fork of Grays River, upstream from the mouth of Beaver Creek. Poorly preserved vesicomyid bivalves have been found, along with possible solemyid and thyasirid fragments (J.L. Goedert and S. Kiel 2010).

Large limestone blocks displaying wavy, laminated structure and containing fossils of modiolid and solemyid bivalves, small gastropods and worm tubes are exposed on the beach along the shore of the Columbia River near the townsite of Knappton (Goedert and Benham 2003; Plate 32a). These limestone blocks are apparently derived from Eocene rocks called "siltstone of Shoalwater Bay" by Wells (1989).

Within rocks of the Olympic Structural Complex is a limestone lense, poorly exposed and probably Oligocene in age, near Mt. Appleton. Fossils of bivalves were collected in 1940 and the assemblage contains solemyids, thyasirids, and vesicomyids according to the late Ralph Stewart, U. S. Geological Survey (Harvey 1959). This locality was visited briefly by J. L. Goedert in 1994 and only a few fossil bivalves were collected, but they are too poorly preserved to be identified. A sample of the limestone was analyzed by J. Peckmann and he confirmed that the site is an ancient seep site. Another seep site from the Olympic Structural Complex was described by Campbell (1992) from the Mio-Pliocene Quinault Formation, and includes solemyid and thyasirid clams, and as-yet unidentified modiolin mussels.

14.10 Conclusions and Outlook

The rich and well-preserved record of fossil seep and other chemosynthetic communities make western Washington an excellent starting point for evolutionary and ecologic studies on geologic timescales. Seep communities with a stratigraphic range from the middle Eocene (~40–45 million years) to the Mio-Pliocene (~5 million years) have produced many oldest records of extant vent and seep taxa, including major groups like vesicomyids and bathymodiolins, galatheid crabs, and also many minute gastropods. The relatively robust paleobathymetric framework based on extensive studies of benthic foraminifera provides insights into the role that water depth had on shaping seep communities in the geologic past: there is a general trend of an increasing proportion of background species, non-symbiotic bivalves and predators in particular, towards the shallower sites (Kiel 2010). Apart from large-scale evolutionary and ecologic trends, ecologic differentiation can be seen in individual sites. Distinct communities within larger seep sites are known for example in the Humptulips Formation (Goedert and Squires 1990). The potential geochemical reasons for such differentiation, as shown in a pioneering study of a Cretaceous seep site in Japan (Jenkins et al. 2007), makes an interesting focus of future studies. This is particularly the case because molecular fossils (biomarkers) are usually well-preserved and abundant in the seep deposits of western Washington (Peckmann et al. 2002; Goedert et al. 2003; Hoffmann 2006); however, many more sites await detailed analysis. More than 65 invertebrate species from around 30 seep sites in western Washington have been dealt with in about 25 papers, and more new species have been recognized but await description (Kiel 2010). Much more work remains to be done, even in this region where ancient seep sites are well studied.

Acknowledgments I am grateful to Jim Goedert for introducing me to the deep-water faunas of western Washington, for guiding me there on several field trips, and for providing insights, unpublished data, images, and constructive criticism for this chapter. I would also like to thank Jörn Peckmann for images.

References

Amano K, Kiel S (2007) Fossil vesicomyid bivalves from the North Pacific region. Veliger 49: 270–293

Babcock RS, Burmester RF, Engebretson DC, Warnock AC et al (1992) A rifted margin origin for the Crescent basalts and related rocks in the northern Coast Range volcanic province, Washington and British Columbia. J Geophys Res 97:6799–6821

Baco AR, Rowden AA, Levin LA, Smith CR et al (2010) Initial characterization of cold seep faunal communities on the New Zealand Hikurangi margin. Mar Geol 272:251–259

Brandon MT, Calderwood AR (1990) High-pressure metamorphism and uplift of the Olympic subduction complex. Geology 18:1252–1255

Brandon MT, Vance JA (1992) Tectonic evolution of the Cenozoic Olympic subduction complex, Washington State, as deduced from fission track ages for detrital zircons. Am J Sci 292:565–636

Campbell KA (1992) Recognition of a Mio-Pliocene cold seep setting from the Northeast Pacific Convergent Margin, Washington, U.S.A. Palaios 7:422–433

Campbell KA (2006) Hydrocarbon seep and hydrothermal vent paleoenvironments and paleontology: past developments and future research directions. Palaeogeogr Palaeoclimatol Palaeoecol 232:362–407

Campbell KA, Bottjer DJ (1993) Fossil cold seeps. Natl Geogr Res Explor 9:326–343

Corliss JB, Dymond J, Gordon LI, Edmond JM et al (1979) Submarine thermal springs on the Galápagos Rift. Science 203:1073–1083

Danner WR (1966) Limestone resources of western Washington. State Wash Div Mines Geol Bull 52:1–474

Goedert JL, Benham SR (2003) Biogeochemical processes at ancient methane seeps: the Bear River site in southwestern Washington. Geol Soc Am Field Guide 4:201–208

Goedert JL, Campbell KA (1995) An Early Oligocene chemosynthetic community from the Makah Formation, northwestern Olympic Peninsula, Washington. Veliger 38:22–29

Goedert JL, Peckmann J (2005) Corals from deep-water methane-seep deposits in Paleogene strata of Western Oregon and Washington, U.S.A. In: Freiwald A, Roberts JM (eds) Cold-water corals and ecosystems. Springer, Berlin, Heidelberg, pp 27–40

Goedert JL, Squires RL (1990) Eocene deep-sea communities in localized limestones formed by subduction-related methane seeps, southwestern Washington. Geology 18:1182–1185

Goedert JL, Squires RL (1993) First Oligocene record of *Calyptogena* (Bivalvia: Vesicomyidae). Veliger 36:72–77

Goedert JL, Squires RL, Barnes LG (1995) Paleoecology of whale-fall habitats from deep-water Oligocene rocks, Olympic Peninsula, Washington State. Palaeogeogr Palaeoclimatol Palaeoecol 118:151–158

Goedert JL, Peckmann J, Reitner J (2000) Worm tubes in an allochthonous cold-seep carbonate from lower Oligocene rocks of western Washington. J Paleontol 74:992–999

Goedert JL, Thiel V, Schmale O, Rau WW et al (2003) The late Eocene 'Whiskey Creek' methane-seep deposit (western Washington State) Part I: geology, palaeontology, and molecular geobiology. Facies 48:223–240

Greinert J, Bohrmann G, Suess E (2001) Gas hydrate-associated carbonates and methane-venting at Hydrate Ridge: classification, distribution, and origin of authigenic lithologies. In: Paull CK, Dillon WP (eds) Natural gas hydrates: occurrence, distribution, and detection. American Geophysical Union, Washington, DC, pp 99–113

Harvey JL (1959) Geologic reconnaissance, S. W. Olympic Peninsula. Master's thesis, University of Washington, Seattle, WA

Hoffmann V-ERK (2006) Biosignaturen in känozoischen Cold-Seeps des Pazifischen Nordwestens der USA. Diplom thesis, Georg-August-Universität, Göttingen

Jenkins RG, Kaim A, Hikida Y, Tanabe K (2007) Methane-flux-dependent lateral faunal changes in a Late Cretaceous chemosymbiotic assemblage from the Nakagawa area of Hokkaido, Japan. Geobiology 5:127–139

Kiel S (2006) New records and species of mollusks from Tertiary cold-seep carbonates in Washington State, USA. J Paleontol 80:121–137

Kiel S (2008a) Fossil evidence for micro- and macrofaunal utilization of large nekton-falls: examples from early Cenozoic deep-water sediments in Washington State, USA. Palaeogeogr Palaeoclimatol Palaeoecol 267:161–174

Kiel S (2008b) An unusual new gastropod genus from an Eocene hydrocarbon seep in Washington State, USA. J Paleontol 82:188–191

Kiel S (2010) On the potential generality of depth-related ecologic structure in cold-seep communities: Cenozoic and Mesozoic examples. Palaeogeogr Palaeoclimatol Palaeoecol 295:245–257

Kiel S, Goedert JL (2006a) Deep-sea food bonanzas: early Cenozoic whale-fall communities resemble wood-fall rather than seep communities. Proc R Soc B 273:2625–2631

Kiel S, Goedert JL (2006b) A wood-fall association from Late Eocene deep-water sediments of Washington State, USA. Palaios 21:548–556

Kiel S, Goedert JL (2007) Six new mollusk species associated with biogenic substrates in Cenozoic deep-water sediments in Washington State, USA. Acta Palaeontol Pol 52:41–52

Kiel S, Campbell KA, Gaillard C (2010) New and little known mollusks from ancient chemosynthetic environments. Zootaxa 2390:26–48

Levin LA, Mendoza GF (2007) Community structure and nutrition of deep methane-seep macrobenthos from the North Pacific (Aleutian) Margin and the Gulf of Mexico (Florida Escarpment). Mar Ecol 28:131–151

Lindberg DR, Hedegaard C (1996) A deep water patellogastropod from Oligocene water-logged wood of Washington State, USA (Acmaeoidea: *Pectinodonta*). J Mollus Stud 62:299–314

Nesbitt EA, Campbell KA, Goedert JL (1994) Paleogene cold seeps and macrinvertebrate faunas in a forearc sequence of Oregon and Washington. In: Swanson DA, Haugerud RA (eds) Geologic field trips in the Pacific Northwest. Geological Society of America, Boulder, pp 1–11

Olu-Le Roy K, Caprais J-C, Fifis A, Fabri M-C et al (2007) Cold-seep assemblages on a giant pockmark off West Africa: spatial patterns and environmental control. Mar Ecol 28: 115–130

Paull CK, Hecker B, Commeau R, Freeman-Lynde RP et al (1984) Biological communities at the Florida Escarpment resemble hydrothermal vent taxa. Science 226:965–967

Peckmann J, Reimer A, Luth U, Luth C et al (2001) Methane-derived carbonates and authigenic pyrite from the northwestern Black Sea. Mar Geol 177:129–150

Peckmann J, Goedert JL, Thiel V, Michaelis W et al (2002) A comprehensive approach to the study of methane-seep deposits from the Lincoln Creek Formation, western Washington State, USA. Sedimentology 49:855–873

Peckmann J, Goedert JL, Heinrichs T, Hoefs J et al (2003) The late Eocene 'Whiskey Creek' methane-seep deposit (western Washington State) Part II: petrology, stable isotopes, and biogeochemistry. Facies 48:241–254

Peckmann J, Senowbari-Daryan B, Birgel D, Goedert JL (2007) The crustacean ichnofossil *Palaxius* associated with callianassid body fossils in an Eocene methane-seep limestone, Humptulips Formation, Olympic Peninsula, Washington. Lethaia 40:273–280

Prothero DR (2001) Chronostratigraphic calibrations of the Pacific Coast Cenozoic: a summary. In: Prothero DR (ed) Magnetic stratigraphy of the Pacific coast Cenozoic. The Pacific Section SEPM, Book 91, Fullerton. Society for Sedimentary Geology pp 377–394

Rigby JK, Goedert JL (1996) Fossil sponges from a localized cold-seep limestone in Oligocene rocks of the Olympic peninsula, Washington. J Paleont 70:900–908

Rigby JK, Jenkins DE (1983) The Tertiary sponges *Aphrocallistes* and *Eurete* from western Washington and Oregon. Nat Hist Mus LA County Contrib Sci 344:1–13

Sahling H, Rickert D, Lee RW, Linke P et al (2002) Macrofaunal community structure and sulfide flux at gas hydrate deposits from the Cascadia convergent margin, NE Pacific. Mar Ecol Prog Ser 231:121–138

Schweitzer CE, Feldmann RM (2008) New Eocene hydrocarbon seep decapod crustacean (Anomura: Galatheidae: Shinkaiinae) and its paleobiology. J Paleontol 82:1021–1029

Snavely PDJ, MacLeod NS (1974) Yachats Basalt-An upper Eocene differentiated volcanic sequence in the Oregon Coast range. US Geol Surv J Res 2:395–403

Squires RL (1995) First fossil species of the chemosynthetic-community gastropod *Provanna*: localized cold-seep limestones in Upper Eocene and Oligocene rocks, Washington. Veliger 38:30–36

Squires RL, Goedert JL (1991) New Late Eocene Mollusks from localized limestone deposits formed by subduction-related methane seeps, southwestern Washington. J Paleontol 65:412–416

Squires RL, Goedert JL, Barnes LG (1991) Whale carcasses. Nature 349:574

Stewart RJ, Brandon MT (2004) Detrital-zircon fission-track ages for the "Hoh Formation": implications for late Cenozoic evolution of the Cascadia subduction wedge. Geol Soc Am Bull 116:60–75

Thurber AR, Kröger K, Neiraa C, Wiklund H et al (2009) Stable isotope signatures and methane use by New Zealand cold seep benthos. Mar Geol 272:260–269

Vacelet J, Boury-Esnault N, Fiala-Médoni A, Fisher CR (1995) A methanotrophic carnivorous sponge. Nature 377:296

Warén A, Bouchet P (2001) Gastropoda and Monoplacophora from hydrothermal vents and seeps; new taxa and records. Veliger 44:116–231

Weaver CE (1937) Tertiary stratigraphy of western Washington and northwestern Oregon. U Wash Publ Geol 4:1–266

Weaver CE (1942) Paleontology of the marine Tertiary formations of Oregon and Washington. U Wash Publ Geol 5:1–789

Wells RE (1989) Geologic map of the Cape Disappointment – Naselle River area, Pacific and Wahkiakum Counties, Washington. U S Geological Survey, Miscellaneous Investigations Map, p I–1832

Wolfe EW, McKee EH (1968) Geology of the Grays River Quadrangle, Wahkiakum and Pacific counties, Washington. State of Washington Department of Natural Resources, Division of Mines and Geology, Geologic Map, p GM–4

Wolfe EW, McKee EH (1972) Sedimentary and igneous rocks of the Grays River quadrangle, Washington. US Geol Surv Bull 1335:1–70

Appendix 1

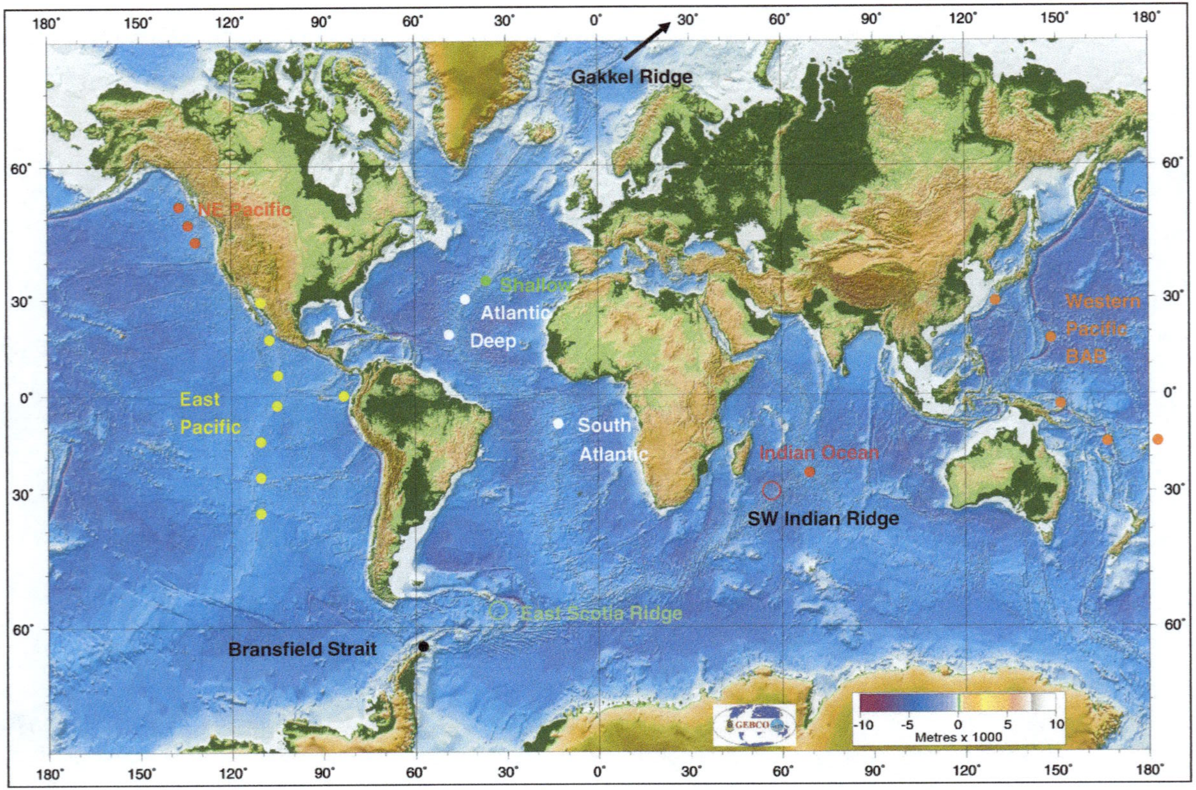

Plate 1 (Chapter 1) Putative biogeographic provinces of hydrothermal vents in the global ocean. Individual vents are not marked. Open circles represent sites of known hydrothermal activity but vent ecosystems are yet to be found (as of November 2008). Bransfield Strait (BS) is yet to be assigned to a biogeographic province

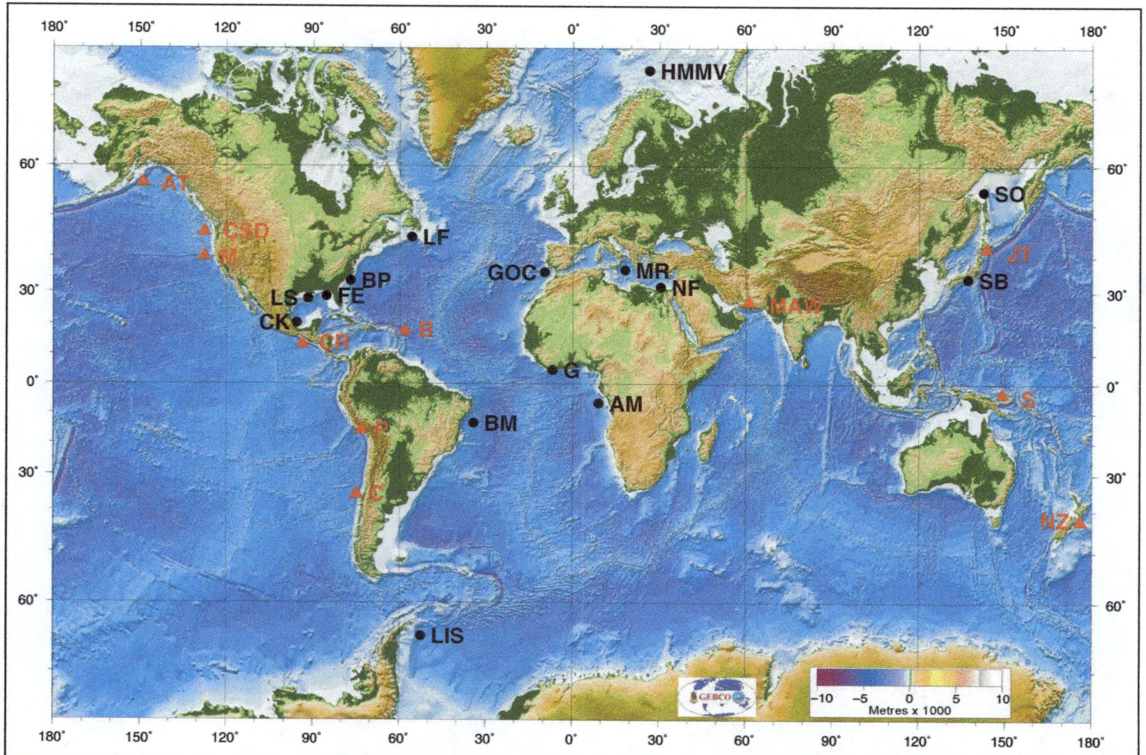

Plate 2 (Chapter 1) Known cold seeps in the global ocean. Seeps on active margins (red triangles) include: Aleutian Trench (AT), Barbados Accretionary Prism (B), Chile (C), Cascadia Margin/Oregon (CSD), Costa Rica (CR), Japan Trench (JT), Monterey Canyon (strictly a strike slip margin)(M), Makran Accretionary Wedge (MAW), New Zealand (NZ), Peru (P), Sissan (S). Seeps on passive margins (Black circles): Angola Margin (AM), Brazil Margin (BM), Blake Plateau (BP), Haakon Mosby Mud Volcano (HMMV), Campeche Knolls (CK), Florida Escarpment (FE), Guinea (G), Gulf of Cadiz (GOC), Laurentian Fan (LF), Larsen Ice Shelf (LIS), Louisiana Slope (LS), Mediterranean Ridge (MR), Nile Fan (NF), Sagami Bay (SB), Sea of Okhotsk (SO)

Plate 3 (Chapter 2) Chemosynthetic taxa found at hydrothermal vents and seeps. **a.** *Beggiatoa* mat growing on sulfides in the Gulf of California. **b.** *Rimicaris* shrimp swarming at a hydrothermal vent on Central Indian Ridge. **c.** *Calyptogena magnifica* clams clustering in basaltic cracks at 21°N latitude on the East Pacific Rise. **d.** *Riftia pachyptila* cluster growing at the base of a hydrothermal vent chimney in the Gulf of California. **e.** A tubeworm cluster composed of *Escarpia spicata* and *Lamellibrachia barhami* growing in soft sediments at cold seeps in the Gulf of California. **f.** A cluster of *Bathymodiolus azoricus* mussels (courtesy of Cindy Van Dover) from the Snake Pit vent locality on the Mid-Atlantic Ridge

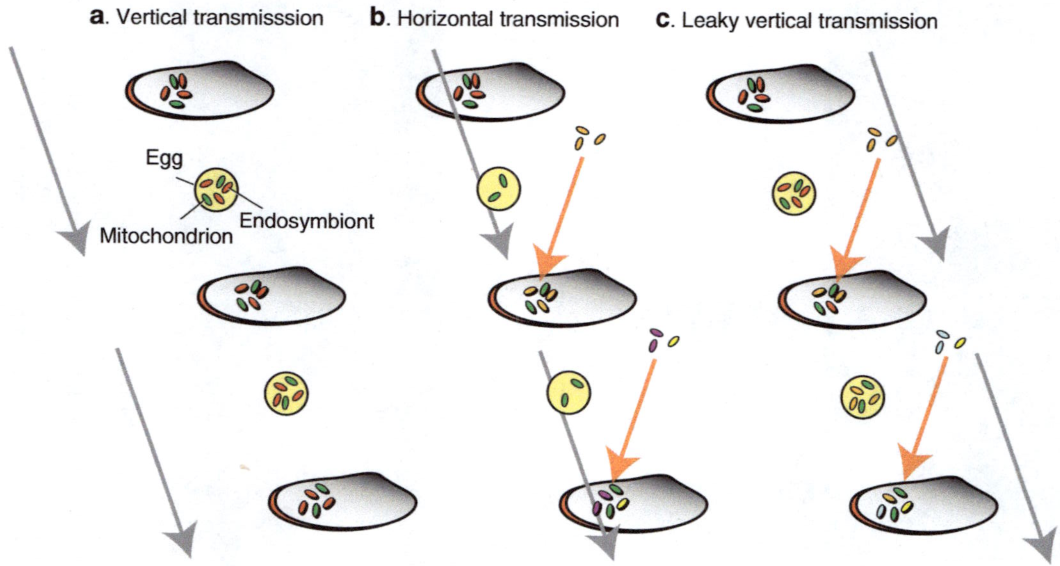

Plate 4 (Chapter 2) Transmission modes that have been documented or inferred in various chemosynthetic organisms, illustrated for a hypothetical bivalve mollusc. The green ovals represent mitochondria. The red and other color ovals represent endosymbiotic bacteria. **a.** Under obligately vertical transmission, the symbiont is inherited maternally along with the mitochondria, leading to genetic coupling of mitochondrial and endosymbiont genotypes. **b.** Under horizontal transmission, the bacteria are acquired anew in each generation, decoupling the cytoplasmic organelles from bacterial genotypes. Furthermore, infections involving environmental bacteria will likely lead to mixed symbiont genotypes (strains) within a host. **c.** Leaky vertical transmission is predominantly vertical with occasional environmental acquisition, or vertical with massive environmental swamping. In either case, the vertical transmission component will create a small lag-time in the decoupling of host mitochondrial and symbiont genotypes. The horizontal component, depending how prevalent, can create mixed-strain infections

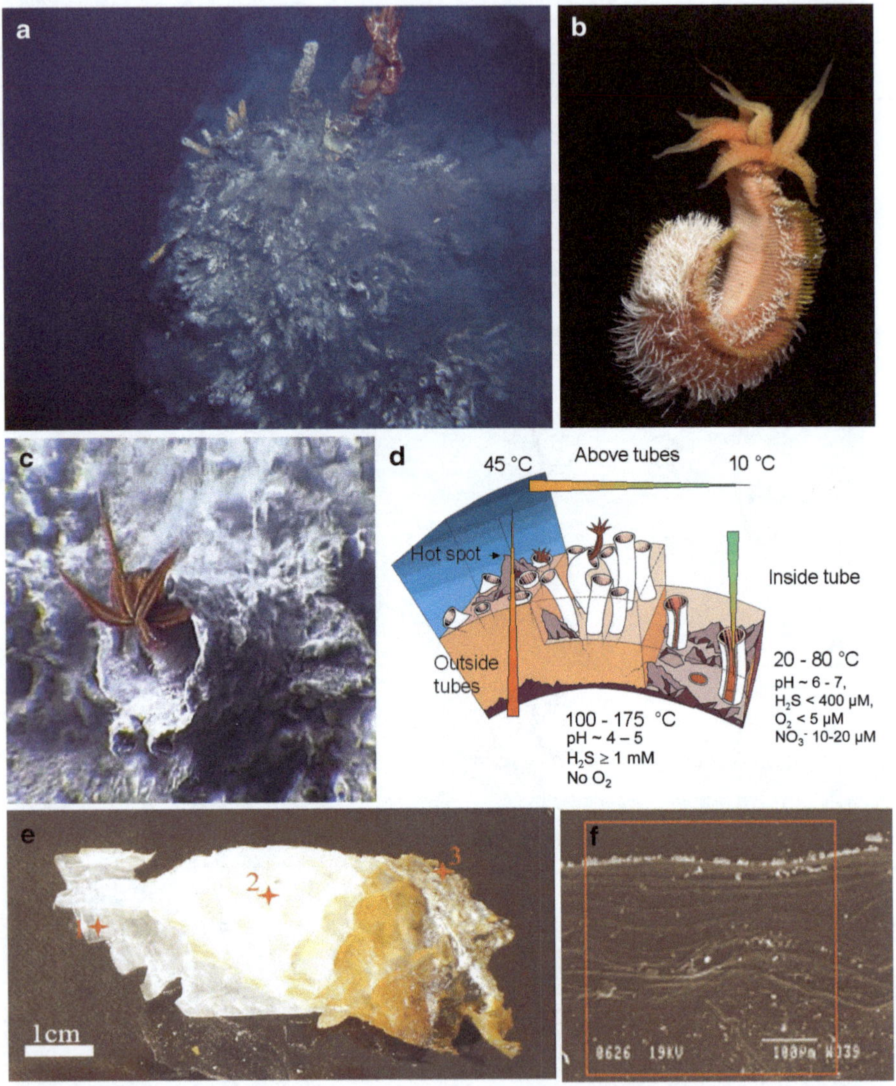

Plate 5 (Chapter 3) *Alvinella pompejana*. **a.** Pompeii worm colony on the wall of a hydrothermal smoker (EPR 13°N, PHARE cruise). **b.** View of an individual and its dorsal filamentous epibiosis. **c.** Brief appearance of a worm at tube opening. **d.** Summary of temperature ranges measured at colony and animal scales. **e**. Tube: 1. Newly secreted portion at tube opening, 2. White central portion, 3. Mineralized aged portion (initially black, then rusty as the minerals oxidize in contact with air). **f**. Thin sections of the tube wall showing successive organic layers and mineral deposits on the outside of the tube

Plate 6 (Chapter 3) *Rimicaris exoculata*. **a.** Swarm on smoker walls (Rainbow, ATOS cruise). **b.** Positioning of shrimps in the thermal gradient, **c.** A *Rimicaris exoculata* exuviate surrounded by live individuals from another species of shrimp feeding on it. Accumulated iron oxides in the branchial chamber are visible through the transparent carapace, **d.** Semi-thin section of the cuticle covered by the epibiotic bacterial mat and iron oxide deposits. **e.** TEM image showing a fraction of a bacterial filament (1) and a rod-shaped bacterium coated with iron oxides (2). **f.** Scheme of the *Rimicaris exoculata* branchial chamber: lower pre-branchial compartment (1), true gill compartment in which respiration takes place (2), upper post-branchial compartment where iron oxides mostly accumulate on Rainbow individuals (3). The exopodite (Ex) delineate the lower and upper compartments. Scaphognatite (Sc) beats drive the water circulation along the three compartments (blue arrow)

Plate 7 (Chapter 4) Image of a black smoker in the Turtle Pits Field on the Mid-Atlantic Ridge. (Photo courtesy of MARUM, University of Bremen and Richard Seifert, University of Hamburg)

Plate 8 (Chapter 4) Cold seep microbial reefs in the Black Sea (GHOSTDABS-field; 230 m water depth). Small Black Sea reef structure before (**a**) and after (**b**) sampling with a submersible showing the different microbial zones. The outer areas are mainly composed of black microbial mats whereas the inner parts are predominantly formed by orange mats (photo courtesy of Walter Michaelis and Richard Seifert (University of Hamburg) and Karen Hissmann (JAGO team))

Plate 9 (Chapter 5) **a.** *Solemyarina velesiana* Iredale, 1931. Moreton Bay, Queensland, Australia. (Image courtesy of Gonzalo Giribet). **b.** *Acharax* sp. off Costa Rica 2274 m. (Image by J Turner, National Museum of Wales). **c.** *Acharax* sp. off Costa Rica 1013 m. (Image by J Turner, National Museum of Wales). **d.** *Nucinella* sp. Panglao Island, Philippines, 0-3 m, soft bottom with seagrass. (Image courtesy of Pierre Lozouet, Museum National d'Histoire Naturelle, Paris. Panglao Marine Biodiversity Project 2004). **e.** *Conchocele bisecta* (Conrad, 1849) Off Vancouver Island, Canada, 366 m (BMNH). **f.** "*Calyptogena*" *magnifica* from vents on the East Pacific Rise

Plate 10 (Chapter 5) Lucinidae from vent and seeps. **a.** *Bathyaustriella thionipta* Glover, Taylor & Rowden 2004, Kermadec Ridge, New Zealand (holotype NIWA3277 H-838). **b.** *Jorgenia louisiana* Taylor & Glover 2009, Louisiana Slope, USA (holotype USNM 1116114). **c.** *Jorgenia luteophila* Taylor & Glover, 2009, Louisiana Slope, 610-850 m (paratype Emilio Garcia collection EFG 27398). **d.** *Lucinoma atlantis* (McLean, 1936) Louisiana Slope, USA (Emilio Garcia collection 24039). **e**. *Lucinoma aequizonata* (Stearns, 1890), Santa Barbara Channel 505 m (syntype USNM 104044). **f**. *Lucinoma gagei,* Oliver & Holmes, 2006, off S. India, 786 m (BMNH). **g**. *Lucinoma myriamae* Cosel, 2005 off Angola 360 m (BMNH). **h**. *Meganodontia acetabulum* Bouchet & Cosel, 2004, off North Taiwan 370m (paratype MNHN)

Plate 11 (Chapter 6) Distribution of sites where bathymodioline mussels were collected and their symbioses studied. Red dots refer to hydrothermal vent sites, green to cold seeps, brown to organic falls

Plate 12 (Chapter 6) Bathymodiolin mussels, their morphology, habitats and evolutionary patterns. **a.** *Idas* sp. Med attached to a carbonate crust recovered from cold seeps in the eastern Mediterranean (MEDECO cruise, 2007, photo by O. Gros). Note the two whitish siphons in the posterior part. Shell length 1 cm. **b.** *Bathymodiolus azoricus* specimens attached to a fluid-emitting edifice at the Rainbow hydrothermal vent site (Mid Atlantic Ridge, 2277m depth). Notice the iron deposits on chimneys and shells, and the presence of *Rimicaris exoculata* shrimps on the right of the image. Image © Ifremer, MOMARDREAM cruise, 2008 (chief scientist: J. Dyment). **c.** Dense bed of the mussel *Bathymodiolus* sp. at a cold seep site situated in a depression on top of a diapir structure in the lower Congo Basin (2700 m depth). The area displayed gas bubble sources. Other metazoans include tubeworms (*Lamellibrachia* sp., in the back), shrimps, and limpets (small white dots visible on mussel shells). Image © Marum, University of Bremen, GUINECO M76/3a cruise, 2008 (chief scientist: A. Boetius). **d.** *Idas* sp. Med, opened valves. Note the white foot, brownish visceral mass below and gills in the left and right valves

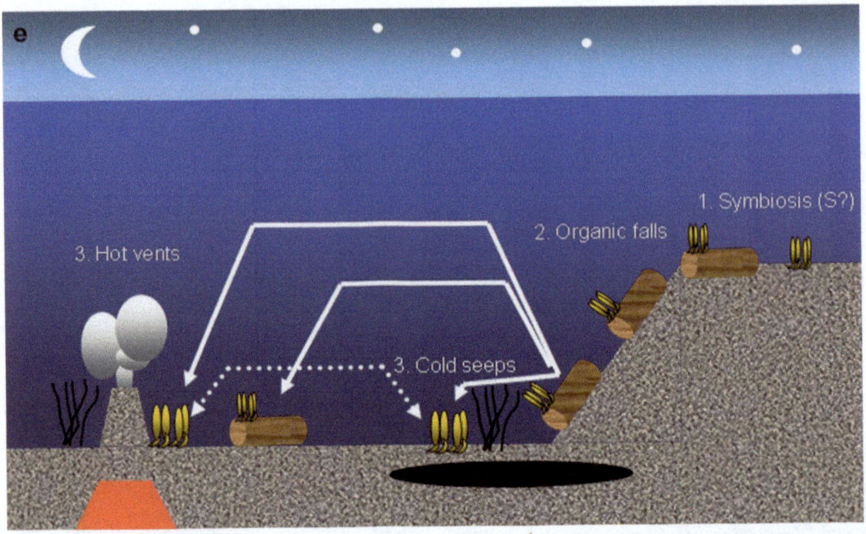

Plate 12 (continued) **e.** The 'stepping stone' hypothesis. Ancestors of bathymodiolines were shallow species (on the right), which might have acquired the ability to associate with bacteria (1), most likely sulphur oxidizers. Mussels able to derive nutrition from both filter-feeding and bacterial chemoautotrophs colonized organic habitats such as wood and whale falls occurring at various depths (2). From such habitats, mussels then colonized hydrothermal vents at ridges and cold seeps on margins (3), giving rise to large *Bathymodiolus*-like species. This event probably occurred several times during the evolution of the group. Shifts between vent and seep probably occur, as evidenced by the existence of species colonizing both habitats (dotted arrow)

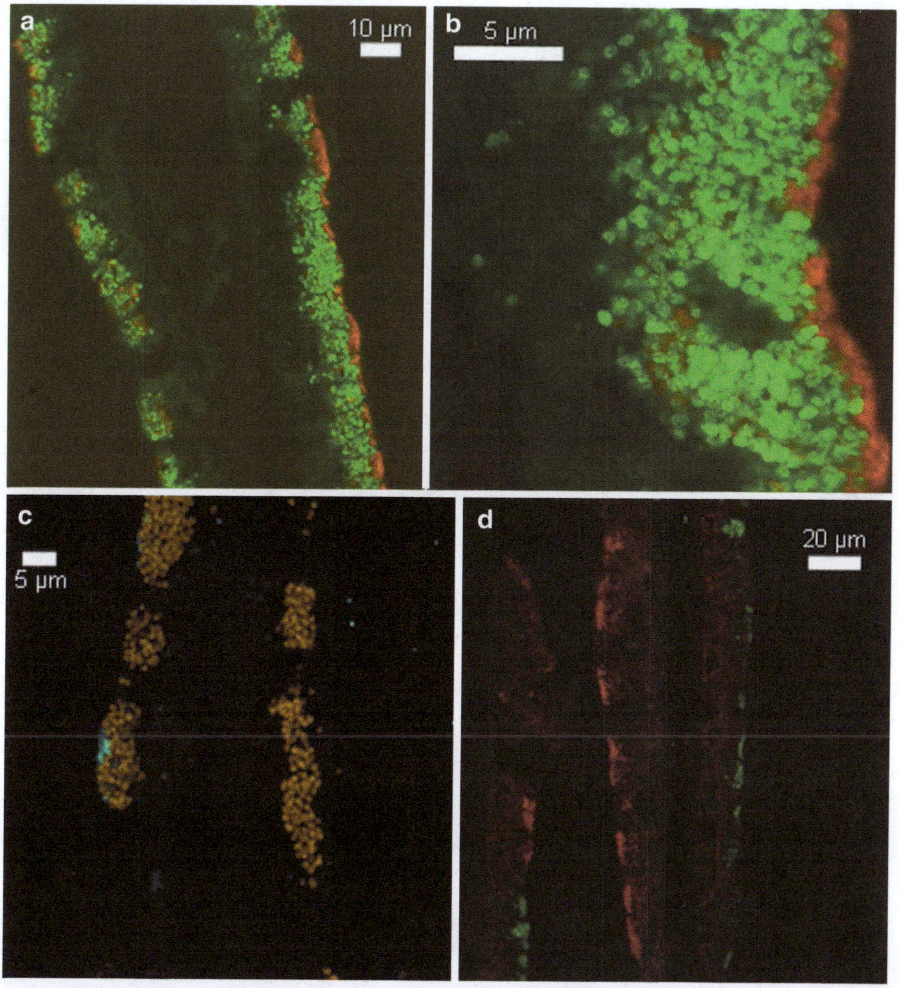

Plate 13 (Chapter 6) Fluorescence *in situ* hybridizations performed on transverse sections of gill filaments using fluorescence-labelled, symbiont-specific probes. **a.** Gill of *Bathymodiolus* aff. *boomerang* from the Gulf of Guinea (collected during the Biozaire 2 cruise, 2001). Methanotrophs in green, sulphur-oxidizers in red. Scale bar = 10 mm. **b.** Higher magnification of a region of the gill epithelium of the same species. **c.** Gill from *Bathymodiolus heckerae* (Gulf of Mexico, Florida Escarpment). Methane oxidizers in orange, *Methylophaga*-related symbionts in blue. Note the close proximity between both symbiont types; scale bar = 5 mm. **d.** Two filaments from *B. heckerae* displaying the two sulphur-oxidizers S1 and S2 (green and red). Symbionts rarely co-occur within a single bacteriocyte but rather occupy distinct areas of the gill epithelium; scale bar = 20 mm

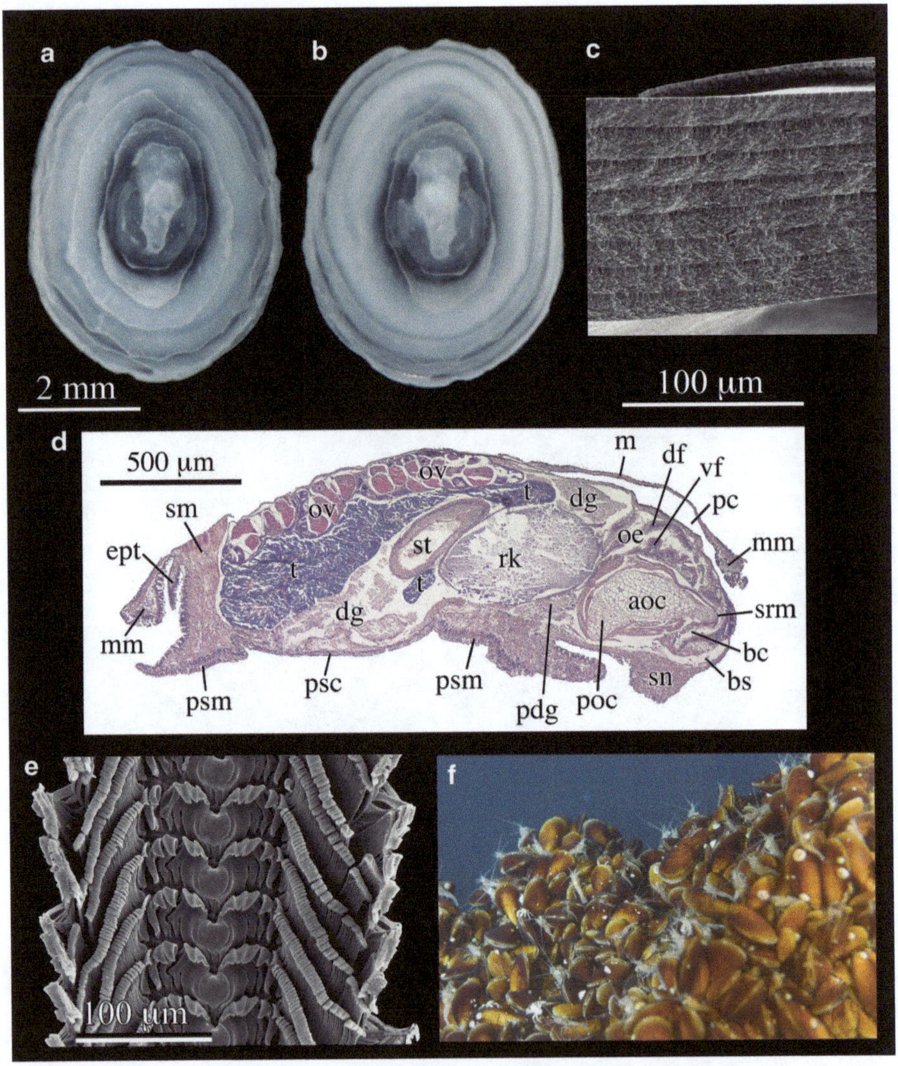

Plate 14 (Chapter 7) Pyropeltidae. *Pyropelta ryukyuensis* Sasaki, Okutani and Fujikura, 2008. **a.** Exterior of shell. **b.** Interior of shell. **c.** Longitudinal section of animal. **d.** Radula. Hatoma Knoll, Okinawa Trough, 1451 m, vent. A-B. Holotype UMUT RM29412. **e.** Paratype UMUT RM29246. **f.** Paratype UMUT RM29418. **g.** Habitat on *Bathymodiolus platifrons*. Hatoma Knoll, ca. 1480 m. *Shinkai 2000* Dive 1361. Abbreviations: aoc, anterior odontophoral cartilage; bc, buccal cavity; bs, buccal sinus; df, dorsal fold of oesophagus; dg, digestive glands; e, oesophagus; ept, epipodial tentacle; i, intestine; m, mantle; mm, mantle margin; ov, ovary; pc, pallial cavity; pdg, pedal ganglion; poc, posterior odontophoral cartilage; psc, central part of pedal sole; psm, marginal part of pedal sole; sm, shell muscle; sn, snout; srm, subradular membrane; st, stomach; t, testis; vf, ventral fold of oesophagus. [**a-e**: Sasaki et al. 2008]

Plate 15 (Chapter 7) Neomphalina. **a-b.** *Neomphalus fretterae* McLean, 1981. SMNH43346. East Pacific Rise (EPR), 13°N, 2630 m, vent. **c.** *Cyathermia naticoides* Warén and Bouchet, 1989. SMNH21160. EPR, 13°N, 2630 m, vent. **d.** Scaly-foot gastropod. Kairei Field, Indian Ocean, 2422 m, vent. UMUT RM30208. **e.** *Melanodrymia aurantiaca* Hickman, 1984. SMNH43311, EPR, 13°N, 2632 m, vent. **f.** *Peltospira operculata* McLean, 1989. SMNH43168. EPR, 13°N, 2649 m, vent. **g.** *Nodopelta heminoda* McLean, 1989. SMNH43316. EPR, 13°N, 2630 m, vent. **h.** *Pachydermia laevis* Warén and Bouchet, 1989. SMNH43130. EPR, 13°N, 2650 m, vent. [**a-b, e-h**: Photo taken by T. Sasaki at SMNH]

Plate 16 (Chapter 7) Provannidae. **a-b.** *Alviniconcha hessleri* Okutani and Ohta, 1988. Holotype, NSMT-Mo 64489, Alice Springs site, Mariana Back-Arc Basin, 3630-3655 m, vent. **c.** *Alviniconcha aff. hessleri* Okutani and Ohta, 1988. Kairei Field, Indian Ocean, 2422 m, vent. JAMSTEC. **d.** *Ifremeria nautilei* Bouchet and Warén, 1991. Paratype, NSMT-Mo 29992, Lau Basin, 1750 m, vent. **e-f.** *Provanna glabra* Okutani, Tsuchida and Fujikura, 1992. Hatsushima seep site, Sagami Bay, Japan, seep. **e.** Shell. UMUT RM30209. **f.** Habitat at *Calyptogena* bed. Off Hatsuhima, 1173-1175 m, *Hyper-Dolphin* Dive 525. Abbreviations: aj, periostracum of *Acharax johnsoni*; ms, *Margarites shinkai*; pg, *Provanna glabra*. [**a-b, d**: Photos taken by T. Sasaki at NSMT; **c**: Okutani et al. 2004; **f**: Sasaki et al. 2007]

Plate 17 (Chapter 7) Habitats of conid gastropods at Hatsushima seep site in Sagami Bay, Japan. **a-b.** *Phymorhynchus buccinoides* Okutani, Fujikura and Sasaki, 1993. *Hyper-Dolphin* Dive 524, 1180 m. **a.** Outcrop covered by *Bathymodiolus* spp. Rectangle is enlarged in B. **b.** *P. buccinoides* extending siphon at base of outcrop. **c-d.** *Oenopota sagamiana* Okutani and Fujikura, 1992. *Hyper-Dolphin* Dive 525, 1173 m. **c.** Large colony of *O. sagamiana*. **d.** Enlarged view of C. [**a-d**: Sasaki et al. 2007]

Plate 18 (Chapter 9) Field shots of some ancient hydrocarbon seep localities from the USA, France, Germany, Ukraine and Morocco. **a.** Isolated carbonate in the hillside, suspected hydrocarbon seep deposit, Santa Ana Mountains. Poison Oak encircling this particular outcrop made access impossible. **b.** Collecting from *Sulcirostra*-bearing carbonate dipping from upper left to lower right of photograph, Seneca area, Oregon. Photograph courtesy of Jörn Peckmann. **c.** Block bearing *Anarhynchia*, Bedford Canyon, California. Photograph courtesy of Kathy Campbell. **d.** *Peregrinella* Bed, Rottier, Drome, France (locality in Thieuloy 1972). **e.** Block with *Ibergirhynchia*, Iberg Reef, Harz Mountains, Germany. Largest brachiopod specimens approximately 1 cm across. Photography courtesy of Steffen Kiel. **f.** *Peregrinella* Bed, near Planerskoje, Crimea, Ukraine. Photograph courtesy of Steffen Kiel. **g, h.** *Dzieduszyckia* bed Sidi Amar locality, Morocco. Photograph courtesy of Jörn Peckmann

Plate 19 (Chapter 9) Selected Mesozoic and Palaeozoic brachiopod genera discussed herein considered to be associated with ancient hydrocarbon seeps. **a.** *Peregrinella multicarinata* (Lamarck) Neocomian, Châtillon, France, USNM 75595, width of specimen 60.8 mm. **b.** *Peregrinella whitneyi* (Gabb), Cretaceous, Wilbur Springs, California, width = 56.0 mm. **c.** *Dzieduszyckia*, Devonian, Morocco, width of specimen estimated at 8 cm. Specimen photographed during field work, photograph courtesy of Jörn Peckmann. **d.** *Halorella*, Triassic, Morgan Mountain, Grant Co., Oregon, width = 56.8 mm. **e.** *Peregrinella whitneyi* (Gabb), Cretaceous, Wilbur Springs, California, USNM 23264, width = 20.7 mm. **f.** *Peregrinella whitneyi* (Gabb), Cretaceous, Wilbur Springs, California, width = 18.4 mm

Plate 19 (continued) **g.** *Anarhynchia gabbi* Ager, Jurassic, Bedford Canyon, California, Ager L = 35.8 mm. **h.** *Peregrinella chisana* Sandy and Blodgett, Cretaceous, Alaska, length = 29.2 mm. **i.** *Cooperrhynchia schucherti* (Stanton), Tithonian, Jurassic, Paskenta, collected Kathy Campbell, W = 15.1 mm. **j.** *Cooperrhynchia* Knoxville Formation, Jurassic, St. Helena Quadrangle, Napa County, California, UCMP A-4312, width = 21.4 mm. **k.** *Ibergirhynchia contraria*, Early Carboniferous (latest Visean), Iberg Reef, Germany, width = 19.7 mm. Specimen collected E Gischler. **l.** *Sulcirostra*, Jurassic, Seneca, Oregon, width = 26.8 mm. **m.** *Sulcirostra*, Jurassic, Seneca, Oregon, width = 32.3 mm. **n.** *Beecheria*, Carboniferous, Newfoundland, length = 20.2 mm. **o.** *Beecheria*, Carboniferous, Newfoundland, length = 21.8 mm

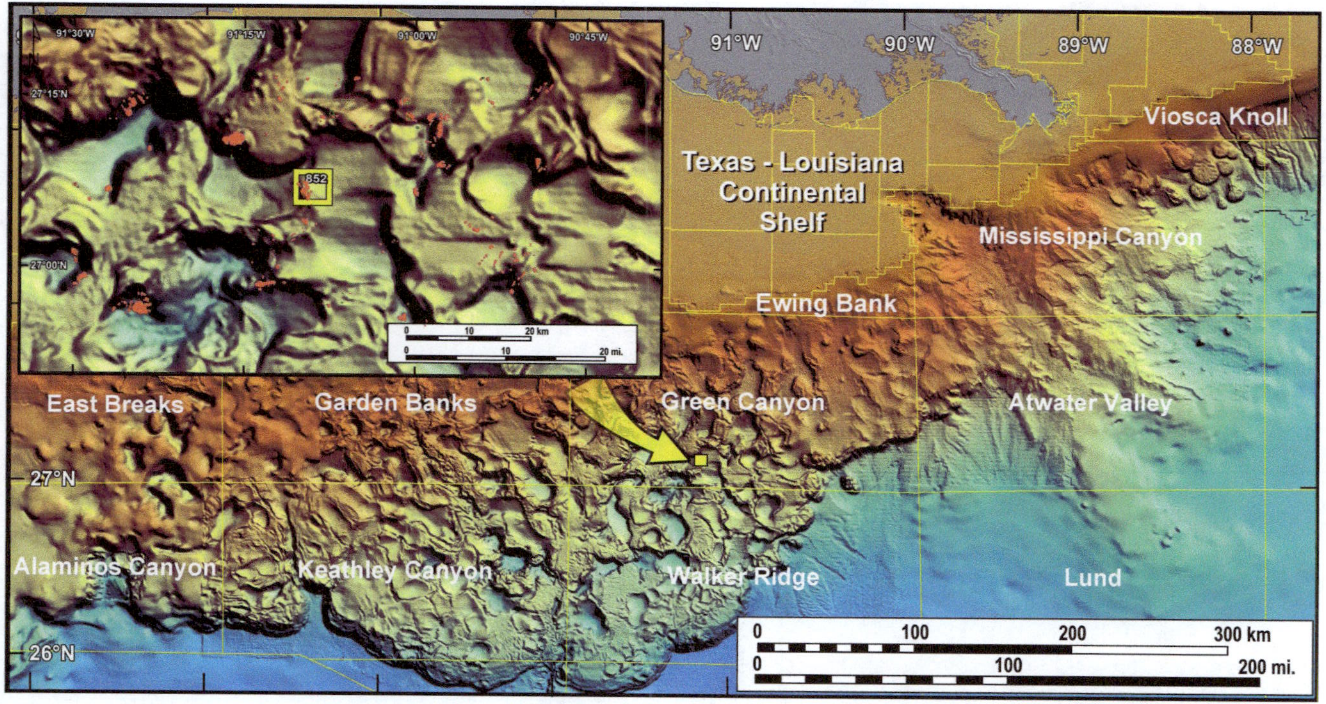

Plate 20 (Chapter 10) Multibeam bathymetry of the northern Gulf's continental slope, which has been enhanced for a 3D-effect, illustrates the complex of intraslope sedimentary basins (smooth areas) and flanking rough topography comprised of domes and ridges. Sites of hydrocarbon seepage and venting are superimposed upon this regional-scale topography (red areas of inset)

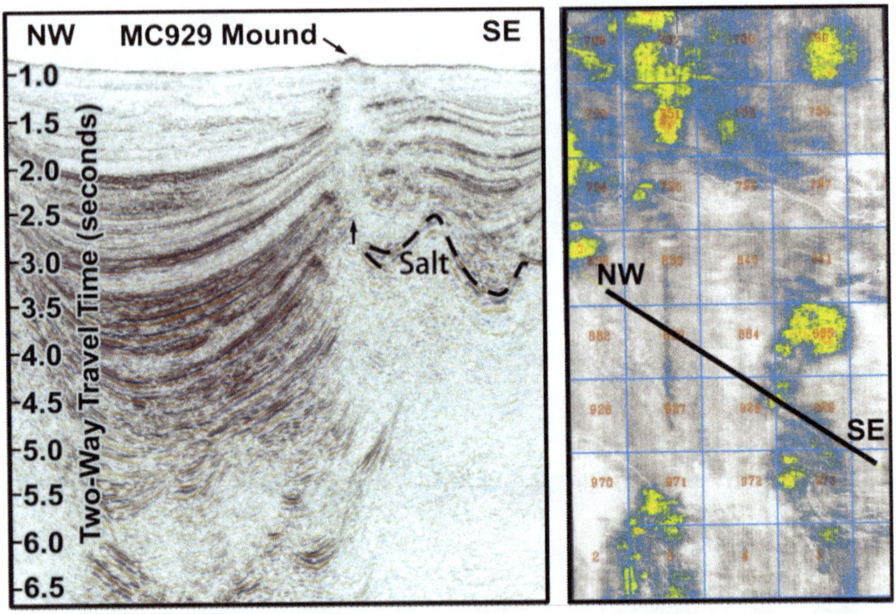

Plate 21 (Chapter 10) The 3D-seismic surface reflectively map of this figure illustrates a series of seafloor "bright spots" that represent hydrocarbon seep sites. These surface reflectively anomalies are arranged at the flanks of sedimentary basins. The NW-SE oriented seismic cross section illustrates a thick sedimentary basin and the MC 929 mound on the flank of the basin. A distinct vertical fluid-gas migration pathway is shown as an acoustically amorphous zone from the edge of subsurface salt to the seafloor. The MC 929 mound has resulted from the expulsion process

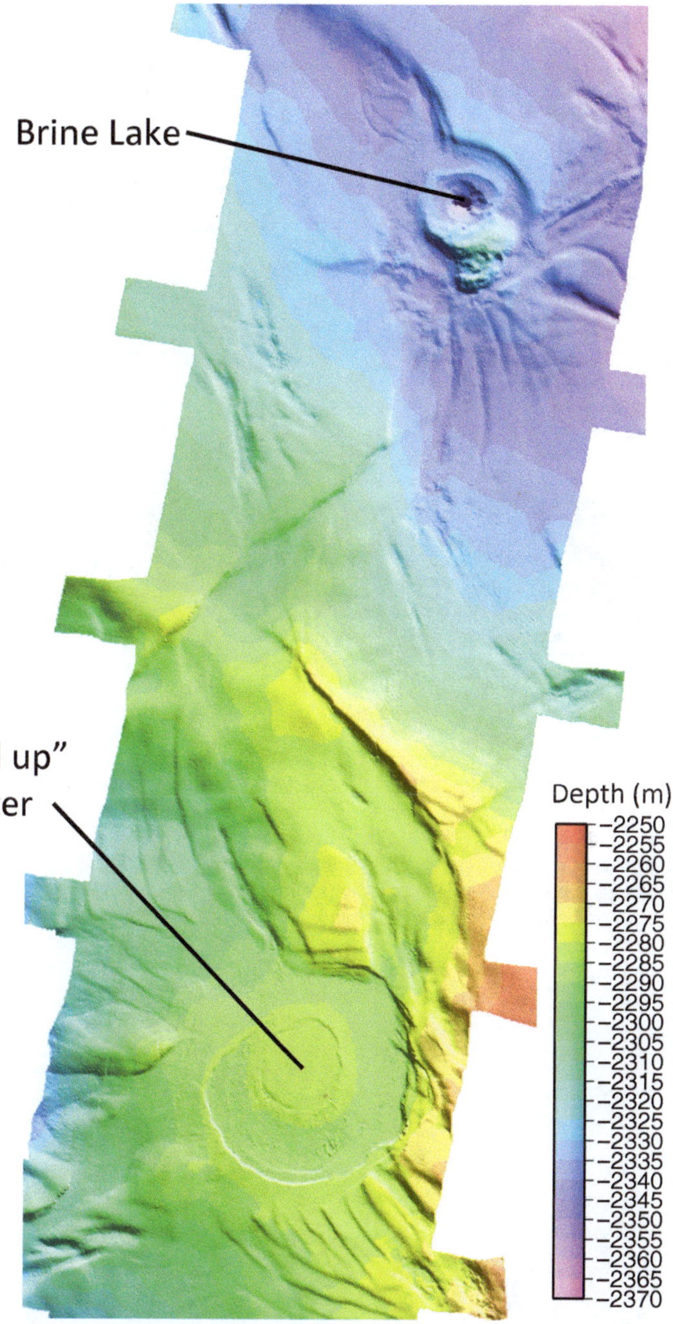

Plate 22 (Chapter 10) Multibeam bathymetry of the Alaminos Canyon 601 lease block. Bathymetry data acquired by the Hugin AUV operated by C&C Technologies as part of the Expedition to the Deep Slope study funded by the U.S. Minerals Management Service and NOAA Office of Ocean Exploration. In the northern section of the image, the well-defined depression of the brine lake is apparent. In the southern section of the image, the outline of the dried up brine lake can be seen along with possible flows from large expulsion events or drainage of the brine from the crater at the top of the feature

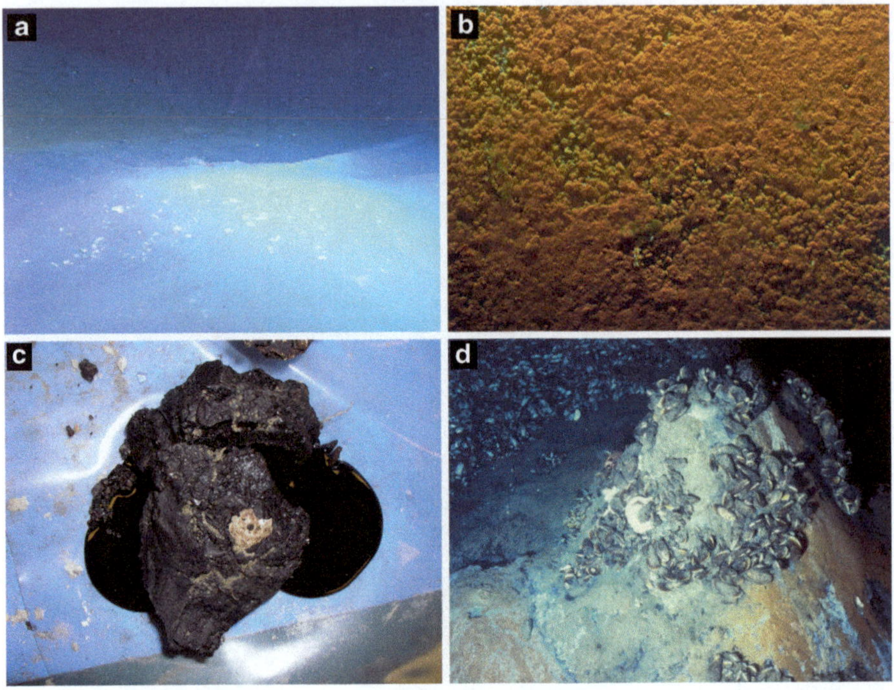

Plate 23 (Chapter 10) Geologic anomalies from the Gulf of Mexico. **a.** Internal waves breaking on the "shoreline" of the Brine Lake in Alaminos Canyon 601. The waves were caused by the impact of the submersible on the surface of the Lake. **b.** An apparent iron-oxide crust formed in the center of the "dried up" crater in the southern feature in Alaminos Canyon 601. **c.** An asphalt nodule that cracked open as the internal temperature rose in the hours following its collection from 950 m depth in Garden Banks 647. **d.** Bathymodiolin mussels inhabit the flanks and tops of the cones as well as the flows and crust areas away from the cones at the top of the mound in Mississippi Canyon 929. Images A and B courtesy of the Expedition to the Deep Slope investigators, C courtesy of Erik Cordes, D courtesy of Harry Roberts

Plate 24 (Chapter 10) Fauna of the Gulf of Mexico seeps. **a.** Habitat formed by the vestimentiferan *Escarpia laminata* and the mussel *Bathymodiolus brooksi* from Alaminos Canyon 818 at approximately 2750 m depth. Also shown are the shrimp *Alvinocaris muricola*, the galatheid crab *Munidopsis* sp., and the holoturian *Chirodota heheva*. **b.** A small portion of the 3000 m^2 mussel bed at the southern crater in Alaminos Canyon 601. The mussels are primarily *Bathymodiolus brooksi* and a worm-like individual of *Chirodota heheva* is present in the upper right corner. This part of the mussel bed lies at the shoreline of the crater where there are still pools of brine. **c.** The iceworm *Hesiocaeca methanicola* on an outcropping piece of methane hydrate at 525 m depth in Green Canyon 234. **d.** The tubeworm-associated clam *Acesta oophagia* enclosing the tips of the tubeworm *Lamellibrachia luymesi* at approximately 460 m in Mississippi Canyon 751. Images A and B courtesy of the Expedition to the Deep Slope investigators, image C courtesy of Chuck Fisher and Erik Cordes, image D courtesy of the Lophelia II: Reefs, Rigs, and Wrecks investigators

Plate 25 (Chapter 10) Fauna of the Gulf of Mexico seeps. **a.** *Bathymodiolus brooksi* mussels heavily colonized by *Ophioctinella acies* brittle stars at 2200 m depth in the Atwater Valley section of the Gulf of Mexico. Image courtesy of Erik Cordes, the Expedition to the Deep Slope 2007, and Aquapix. **b.** *Sarsiaster greigi* burrowing through sediments at 2200 m depth in the Atwater Valley section of the Gulf of Mexico. Image courtesy of Erik Cordes and the Expedition to the Deep Slope 2007

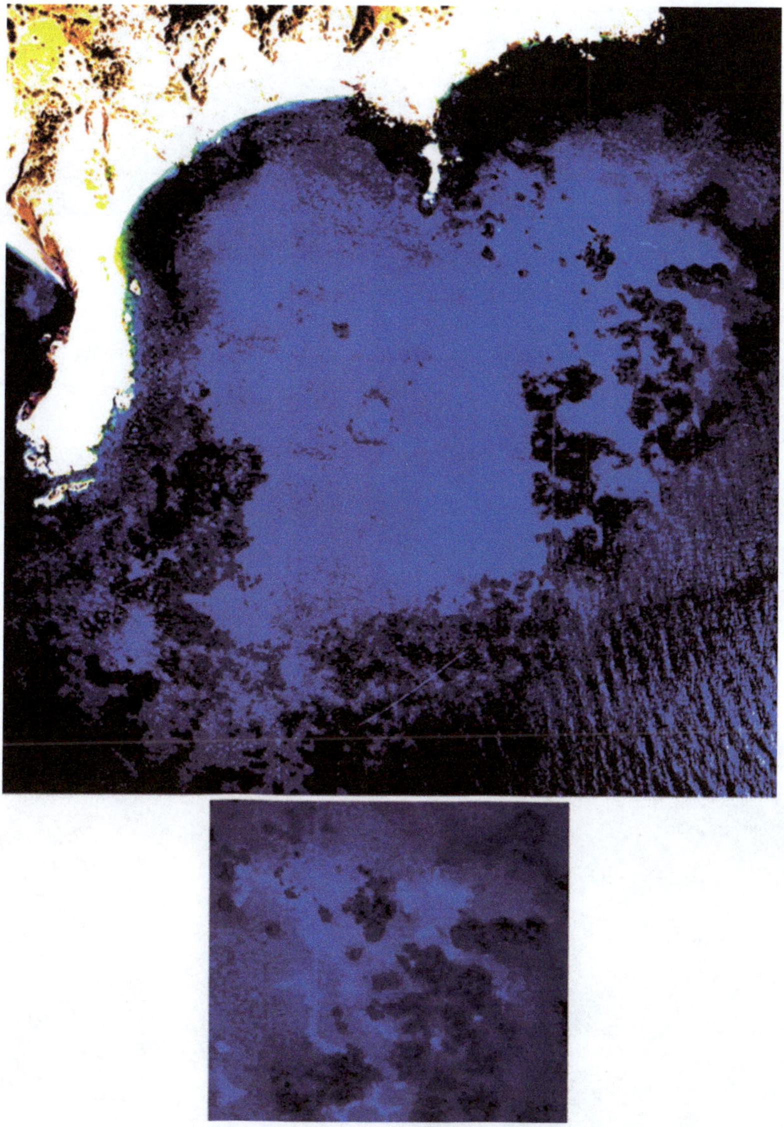

Plate 26 (Chapter 11) Aerial view, approximately 1 km square, of one of the hydrothermal brine seep areas off SE Miles, to the east of Palaeochori Bay. Upper photograph – the dark areas are seagrass beds fringing the minero-bacterial mats over the brine seeps (pale background). Lower photograph – section of the above after adjusting the contrast. The white mats overly the fractures through which the brine seeps to the surface. Photograph acquired by the NERC Airborne Remote Sensing Facility under Project MC04/10

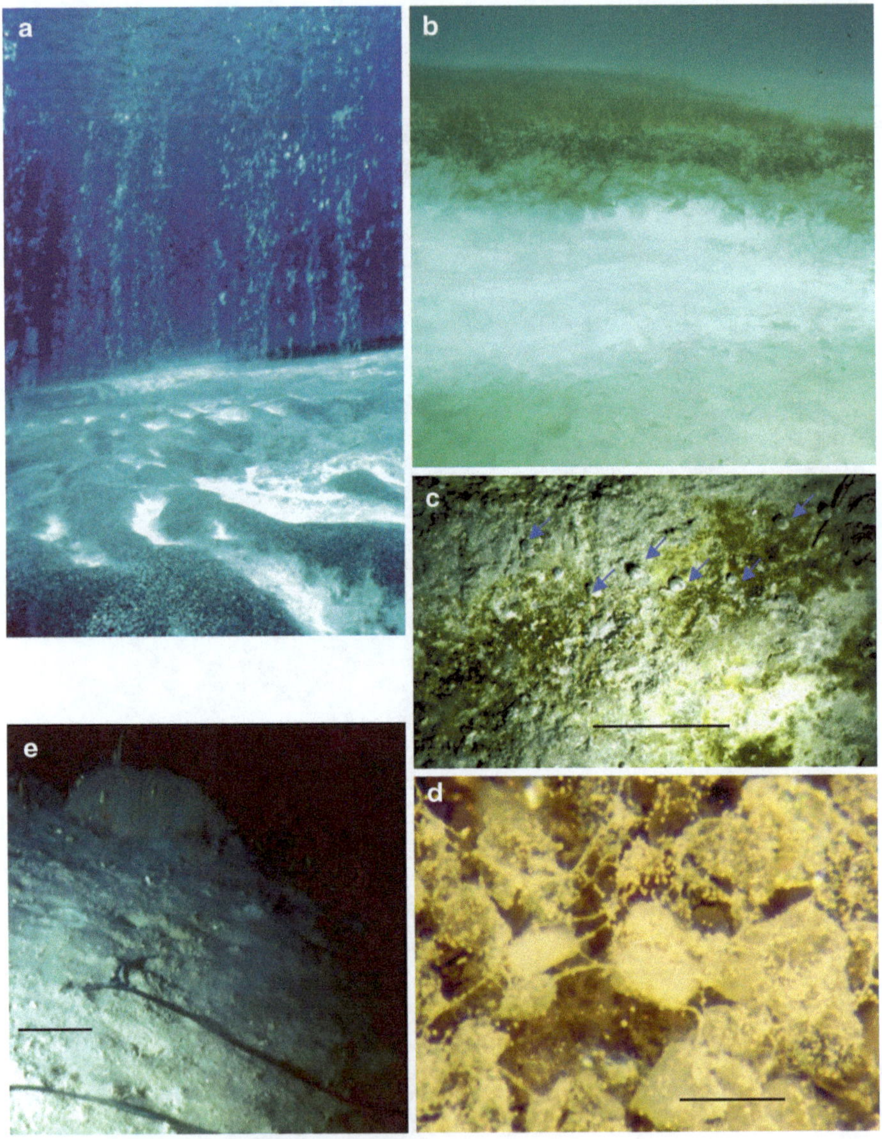

Plate 27 (Chapter 11) Close-up views of the Milos hydrothermal brine seeps. **a.** Gas escape on the fringes of the brine seep: the reflective material in the white mats is sulphur and silicate. **b.** Brine pool (foreground) in a depression at the base of a bed of seagrass (*Cymodocea nodosa*), the white floc at the interface is a mixture of bacteria and elemental sulphur. **c.** Gastropods (*Cyclope neritea*), grazing on the algal-bacteria mats. **d.** Globular sulphur-oxidising bacteria (*Achromatium volutans*) on sand grains over the brine seep: the strands are of elemental sulphur. **e.** Proboscis of an echiuran (*Bonellia* sp.) feeding on bacterial mat. Scale bars: (C) 10 cm, (D) 1 mm, (E) 10 cm. Photographs: A, B and C: JA Hughes & I. Akoumianaki, D: AJ Southward, copyright EC Southward, E: CJ Smith, HCMR

Plate 28 (Chapter 12) Photographs of vent and seep communities around Japan. **a.** The vesicomyids bivalve *Calyptogena phaseoliformis* in the Japan Trench. **b.** The vesicomyids bivalves *Calyptogena okutanii* and *C. soyoae* in Sagami Bay. **c.** Tubeworms of *Lamellibrachia* sp. L1 and *Alaysia* sp. in Sagami Bay. **d.** The vesicomyid bivalve *Calyptogena similaris* in the Nankai Trough

Plate 29 (Chapter 12) Photographs of vent and seep communities around Japan. **a.** Barnacles of the genus *Neoverruca* on a hydrothermal chimney in the Izu-Ogasawara Trench. **b.** The mussel *Bathymodiolus septemdielum* in the Izu-Ogasawara Arc. **c.** The mussel *Bathymodiolus platifrons* and the galatheid crab *Shinkaia crosnieri* in the Okinawa Trough, **d.** *Bathymodiolus platifrons* in the Okinawa Trough

Plate 30 (Chapter 13) Colonization experiments at hydrothermal vents. **a-c.** Artificial tube aggregations deployed in high (**a**), intermediate (**b**), and low (**c**) productivity habitats at "Tica" vent, on the East Pacific Rise (9°50'N). **d.** Colonization by filamentous bacteria on an artificial aggregation, deployed on the basalt near natural aggregations of *Riftia pachyptila*; image taken in the laboratory on board the ship, immediately after recovery

Plate 31 (Chapter 14) Field images of a seep deposit in the Lincoln Creek Formation, cropping out along the bank of the Satsop River (site SR2 of Peckmann et al. 2002). Note a structure called 'inverted stromatactis' by Peckmann et al. (2002) in the block just below the hammer in figure **b.** Images courtesy of Jörn Peckmann

Plate 32 (Chapter 14) Seep fossils from Washington State, USA. **a.** *Acharax dalli* from an undescribed late Eocene seep deposit found float on the bank of the Columbia River. **b.** Late Oligocene seep carbonate block from near the ghost town Knappton with several bathymodiolins; note that all specimens are more-or-less aligned, which is very unusual for fossil bathymodiolins. **c.** *Conchocele bisecta* from the early Miocene near Frankfort, found float on the banks of the Columbia River. **d.** Undescribed solemyid from an early Miocene seep carbonate block found near Frankfort on the Columbia River. **e.** The sponge *Hexactinella*(?) *conica* from an Oligocene section of the Lincoln Creek Formation on the Canyon River

Plate 32 (continued) **f.** So-called 'seep doughnut', being the calcified end of a fluid conduit, found just upstream from where the sponge of Fig. e was found. **g.** Cluster of vesicomyid bivalves from a seep deposit in the Canyon River. Figs. e and f courtesy of Jim Goedert, Fig. g courtesy of Jörn Peckmann

Index

V

W

Kolibri 360 GmbH
Ferdinand-Jühlke-Straße 7
99095 Erfurt, Deutschland
produktsicherheit@kolibri360.de